AF842552

# Inorganic Pollutants in Wastewater
## Methods of Analysis, Removal and Treatment

Edited by
Inamuddin
Ali Mohammad
Abdullah M. Asiri

This book, 'Inorganic Pollutants in Wastewater: Methods of Analysis, Removal and Treatment' extensively investigates the most recent improvements in the area of inorganic pollutants analysis, removal and treatment of wastewater by utilizing different materials such as natural polymers, husks, graphene and carbon nanotube composites, fruit cortex etc. It covers photocatalysis, adsorption, desalination and electrochemical technologies used for the analysis and treatment of inorganic pollutants.

The book is the result of the significant commitment of specialists from different interdisciplinary fields of science. It comprehensively investigates the most abundant, top to bottom, and avant-garde research and reviews.

# Inorganic Pollutants in Wastewater
## Methods of Analysis, Removal and Treatment

Edited by

**Inamuddin[1,2], Ali Mohammad[2] and Abdullah M. Asiri[1]**

[1] Chemistry Department, Faculty of Science, King Abdulaziz University, Jeddah 21589, Saudi Arabia

[2] Department of Applied Chemistry, Faculty of Engineering and Technology, Aligarh Muslim University, Aligarh-202 002, India

Published by **Materials Research Forum LLC**
Millersville, PA 17551, USA



Published as part of the book series
**Materials Research Foundations**
Volume 16 (2017)
ISSN 2471-8890 (Print)
ISSN 2471-8904 (Online)

Print ISBN 978-1-945291-34-0
ePDF ISBN 978-1-945291-35-7



Distributed worldwide by

**Materials Research Forum LLC**
105 Springdale Lane
Millersville, PA 17551
USA
http://www.mrforum.com

Manufactured in the United States of America
10 9 8 7 6 5 4 3 2 1

# Table of Contents

# Preface

Water pollution is the defilement of water bodies e.g. lakes, streams, seas, aquifers, and groundwater. This type of ecological debasement happens when pollutants are straightforwardly or in a roundabout way released into water bodies without sufficient treatment to evacuate destructive compounds. Water pollution influences the whole biosphere–plants and living organisms living in these waterways. Water contamination is a major worldwide issue which requires continuous assessment and amendment of water resource policy at all levels (universal down to singular aquifers and wells). It has been proposed that water contamination is the main overall reason for death and infections. Water is commonly alluded to as contaminated when it is impaired by anthropogenic contaminants such as inorganic as well as organic substances and either does not bolster a human utilize, for example, drinking water, or experiences a stamped move in its capacity to help its constituent biotic groups, for example, animals. Natural phenomena, for example, volcanoes, algal blooms, tempests, and seismic tremors additionally cause real changes in water quality and the natural status of water. A number of wastewater analysis and treatment processes have been developed over the past few years. This edition covers the photocatalysis, adsorption, desalination and electrochemical technologies used for the analysis and treatment of inorganic pollutants.

This book, 'Inorganic Pollutants in Wastewater: Methods of Analysis, Removal and Treatment' is intended to extensively investigate the most recent improvements in the area of inorganic pollutants analysis, removal and treatment of wastewater by utilizing different materials such as natural polymers, husks, graphene and carbon nanotube composites, fruit cortex etc.

The book is the result of the significant commitment of specialists from different interdisciplinary fields of science. It comprehensively investigates the most abundant, top to bottom, and avant-garde research and reviews. We are grateful to all the contributing authors and their co-authors for their esteemed contribution. We might likewise want to thank all distributors, authors, and others people who conceded consent to utilize their figures, tables, and schemes. Albeit each exertion has been made to acquire the copyright authorizations from the individual proprietors to incorporate reference to the replicated materials, we might want to offer our earnest statements of regret to any copyright holder if accidentally their privilege is being encroached.

*Inamuddin*[1,2], *Ali Mohamma*[2] *and Abdullah M. Asiri*[1]

[1] Chemistry Department, Faculty of Science, King Abdulaziz University, Jeddah 21589, Saudi Arabia

[2] Department of Applied Chemistry, Faculty of Engineering and Technology, Aligarh Muslim University, Aligarh- 202 002, India

# Chapter 1

# Photocatalysis: Present, past and future

Thillai Sivakumar Natarajan[1,2] and Rajesh J. Tayade*[1]

[1]Inorganic Materials and Catalysis Division, CSIR-Central Salt and Marine Chemicals Research Institute (CSIR-CSMCRI), Council of Scientific and Industrial Research (CSIR), Gijubhai Badheka Marg, Bhavnagar 364 002, Gujarat, India

[2]UCD School of Chemical and Bioprocess Engineering, UCD Engineering and Materials Science Centre, University College Dublin (UCD), Belfield, Dublin 4, Ireland

thillai.natarajan@ucd.ie, organthillai@gmail.com; *tayade@csmcri.res.in, rtayade@gmail.com

**Abstract**

Photocatalysis is one of the effective catalytic processes with potential applications in solving the environmental pollution and energy crisis problems by degrading the pollutants, reducing the $CO_2$ levels and production of sustainable hydrogen ($H_2$) fuel, respectively, using semiconductors as a photocatalyst under irradiation of light. Various types of photocatalytic materials, reactors, and processes with efficient photocatalytic performance have been studied to fulfill the practical requirements. In this chapter, initially, we will briefly discuss the principle of photocatalysis, its application in pollutants abatements and energy production along with the basic requirements. Simultaneously, we will discuss the past and present state of art of the utilization of this potential technique in these applications. Finally we describe future viewpoint of this technique in these fields as well.

**Keywords**

Photocatalysis, Pollutants Degradation, Water Splitting, $H_2$ Production, $CO_2$ Reduction, Irradiation

## Contents

## 1. Introduction

Environmental problems and energy crisis are the two major issues the world is facing today. The continued rise in the world population with industrialization produces a significant amount of contaminants to the water, soil and air environments. Several types of hazardous pollutants such as dyes, organic compounds, pharmaceuticals, heavy metals, agricultural chemicals, radioactive materials and volatile organic compounds (VOCs) etc., have been detected in the water and air. For example, dyes are majorly utilized in different industries like paper, textile, leather and cosmetic industries and approximately 15% of dyes are being lost during the dyeing process. World Health Organization (WHO) reported that diseases transmitted through dirty water or unhygienic water resources are the second leading cause of children's death worldwide, U.N. world water development has reported that by 2050, at least one in four peoples is likely to live in a country affected by chronic or recurring shortages of freshwater [1, 2]. The unstoppable liberation of contaminants to the environments has made a significant interest to the scientific community to develop energy efficient technologies for purifying the environment. In addition, the worldwide energy requirements mainly rely on fossil fuels. However, the emission of a large amount of carbon dioxide ($CO_2$), which is one of the major greenhouse gasses contributing to global warming causes, is threatening to the life of mankind and animals. The $CO_2$ concentration in the atmosphere was 404 pp min December 2016, and it is increasing day by day [3]. The significant dependency on fossil fuels and other growing anthropogenic emission of greenhouse gases result in the emission of about 30.4 Gt of $CO_2$ to the atmosphere, it may be expected to increase to 36-43 Gt by 2035 [4, 5]. Therefore, it is highly emphasized to develop a clean and sustainable route to generate a renewable fuel to mitigate the level of $CO_2$ emission or a technology to reduce $CO_2$ into useful products.

To address the environmental pollution various conventional technologies like physical, biological, thermal, and chemical treatments have been developed. However, these are energy intensive, transforming the pollutants from one phase to another which creates secondary pollution, high operating cost over long-term and are economically not viable [6-8]. As an alternative, advanced oxidation processes (AOPs) has been developed, it uses the *in-situ* formed hydroxyl radicals ($^{\bullet}OH$) as a potential oxidizing agent to degrade the environmental pollutants [9-11]. Among the AOPs, photocatalysis is a promising and environmentally benign technology for degrading the pollutants in the presence of light irradiation using semiconductors as a photocatalyst. The pioneering work of photoelectrochemical water splitting using $TiO_2$ electrode by Fujishima and Honda in 1972 [12] has enhanced the research interest in the field of photocatalysis and degradation of a wide range of pollutants into less harmful substances at ambient pressure and temperature, and sustainable solar light also used for irradiation. Subsequently, different semiconductors ($TiO_2$, ZnO, $WO_3$, $ZrO_2$, etc.) used for degradation of different types of organic and inorganic pollutants under UV light irradiation led to the development of different geometry of slurry and immobilized type photocatalytic reactors [13-21]. These semiconductors are activated under UV light; possess wide band gap energy (uses 4% of solar spectrum) and lower surface area, the high recombination rate of photogenerated charge carriers, and catalyst recovery problem in slurry reactors limit their practical applicability. These semiconductors have been doped with metals, non-metals, and anions, sensitized with dye molecules, prepared magnetic photocatalyst, composites with other semiconductors and mixed with high surface area materials to enhance the adsorption capacity leading to the improved degradation efficiency [22-30]. However, their stability and reproducibility of degradation efficiency were poor, led to the development of direct visible light responsive low band gap energy sulfides (CdS, ZnS), $Cu_2O$, oxynitrides and oxyhalides photocatalyst but these were not stable under light irradiation. However, their low band gap energy with high visible light response triggered the research to improve their stability, also gave impetus input to develop different types of visible light active photocatalysts. Bismuth ($Bi_2O_3$, $Bi_2WO_6$, BiOX (X-Cl, Br,I) etc.), g-$C_3N_4$, $MS_2$ (M-Mo, W, Sn), $AB_2X_4$, silver ($Ag_3PO_4$, $Ag_2O$), ferrites and metal-organic frameworks (MOFs) etc., based visible light photocatalysts have been developed for pollutants degradation [31-36]. Nevertheless, their efficiency was poor because the redox potential of photogenerated electrons and holes on the CB and VB was lower than the potential required for $^{\bullet}OH$ and superoxide radical anions ($O_2^{\bullet-}$)generation and the poor charge carriers transfer rate. As a result, dual semiconductors with solid state electron mediators and redox- or solid state electron mediators free and ternary composite photocatalyst systems have been developed for degradation of organic and inorganic pollutants and still the research is underway to develop various photocatalytic systems [37-40].

Similarly, the $CO_2$ mitigation was performed by different carbon capture and storage (CCS) techniques such as a post or pre-combustion capture and oxyfuel combustion. Post-combustion involves the capturing of $CO_2$ from flue gas constituents produced by combustion, in pre-combustion, it is removed from the fuel before combustion, and in oxyfuel combustion, the fuel is burned in the presence of oxygen stream that comprises low or no nitrogen. The captured $CO_2$ stored on the natural sites like deep ocean sequestration and geological storage. However, the long-term storage leads to $CO_2$ leakage which creates negative impact on earth surface and is economically not feasible [41-43]. Subsequently, electricity, hydrogen, battery technology, biofuels and conversion of $CO_2$ into useful chemicals such as polymer, fuels, minerals and other chemical commodities are considered as possible alternative sources to replace fossil fuels. Battery technology suffers from low gravimetric and volumetric energy densities and biofuel lack to fulfill the global energy supply [40, 41]. However, the energy required for these processes is relatively high and these are short-term solutions for $CO_2$ emission. Conversion of $CO_2$ into useful chemicals by catalytic route is an alternative method to decrease the $CO_2$ concentration, however, these processes use high temperature and pressure and also requires an external $H_2$ source [44-47]. Therefore, for the production of solar fuels, hydrogen ($H_2$) is a clean and sustainable fuel; it can be produced from nonfossil sources like water and solar light as the energy source. However, the problem in the storage of produced $H_2$ makes this technology not accessible for immediate practical usability. So, the reduction of $CO_2$ into liquid carbonaceous fuels such as methanol and methane is a possible technology to eliminate the $CO_2$ emission, but these being high energy processes would be studied further. However, this technology has potential advantages over the catalytic routes and other $CO_2$ emission reduction processes. Thus, the production of hydrogen by water splitting and liquid carbonaceous fuels by $CO_2$ reduction using renewable energy sources has been considered as the key strategies to fulfill the increasing global energy demands. Since the Fujishima and Honda discovery, different semiconductor photocatalytic materials have been developed to produce $H_2$ and $CO_2$ reduction. However, the photocatalytic efficiency mainly relies on the conduction (CB) and valence band (VB) edge positions of the respective semiconductors[48-55]. Therefore, the development of an efficient photocatalytic system to degrade the pollutants, produce $H_2$ and reduce the $CO_2$ is highly challenging and even more complex for practical implementations. In the following paragraphs, we aimed to describe the principles of photocatalysis in degradation of pollutants, water splitting, and $CO_2$ reduction. The status of this energy efficient and benign photocatalytic system in environmental remediation and energy application also is discussed.

## 2. Principle of photocatalysis

Irradiation of semiconductor photocatalysts in the presence of light energy with equal to or greater than that of their band gap energy generates electrons and holes in their CB and

VB, respectively. Subsequently, holes and electrons in the VB and CB react with surface adsorbed water molecule ($H_2O$) or hydroxyl group (-OH) and oxygen ($O_2$) to generate reactive $^{\bullet}OH$ and $O_2^{\bullet -}$ radicals. These reactive radicals react with different water or air pollutants and degrade them into carbon dioxide, water, and other less toxic inorganic ions. In some cases, holes in the VB directly oxidize the pollutants adsorbed on the photocatalyst surface and the electrons in the CB indirectly degrade the pollutants using $^{\bullet}OH$ radicals formed by cleavage of *in-situ* produced hydrogen peroxide ($H_2O_2$). On the other hand, the surface adsorbed dye excites under light irradiation and injects an electron into the CB of the catalyst which reacts with oxygen to yield hydroxyl ($^{\bullet}OH$) radicals indirectly via formation of $H_2O_2$.

$$\text{Photocatalyst} + h\upsilon\ (\lambda \geq \text{band gap}) \rightarrow \text{Photocatalyst}\ (h+_{VB}) + \text{Photocatalyst}\ (e^-_{CB})$$

$$\text{Photocatalyst}\ (h+_{VB}) + H_2O \text{ or } -OH \rightarrow H^+ + {}^{\bullet}OH$$

$$\text{Photocatalyst}\ (e^-_{CB}) + O_2 \rightarrow O_2^{\bullet -}$$

$$OH + O_2^{\bullet -} + \text{water or air pollutants} \rightarrow \text{intermediate products} \rightarrow CO_2 + H_2O + \text{inorganic ions}$$

Photocatalytic degradation of pollutants is generally described as a thermodynamically downhill reaction which occurs spontaneously with negative Gibbs energy change (i.e. $\Delta G < 0$). However, the photocatalytic water splitting reaction is different than the water pollutants degradation; in the case of water splitting, the photo produced electrons and holes act as reducing and oxidizing agents to generate hydrogen ($H_2$) and oxygen ($O_2$), respectively. This reaction involved a high positive change in the Gibbs free energy (i.e. $\Delta G > 0$, 237 kJ/mol) and hence it is a thermodynamically uphill reaction. This reaction resembles natural photosynthesis in green plants and is highly challenging in chemistry, therefore it is often called as the "Holy Grail" of Chemistry.

$$H_2O \rightarrow \tfrac{1}{2} O_2 + H_2 \qquad \Delta G = +237 \text{ kJ/mol}$$

Similarly, the photocatalytic reduction of $CO_2$ is also challenging because $CO_2$ molecules are relatively inert and stable compounds. Thus, the $CO_2$ reduction is also an uphill reaction requiring huge energy input ($\Delta G > 0$), which is offered by the incident photons. The high negative potential (-1.90 V) is required to reduce the $CO_2$ to $CO_2^{\bullet -}$ anion radical by single electron and hence makes this reaction highly implausible. Thus,

$$CO_2 + 1\ e^- \rightarrow CO_2^{\bullet -} \qquad -1.90 \text{ V}$$

$CO_2$ reduction occurs by proton-assisted transfer of multiple electrons. In the presence of light irradiation, $H_2$ is produced by the photocatalytic splitting of water and then the produced hydrogen hydrogenates $CO_2$ to formic acid (HCOOH), carbon monoxide (CO), formaldehyde (HCHO), methane ($CH_4$), methanol ($CH_3OH$), and other products, as listed below.

$2H^+ + 2e^- \rightarrow H_2$ -0.41 V

$CO_2 + 2H^+ + 2e^- \rightarrow HCOOH$ -0.61 V

$CO_2 + 2H^+ + 2e^- \rightarrow CO + H_2O$ -0.53 V

$CO_2 + 4H^+ + 4e^- \rightarrow HCHO + H_2O$ -0.48 V

$CO_2 + 6H^+ + 6e^- \rightarrow CH_3OH + H_2O$ -0.38 V

$CO_2 + 8H^+ + 8e^- \rightarrow CH_4 + H_2O$ -0.24 V

Unlike photocatalytic degradation, $H_2$ production and $CO_2$ reduction are an uphill reaction and highly challenging, hence, semiconductor photocatalyst should possess band edge positions with suitable redox potentials which satisfy the potential required to produce hydrogen and different $CO_2$ reduction products.

## 3. Overview of development in photocatalysis

The pollutants degradation using light irradiation and semiconductors in $19^{th}$ century triggered researchers to prepare semiconductors with efficient degradation efficacy. The discovery by Fujishima and Honda has offered further indispensable inputs for the development of photocatalysis and synthesizing $TiO_2$ based photocatalysts. Afterwards, $TiO_2$ has been efficiently studied for the degradation of various pollutants (Fig. 1) due to its unique properties. Furthermore, different types of slurry photocatalytic reactors have been developed, however, the difficulty in catalyst regeneration led to designing of immobilized photocatalytic reactors. The low surface area and higher band gap energy of $TiO_2$ have restricted its use into UV light. Therefore, $TiO_2$ was decorated with metals and non-metal ions (Fig. 2), sensitized with dye molecules, prepared composite of $TiO_2$ with other metal oxides and synthesized different morphologies of $TiO_2$ using different synthesis methodologies (Fig. 3a) to improve photocatalysts visible light absorption capability and surface area. However, the stability and reusability of modified $TiO_2$ photocatalyst were poor and synthesis methodologies comprised of complicated and energy intensive steps which guided to develop variety of visible light responsive low band gap energy non-$TiO_2$ based photocatalysts. The surface area of $TiO_2$ was also enhanced by supporting them on photoactive and non-photoactive high surface area adsorbents. In addition, the recombination of charge carriers was drastically reduced by synthesizing binary and ternary composites, dual semiconductors with and without redox-mediator based photocatalysts (Fig. 3b). The difficulties associated with UV and solar light-based reactions led to use of light emitting diode (LED) as an irradiation source for designing of energy efficient and portable reactors. Like pollutants degradation, simultaneously these processes and photocatalytic systems have been utilized to fulfill the energy requirement by producing sustainable $H_2$ fuel and mitigate the $CO_2$ pollution by converting them into value-added products. The high energy requirement of these processes challenges the researcher to develop potential photocatalytic systems with appropriate band edge potentials for sustainable fuel production and $CO_2$ reduction (Fig. 4). Z-scheme type photocatalysts with appropriate CB and VB potentials which mimic the natural photosynthesis have been developed, however, their efficiencies were insufficient to fulfill the commercial technology sectors. Therefore, the research keeps on moving forward to design a photocatalytic system to achieve the practical sector requirement for efficient pollutants degradation, energy production and reducing $CO_2$ into value added products. As discussed above, various types of processes and techniques have been implemented in different years to improve the efficiency of photocatalysis. Initially, this has been majorly utilized to decontaminate the pollutants present in water and air environments with less concentration on fuel production. However, the continuous consumption of fossil fuels and $CO_2$ emission led us to utilize this photocatalytic processes to produce sustainable and renewable $H_2$ fuel and reduce the $CO_2$ concentration by converting them into value-added products. Therefore, the forthcoming section describes the detailed progress of each of these applications and their journey towards scientific advancement.

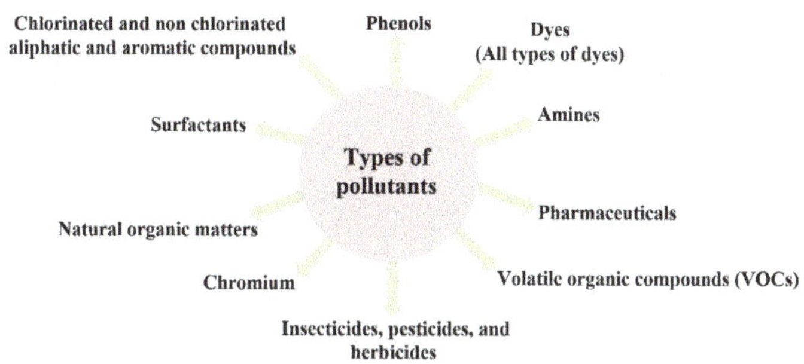


*Fig. 1. Different types of pollutants used to study the photocatalytic activity of the catalysts.*

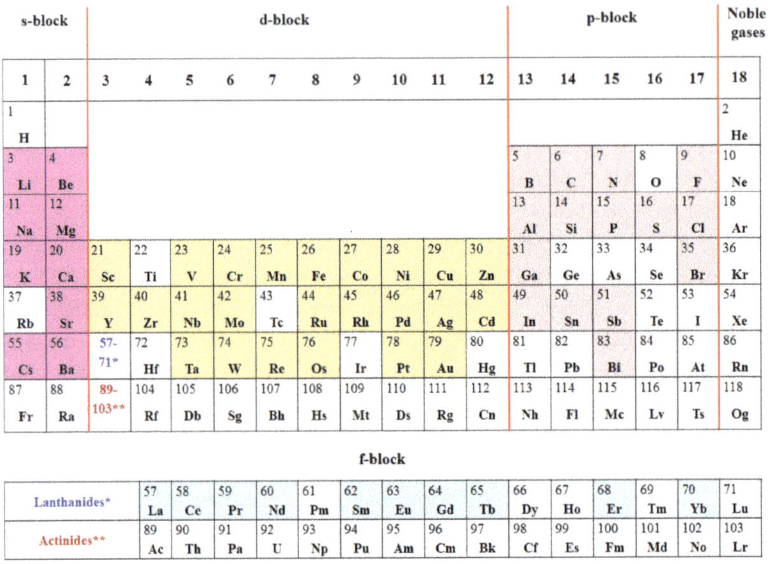


*Fig. 2. Periodic table shows the elements used for doping over $TiO_2$ for pollutants degradation, $H_2$ generation and $CO_2$ reduction reactions.*

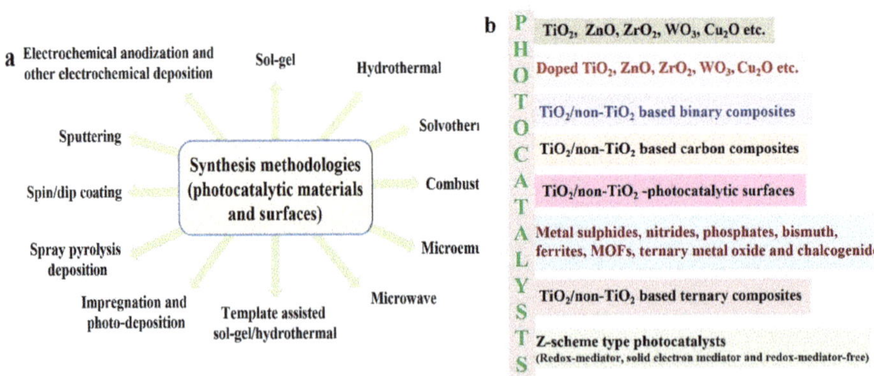


*Fig. 3. (a) Synthesis methodologies for development of photocatalytic materials and surfaces and (b) Different types of photocatalysts used so far.*

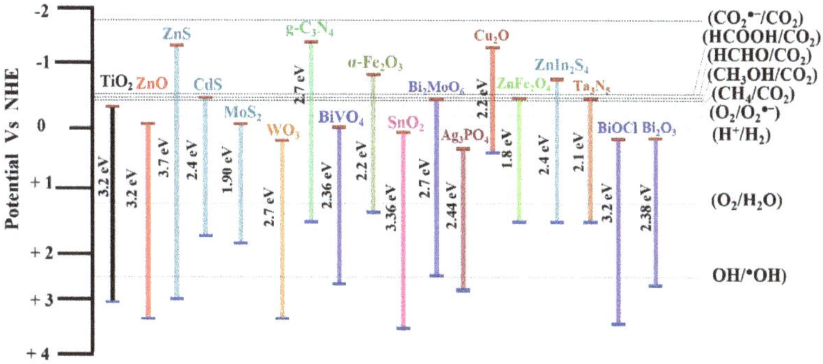


*Fig. 4. Band-edge positions and band gap energies of selected semiconductors.*

## 4. Photocatalysis in pollutants degradation

In 1937, Goodeve studied the absorption spectra and photosensitizing activity of the white ZnO and $TiO_2$ pigments and used them for oxidation of oil vehicles and bleaching of adsorbed dye stuffs [56]. In that article, it was mentioned that in 1929, both Keidel and Wagner separately studied the action of $TiO_2$ on discolouring of dye stuffs. In 1955, Markham studied the photocatalytic properties of ZnO, $TiO_2$ and $Sb_2O_3$ oxides in oxidation of organic compounds under UV light. In addition, various phenomena on these metal oxides such as photoconductivity, phosphorescence or fluorescence, phototropy and photolysis have been studied under UV light exposure. A summary of these effects provided a valuable input for interpretation of effect of UV light and observable changes in these properties by the effects of external parameters [57]. The discovery of photoelectrochemical splitting of water using $TiO_2$ and Pt electrode by Fujishima and Honda in 1972 [12] increased the researcher's interest and triggered the research on the use of semiconductor as a photocatalyst for the degradation of environmental pollutants in the presence of light irradiation. With this invention, in 1976, Carry et al. used $TiO_2$ as a photocatalyst for photodechlorination of polychlorobiphenyls [58], followed by in 1977, Frank and Bard studied the photocatalytic oxidation of cyanide and sulphite ion in water using $TiO_2$, ZnO, CdS, $Fe_2O_3$ and $WO_3$ semiconductors under xenon light irradiation [59, 60]. Successively, other semiconductors such as $TiO_2$, ZnO, $ZrO_2$, $WO_3$, $V_2O_5$, $SnO_2$, $CeO_2$, $Cu_2O$, CdS and $Fe_2O_3$ etc., have also been intensively studied for the degradation of dyes, organics, pharmaceuticals, air pollutants, and inorganic hazardous pollutants to innocuous end-products [13-21, 61-66]. Among these, $TiO_2$ is a potential photocatalyst because of its high chemical and biological stability, rapid degradation ability, none corrosiveness under light irradiation, and environmental acceptability. Subsequently, in late 1980, frenzied series of studies have been reported on the photocatalytic degradation of pollutants (water or air) in the presence of light irradiation using $TiO_2$ as a photocatalyst. Moreover, $TiO_2$ exhibits three phases such as anatase, rutile and brookite; anatase exhibited higher photocatalytic performance. The influence of different operational parameters such as catalyst concentration, initial concentration of pollutants, pH of reaction solution, light intensity, co-presence of other ions, temperature and quantum yield measurement, have been studied on the degradation efficiency of $TiO_2$. Continuously, various geometry of $TiO_2$ based slurry photocatalytic reactors have been developed for the degradation of pollutants under UV light irradiation and it has been reviewed in several research articles [13-21, 67-81]. However, the presence of $TiO_2$ photocatalyst nanomaterials in the slurry reactors tends to aggregate and scatter the incident light which decreases the amount of light reaching the catalyst surface and their subsequent production of electrons and holes. The recovery and reuse of $TiO_2$ from the slurry reaction medium need a proper post-treatment which makes these slurry photocatalytic reactors practically unsuitable for large scale application. This has led to immobilization of $TiO_2$ nanoparticles on different solid supports using various coating techniques and development of various immobilized photocatalytic reactors [19, 76, 79, 82-91]. However, these reactors have showed lower photocatalytic performance than the slurry counterparts because of the reduced surface area, less porosity and increase in the coating thickness (limited exposure of the photocatalyst to light). To overcome the problem in both the type of reactors, magnetically separable photocatalytic materials have also been developed to have easy separation of the catalyst particles. Magnetic materials such as ferrites ($MFe_2O_4$, M-Zn, Ni, Co, etc.,), $Fe_3O_4$ and $Fe_2O_3$ etc., were coupled with non-magnetic photocatalysts in order to induce the magnetic property along with the enhancement in their reusability. It is also understood that, some magnetic materials itself demonstrated good visible light response and degradation ability [92-96]. Furthermore, most of the photocatalytic reactions have been performed under traditional UV irradiation (low/medium/high pressure mercury vapor lamp), which possess serious drawbacks such as low photonic efficiency, requirement of cooling arrangement, less lifetime (1000 h), need of high voltage at initial stage and use of hazardous mercury metal. To conquer this constraint, energy efficient ultraviolet-light emitting diodes (UV-LED) has been used as alternative for degradation of pollutants. It offers more advantages like high robustness, long lifetime (100000 h), little heat production, good linearity of the emitted light intensity with current, no cooling requirements and suitability for designing of compact and portable photocatalytic reactors [97-111]. Chen et al., investigated the photocatalytic oxidation of perchloroethylene under UV-LED reactor using P25 coated fiber surface [98]. Wang and Ku studied the degradation efficiency of Degussa P-25 $TiO_2$ (P25) coated quartz plate by degradation of an aqueous solution of reactive red 22 dye under UV-LED irradiation [99]. Subsequently, various types of UV-LED based slurry and immobilized photocatalytic reactors have been developed for the degradation of different organic and inorganic pollutants [100-111]. We have also developed UV-LED source based slurry and immobilized photocatalytic reactors for degradation of methylene blue (MB), rhodamine B (RhB), congo red (CR) and malachite green (MG) dyes using P25, $TiO_2$ nanotube array and $TiO_2$ coated quartz tube surfaces as a photocatalyst [100, 101, 103, 105, 108].

In addition, the photoactivity of $TiO_2$ is limited to UV light irradiation because of its wide band gap energy (3.2 eV). The problems with UV sources make these $TiO_2$ based processes impractical for commercialization. Besides, $TiO_2$ possesses a lower surface area and its high electron-hole pair recombination resulted in low electron transfer rate leads to deterioration of the photocatalytic activity of $TiO_2$ based reactions. The surface area of $TiO_2$ has been increased by synthesizing $TiO_2$ through template assisted methods, different morphologies of $TiO_2$ like nanotube, nanoflower, nanosheets, nanorods, etc., (Fig. 3) and making composite with zeolite, carbon, graphene, clays and other porous materials [112-116]. We have prepared $TiO_2$ coated NaY and HY zeolites, demonstrated that 1%-$TiO_2$ loaded NaY and HY zeolites exhibited higher activity in degradation of MB dye, which may be because of their higher surface area, leads to enriched adsorption percentage than the bare $TiO_2$ [117]. Similarly, the visible light response of $TiO_2$ and charge transfer rate has also been improved by doping with metals, non-metals and dye-sensitization. In 1981 platinized $TiO_2$ materials with improved light absorbance in the visible range were developed for splitting of water and organic transformation reactions, but, their stability and reusability were extremely poor [118]. In 1984, Serpone et al. reported two semiconductor systems ($CdS/TiO_2$) by coupling $TiO_2$ with CdS and revealed that photogenerated electron and holes were vectorially separated through the interparticle electron transfer pathway and improved the visible light response [119]. To prove this, in 1995 they developed different double semiconductors comprised of $TiO_2$, CdS, ZnO, $SnO_2$, $WO_3$ and $Fe_2O_3$ ($CdS/TiO_2$, $ZnO/TiO_2$, $CdS/WO_3$, $ZnO/Fe_2O_3$ etc.,) for the photocatalysed oxidation of phenol, 2-chlorophenol and pentachlorophenol reactions. As a follow up, different semiconductor composite photocatalytic systems have been studied so far to decrease the photogenerated electron hole recombination rate and enhance the visible light photocatalytic performance [120]. Similarly, in 1994 different metals ($Fe^{3+}$, $Mo^{5+}$, $Ru^{3+}$, $Os^{3+}$, $Re^{5+}$, $V^{4+}$, $Co^{3+}$, $Al^{3+}$and $Rh^{3+}$) doped $TiO_2$ have been designed for enhancing the visible light response of $TiO_2$ [22]. Not only these, various other transition metal ions doped $TiO_2$ photocatalytic materials have also been prepared for the degradation of various organic and inorganic pollutants [24, 25, 27,121-127]. In metal doped $TiO_2$ photocatalyst, the photogenerated electrons are trapped by metal which leads to enhanced charge carrier transfer rate and visible light photocatalytic degradation efficiency. In 2001, Asahi et al., reported nitrogen doped $TiO_2$ which extended the $TiO_2$ response to visible light and enhanced its photocatalytic efficiency by degradation of methylene blue and gaseous acetaldehyde under visible light irradiation [23]. In 2002, visible active chemically modified $TiO_2$ (i.e. n-$TiO_{2-x}C_x$) with band gap energy of 2.32 eV was prepared by Khan et al., for water-splitting into hydrogen and oxygen [128]. Yu et al., reduced the band gap energy of $TiO_2$ by doping with $F^-$ ions and used for oxidation of acetone [129]. Luo et al., improved the visible light response of $TiO_2$ by co-doping with bromine and chlorine and examined their efficiency for water splitting application [130]. Subsequently, different non-metals (B, C, S, P, F etc.,) doped $TiO_2$ photocatalytic materials have been synthesized to enhance the visible light photocatalytic degradation efficiency. Consequently, several articles reports appeared on the different synthesis methodologies of $TiO_2$, modification of $TiO_2$, the mechanistic study of degradation and influence of different operational parameters on the degradation efficiency [20, 26, 27, 125-127]. The modification involves intricate synthesis procedure and the reproducibility of doping level needs an attention. Moreover, in the presence of light irradiation, the modified photocatalyst undergo corrosion and long-term reactions lead to leaching of doped metal ions gradually results in a decrease in the photocatalytic degradation efficiency.

Non-$TiO_2$ sulphide based visible light photocatalysts like CdS, ZnS, $Cu_2O$ have been developed for the degradation of pollutants. The stability of these photocatalysts is poor and these are oxidized by holes in the VB to liberate metal ions which are hazardous to our health [31]. Consequently, bismuth-based photocatalysts such as $Bi_2O_3$, $Bi_2WO_6$, $Bi_2MTaO_7$ (M = In, Ga, and Fe), $BiVO_4$, BiOX (X-Cl, Br, I), $Bi_2MNbO_7$ (M = Al, Fe, In, Sm) and $Bi_2MoO_6$ etc., have been developed to improve the mobility of photogenerated carriers. The activity of photocatalyst depends upon the mobility of photogenerated electron-hole pairs and positions of the VB and CB. The VB of metal oxide photocatalysts is composed of O 2p, nevertheless, the VB of Bi-based photocatalytic materials consists of O 2p and Bi 6s hybrid orbitals and the CB consists of Bi 6p. So, the broad VB increases the mobility of the charge carriers and improved the degradation efficiency. Subsequently, it has also been combined with the $TiO_2$ to improve the photocatalytic degradation efficiency [32, 33, 131]. In 2009, Wang et al., prepared visible light responsive metal-free graphitic-like carbon nitride (g-$C_3N_4$) photocatalyst by heating cyanamide between 673 and 873 K for water splitting reactions. The absorption edge of g-$C_3N_4$ is in longer wavelengths, leading to decrease in the bandgap (2.7 eV) was obtained. Unlike the sulfide photocatalysts, the g-$C_3N_4$ is stable under light irradiation because of strong covalent bonds between carbon and nitride atoms [34]. Yan et al., synthesized g-$C_3N_4$ by heating low-cost melamine and examined their degradation performance [132]. In 2010, Yi et al., reported visible light active silver orthophosphate ($Ag_3PO_4$) for water splitting and degradation of MB and RhB dyes [35]. Similarly, various visible light responsive photocatalysts such as $MS_2$ (M-Mo, W, Sn), $AB_2X_4$ (A =Cu, Zn, Cd, Ca, etc.; B = In, Ga, Al; X = S, Se, Te) and ferrites etc., have been developed for the degradation of pollutants [95, 96, 133-135]. For an efficient visible light photocatalysis, semiconductors should possess lower band gaps, lesser rates of charge recombination, more positive VB and negative CB potentials. However, a single visible light responsive photocatalyst with lower band gap cannot hold less recombination and high redox ability and vice-versa. To overcome these, recently $TiO_2$ and non-$TiO_2$ based heterojunction photocatalytic systems that consist of two semiconductors are developed to improve the visible light absorption, separation efficiency of charge carrier and their redox ability. In heterojunction-like composites, the charge carriers are spatially separated by transferring the electron from CB of semiconductor 1 (SC1) to CB of semiconductor 2 (SC2) and holes are transferred from VB of semiconductor 2 (SC2) to VB of semiconductor 1 (SC1) (Fig. 5). Subsequently, different $TiO_2$ and non-$TiO_2$ heterojunction-like composites were synthesized for degradation of pollutants. However, the band edge potential of semiconductors is most essential for efficient photocatalytic activity. The band gap and band edge potential of some semiconductor photocatalysts are listed in Table 1. However, the redox ability of spatially separated photogenerated electrons and holes on the CB and VB of the two different semiconductors in the heterojunction-like composites are lower, because the redox potential of accumulated holes on the VB is lower than the potential required for $^{\bullet}OH$ generation ($H_2O$ or $-OH/^{\bullet}OH$ = +2.4 eV vs. NHE). In some cases, the accumulated holes on the VB could directly degrade the pollutants or the CB electrons indirectly degrade the pollutants using $^{\bullet}OH$ radicals formed by cleavage of *in-situ* produced hydrogen peroxide ($H_2O_2$); nevertheless, their efficiency does not fulfill the practical requirements. For example, Ge et al., developed g-$C_3N_4$/$Bi_2WO_6$ composite photocatalyst which exhibited efficient visible light photocatalytic activity than the pristine g-$C_3N_4$ and $Bi_2WO_6$ in degradation of MO. This may be due to the effective separation of photogenerated electron and holes by transferring electrons from the CB of g-$C_3N_4$ to the CB of $Bi_2WO_6$, and holes in the VB of $Bi_2WO_6$ moved to the VB of g-$C_3N_4$. However, accumulated holes in the VB of g-$C_3N_4$ may not produce $^{\bullet}OH$ by water oxidation because redox potential of VB edge of g-$C_3N_4$ (+1.57) is lower than the potential required for $^{\bullet}OH$ generation [136]. Fu et al., synthesized g-$C_3N_4$/CdS organic-inorganic heterojunctions photocatalysts, $0.7C_3N_4$–0.3CdS ratio exhibited higher activity than another ratio of composites, bare g-$C_3N_4$ and CdS in degradation of MO (5 mg/L) and 4-aminobenzoic acid (4-ABA, 25 mg/L). Under visible light irradiation, $g\text{-}C_3N_4$ and CdS are excited, charge carriers are well separated, nevertheless the holes gathered on the VB of $g\text{-}C_3N_4$ may not produce $^{\bullet}OH$, which decreases the final degradation efficiency.

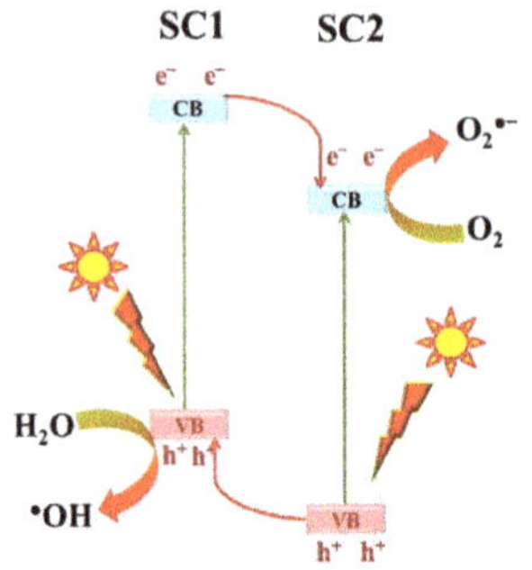


*Fig 5. Schematic representation of photogenerated holes and electron separation mechanism on heterojunction-like photocatalyst.*

In some cases, dye sensitization also accompanied with the heterojunction-like carrier separation mechanism which leads to improved degradation efficiency [137]. Sun et al., prepared BiOCl/$SnO_2$ heterojunction photocatalyst for degradation of RhB. It was revealed that dye photosensitization, defects states, and the heterojunction are mainly responsible for enriched degradation efficiency [162]. As described above the separated holes may directly decompose the pollutants, nevertheless, their degradation mechanism is still unclear.

*Table 1. Band gap and band edge position of some semiconductor photocatalysts.*

| Semiconductor | Potential (eV) | | Bandgap (eV) | Reference |
|---|---|---|---|---|
| | **VB** | **CB** | | |
| $TiO_2$ (anatase) | + 2.97 | -0.23 | 3.2 | [138] |
| $TiO_2$ (rutile) | + 2.50 | -0.50 | 3.0 | [139] |
| g-$C_3N_4$ | +1.57 | -1.13 | 2.7 | [140] |
| TaON | +2.12 | -0.17 | 2.29 | [141] |
| $Ta_3N_5$ | +1.70 | -0.40 | 2.1 | [142] |
| $CeO_2$ | +2.44 | -0.32 | 2.76 | [143] |
| $WO_3$ | + 3.35 | + 0.63 | 2.72 | [144] |
| ZnO | +3.10 | -0.10 | 3.2 | [145] |
| CdS | +1.75 | -0.39 | 2.4 | [137] |
| ZnS | +2.66 | -1.12 | 3.78 | [146] |
| $Bi_2O_3$ | +2.67 | +0.29 | 2.38 | [147] |
| BiOI | +2.31 | +0.58 | 1.73 | [148] |
| BiOBr | +3.06 | +0.29 | 2.77 | [149] |
| BiOCl | +3.52 | +0.20 | 3.20 | [150] |
| $Bi_2WO_6$ | +2.94 | +0.24 | 2.70 | [149] |
| $Bi_2MoO_6$ | +2.44 | -0.32 | 2.70 | [151] |
| $BiVO_4$ | +2.36 | 0.00 | 2.36 | [138] |
| $BiFeO_3$ | +2.63 | +0.54 | 2.09 | [152] |
| $Bi_2O_2CO_3$ | +3.44 | +0.23 | 3.21 | [153] |
| $BiPO_4$ | +4.28 | +0.43 | 3.85 | [154] |
| $Bi_2S_3$ | +1.43 | +0.13 | 1.30 | [155] |
| AgBr | +2.60 | 0.00 | 2.60 | [154] |
| AgI | +2.37 | -0.41 | 2.78 | [156] |
| $Ag_3PO_4$ | +2.90 | +0.46 | 2.44 | [151] |
| $Ag_2S$ | +1.10 | 0.00 | 1.1 | [157] |
| $Ag_2CrO_4$ | +2.20 | +0.45 | 1.75 | [158] |

| | | | | |
|---|---|---|---|---|
| $Ag_2CO_3$ | +2.60 | +0.43 | 2.17 | [159] |
| $MoS_2$ | +1.78 | -0.12 | 1.90 | [160] |
| $SnS_2$ | +1.19 | -0.98 | 2.17 | [161] |
| $SnO_2$ | +3.40 | +0.04 | 3.36 | [162] |
| $NaNbO_3$ | +2.57 | -0.69 | 3.26 | [163] |
| $CoFe_2O_4$ | +1.76 | +0.43 | 1.33 | [164] |
| $ZnFe_2O_4$ | +1.49 | -0.39 | 1.88 | [165] |
| $Nb_2O_5$ | +2.53 | -0.47 | 3.0 | [166] |
| $CuBi_2O_4$ | +2.35 | +0.87 | 1.48 | [167] |
| $CaIn_2S_4$ | +0.88 | -1.10 | 1.98 | [168] |
| $ZnIn_2S_4$ | +1.52 | -0.85 | 2.4 | [169] |
| $In_2S_3$ | +1.26 | -0.86 | 2.12 | [170] |
| $In_2O_3$ | +2.17 | -0.63 | 2.8 | [171] |
| $\alpha$-$Fe_2O_3$ | +1.38 | -0.82 | 2.2 | [172] |
| $CaWO_4$ | +4.59 | +0.50 | 4.09 | [173] |
| $BaTiO_3$ | +2.34 | -0.84 | 3.18 | [157] |
| CuO | +2.16 | +0.46 | 1.70 | [174] |
| $Cu_2O$ | +0.40 | -1.60 | 2.2 | [138] |

To improve the separation and degradation efficiency further, different dual semiconductors photocatalysts with solid-state electron mediators (metals, RGO and carbon) and redox-mediator-free system have been designed for degradation of pollutants. However, they discussed that photogenerated electron-hole pairs are efficiently separated through "Z-scheme" mechanism and named as Z-scheme photocatalysts with solid-state electron mediators and direct Z-scheme photocatalysts. The experimental evidence for this mechanism is ambiguous and not investigated in the deepest manner. Generally, Z-scheme term resembles the natural photosynthesis which is mainly used for thermodynamically uphill reaction (example, splitting water into hydrogen and oxygen) such as high positive change in the Gibbs free energy (i.e. $\Delta G > 0$). However, photocatalytic degradation of pollutants is a thermodynamically downhill reaction which occurs spontaneously with negative Gibbs energy change (i.e. $\Delta G < 0$).

So, the usage of "Z-scheme" in photocatalytic degradation of pollutants leads to the misconception and ambiguity [175]. Therefore, in this section of chapter we have dictated Z-scheme photocatalysts system with solid-state electron mediators and without redox-mediators (direct) as dual semiconductor system with solid-state electron mediators and without redox-mediators. In these cases, the photogenerated electrons in the CB of the SC 2 travelled to VB of the semiconductor 1 and recombines with the holes on the SC 1. Concurrently, the electrons and holes present in the CB and VB of the SC 1 and 2 are separated and improved the degradation efficacies [38, 39]. Metal nanoparticles (Ag, Au, Cu), reduced graphene oxide (RGO) and carbon materials are largely used as a solid-state electron mediators. In 2006, Tada et al., prepared gold mediator based $CdS/Au/TiO_2$ dual semiconductors photocatalytic system for methylviologen ($MV^{2+}$) reduction under visible light irradiation. Charge carriers were separated through vectorial electron transfer ($TiO_2$ → Au → CdS) such as electron on the CB of $TiO_2$ migrated to Au, then to the VB of the CdS where they recombined with holes on the VB. Afterwards the separated electrons and holes on the CB and VB of the CdS and $TiO_2$ efficiently participated in the reaction and displayed higher photocatalytic activity than the mono and bi photocatalytic systems [176]. Zhang et al., prepared non-$TiO_2$ based AgBr-Ag-$Bi_2WO_6$ system where charge carriers separated via $Bi_2WO_6$→Ag→AgBr vector-like transfer. It results in improved visible light activity in degradation of procion red MX-5B (MX-5B) and pentachlorophenol (PCP) compounds [38]. Similarly, RGO mediator system like g-$C_3N_4$/RGO/$Bi_2WO_6$ was developed by Ma et al., for degradation of 2, 4, 6-trichlorophenol (TCP) under visible-light irradiation [177]. RGO systems possess large specific surface area and their high electrical conductivity facilitates the transfer of charge carriers. Successively, various dual semiconductor photocatalysts with solid-state electron mediators have been prepared for the degradation of dyes and organic compounds. However, the drawbacks like absorption of visible light by metal particles decreases the light reaching the catalytic surface by surface plasmon resonance (SPR) effect and the intricate synthesis procedure for carbon and graphene limits their practical implementation. Consequently, electron/redox-mediator-free direct dual semiconductors photocatalytic systems have been developed to circumvent these drawbacks. For example, Arai et al., first prepared a $CuBi_2O_4/WO_3$ composite photocatalyst by mechanical mixing and studied their visible light activity by degradation of acetaldehyde ($CH_3CHO$). In the presence of visible light irradiation, the electrons and holes in the n-type $WO_3$ and p-type $CuBi_2O_4$ electrodes travel in the porous semiconductor films where recombination occurred. Afterward, the holes and electrons in the $WO_3$ and $CuBi_2O_4$ possess strong oxidation and reduction potential, which completely degrades the $CH_3CHO$ within 60 min [178]. Likewise, various redox-mediator-free direct dual semiconductor systems such as $NaNbO_3/WO_3$, g-$C_3N_4/TiO_2$, g-$C_3N_4/BiVO_4$, g-$C_3N_4$/N-

ZnO, $Ag_4V_2O_7/Ag_3VO_4$, $Bi_2MoO_6/Ag_3PO_4$, $MoS_2/Ag_3PO_4$, and $Bi_2MoO_6$/BiOBr etc., have been synthesised for the degradation of different types of organic and inorganic pollutants under visible light irradiation. We have also developed the redox-mediator-free g-$C_3N_4/TiO_2$, $CaIn_2S_4/TiO_2$ and $ZnIn_2S_4/Bi_2WO_6$ dual semiconductors for the degradation of isoniazid and metronidazole pharmaceuticals. The results revealed that dual semiconductors exhibited enriched degradation efficiency than the respective bare semiconductors. Under light irradiation, the generated electrons in the CB of $TiO_2$ and $Bi_2WO_6$ travelled to the VB of g-$C_3N_4$, $CaIn_2S_4$ and $ZnIn_2S_4$ and recombined with holes on the respective semiconductors. Subsequently, the electrons on the CB of g-$C_3N_4$ (-1.13 eV vs. NHE), $CaIn_2S_4$(-1.1 eV vs NHE) and $ZnIn_2S_4$(-0.85 eV vs NHE), and the holes in the VB of $TiO_2$ (+2.91 eV vs NHE) and $Bi_2WO_6$ (+2.94 eV vs NHE) were well separated, possess high negative and positive potentials for reduction of $O_2$ to $O_2^{\bullet -}$ and oxidation of $H_2O$ or -OH to $^{\bullet}$OH which lead to enhancement in the degradation efficiency. This mechanism was confirmed by reactive radicals quenching study, photoluminescence, hydroxyl radicals determination by fluorescence and electron spin resonance (ESR) spectroscopy analysis [40, 168, 179]. Recently, the visible light response, charge transfer rate and degradation efficiency were improved by synthesizing ternary composite photocatalyst and redox-mediator-free double direct dual semiconductor photocatalyst systems. For example, Zamiri, et al., developed ZnO-ZnS-$Ag_2S$ ternary nanostructures for NOx degradation and reported that 45.9, 47.7 and 53.1% of initial concentrations were degraded after 60min using ZnO, ZnO-ZnS and ternary systems, respectively. The mono and binary systems were not stable under reaction atmosphere due to photo-corrosion effects, hence dropped its activity. However, ternary system was stable up to third run of reaction [180]. Yuan et al., prepared $BiVO_4$-$Cu_2O$-$TiO_2$ ternary system for degradation of RhB dye under 9W energy-saving fluorescent lamp. Ternary composite was capable to degrade 97.8% of RhB after 8h irradiation while it was 80.7 and 31.6% for $Cu_2O$-$TiO_2$ and $Cu_2O$ systems, respectively. The enhanced photocatalytic activity of ternary system may be due to the formation of heterojunction among them, matched band edge positions of semiconductors and higher charge transfer rate, respectively [138]. Successively, graphene-$CoFe_2O_4$/CdS, $MoS_2$ quantum dots-graphene-$TiO_2$, $Fe_3O_4$/ZnO/$CoWO_4$, ZnO/$Ag_3VO_4/Fe_3O_4$, ZnO/AgI/$Ag_2CO_3$, ZnO-$Ag_2O$/porous g-$C_3N_4$, $Ag_3PO_4$/AgBr/Ag-HKUST-1-MOF and g-$C_3N_4$/CdS/$TiO_2$ nano-sheets etc., have been reported for the degradation of organic and inorganic pollutants. Likewise, Li et al., studied the visible light photoelectrocatalytic degradation of MO and phenol using redox-mediator-free double direct g-$C_3N_4/WO_3/Bi_2WO_6$ dual semiconductor systems. The produced electron-hole pairs in the CB and VB of $WO_3$, g-$C_3N_4$ and $Bi_2WO_6$ semiconductors were separated but recombination occurred between the electrons and holes in the CB and VB of $WO_3$ and g-$C_3N_4$. Similarly, electrons in the CB of g-

$C_3N_4$ combined with the holes in the VB of $Bi_2WO_6$. Thus, the presence of electrons in the CB of $Bi_2WO_6$ and holes in the VB of $WO_3$ were well separated, confirmed by enhancement in the photocurrent leading to the production of higher concentration of reactive radicals and improved the degradation efficiency[181, 182].We have also prepared redox-mediator-free double direct $ZnIn_2S_4$/g-$C_3N_4$/$BiVO_4$ nanorod-based ternary semiconductors which demonstrated higher degradation efficiency for CR dye and metronidazole pharmaceutical degradation than the bare semiconductors [183]. Most of the existed visible light photocatalytic systems are used high power artificial solar lights such as xenon, tungsten and halogen lamps as a visible light source, however due to their inherent drawbacks practical implementation are limited. Consequently, energy efficient visible light emitting diodes (VLED) have been used for the degradation of pollutants. For example, Yin et al., studied the photocatalytic deNOx abilities of N-doped $TiO_2$, Fe and Pt loaded-N doped $TiO_2$ catalyst under the irradiation of UV, blue, green, and red LED light. The results demonstrated that Pt loaded-N doped $TiO_2$ showed higher deNOx ability under all LED irradiation [184]. Sun et al., has used blue LED as an irradiation source for degradation of RhB and phenol using $Bi_2MoO_6$ photocatalyst [185]. We have also investigated the degradation behaviour of RhB dye using BiOX (X-Cl, Br, I) and $TiO_2$/$ZnFe_2O_4$ photocatalysts under blue, green and red LEDs irradiation [95,186]. LED could be used to degrade pollutants only as it is at lower concentration ineffective for higher concentration or need large number of LEDs.

In addition, metal organic frameworks (MOFs), a new class of organic visible light responsive semiconductors have been used for degradation of pollutants. MOFs are highly porous crystalline coordination polymers prepared using metal ions and polyfunctional organic ligands. MOFs are different from other high surface area materials like zeolites, other microporous and mesoporous materials because they contain light absorbing organic chromophores. This high surface area material is highly useful to enhance adsorption percentage of the pollutants, so MOFs have also been used in gas storage and separation, catalysis and drug delivery applications, [36, 187-191]. In 2007, Garcia et al., examined the semiconducting property of MOF-5 (Zn-based MOFs) and used it as a photocatalyst for degradation of phenol and 2, 6-di-tert-butylphenol. They found that MOF-5 showed better photocatalytic activity compared to P-25 $TiO_2$ and ZnO. However, MOF-5 shows poor stability under light irradiation [192, 193]. Mahata et al., prepared Co, Ni and Zn-based MOFs and studied their photocatalytic activity for degradation of orange G, RhB, Remazol Brilliant Blue R and MB dyes [194]. Subsequently, iron, aluminum, chromium, titanium based MOFs and composites of metal oxide with MOFs were developed [187-191] and still different studies are underway for designing of novel MOFs photocatalysts for degradation of organic and inorganic pollutants. In addition, the MOFs materials were also used as a template to synthesize corresponding metal oxide photocatalysts with the high surface area by pyrolysis of MOFs and used them for pollutants degradation [195, 196]. Though the MOFs are effective visible light responsive photocatalyst, nevertheless difficulties in their synthesis procedure and stability (reusability of results) not paving the way for them to move forward to implement the practical technology. As discussed above, till date different types of semiconductor materials have been used as a photocatalyst for the degradation of various organic and inorganic pollutants in the presence of UV, visible, LED and direct sunlight irradiation. However, all the photocatalytic systems have their own limitations, which put restrictions on this technology for being effectively used in the commercial sectors. Thus, to make use of the photocatalytic system for pollutants degradation at industrial and government sectors, efforts for better developments are needed.

## 5. Photocatalysis in $H_2$ production and $CO_2$ reduction

Photocatalytic production of $H_2$ fuel and $CO_2$ reduction reactions are the energy efficient alternative routes to the fossil fuel consumption and reduction of the $CO_2$ emission to the atmosphere. As discussed earlier in section 2, the $H_2$ production by water splitting and $CO_2$ reduction to valuable chemicals such as HCOOH, HCHO, $CH_3OH$, and $CH_4$ etc., are completely different from the pollutants degradation. Both the $H_2$ production and $CO_2$ reduction are thermodynamically uphill reactions with a high positive change in the Gibbs free energy. Considerable efforts have been made in respect of $H_2$ production and $CO_2$ reduction till date. It is, therefore, worthwhile, to sum up the journey of photocatalytic technology, its applications, and short comings in the following section.

After the discovery by Fujishima and Honda, the production of $H_2$ using semiconductor-based photocatalytic water splitting method under solar light was considered as one of the efficient technologies to solve the world energy crisis. For efficient $H_2$ production reaction, the bottom of CB level must be more negative than the reduction potential of $H^+/H_2$ (0 V), while the top of VB level must be lower than the oxidation potential of $O_2/H_2O$ (1.23 V), as illustrated in Fig. 4. $TiO_2$ material possesses sufficient VB (+2.97) and CB (-0.23) edge potentials to oxidize and reduce the water molecules and hence significant studies have been performed on $TiO_2$ for $H_2$ production. However, there are some other factors such as wide band gap energy (3.2eV), faster rate of photogenerated electrons and holes recombination and the low transfer rate of charge carriers to the surface which affect visible light photocatalytic activity. Moreover, the quantum efficiency of the reaction is also very low. To improve these, the photocatalytic reactions were performed in the presence of sacrificial agents such as aqueous solution of alcohols, organic acids, sulfite and iodate etc. Sacrificial agents or electron donors are irreversibly oxidized by photogenerated holes resulting in the enrichment of the concentration of electrons and thereby improving the $H_2$ production efficiency[197-200]. Besides, the added sacrificial agents prevent the back reaction between the generated $H_2$ and $O_2$ to form $H_2O$. Moreover, the use of $TiO_2$ with co-catalysts such as metal ions (Fe, Cu, Ni, Pt, Au, Ru, Rh, Pd, Ag, Au, lanthanides, alkali/alkali earth metals etc.), non-metal ions (B, N, C, S), and co-doping elements (Sn-Eu, Au-Rh, Au-Pt, Cu-Pt, B-N etc.), (Fig. 2) effectively improves the visible light photocatalytic $H_2$ production activity [48-51, 118, 130,198-200,201-205]. The introduction of dopants creates the intra-band states between the VB and CB of the semiconductor which provides electronic states and improves the visible light response, decreases the band gap energy and also offer a large amount of active sites for water reduction reaction. The $H_2$ production efficiency was also improved by declining the back reaction between formed $H_2$ and $O_2$ by performing the reaction in the gas phase using NaOH coated metal loaded $TiO_2$. NaOH serves to hold a higher amount of water on the catalyst surface, which enhances the photolysis of water in gas phase than the liquid phase [206, 207]. Moreover, the $TiO_2$ combined with other metal oxides ($ZrO_2$, ZnO,$SrTiO_3$, $SiO_2$, CuO etc.) and other mesoporous materials, showed higher $H_2$ production efficiency than the bare catalyst [208]. Like $TiO_2$, in 1990 titanates ($K_2Ti_4O_9$, $K_2Ti_2O_5$, $Na_2Ti_3O_7$, etc.,) and tantalates have also been used as efficient photocatalysts for $H_2$ production. The charge carriers recombination rate on these modified photocatalysts was dramatically decreased and the $H_2$ production efficiency and the quantum yield of these modified photocatalysts were improved [204, 209, 210]. Furthermore, the stability and reusability of the single and modified $TiO_2$ materials are very poor, because stability is the most important factor suggesting that the $H_2$ production efficiency of the photocatalyst should not significantly decrease with time.

To improve the visible light absorption further, non-$TiO_2$ based low band gap energy photocatalysts have been prepared to enhance the $H_2$ production efficiency. Initially, visible light responsive CdS or CdSe and other chalcogenides with low band gap energy were designed which demonstrated sufficient CB edge potential for efficient $H_2$ production [211]. In 1981, Harbour et al., studied the $H_2$ production efficiency of Pt/CdS and CdS particles using ethylenediaminetetraacetic acid (EDTA) as a sacrificial agent, they found that Pt/CdS-EDTA provided higher $H_2$ amount than the CdS-EDTA system [212]. However, CdS and CdSe were not stable in the reaction because the $S^{2-}$ and $Se^{2-}$ anions were more susceptible to oxidation by photogenerated holes than water, causing the CdS or CdSe itself to be oxidized and degraded and the quantum efficiency of these photocatalytic systems was also lower. CdS was modified by other metal oxides, sulphides, solid solutions and different nanostructures of CdS were also prepared to enhance the $H_2$ production efficiency [213, 214]. Kalyanasundaram et al., studied visible light driven water cleavage process using CdS with the addition of Pt and suppressed its photo-corrosion by addition of $RuO_2$. The photogenerated hole on the VB of CdS trapped by $RuO_2$ does not degrade the CdS and enriches the $H_2$ production efficiency [215]. Later in 1992, Kobayakawa et al., studied the $H_2$ production ability of $CuInS_2$ semiconductor with band gap range of 1.5-1.9 eV. It enhances the visible light absorption and possesses suitable CB edge potential for efficient $H_2$ evolution [216]. In 1998, $BiVO_4$ was identified as one of the visible light active photoanodes with band gap of 2.4 eV for $H_2$ production. However, the CB edge of unmodified $BiVO_4$ is located just below the $H_2$ production potential and hence suffers from charge carrier recombination, poor charge transport and water reduction properties [217]. Thus, various modifications have been carried out on the $BiVO_4$ to alleviate these limitations. In 2000, (oxy) nitrides ($Ta_3N_5$, TaON, $BaNbO_2N$, $SrTaO_2N$ etc.) have been utilized as photocatalysts for $H_2$ production under visible light.(Oxy)nitrides shows absorption bands in 500-750 nm, corresponding to the narrow bandgap energy in the range of 1.9-2.5 eV and possess appropriate VB and CB position for water oxidation and reduction [142, 218, 219]. However, (oxy)nitrides usually suffer from poor stability in water medium by self-oxidative deactivation and the photocatalytic activity in water over a long period cannot be maintained. This self-oxidative degradation has been suppressed by co-catalyst loading, but could not suppress completely [220, 221]. In addition, these (oxy)nitrides photocatalysts are majorly used as an electrode material to produce $H_2$ under visible light irradiation, hardly been used in the suspension based photocatalytic reaction. In 2003, Lei et al., synthesized a new ternary chalcogenide $ZnIn_2S_4$ with band gap energy of 2.3 eV which belonged to the family of ternary $AB_2X_4$ compounds and showed high activity for water reduction under visible light. No photo-corrosion was observed after 150 h of photocatalytic reaction, thus, paved the way for development of stable visible light active ternary metal sulphides for $H_2$ production [222].In 2005, Nørskov et al., prepared efficient $MoS_2$ catalyst for $H_2$ production. It has low band gap energy and high visible light absorption [223]. In 2009, Wang et al., developed a metal-free polymeric graphitic carbon nitride (g-$C_3N_4$) for $H_2$ production under visible light irradiation [34]. Followed by other catalysts, $In_2O_3$, $In_2S_3$, ferrites, MOFs etc., were investigated for $H_2$ production under visible light irradiation. The narrow band gap energy of these photocatalysts leads to strong absorption in the visible light range, however, single system using these photocatalysts suffers from poor charge transport and stability and some systems have low CB edge position than the $H_2$ production potential. To conquer these constraints, a non-$TiO_2$ based visible light active catalyst are combined with other non-$TiO_2$ or $TiO_2$ materials to make various binary and ternary composite photocatalysts with sufficient CB edge position and enhanced the charge transport rates for improved $H_2$ production [204, 224-228]. Furthermore, $H_2$ production efficiency of these composites was further enhanced by addition of co-catalysts, but still these systems suffer from poor stability and low quantum efficiency. The graphene material has been combined with these photocatalysts to enhance their photocatalytic efficiency by improving the charge carrier separation and transfer rates, stability and quantum efficiency. Graphene possess high specific surface area, thermal conductivity and excellent mobility of charge carriers at room temperature which make it an ideal support for these photocatalysts to enhance their $H_2$ production efficiency. The enhancement of photocatalytic performance may be because of efficient suppression of the recombination of the photogenerated charge carriers, decrease in the band gap of parent semiconductor, increase in catalytic sites, high surface area and acting as a photosensitizer and co-catalyst to catalyse the $H_2$ production [229, 230].

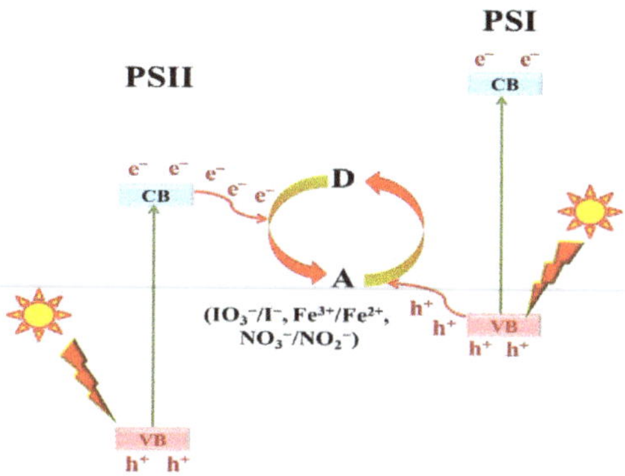


*Fig. 6. Schematic representation of photogenerated charge carrier separation mechanism in Z-scheme photocatalyst with shuttle redox-mediators.*

Likewise, another visible light active photocatalytic system such as Z-scheme which mimics the natural photosynthesis process, (photosystem II and photosystem I, PS II and PSI), has been developed for water splitting applications. In Z-scheme, two different semiconductors are combined, in which one has a higher VB potential and another having a higher CB potential for water oxidation and reduction using as appropriate shuttle redox-mediators [231]. Z-scheme system utilizes higher visible light than the single water splitting photocatalytic system which led to efficient separation of photogenerated charge carriers in two different semiconductors (Fig. 6) and the potential is kept higher than the required $H_2$ evolution potential in order to enhance the $H_2$ production efficiency. Abe et al., used Z-scheme system that consists of Pt-$TiO_2$ anatase and $TiO_2$ rutile for $H_2$ and $O_2$ evolution using $IO_3^-/I^-$ shuttle redox mediator. It was revealed that 180 and 90 μmol/h of

$H_2$ and $O_2$ were produced from a basic aqueous NaI solution under UV irradiation [231]. The photogenerated electrons in the CB of $BiVO_4$ recombine with the photogenerated holes in the VB of Ru/$SrTiO_3$: Rh, the electrons and holes in the CB and VB of Ru/$SrTiO_3$: Rh and $BiVO_4$ are used to reduce and oxidize water into $H_2$ and $O_2$, respectively [232]. Kato et al., reported Pt/$SrTiO_3$: Rh as an $H_2$-evolving photocatalyst, and $BiVO_4$, $Bi_2MoO_6$, and $WO_3$ as $O_2$-evolving photocatalysts using $Fe^{3+}/Fe^{2+}$ mediator [233]. Subsequently, different Z-scheme photocatalysts like $SrTiO_3$: Cr/Ta, $WO_3$ and Ru loaded $SrTiO_3$:Rh, $BiVO_4$with $IO_3^-$ /$I^-$ and $Fe^{3+}/Fe^{2+}$ were developed [234, 235]. Similarly, exhaustive water-splitting systems constructed in the last decade have been reported in various review articles. Z-scheme photocatalyst with shuttle redox-mediators holds major drawbacks such as the presence of redox mediators which lead to backward reactions, difficulty to achieve simultaneous production of $H_2$ and $O_2$ and the redox mediators strongly absorb visible light which reduces the light reaching to the photocatalyst surface. Furthermore, this system applicable to aqueous-phase only and efficiency relies on the suitable and stable redox-mediators. To overcome these drawbacks, solid-state electron mediators and redox-mediator-free direct Z-scheme photocatalytic systems have been developed for water splitting applications (Fig. 7a and 7b). In such systems, two photosystems are connected through a solid-state electron mediators (metal, carbon, GO) or direct intimate contact (shuttle redox-mediator-free). Ma et al., prepared sandwich-structured g-$C_3N_4$/Au/CdZnS photocatalyst for $H_2$ production from an aqueous mixture of $Na_2S$ and $Na_2SO_3$ solutions. It was revealed that 18.3 mmol/g of $H_2$ was produced using g-$C_3N_4$/CdZnS, while it was enhanced to 24.6 mmol/g for g-$C_3N_4$/Au/CdZnS catalysts. Under light irradiation, the electrons in the CB of CdZnS travelled to metallic Au and combined with the holes in VB of g-$C_3N_4$. The electrons in the CB of g-$C_3N_4$ and holes in the VB of CdZnS were efficiently separated, thus electrons in the CB of g-$C_3N_4$ held high negative potential and easily reduces the $H^+$ to $H_2$[236]. Lu et al., prepared Z-Scheme g-$C_3N_4$/Ag/$MoS_2$ternary photocatalysts for $H_2$ production from aqueous solutions with 15vol % of triethanolamine (TEOA). About 10.4 μmol/h of $H_2$ was produced using ternary Z-scheme, which is about 3.51 times higher than the bare g-$C_3N_4$ (2.963 μmol/h) [237]. Likewise, different solid-state electron mediators based Z-scheme photocatalysts have been designed for $H_2$ production but still, research is under progress in this area. The solid-state electron mediators possess serious problems which decrease the $H_2$ production efficiency and practical implementation. Therefore, shuttle redox-mediator-free direct Z-scheme photocatalytic systems with intimate contact have been developed for $H_2$ production. Sasaki et al., designed Ru/$SrTiO_3$:Rh/$BiVO_4$ direct Z-scheme powder photocatalysts, where Ru/$SrTiO_3$:Rh powder for $H_2$ evolution and $BiVO_4$ powder for $O_2$ evolution were used [238]. Zhao et al., designed redox-mediator-free g-$C_3N_4$/$WO_3$ Z-scheme photocatalyst for $H_2$ production in water/TEOA solution, water splitting efficiency was improved by the addition of Pt and RGO. The highest $H_2/O_2$ evolution rates of 2.84 and 1.46 μmol/h, with a quantum yield of 0.9% were obtained [239]. Zhou et al., prepared CdS nanoparticles/$BiVO_4$ nanowires Z-scheme photocatalyst for $H_2$ production using 20 vol% lactic acid solution. CdS/$BiVO_4$ with 1:2 wt ratio showed 3.65 times (124 μmol/h) higher photocatalytic activity than the bare CdS (34 μmol/h). The Pt loading further enhanced the $H_2$ production efficiency, Pt-CdS/$BiVO_4$ (1253 μmol/h) displayed 2.03 times higher activity than the Pt/CdS (568 μmol/h) [240]. Similarly, a various number of direct Z-scheme photocatalysts have been developed. The higher $H_2$ production efficiency was obtained as compared to the single-step photocatalytic system. However, the quantum yield obtained so far are not significantly higher.

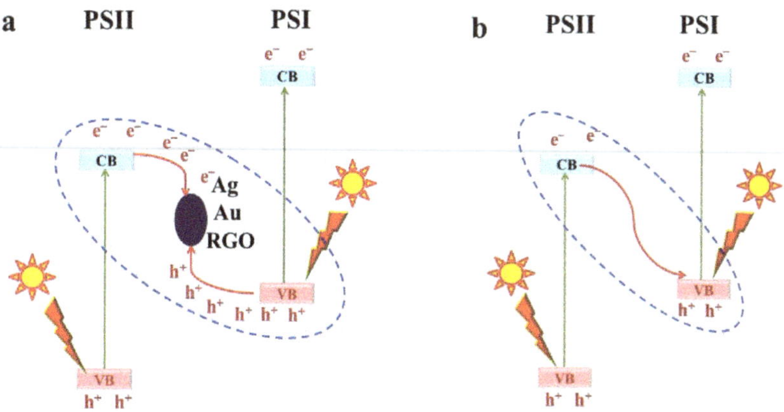


*Fig. 7. Schematic representation of photogenerated charge carrier separation mechanism in Z-scheme photocatalyst with (a) solid-state electron mediators and (b) redox-mediator-free (direct Z-scheme).*

In the above-mentioned reactions, precious metals have been majorly used as co-catalyst for enhancing the $H_2$ production efficiency, but their preciousness it cannot use as a co-catalyst for large-scale energy production. These precious co-catalysts have been replaced by inexpensive metals such as Fe, Co, Ni, Cu, their oxides, hydroxides ($Co(OH)_2$, $Ni(OH)_2$, $Cu(OH)_2$) and sulphides (FeS, CoS, NiS, $NiS_2$, CuS) to enhance the $H_2$ production efficiency [241-246]. Though, the loading of these co-catalysts enhances the $H_2$ production efficiency of pristine semiconductors, however, still their performance has not reached to the efficiency level as obtained by precious metal loading. Recently, transition metal disulfides (like $WS_2$, $MoS_2$) based photocatalysts have emerged as inexpensive co-catalysts for $H_2$ production than the precious metals. Among these, $MoS_2$ is a typical graphene-like layered transition-metal dichalcogenide low-cost co-catalysts, but bulk $MoS_2$ cannot be used as a long-term $H_2$ production catalyst. $MoS_2$ nanosheets which provide higher specific surface area, more active sites, adjustable bandgap energy, band position and controllable phase transformation for efficient $H_2$ production [247, 248].

All the above photocatalytic systems are majorly used as sacrificial agents enhance the $H_2$ conversion rate. To decrease the water pollution and make the sustainable $H_2$ production system, recently pollutants containing wastewater has been used as a source for $H_2$ production and simultaneously degrades the pollutants in the presence of UV and visible light irradiation [249-255]. Choi et al. produced $H_2$ through photocatalytic degradation of 4-chlorophenol and bisphenol pollutants under artificial and natural solar light irradiation. They revealed that $H_2$ production decreased with increase in the degradation percentage of pollutants [251]. The organic substances serve as a hole scavenger that decreases the recombination of photogenerated electron-holes pairs and enhances the $H_2$ production efficiency. In recent days' researchers are working on this field to develop better photocatalytic systems for simultaneous degradation of pollutants and $H_2$ production. The dual purpose photocatalytic systems are supposed to be useful because of a decrease in energy requirement and overall cost of wastewater treatment.

Like $H_2$ production, the $CO_2$ reduction is an effective route to decline $CO_2$ emission and concentration in the atmosphere but quite challenging process because $CO_2$ molecules are highly inert and stable compounds, their reduction is much more difficult than the proton reduction. In addition, it involves carbon reduction, proton transfer, and hydrogenation processes as well as the analysis of a variety of $CO_2$ reduction products which make more complicated process than the water splitting [54, 55]. In 1978, Halmann, photoelectrochemically reduced the aqueous $CO_2$ to HCOOH, HCHO, and $CH_3O$ Husing p-type GaP electrode in the presence of high-pressure mercury (Hg) and halogen lamp [256]. Hemminger et al., reduced $CO_2$ molecules to $CH_4$ in the gas phase with water vapor on the $SrTiO_3$ surface under the irradiation of 500 W high-pressure mercury vapor lamp [257]. Inoue et al., in 1979, first time performed the solid-liquid phase reduction reaction of $CO_2$ to HCHO and $CH_3OH$ using $TiO_2$, CdS, ZnO, GaP, SiC and $WO_3$ semiconductors as a photocatalyst in the presence of Xe or high-pressure Hg arc lamp (500W) [52]. Halmann et al., photocatalytically reduced $CO_2$ to HCOOH, HCHO, $CH_3OH$, $CH_3CHO$ and ethanol (EtOH) using $TiO_2$ doped with $RuO_2$ loading under a high

pressure mercury immersion lamp [53]. Tennakone et al., used Pt, Au, Ag, Co, Pb, and Hg coated $TiO_2$ powder for reduction of $CO_2$ to HCHO. Amongst these, Hg-coated $TiO_2$ showed the highest yield of HCHO [258]. In later years, many researchers have worked on the photocatalytic reduction of $CO_2$ in gas or water phase using various semiconductors photocatalysts ($TiO_2$, ZnO, CdS, metal-doped $TiO_2$ etc.). The $CO_2$ reduction efficiency is highly depended on the pH of the reaction medium because the acid medium leads to the low solubility of $CO_2$ in a water medium. The reduction efficiency was improved by performing the reaction in organic solvents. Liu et al., studied the effect of solvents (2-propanol, carbon tetrachloride, acetonitrile, dichloromethane etc.,) on the photocatalytic $CO_2$ reduction efficiency using $TiO_2$ nanoparticle embedded $SiO_2$ matrix. The results revealed that $CO_2$ solubility and reduction efficiency were greatly influenced by solvents, and the major reduction products such as formate and carbon monoxide were increased by increasing the dielectric constant of the solvent [259]. Different other solvents triethanolamine (TEA), trimethylamine (TMA), and *N*, *N*-dimethyl formamide (DMF) etc., were used to enhance the $CO_2$ solubility and reduction efficiency. The $CO_2$ solubility was further improved by changing pH of the reaction medium and the reduction reaction was also carried out in sodium hydroxide (NaOH), potassium hydroxide (KOH) or sodium bicarbonate ($NaHCO_3$) solution [260-262]. In addition, the low adsorption of $CO_2$ on the catalyst surface, competition between the $H_2O$ and $CO_2$ to capture a photogenerated electron from the semiconductor and the selectivity in the product formation decrease the photocatalytic $CO_2$ reduction efficiency. To enhance the adsorption quantity and improve product selectivity, the $TiO_2$ photocatalyst was embedded on the zeolites, $SiO_2$, Ti-containing zeolites (TS-1, Ti–Beta), mesoporous molecular sieves (Ti-MCM-41, Ti-MCM-48, Ti-HMS, Ti-FSM), and porous silica glass (Vycor) materials [263-268]. For instance, Anpo et al., included $TiO_2$ catalyst within the framework of mesoporous zeolites (Ti-MCM-41 and Ti-MCM-48), studied selectivity in formation of $CH_3OH$ in the photocatalytic reduction of $CO_2$ in gaseous $H_2O$ and compared with Ti-microporous zeolites (TS-1). The results revealed that selectivity for the $CH_3OH$ formation was highly depended on the photocatalyst. Ti-containing zeolites led to the considerable amount of $CH_3OH$, while in the case of bulk $TiO_2$,$CH_4$ was the major product [265]. The enrichment in the reduction efficiency was due to the larger effective surface area, reduced rate of charge recombination and small size of the anatase crystallites. Moreover, the $CO_2$ adsorption capacity on the $TiO_2$ photocatalysts was enhanced by introducing the basic site on that surface. Xie et al., studied the reduction of $CO_2$ to $CH_4$ with $H_2O$ using Pt-$TiO_2$ catalyst with the addition of basic oxides such as MgO, CaO, SrO, BaO, $La_2O_3$ and $Lu_2O_3$, the enhanced $CO_2$ adsorption were obtained [269]. Liu et al., prepared MgO-$TiO_2$ composites for $CO_2$ reduction with $H_2O$ vapor and they observed good adsorption capacity and increased $CO_2$ adsorption after MgO addition [270]. As mentioned earlier, the lower lifetime of photogenerated electron-hole pairs and poorly visible light response of $TiO_2$ was responsible for lowering the visible light photocatalytic $CO_2$ reduction efficiency. Like pollutants degradation and $H_2$ production, the $TiO_2$ was loaded with metals (Cu, Ru, Pd, Ag, Au, Rh, Pt, Fe, In), and sensitization (dyes, quantum dots) for improving the $CO_2$ reduction efficiency and selectivity (Fig. 2) [55, 261,271-276]. Hirano et al., synthesized copper nanoparticles loaded $TiO_2$ which effectively trapped electron, reduced charge recombination and produced mainly $CH_3OH$ [271]. $TiO_2$ also combined with other low band gap semiconductors (CdS, $Cu_2O$, AgBr, $FeTiO_3$, $CeO_2$ etc.,) graphene, carbon nanotube, other carbons and clays, yielded higher $CO_2$ reduction efficiency under visible light, because of defects in graphene and carbon structure which improve the visible light response of $TiO_2$ and charge transfer rate[55, 274-276, 277-279].

As described above, different modifications have been performed on the $TiO_2$ photocatalyst to improve the $CO_2$ reduction efficiency; however, their poor reusability and stability under light irradiation diminish their practical usability. In view of this, different non-$TiO_2$ based semiconductors such as $Cu_2O$, $ZrO_2$, $Ta_2O_5$, $CaFe_2O_4$, $Zn_2GeO_4$, CdS, $InTaO_4$, bismuth catalyst, g-$C_3N_4$, $AB_2X_4$, silver-based catalyst etc., have been developed to enhance visible light utilization and $CO_2$ reduction efficiency [273-275, 280-286]. In 1989, Tennakone et al., selectively reduced the $CO_2$ to $CH_3OH$ using aqueous suspension of hydrous cuprous oxide ($Cu_2O \cdot xH_2O$). The higher selectivity in the $CH_3OH$ formation is mainly due to the higher flat-band potential, multi electron transfer and strong chemisorption of $CO_2$, but the stability $Cu_2O$ is very poor [280]. In 1994, Matsumoto et al., studied the $CO_2$ reduction behavior of $CaFe_2O_4$ catalyst in 0.01 M NaOH solution,$NaH_2PO_2$ and $Fe^{2+}$ ion were added as reducing agent in the solution, HCHO and $CH_3OH$ were the major reduction products, however, this catalyst utilizes little amount of visible light [281]. Liu et al., selectively produced the EtOH by reducing the $CO_2$ in water using $BiVO_4$. It was stable up to five cycles of reduction experiments; nevertheless, the CB band edge potential was lower than the potential required for $CO_2$ reduction [282]. In 2011, Zhou et al., and Cheng et al., studied the reduction behavior of ultrathin and uniform $Bi_2WO_6$ square nanoplates. The $Bi_2WO_6$ utilizes higher amounts of visible light, the ultrathin geometry of $Bi_2WO_6$ efficiently decreased the charge carrier recombination rate, ultimately improved the $CO_2$ reduction efficiency [283, 284]. Jiang et al., synthesized $CdIn_2S_4$ microsphere with different low band gap energy (1.68, 2.09 and 2.05 eV) with respect to the different sulfur precursor (L-cysteine, thioacetamide, thiourea), effectively reduced the $CO_2$ to dimethoxymethane and methyl formate in $CH_3OH$ solution. The photocatalyst absorbs high concentration of visible light, however, suffers from higher recombination rate of charge carriers [286]. Similarly, other low band gap energy visible light responsive semiconductors are reported for production of solar fuels by reduction of $CO_2$.

To improve the catalyst stability, charge carrier separation, and $CO_2$ reduction, the photocatalytic materials have been combined with other semiconductors, graphene and co-catalysts to make the binary and ternary composite materials. Wang et al., designed an optical fiber photocatalytic reactor, studied both aqueous and vapor-phase $CO_2$ reduction reactions using NiO/$InTaO_4$ photocatalyst and thin film under visible light and real sunlight irradiation. Around 11.1 μmol/g/ h of $CH_3OH$ was produced under visible light, while it was 11.30 μmol/g/h for real sunlight irradiation. The quantum efficiency was 14 times higher than that of the aqueous-phase reaction because NiO/$InTaO_4$ thin film had greater light energy utilization [287]. Li et al., synthesized CdS or $Bi_2S_3$/$TiO_2$ nanotube photocatalyst for reduction of $CO_2$ to $CH_3OH$. The visible light response of $TiO_2$ nanotube was significantly improved after CdS or $Bi_2S_3$ loading. $Bi_2S_3$/$TiO_2$ nanotube showed higher reduction efficiency (224. 6 μmol/L of $CH_3OH$) than the CdS/$TiO_2$ nanotube (159.5 μmol/L of $CH_3OH$), because of a higher amount of $CO_2$ adsorption and enhanced the visible light response [288]. Lv et al., synthesized RGO modified $NiO_x$-$Ta_2O_5$ composite, yielded a higher amount of $H_2$ and $CH_3OH$ because of the effective lengthening the lifetime of the photogenerated electron-hole pairs [289]. Liu et al., developed visible light responsive $CdIn_2S_4$/mesoporous g-$C_3N_4$ hybrids for reduction of $CO_2$ to $CH_3OH$. About 42.7 μmol/g/ h of $CH_3OH$ was formed after 4h reaction in 0.1 M NaOH solution on use continuous-flow reactor system. The photocatalyst was stable up to 4 cycles and the charge carriers were separated by transferring the electrons from CB of g-$C_3N_4$ to CB of $CdIn_2S_4$ whereas holes in VB of $CdIn_2S_4$ migrated to the surface of mesoporous g-$C_3N_4$[290]. Likewise, various visible light responsive binary and ternary composite materials have been designed for enhancing the lifetime of charge carriers and $CO_2$ reduction efficiency [291-293]. Moreover, like $H_2$ production, the Z-scheme type photocatalysts have also been prepared to increase the visible light absorption, charge carrier's separation and transfer rate, $CO_2$ reduction efficiency, and selectivity. Sato et al., designed a hybrid Z-scheme photocatalyst comprising of semiconductors (N-$Ta_2O_5$, InP, GaP, Ni-ZnS, reduced $SrTiO_3$) and metal complex electrocatalyst (ruthenium complex polymer) to improve the conversion efficiency. This system mimics the natural photosynthesis, efficiently separate the charge carriers and enhance the $CO_2$ to HCOOH reduction rate [294-296]. Sekizawa et al. developed Ag-tantalum oxynitride (TaON) and Ru(II) dinuclear complex based system for reduction of $CO_2$ into HCOOH using methanol as a reducing agent [297]. However, the poor stability of metal complexes under irradiation is not suitable for practical implementation. Therefore, visible light responsive semiconductors with suitable band edge position have been utilized to develop efficient and robust Z-scheme photocatalytic materials. Yu et al., selectively produced $CH_3OH$ using g-$C_3N_4$/ZnO direct Z-scheme photocatalyst, exhibited 2.3 times higher activity than the bare g-$C_3N_4$, and attributed to the efficient separation of charge carriers at the intimate contact interface between the g-$C_3N_4$ and ZnO by following the Z-scheme transfer mechanism [298]. Various other Z-scheme systems such as CdS/$WO_3$ [299], $Ag_3PO_4$/g-$C_3N_4$ [300], α-$Fe_2O_3$/$Cu_2O$ [301], BiOI/g-$C_3N_4$ [302], etc., have been developed for $CO_2$ reduction and still the research is underway to develop different Z-scheme type photocatalysts.

Similarly, MOFs are other efficient visible light responsive high surface area photocatalytic materials. Due to their unique properties, these have been mostly used for improving the $CO_2$ adsorption capacity and enhancing reduction efficiency. Fu et al., synthesized amine functionalized titanium MOFs ($NH_2$-MIL-125 (Ti)) for reduction of $CO_2$ into HCOOH under visible light irradiation. Maximum $CO_2$ adsorption was observed for $NH_2$-MIL-125 (Ti) (132.2 $cm^3$/g) which was 1.3 times higher than the MIL-125 (Ti) (98.6 $cm^3$/g)and this higher conversion of $CO_2$ to HCOOH (8.4 μmol) was obtained [303]. Sun et al., investigated $CO_2$ reduction efficiency using zirconium (Zr) based MOFs, $NH_2$-Uio-66(Zr) in acetonitrile solution and TEOA as a sacrificial agent. These MOFs possess higher negative redox potential than the Ti-MOFs which led to higher concentration of HCOOH (13.2 μmol) was obtained [304]. Subsequently, several MOFs materials and their modification by loading with metal nanoparticles, other semiconductors, and graphene oxide have been designed for $CO_2$ reduction. However, the poor thermal and photostability of these materials make this process economically and environmentally not viable. Therefore, still, research work is underway to develop highly stable MOF photocatalysts for efficient reduction of $CO_2$. As described above, different photocatalytic processes have been utilized for the degradation of pollutants, $H_2$ production, and $CO_2$ reduction reactions. Nevertheless, the problems highlighted in the above sections are highly challenging and require an additional improvement in the designing of efficient photocatalytic systems for alleviating these drawbacks.

## 6. Future outlook

As discussed, the photocatalysis is one of the efficient processes to decontaminate the pollutants present in the environment, produce sustainable $H_2$ fuel and reduce $CO_2$ into value added products. However, the poor separation and transport efficiency of photogenerated electron-hole pairs, poor solar-to-hydrogen and $CO_2$ conversion efficiency, the back reaction of generated $H_2$ and $O_2$, the lower lifetime and poor reusability of these photocatalytic systems in real scale have created hurdles for their commercialization.

Presently, a variety of single, binary and ternary composite photocatalysts are available, still, research is underway to design other novel catalysts for degradation of pollutants under visible light, and their practical usability. In addition, the visible light degradation activity of developed systems is mainly performed in the presence of artificial visible light, thus, still the direct solar light utilization is poor and it is a major challenge ahead for the development of practical systems. Moreover, the future lab scale photocatalytic system might be applied to work on the degradation of real wastewater, thus it can be easily utilized for practical implementation.

In the case of $H_2$ fuel production and $CO_2$ reduction, the preparation of Z-scheme photocatalytic systems using low band gap energy semiconductors with appropriate band edge potential are assumed to improve the charge carrier generation, separation, and transport efficiency. The improved charge transport enhances the solar-to-hydrogen and $CO_2$ reduction efficacy. Further studies devoted to using the *in-situ* characterization techniques for proving the Z-scheme type transfer of charge carrier at the interfaces of two closely connected semiconductors are needed. Moreover, the research efforts have to be directed to design systems which simultaneously yield the $H_2$ fuel and degrade the pollutants under artificial visible light and direct solar light irradiation. The simultaneous processes can decrease the total energy consumption which will be highly useful to develop practical technology for $H_2$ fuel generation from real wastewater with simultaneous purification which enhances solar energy conversion efficiency. The quantum efficiency obtained so far is not up to the mark and further studies are needed for the development of multicomponent composite photocatalysts to extend their light absorption range, tune the band gap energy, facilitate the charge carrier generation, migration and utilization efficiency and increase the quantum efficiency. Besides, significant improvement is needed on the development of highly active, robust, inexpensive and earth-abundant cocatalysts for $H_2$ evolution reactions. Future studies need to be focused on developing dye-sensitized photocatalytic systems to enhance solar-to-hydrogen conversion efficiency. Adsorption and activation of $CO_2$ onto the surface of photocatalysts are the vital steps in defeating the competitive reduction to enhance the selectivity towards carbonaceous products. Therefore, future research will be focused on the development of efficient single or composite photocatalysts with suitable CB edge potential, changing the pH of reaction mixture, modification of catalyst surface by introducing the hydrophobicity nature, functionalized with basic sites or some organic semiconductors and supporting on high surface area materials to enhance the $CO_2$ adsorption capacity and selectivity. Like $H_2$ production, the loading of cocatalysts is needed to improve water oxidation and selectivity in the $CO_2$ reduction products formation. Furthermore, the methodical mechanistic study is required to prove the properties of cocatalysts, their relationship in enhancing the carrier separation and photocatalytic performance in either $H_2$ generation or $CO_2$ reduction. In addition, the in-depth understanding of the reaction mechanism of $CO_2$ reduction at the molecular level through *in-situ* spectroscopical studies is needed. Finally, it is expected that the future research should be dedicated to develop robust and effective photocatalytic systems based on the above-described ideas for environmental remediation and energy application using proper technologies and promote them to apply simultaneously in all these applications.

**Acknowledgments**

CSIR-CSMCRI Communication No. PRIS 040/2017.

**References**

[1] M.A. Montgomery, M. Elimelech, Water and sanitation in developing countries: Including health in the equation, Environ. Sci. Technol.41 (2007) 17-24. https://doi.org/10.1021/es072435t

[2] Information on http://www.un.org/waterforlifedecade/background.shtml

[3] Information on https://www.co2.earth/

[4] G. Centi, S. Perathoner, Towards solar fuels from water and $CO_2$, ChemSusChem 3 (2010) 195-208. https://doi.org/10.1002/cssc.200900289

[5] E.J. Maginn,What to do with $CO_2$, J. Phys. Chem. Lett.24 (2010) 3478-3479. https://doi.org/10.1021/jz101582c

[6] A. Ahmad, S.H. Mohd-Setapar, C.S. Chuong, A. Khatoon, W.A. Wani, R. Kumar, M. Rafatullah, Recent advances in new generation dye removal technologies: novel search for approaches to reprocess wastewater, RSC Adv. 5 (2015) 30801-30818. https://doi.org/10.1039/C4RA16959J

[7] K. Kumari, T.E. Abraham, Biosorption of anionic textile dyes by nonviable biomass of fungi and yeast, Bioresour. Technol. 98 (2007) 1704-1710. https://doi.org/10.1016/j.biortech.2006.07.030

[8] E. Khan, M. Li, C. P. Huang, Hazardous waste treatment technologies, Water Environ. Res. 79 (2007) 1858-1902. https://doi.org/10.2175/106143007X218601

[9] W.H. Glaze, J.W. Kang, D.H. Chapin, The chemistry of water treatment processes involving ozone, hydrogen peroxide and ultraviolet radiation, Ozone Sci. Eng. 9 (1987) 335-352. https://doi.org/10.1080/01919518708552148

[10] E.M. Aieta, K.M. Regan, J.S. Lang, L. McReynolds, J.W. Kang, W.H. Glaze, Advanced oxidation processes for treating groundwater contaminated with TCE and PCE: pilot scaleevaluations,J. Am. Water Works Association 80 (1988) 64-72.

[11] R. Andreozzi, V. Caprio, A.Insola, R. Marotta, Advanced oxidation processes (AOP) for water purification and recovery, Catal. Today 53(1999) 51-59. https://doi.org/10.1016/S0920-5861(99)00102-9

[12] A. Fujishima, K. Honda, Electrochemical photolysis of water at a semiconductor electrode, Nature 238(1972) 37-38. https://doi.org/10.1038/238037a0

[13] F. Steinbach, Heterogeneous photocatalysis, in:Catalysis, Springer-Verlag Berlin Heidelberg, Germany, 1972, pp. 117-154. https://doi.org/10.1007/bfb0051372

[14] M. Sciavello (Ed.), Photocatalysis and Environment: Trends and Applications, Kluwer Academic Publishers, Dordrecht,1988.

[15] N. Serpone, E. Pelizzetti (Eds.), Photocatalysis: Fundamentals and Applications, Wiley, New York, 1989.

[16] J.-M. Herrmann, C. Guillard, P. Pichat, Heterogeneous photocatalysis: an emerging technology for water treatment, Catal. Today 17 (1993) 7-20. https://doi.org/10.1016/0920-5861(93)80003-J

[17] M.A. Fox, M.T. Dulay, Heterogeneous photocatalysis, Chem. Rev. 93 (1993) 341-357. https://doi.org/10.1021/cr00017a016

[18] A. Mills, R. H. Davies, D. Worsley, Water purification by semiconductor photocatalysis, Chem. Soc. Rev. 22 (1993) 417-425. https://doi.org/10.1039/cs9932200417

[19] A.L. Linsebigler, G. Q. Lu, J. T. Yates, Photocatalysis on $TiO_2$ surfaces: principles, mechanisms, and selected results, Chem. Rev. 95 (1995) 735-758. https://doi.org/10.1021/cr00035a013

[20] X. Chen, S.S. Mao, Titanium dioxide nanomaterials: Synthesis, properties, modification, and applications, Chem. Rev. 107 (2007) 2891-2959. https://doi.org/10.1021/cr0500535

[21] S. Sarkar, R. Das, H. Choi, C. Bhattacharjee, Involvement of process parameters and various modes of application of $TiO_2$ nanoparticles in heterogeneous photocatalysis of pharmaceutical wastes- a short review, RSC Adv. 4 (2014) 57250-57266. https://doi.org/10.1039/C4RA09582K

[22] W. Choi, A. Termin, M.R. Hoffmann, The role of metal ion dopants in quantum-sized $TiO_2$: correlation between photoreactivity and charge carrier recombination dynamics, J. Phys. Chem. 98 (1994) 13669-13679. https://doi.org/10.1021/j100102a038

[23] R. Asahi, T. Morikawa, T. Ohwaki, K. Aoki, Y. Taga, Visible-light photocatalysis in nitrogen-doped titanium oxides, Science 293 (2001) 269-271. https://doi.org/10.1126/science.1061051

[24] R.J. Tayade, R.G. Kulkarni, R.V. Jasra, Transition metal ion impregnated mesoporous $TiO_2$ for photocatalytic degradation of organic contaminants in water, Ind. Eng. Chem. Res. 45 (2006) 5231-5238. https://doi.org/10.1021/ie051362o

[25] S. Malato, P. Fernandez-Ibanez, M.I. Maldonado, J. Blanco, W. Gernjak, Decontamination and disinfection of water by solar photocatalysis: Recent overview and trends, Catal. Today 147 (2009) 1-59. https://doi.org/10.1016/j.cattod.2009.06.018

[26] L.G. Devi, R. Kavitha, A review on non-metalion doped titania for the photocatalytic degradation of organic pollutants under UV/solar light: Role of photogenerated charge carrier dynamics in enhancing the activity, Appl. Catal. B Environ. 140-141 (2013) 559-587. https://doi.org/10.1016/j.apcatb.2013.04.035

[27] R. Daghrir, P. Drogui, D. Robert, Modified TiO2 for environmental photocatalytic applications: A review, Ind. Eng. Chem. Res. 52 (2013) 3581-3599. https://doi.org/10.1021/ie303468t

[28] K. Mogyorosi, I. Dekany, J.H. Fendler, Preparation, characterization of clay mineral intercalated titanium dioxide nanoparticles, Langmuir 19 (2003) 2938-2946. https://doi.org/10.1021/la025969a

[29] G.L. Puma, A. Bono, D. Krishnaiah, J.G. Collin, Preparation of titanium dioxide photocatalyst loaded onto activated carbon support using chemical vapour deposition: a review paper, J. Hazard. Mater. 157 (2008) 209-219. https://doi.org/10.1016/j.jhazmat.2008.01.040

[30] E.P. Reddy, L. Davydov, P. Smirniotis, $TiO_2$-loaded zeolites and mesoporous materials in the sonophotocatalytic decomposition of aqueous organic pollutants: the role of the support, Appl. Catal. B: Environ. 42 (2003) 1-11. https://doi.org/10.1016/S0926-3373(02)00192-3

[31] N. Soltani, E. Saion, M.Z. Hussein, M. Erfani, A. Abedini, G. Bahmanrokh, M. Navasery, P. Vaziri, Visible light-induced degradation of methylene blue in the

presence of photocatalytic ZnS and CdS nanoparticles, Int. J. Mol. Sci. 13 (2012) 12242-12258. https://doi.org/10.3390/ijms131012242

[32] L. Zhang, W. Wang, Z. Chen, L. Zhou, H. Xu, W. Zhu, Fabrication of flower-like Bi2WO6 superstructures as high performancevisible-light drivenphotocatalysts, J. Mater. Chem. 17 (2007) 2526-2532. https://doi.org/10.1039/b616460a

[33] A. Huizhong, D. Yi, W. Tianmin, W. Cong, H. Weichang, Z. Junying, Photocatalytic properties of BiOX (X = Cl, Br, and I), Rare Metals 27 (2008) 243-250. https://doi.org/10.1016/S1001-0521(08)60123-0

[34] X.C. Wang, K. Maeda, A. Thomas, K. Takanabe, G. Xin, J.M. Carlsson, K. Domen, M. Antonietti, A metal-free polymeric photocatalyst for hydrogen production from water under visible light, Nat. Mater. 8 (2009) 76-80. https://doi.org/10.1038/nmat2317

[35] Z. Yi, J. Ye, N. Kikugawa, T. Kako, S. Ouyang, H. S. Williams, H. Yang, J. Cao, W. Luo, Z. Li, Y. Liu, R. L. Withers, An orthophosphate semiconductor with photooxidation properties under visible-light irradiation, Nat. Mater. 9 (2010) 559-564. https://doi.org/10.1038/nmat2780

[36] C.G. Silva, A. Corma, H. Garcia, Metal-organic frameworks as semiconductors, J. Mater. Chem. 20 (2010) 3141-3156. https://doi.org/10.1039/b924937k

[37] S.C. Yan, S.B. Lv, Z.S. Li, Z.G. Zou, Organic–inorganic composite photocatalyst of g-$C_3N_4$ and TaON with improved visible light photocatalytic activities, Dalton Trans.39 (2010) 1488-1491. https://doi.org/10.1039/B914110C

[38] L. Zhang, K.-H. Wong, Z. Chen, J.C. Yu, J. Zhao, C. Hu, C.-Y. Chang, P.-K. Wong, AgBr-Ag-$Bi_2WO_6$ nanojunction system: A novel and efficient photocatalyst with double visible-light active components, Appl. Catal. A. Gen. 363 (2009) 221-229. https://doi.org/10.1016/j.apcata.2009.05.028

[39] J. Yu, S. Wang, J. Low, W. Xiao, Enhanced photocatalytic performance of direct Z-scheme g-$C_3N_4$-$TiO_2$ photocatalysts for the decomposition of formaldehyde in air, Phys. Chem. Chem. Phys. 15 (2013) 16883-16890. https://doi.org/10.1039/c3cp53131g

[40] W.K. Jo, T.S. Natarajan, Influence of $TiO_2$ morphology on the photocatalytic efficiency of direct Z-scheme g-$C_3N_4$/$TiO_2$ photocatalysts for isoniazid degradation, Chem. Eng. J 281 (2015) 549-565. https://doi.org/10.1016/j.cej.2015.06.120

[41] J.D. Figueroa, T. Fout, S. Plasynski, H. McIlvried, R.D. Srivastava, Advances in $CO_2$ capture technology-The U.S. department of energy's carbon sequestration program, Inter. J Greenhouse Gas Control2 (2008) 9-20. https://doi.org/10.1016/S1750-5836(07)00094-1

[42] N. MacDowell, N. Florin, A. Buchard, J. Hallett, A. Galindo, G. Jackson, C. S. Adjiman, C. K. Williams, N. Shah, P. Fennell, An overview of $CO_2$ capture technologies, Energy Environ. Sci. 3 (2010) 1645-1669. https://doi.org/10.1039/c004106h

[43] D.Y.C. Leung, G. Caramanna, M.M. Maroto-Vale, An overview of current status of carbon dioxide capture and storage technologies, Renew. Sustainable Energy Rev. 39 (2014) 426-443. https://doi.org/10.1016/j.rser.2014.07.093

[44] M.Z. Jacobson, Review of solutions to global warming, air pollution, and energy security, Energy Environ. Sci. 2 (2009) 148-173. https://doi.org/10.1039/B809990C

[45] X. Xiaoding, J.A. Moulijn, Mitigation of $CO_2$ by chemical conversion: plausible chemical reactions and promising products, Energy & Fuels10(1996) 305-325. https://doi.org/10.1021/ef9501511

[46] T. Sakakura, J.-C. Choi, H. Yasuda, Transformation of carbon dioxide, Chem. Rev. 107 (2007) 2365-2387. https://doi.org/10.1021/cr068357u

[47] S. Sankaranarayanan, S. Kannan, Carbon dioxide-a potential raw material for the production of fuel, fuel additives and bio-derived chemicals, Indian J. Chem. 51A (2012) 1252-1262.

[48] A.J. Bard, M.A. Fox, Artificial photosynthesis: solar splitting of water to hydrogen and oxygen, Acc. Chem. Res. 28 (1995) 141-145. https://doi.org/10.1021/ar00051a007

[49] J.A. Turner, Sustainable hydrogen production, Science 305 (2004) 972-974. https://doi.org/10.1126/science.1103197

[50] L. Rovelli, K.R. Thampi, Solar water splitting using semiconductor systems. Chapter 1, in: Jacinto Sa (Eds). Fuel Production with Heterogeneous Catalysis. CRC Press, Taylor & Francis Group, UK, 2014, pp.1-62. https://doi.org/10.1201/b17552-2

[51] K. Natarajan, T.S. Natarajan, R.I. Kureshy, H.C. Bajaj, W.-K. Jo, R.J. Tayade, Photocatalytic $H_2$ production using semiconductor nanomaterials via water splitting-an overview, in: A. Al-Ahmed, M.K. Hossain, M. Afzaal, H.M.

Bahaidarah (Eds.), Recent Advances in Renewable Energy Research, Adv. Mater. Res. Trans Tech Publications, Switzerland, 2015, pp. 130-156.

[52] T. Inoue, A. Fujishima, S. Konishi, K. Honda, Photoelectrocatalytic reduction of carbon dioxide in aqueous suspensions of semiconductor powders, Nature 277 (1979) 637-638. https://doi.org/10.1038/277637a0

[53] M. Halmann, V.Katzir, E. Borgarello, J. Kiwi, Photoassisted carbon dioxide reduction on aqueous suspensions of titanium dioxide, Sol.Energy Mater. 10 (1984) 85-91. https://doi.org/10.1016/0165-1633(84)90010-8

[54] P. Usubharatana, D. McMartin, A. Veawab, P. Tontiwachwuthikul, Photocatalytic process for $CO_2$ emission reduction from industrial flue gas streams, Ind. Eng. Chem. Res. 45 (2006) 2558-2568. https://doi.org/10.1021/ie0505763

[55] S.N. Habisreutinger, L. Schmidt-Mende, J.K. Stolarczyk, Photocatalytic reduction of $CO_2$ on $TiO_2$ and other semiconductors, Angew. Chem. Int. Ed. 52 (2013) 7372-7408. https://doi.org/10.1002/anie.201207199

[56] C.F. Goodeve, The absorption spectra and photosensitising activity of white pigments, Trans. Faraday Soc. 33 (1937) 340-347. https://doi.org/10.1039/tf9373300340

[57] S.M.C. Markham, Photocatalytic properties of oxides, Report of the New England Association of Chemistry Teachers, (1955) 540-543.

[58] J.H. Carey, J. Lawrence, H. M. Tosine, Photodechlorination of PCB's in the presence of titanium dioxide in aqueous suspensions, Bull. Environ. Contam. Toxicol. 16 (1976) 697-701. https://doi.org/10.1007/BF01685575

[59] S.N. Frank, A.J. Bard, Heterogeneous photocatalytic oxidation of cyanide ion in aqueous solutions at titanium dioxide powder, J. Am. Chem. Soc. 99 (1977) 303-304. https://doi.org/10.1021/ja00443a081

[60] S.N. Frank, A.J. Bard, Heterogeneous photocatalytic oxidation of cyanide and sulfite in aqueous solutions at semiconductor powders, J. Phys. Chem. 81 (1977) 1484-1488. https://doi.org/10.1021/j100530a011

[61] G. Al-Sayyed, J.-C. D'Oliveira, P. Pichat, Semiconductor-sensitized photodegradation of 4-chlorophenol in water, J. Photochem Photobiol. A. Chem. 58 (1991) 99-114. https://doi.org/10.1016/1010-6030(91)87101-Z

[62] V. Kandavelu, H. Kastien, K.R. Thampi, Photocatalytic degradation of isothiazolin-3-ones in water and emulsion paints containing nanocrystalline $TiO_2$

and ZnO catalysts, Appl. Catal. B. Environ. 48 (2004) 101-111. https://doi.org/10.1016/j.apcatb.2003.09.022

[63] C.-C. Chen, H.-J. Fan, J.-L. Jan, Degradation pathways and efficiencies of acid blue 1 by photocatalytic reaction with ZnO nanopowder, J. Phys. Chem. C 112 (2008)11962-11972. https://doi.org/10.1021/jp801027r

[64] C.H. Wu, Comparison of azo dye degradation efficiency using UV/single semiconductor and UV/coupled semiconductor systems, Chemosphere 57 (2004) 601-608. https://doi.org/10.1016/j.chemosphere.2004.07.008

[65] B. Neppolian, H.C. Choi, S. Sakthivel, B. Arabindoo, V. Murugesan, Solar/UV-induced photocatalytic degradation of three commercial textile dyes, J. Hazard. Mater. 89 (2002) 303-317. https://doi.org/10.1016/S0304-3894(01)00329-6

[66] C. Karunakaran, S. Senthilvelan, Photocatalysis with $ZrO_2$: oxidation of aniline, J Mol. Catal. A. Chem. 233 (2005) 1-8. https://doi.org/10.1016/j.molcata.2005.01.038

[67] A. Fujishima, T.N. Rao, D.A. Tryk, Titanium dioxide photocatalysis, J. Photochem. Photobio. C: Photochem. Rev. 1 (2000) 1-21. https://doi.org/10.1016/S1389-5567(00)00002-2

[68] S. Parra, S.E. Stanca, I. Guasaquillo, K.R. Thampi, Photocatalytic degradation of atrazine using suspended and supported $TiO_2$, Appl. Catal. B: Environ. 51 (2004) 107-116. https://doi.org/10.1016/j.apcatb.2004.01.021

[69] K. Kabra, R. Chaudhary, R.L. Sawhney, Treatment of hazardous organic and inorganic compounds through aqueous-phase photocatalysis: A review, Ind. Eng. Chem. Res. 43 (2004) 7683-7696. https://doi.org/10.1021/ie0498551

[70] K. Wang, J. Zhang, L. Lou, S. Yang, Y. Chen, UV or visible light induced photodegradation of AO7 on $TiO_2$ particles: the influence of inorganic anions, J. Photochem. Photobio. A: Chem. 165 (2004) 201-207. https://doi.org/10.1016/j.jphotochem.2004.03.025

[71] I.K. Konstantinou, T.A. Albanis, $TiO_2$-assisted photocatalytic degradation of azo dyes in aqueous solution: kinetic and mechanistic investigations: A review, Appl. Catal. B: Environ. 49 (2004) 1-14. https://doi.org/10.1016/j.apcatb.2003.11.010

[72] R.J. Tayade, R.G. Kulkarni, R.V. Jasra, Photocatalytic degradation of aqueous nitrobenzene by nanocrystalline TiO2,Ind. Eng. Chem. Res. 45 (2006) 922-927. https://doi.org/10.1021/ie051060m

[73] R.J. Tayade, P.K. Surolia, R.G. Kulkarni, R.V. Jasra, Photocatalytic degradation of dyes and organic contaminants in water using nanocrystalline anatase and rutile $TiO_2$, Sci. Tech. Adv. Mater. 8 (2007) 455-462. https://doi.org/10.1016/j.stam.2007.05.006

[74] M. Lam, J.C. Sin, A.R. Mohamed, Recent patents on photocatalysis over nanosized titanium dioxide, Recent Patents on Chem. Eng. 1 (2008) 209-219. https://doi.org/10.2174/2211334710801030209

[75] P.K. Surolia, M.A. Lazar, R.J. Tayade, R.V. Jasra, Photocatalytic degradation of 3,3′-dimethylbiphenyl-4,4′-diamine (o-Tolidine) over nanocrystalline $TiO_2$ synthesized by sol-gel, solution combustion, and hydrothermal methods, Ind. Eng. Chem. Res.47 (2008)5847-5855. https://doi.org/10.1021/ie800073j

[76] R. Thiruvenkatachari, S. Vigneswaran, I.S. Moon, A review on UV/$TiO_2$ photocatalytic oxidation process, Korean J. Chem. Eng. 25 (2008) 64-72. https://doi.org/10.1007/s11814-008-0011-8

[77] R.S. Thakur, R. Chaudhary, C. Singh, Fundamentals and applications of the photocatalytic treatment for the removal of industrial organic pollutants and effects of operational parameters: A review, J. Renewable Sustainable Energy 2 (2010) 042701-(1-37). https://doi.org/10.1063/1.3467511

[78] A.R. Khataee, M.B. Kasiri, Photocatalytic degradation of organic dyes in the presence of nanostructured titanium dioxide: Influence of the chemical structure of dyes, J. Mol. Catal. A: Chem. 328 (2010) 8-26. https://doi.org/10.1016/j.molcata.2010.05.023

[79] H. Chen, C.E. Nanayakkara, V.H. Grassian,Titanium dioxide photocatalysis in atmospheric chemistry, Chem. Rev. 112 (2012) 5919-5948. https://doi.org/10.1021/cr3002092

[80] M.A. Lazar, R.J. Tayade, H.C. Bajaj, R.V. Jasra, Correlation of surface properties and photocatalytic activity of nanocrystalline $TiO_2$ on the synthesis route, Nano Hybrids 1 (2012) 57-80. https://doi.org/10.4028/www.scientific.net/NH.1.57

[81] J. Schneider, M. Matsuoka, M. Takeuchi, J. Zhang, Y. Horiuchi, M Anpo, D. W. Bahnemann, Understanding $TiO_2$ photocatalysis: mechanisms and materials, Chem. Rev. 114 (2014) 9919-9986. https://doi.org/10.1021/cr5001892

[82] D.D. Dionysiou, G. Balasubramanian, M.T. Suidan, A.P. Khodadoust, I. Baudin, J.-M. Laine, Rotating disk photocatalytic reactor: development, characterization,

and evaluation for the destruction of organic pollutants in water, Water Res. 34 (2000) 2927-2940. https://doi.org/10.1016/S0043-1354(00)00022-1

[83] J. Zhao, X. Yang, Photocatalytic oxidation for indoor air purification: a literature review, Building Environ. 38 (2003) 645-654. https://doi.org/10.1016/S0360-1323(02)00212-3

[84] K.V.S. Rao, M. Subrahmanyam, P. Boule, Immobilized $TiO_2$ photocatalyst during long-term use: decrease of its activity, Appl. Catal. B: Environ. 49 (2004) 239-249. https://doi.org/10.1016/j.apcatb.2003.12.017

[85] C.M. Ling, A.R. Mohamed, S. Bhatia, Performance of photocatalytic reactors using immobilized TiO2 film for the degradation of phenol and methylene blue dye present in water stream, Chemosphere 57 (2004) 547-554. https://doi.org/10.1016/j.chemosphere.2004.07.011

[86] N.M. Mahmoodi, M. Arami, N.Y. Limaee, N.S. Tabrizi, Kinetics of heterogeneous photocatalytic degradation of reactive dyes in an immobilized $TiO_2$ photocatalytic reactor, J. Colloid Interface Sci. 295 (2006) 159-164. https://doi.org/10.1016/j.jcis.2005.08.007

[87] A. Danion, J. Disdier, C. Guillard, O. Paisse, N.J.-Renault, Photocatalytic degradation of imidazolinone fungicide in $TiO_2$-coated optical fiber reactor, Appl. Catal. B: Environ. 62 (2006) 274-281. https://doi.org/10.1016/j.apcatb.2005.08.008

[88] T.V. Gerven, G. Mul, J. Moulijn, A. Stankiewicz, A review of intensification of photocatalytic processes, Chem. Eng. Process. 46 (2007) 781-789. https://doi.org/10.1016/j.cep.2007.05.012

[89] M. Birnie, S. Riffat, M. Gillott, Photocatalytic reactors: design for effective air purification, Int. J. Low-Carbon Tech. 1 (2006) 47-58. https://doi.org/10.1093/ijlct/1.1.47

[90] A.Y. Shan, T.I. Mohd. Ghazi, S.A. Rashid, Immobilisation of titanium dioxide onto supporting materials in heterogeneous photocatalysis: A review, Appl. Catal. A: Gen. 389 (2010) 1-8. https://doi.org/10.1016/j.apcata.2010.08.053

[91] C.S.L. Vegaa, B.S.Rosalesb, H. de Lasa, Energy efficiency limits in photo-CREC-air photocatalytic reactors, Chem. Eng. Sci. 156 (2016) 77-88. https://doi.org/10.1016/j.ces.2016.08.038

[92] F. Chen, J. Zhao, Preparation and photocatalytic properties of a novel kind of loaded photocatalyst of $TiO_2/SiO_2/\gamma$-$Fe_2O_3$, Catal. Lett. 58 (1999) 245-247. https://doi.org/10.1023/A:1019071126825

[93] V. Belessi, D. Lambropoulou, I. Konstantinou, R. Zboril, J. Tucek, D. Jancik, Structure and photocatalytic performance of magnetically separable titania photocatalysts for the degradation of propachlor, Appl. Catal. B Environ. 87 (2009) 181-189. https://doi.org/10.1016/j.apcatb.2008.09.012

[94] T.A. Gad-allah, S. Kato, S. Satokawa, T. Kojima, Treatment of synthetic dyes wastewater utilizing a magnetically separable photocatalyst ($TiO_2/SiO_2/Fe_3O_4$): Parametric and kinetic studies, Desalination244 (2009) 1-11. https://doi.org/10.1016/j.desal.2008.04.031

[95] K. Natarajan, P. Singh, H.C. Bajaj, R.J. Tayade, Facile synthesis of $TiO_2/ZnFe_2O_4$ nanocomposite by sol-gel auto combustion method for superior visible light photocatalytic efficiency, Korean J. Chem. Eng. 33 (2016) 1788-1798. https://doi.org/10.1007/s11814-016-0051-4

[96] G. Mamba, A. Mishra, Advances in magnetically separable photocatalysts: smart, recyclable materials for water pollution mitigation, Catalysts6 (2016) 1-34.

[97] B.D. Johnson, High-power short-wave LED purifies air, Photon. Spectra 37 (2003) 111.

[98] D.H. Chen, X. Ye, K. Li, Oxidation of PCE with a UV LED Photocatalytic Reactor, Chem. Eng. Technol. 28 (2005) 95-97. https://doi.org/10.1002/ceat.200407012

[99] W.Y. Wang, Y. Ku, Photocatalytic degradation of Reactive Red 22 in aqueous solution by UV-LED radiation, Water Res. 40 (2006) 2249-2258. https://doi.org/10.1016/j.watres.2006.04.041

[100] R.J. Tayade, T.S. Natarajan, H.C. Bajaj, Photocatalytic degradation of methylene blue dye using ultraviolet light emitting diodes, Ind. Eng. Chem. Res. 48 (2009) 10262-10267. https://doi.org/10.1021/ie9012437

[101] T.S. Natarajan, M. Thomas, K. Natarajan, H.C. Bajaj, R.J. Tayade, Study on UV-LED/$TiO_2$ process for degradation of rhodamine B dye, Chem. Eng. J169 (2011) 126-134. https://doi.org/10.1016/j.cej.2011.02.066

[102] W.K. Jo, S.-S. Eun, S.-H. Shin, Feasibility of light-emitting diode uses for annular reactor inner-coated with $TiO_2$ or nitrogen-doped $TiO_2$ for control of dimethyl

sulphide, Photochem. Photobiol. 87 (2011) 1016-1023. https://doi.org/10.1111/j.1751-1097.2011.00965.x

[103] T.S. Natarajan, K. Natarajan, H.C. Bajaj, R.J. Tayade, Energy efficient UV-LED source and $TiO_2$ nanotube array based reactor for photocatalytic application, Ind. Eng. Chem. Res. 50 (2011) 7753-7762. https://doi.org/10.1021/ie200493k

[104] L.H. Levine, J.T. Richards, J.L. Coutts, R. Soler, F. Maxik, R.M. Wheeler, Feasibility of ultraviolet-light-emitting diodes as an alternative light source for photocatalysis, J. Air Waste Manag. Assoc. 61 (2011) 932-940. https://doi.org/10.1080/10473289.2011.596746

[105] K. Natarajan, T.S. Natarajan, H.C. Bajaj, R.J. Tayade, Photocatalytic reactor based on UV-LED/$TiO_2$ coated quartz tube for degradation of dyes, Chem. Eng. J 178 (2011) 40-49. https://doi.org/10.1016/j.cej.2011.10.007

[106] R. Sharmin, M.B. Ray, Application of ultraviolet light-emitting diode photocatalysis to remove volatile organic compounds from indoor air, J. Air Waste Manag. Assoc. 62 (2012) 1032-1039. https://doi.org/10.1080/10962247.2012.695760

[107] M. Izadifard, G. Achari, C.H. Langford, Application of photocatalysts and LED light sources in drinking water treatment, Catalysts 3 (2013) 726-743. https://doi.org/10.3390/catal3030726

[108] T.S. Natarajan, K. Natarajan, H.C. Bajaj, R.J. Tayade, Ultraviolet light emitting diode (UV-LED) source-based photocatalytic reactors for environmental remediation, in:Inamuddin (Eds), Advanced Functional Polymers and Composites. Volume 2, Nova Science Publishers, Inc. 2013, pp. 33-91.

[109] K. Dai, L. Lu, G. Dawson, Development of UV-LED/$TiO_2$ device and their application for photocatalytic degradation of methylene blue, J. Mater. Eng. Perfor. 22 (2013) 1035-1040. https://doi.org/10.1007/s11665-012-0344-7

[110] W.K. Jo, R.J. Tayade, New generation energy-efficient light source for photocatalysis LEDs for environmental applications, Ind. Eng. Chem. Res. 53 (2014) 2073-2084. https://doi.org/10.1021/ie404176g

[111] W.K. Jo, R.J. Tayade, Recent developments in photocatalytic dye degradation upon irradiation with energy-efficient light emitting diodes, Chinese J. Catal. 35 (2014) 1781-1792. https://doi.org/10.1016/S1872-2067(14)60205-9

[112] T. Kasuga, M. Hiramatsu, A. Hoson, T. Sekino, K. Niihara, Formation of titanium oxide nanotube, Langmuir 14 (1998) 3160-3163. https://doi.org/10.1021/la9713816

[113] T.S. Natarajan, H.C. Bajaj, R.J. Tayade, Preferential adsorption behavior of methylene blue dye onto surface hydroxyl group enriched $TiO_2$ nanotube and its photocatalytic regeneration, J. Colloid Interface Sci. 433 (2014) 104-114. https://doi.org/10.1016/j.jcis.2014.07.019

[114] B. Wen, C. Liu, Y. Liu, Depositional characteristics of metal coating on single-crystal $TiO_2$ nanowires, J. Phys. Chem. B 109 (2005) 12372-12375. https://doi.org/10.1021/jp050934f

[115] S. Daothong, N. Songmee, S. Thongtem, P. Singjai, Size-controlled growth of $TiO_2$ nanowires by oxidation of titanium substrates in the presence of ethanol vapor, Scripta Materialia 57 (2007) 567-570. https://doi.org/10.1016/j.scriptamat.2007.06.030

[116] Y. Wang, Y. He, Q. Lai, M. Fan, Review of the progress in preparing nano $TiO_2$: An important environmental engineering material, J. Environ. Sci. 26 (2014) 2139-2177. https://doi.org/10.1016/j.jes.2014.09.023

[117] R.J. Tayade, R.G. Kulkarni, R.V. Jasra, Enhanced photocatalytic activity of $TiO_2$-coated NaY and HY zeolites for the degradation of methylene Blue in water, Ind. Eng. Chem. Res. 46 (2007)369-376. https://doi.org/10.1021/ie060641o

[118] S. Sato, J.M. White,Photocatalytic water decomposition and water-gas shift reactions over NaOH-coated, platinized $TiO_2$, J. Catal. 69 (1981) 128-139. https://doi.org/10.1016/0021-9517(81)90135-4

[119] N. Serpone, E. Borgarello, M. Grätzel, Visible light inducedgeneration of hydrogen from H2S in mixed semiconductor dispersions; improved efficiency through inter-particle electron transfer, J. Chem. Soc. Chem. Commun. (1984) 342-344. https://doi.org/10.1039/C39840000342

[120] N. Serpone, P. Maruthamuthu, P. Pichat, E. Pelizzetti, H. Hidaka, Exploiting the interparticle electron transfer process in the photocatalysed oxidation of phenol, 2-chlorophenol, and pentachlorophenol: chemical evidence for electron and hole transfer between coupled semiconductors, J. Photochem. Photobiol. A. Chem. 85 (1995) 247-255. https://doi.org/10.1016/1010-6030(94)03906-B

[121] J. Zhou, Y. Zhang, X.S. Zhao, A.K. Ray, Photodegradation of benzoic acid over metal-doped $TiO_2$, Ind. Eng. Chem. Res. 45 (2006) 3503-3511. https://doi.org/10.1021/ie051098z

[122] P.K. Surolia, R.J. Tayade, R.V. Jasra, Effect of anions on the photocatalytic activity of Fe(III) salts impregnated $TiO_2$, Ind. Eng. Chem. Res. 46 (2007) 6196-6203. https://doi.org/10.1021/ie0702678

[123] S. Neatu, V.I. Parvulescu, G. Epure, E. Preda, V. Somoghi, A. Damin, S. Bordiga, A. Zecchina, Photo-degradation of yperite over V, Fe and Mn-doped titania-silica photocatalysts, Phys. Chem. Chem. Phys. 10 (2008) 6562-6570. https://doi.org/10.1039/b810200g

[124] R.J. Tayade, H.C. Bajaj, R.V. Jasra, Photocatalytic removal of organic contaminants from water exploiting tuned bandgap photocatalysts, Desalination 275 (2011) 160-165. https://doi.org/10.1016/j.desal.2011.02.047

[125] S.M. Gupta, M. Tripathi, A review of $TiO_2$ nanoparticles, Chinese Sci. Bulletin 56 (2011) 1639-1657. https://doi.org/10.1007/s11434-011-4476-1

[126] S. Bingham, W.A. Daoud, Recent advances in making nano-sized $TiO_2$ visible-light active through rare-earth metal doping, J. Mater. Chem. 21 (2011) 2041-2050. https://doi.org/10.1039/C0JM02271C

[127] S. Banerjee, S.C. Pillai, P. Falaras, K.E. O'Shea, J.A. Byrne, D.D. Dionysiou, New insights into the mechanism of visible light photocatalysis, J. Phys. Chem. Lett. 5 (2014) 2543-2554. https://doi.org/10.1021/jz501030x

[128] S.U.M. Khan, M. Al-Shahry, W.B. Ingler Jr. Efficient photochemical water splitting by a chemically modified n-$TiO_2$, Science 297 (2002) 2243-2245. https://doi.org/10.1126/science.1075035

[129] J.C. Yu, J. Yu, W. Ho, Z. Jiang, L. Zhang, Effects of F- doping on the photocatalytic activity and microstructures of nanocrystalline $TiO_2$ powders, Chem. Mater. 14 (2002) 3808-3816. https://doi.org/10.1021/cm020027c

[130] H. Luo, T. Takata, Y. Lee, J. Zhao, K. Domen, Y. Yan, Photocatalytic activity enhancing for titanium dioxide by co-doping with bromine and chlorine, Chem. Mater. 16 (2004) 846-849. https://doi.org/10.1021/cm035090w

[131] H. Zhao, F. Tian, R. Wang, R, Chen, A Review on bismuth-related nanomaterials for photocatalysis, Rev. Adv. Sci. Eng. 3 (2014) 3-27 https://doi.org/10.1166/rase.2014.1050

[132] S.C. Yan, Z.S. Li, Z.G. Zou, Photodegradation performance of g-$C_3N_4$ fabricated by directly heating melamine, Langmuir 25(2009) 10397-10401. https://doi.org/10.1021/la900923z

[133] K.M. Garadkar, A.A. Patil, P.P. Hankare, P.A. Chate, D.J. Sathe, S.D. Delekar, $MoS_2$: Preparation and their characterization, J. Alloys Compounds 487 (2009) 786-789. https://doi.org/10.1016/j.jallcom.2009.08.069

[134] K.K. Banger, M. H.-C. Jin, J. D. Harris, P.E. Fanwick, A.F. Hepp, A new facile route for the preparation of single-source precursors for bulk, thin-film, and nanocrystallite I–III–VI semiconductors, Inorg. Chem. 42 (2003) 7713-7715. https://doi.org/10.1021/ic034802h

[135] P. Kanhere, Z. Chen, A review on visible light active perovskite-based photocatalysts, Molecules 19 (2014) 19995-20022. https://doi.org/10.3390/molecules191219995

[136] L. Ge, C. Han, J. Liu, Novel visible light-induced g-$C_3N_4$/$Bi_2WO_6$ composite photocatalysts for efficient degradation of methyl orange, Appl. Catal. B Environ. 108-109 (2011) 100-107. https://doi.org/10.1016/j.apcatb.2011.08.014

[137] J. Fu, B. Chang, Y. Tian, F. Xi, X. Dong, Novel $C_3N_4$–CdS composite photocatalysts with organic–inorganic heterojunctions: in situ synthesis, exceptional activity, high stability and photocatalytic mechanism, J. Mater. Chem. A 1 (2013) 3083-3090. https://doi.org/10.1039/c2ta00672c

[138] H. Yuan, J. Liu, J. Li, Y. Li, X. Wang, Y. Zhang, J. Jiang, S. Chen, C. Zhao, D. Qian, Designed synthesis of a novel $BiVO_4$–$Cu_2O$–$TiO_2$ as an efficient visible-light-responding photocatalyst, J. Colloid Interface Sci. 444 (2015) 58-66. https://doi.org/10.1016/j.jcis.2014.12.034

[139] A. Di Paola, M. Bellardita, R. Ceccato, L. Palmisano, F. Parrino, Highly active photocatalytic $TiO_2$ powders obtained by thermohydrolysis of $TiCl_4$ in water, J. Phys. Chem. C 113 (2009) 15166-15174. https://doi.org/10.1021/jp904673e

[140] L. Huang, Y. Li, H.Xu, Y.Xu, J.Xia, K.Wang, H. Li, X. Cheng, Synthesis and characterization of $CeO_2$/g-$C_3N_4$ composites with enhanced visible-light photocatatalytic activity, RSC Adv. 3 (2013) 22269-22279. https://doi.org/10.1039/c3ra42712a

[141] S.P. Adhikari, Z.D. Hood, K.L. More, I. Ivanov, L. Zhang, M. Gross, A. Lachgar, Visible light assisted photocatalytic hydrogen generation by $Ta_2O_5$/$Bi_2O_3$,

TaON/$Bi_2O_3$, and $Ta_3N_5$/$Bi_2O_3$ composites, RSC Adv. 5 (2015) 54998-55005. https://doi.org/10.1039/C5RA06563A

[142] W.-J. Chun, A. Ishikawa, H. Fujisawa, T. Takata, J. N. Kondo, M. Hara, M. Kawai, Y. Matsumoto, K. Domen, Conduction and valence band positions of $Ta_2O_5$, TaON, and $Ta_3N_5$ by UPS and electrochemical methods, J. Phys. Chem. B 107 (2003) 1798-1803. https://doi.org/10.1021/jp027593f

[143] N. Wetchakun, S. Chaiwichain, B. Inceesungvorn, K. Pingmuang, S. Phanichphant, A. I. Minett, J. Chen, $BiVO_4$/$CeO_2$ nanocomposites with high visible-light-induced photocatalytic activity, ACS Appl. Mater. Interfaces 4 (2012) 3718-3723. https://doi.org/10.1021/am300812n

[144] J. Lua, Y. Wang, F. Liu, L. Zhang, S. Chai, Fabrication of a direct Z-scheme type $WO_3$/$Ag_3PO_4$ composite photocatalyst with enhanced visible-light photocatalytic performances, Appl. Surf. Sci. 393 (2017) 180-190. https://doi.org/10.1016/j.apsusc.2016.10.003

[145] S. Rajendran, M. M. Khan, F. Gracia, J. Qin, V.K. Gupta, S. Arumainathan, $Ce^{3+}$-ion-induced visible-light photocatalytic degradation and electrochemical activity of ZnO/$CeO_2$ nanocomposite, Scientific Reports 6:31641 (2016) 1-11. https://doi.org/10.1038/srep31641

[146] R. Tang, H. Su, S. Duan, Y. Sun, L. Li, X. Zhang, S. Zeng, D. Sun, Enhanced visible-light-driven photocatalytic performances using $Bi_2WO_6$/MS (M = Cd, Zn) heterostructures: facile synthesis and photocatalytic mechanisms, RSC Adv. 5 (2015) 41949-41960. https://doi.org/10.1039/C5RA04655F

[147] H. Fan, H. Li, B. Liu, Y. Lu, T. Xie, D. Wang, Photoinduced charge transfer properties and photocatalytic activity in $Bi_2O_3$/$BaTiO_3$ composite photocatalyst, ACS Appl. Mater. Interfaces 4 (2012) 4853-4857. https://doi.org/10.1021/am301199v

[148] D. Jiang, L. Chen, J. Zhu, M. Chen, W. Shi, J. Xie, Novel p–n heterojunction photocatalyst constructed by porous graphite-like $C_3N_4$ and nanostructured BiOI: facile synthesis and enhanced photocatalytic activity, Dalton Trans. 42 (2013) 15726-15734. https://doi.org/10.1039/c3dt52008k

[149] J. Xia, J. Di, S. Yin, H. Xu, J. Zhang, Y. Xu, L. Xu, H. Li, M, Ji, Facile fabrication of the visible-light-driven $Bi_2WO_6$/BiOBr composite with enhanced photocatalytic activity, RSC Adv. 4 (2014) 82-90. https://doi.org/10.1039/C3RA44191A

[150] J. Cao, C. Zhou, H. Lin, B. Xu, S. Chen, Surface modification of m-$BiVO_4$ with wide band-gap semiconductor BiOCl to largely improve the visible light induced photocatalytic activity, Appl. Surf. Sci. 284 (2013) 263-269. https://doi.org/10.1016/j.apsusc.2013.07.092

[151] Y.S. Xu, W.D. Zhang, Monodispersed $Ag_3PO_4$ nanocrystals loaded on the surface of spherical $Bi_2MoO_6$ with enhanced photocatalytic performance, Dalton Trans. 42 (2013) 1094-1101. https://doi.org/10.1039/C2DT31634J

[152] X. Wang, Y. Lin, X. Ding, J. Jiang, Enhanced visible-light-response photocatalytic activity of bismuth ferrite nanoparticles, J. Alloys Compounds 509 (2011) 6585-6588. https://doi.org/10.1016/j.jallcom.2011.03.074

[153] M. Xiong, L. Chen, Q. Yuan, J. He, S.-L. Luo, C.-T. Au, S.-F. Yin, Facile fabrication and enhanced photosensitized degradation performance of the g-$C_3N_4$–$Bi_2O_2CO_3$ composite, Dalton Trans. 43 (2014) 8331-8337. https://doi.org/10.1039/C4DT00486H

[154] H. Xu, Y. Xu, H. Li, J. Xia, J. Xiong, S. Yin, C. Huang, H. Wan, Synthesis, characterization and photocatalytic property of AgBr/$BiPO_4$ heterojunction photocatalyst, Dalton Trans. 41 (2012) 3387-3394. https://doi.org/10.1039/c2dt11969b

[155] M. Lu, G. Yuan, Z. Wang, Y. Wang, J. Guo, Synthesis of $BiPO_4$/$Bi_2S_3$ heterojunction with enhanced photocatalytic activity under visible-light irradiation, Nanoscale Res. Let. 10:385 (2015) 1-7. https://doi.org/10.1186/s11671-015-1092-z

[156] T. Wang, W.Quan, D. Jiang, L. Chen, D. Li, S. Meng, M. Chen, Synthesis of redox-mediator-free direct Z-scheme AgI/$WO_3$ nanocomposite photocatalysts for the degradation of tetracycline with enhanced photocatalytic activity, Chem. Eng. J 300 (2016) 280-290. https://doi.org/10.1016/j.cej.2016.04.128

[157] O. Mehraj, B.M. Pirzada, N.A. Mir, S. Sultana, S. Sabir, $Ag_2S$ sensitized mesoporous $Bi_2WO_6$ architectures with enhanced visible light photocatalytic activity and recycling properties, RSC Adv. 5 (2015) 42910-42921. https://doi.org/10.1039/C5RA05117G

[158] Y. Deng, L. Tang, G. Zeng, J. Wang, Y. Zhou, J. Wang, J. Tang, Y. Liu, B. Peng, F. Chen, Facile fabrication of a direct Z-scheme $Ag_2CrO_4$/g-$C_3N_4$ photocatalyst with enhanced visible light photocatalytic activity, J Mol. Catal. A: Chem. 421 (2016) 209-221. https://doi.org/10.1016/j.molcata.2016.05.024

[159] J. Bao, S. Guo, J.Gao, T. Hu, L. Yang, C. Liu, J. Peng, C.Jiang, Synthesis of $Ag_2CO_3/Bi_2WO_6$ heterojunctions with enhanced photocatalytic activity and cycling stability, RSC Adv. 5 (2015) 97195-97204. https://doi.org/10.1039/C5RA18938A

[160] C. Zhu, L. Zhang, B. Jiang, J. Zheng, P. Hu, S. Li, M. Wu, W. Wu, Fabrication of Z-scheme $Ag_3PO_4/MoS_2$ composites with enhanced photocatalytic activity and stability for organic pollutant degradation, Appl. Surf. Sci. 377 (2016) 99-108. https://doi.org/10.1016/j.apsusc.2016.03.143

[161] H. Meng, T. Wang, X. Yu, Y. Zhu, Y. Zhang, $BiOCl/SnS_2$ hollow spheres for the photocatalytic degradation of waste water, RSC Adv. 5 (2015) 107088-107097. https://doi.org/10.1039/C5RA22526D

[162] M. Sun, Q. Zhao, C. Du, Z. Liu, Enhanced visible light photocatalytic activity in $BiOCl/SnO_2$: heterojunction of two wide band-gap semiconductors, RSC Adv. 5 (2015) 22740-22752. https://doi.org/10.1039/C4RA14187C

[163] C. Shifu, J. Lei, T. Wenming, F, Xianliang, Fabrication, characterization and mechanism of a novel Z-scheme photocatalyst $NaNbO_3/WO_3$ with enhanced photocatalytic activity, Dalton Trans. 42 (2013) 10759-10768. https://doi.org/10.1039/c3dt50699a

[164] Y. Yao, G. Wu, F. Lu, S. Wang, Y. Hu, J. Zhang, W. Huang, F. Wei, Enhanced photo-Fenton-like process over Z-scheme $CoFe_2O_4/g\text{-}C_3N_4$ heterostructures under natural indoor light, Environ. Sci. Pollut. Res. 23 (2016) 21833-21845. https://doi.org/10.1007/s11356-016-7329-2

[165] K.J. McDonald, K.-S. Choi, Synthesis and photoelectrochemical properties of $Fe_2O_3/ZnFe_2O_4$ composite photoanodes for use in solar water oxidation, Chem. Mater. 23 (2011) 4863-4869. https://doi.org/10.1021/cm202399g

[166] R. Shao, X. Zeng, Z. Cao, H. Dong, L. Wang, F. Wang, J. Liu, Z. Li, Q. Liang, A novel $Ag_3PO_4/Nb_2O_5$ fiber composite with enhanced photocatalytic performance and stability, RSC Adv. 5 (2015) 102101-102107. https://doi.org/10.1039/C5RA17555K

[167] Y. Deng, Y. Chen, B. Chen, J. Ma, Preparation, characterization and photocatalytic activity of $CuBi_2O_4/NaTaO_3$ coupled photocatalysts, J. Alloys Compounds 559 (2013) 116-122. https://doi.org/10.1016/j.jallcom.2013.01.069

[168] W.K. Jo, T.S. Natarajan, Facile synthesis of novel redox-mediator-free direct Z-scheme $CaIn_2S_4$ marigold-flower-like/$TiO_2$ photocatalysts with superior

photocatalytic efficiency, ACS Appl. Mater. Interfaces 7 (2015) 17138-17154. https://doi.org/10.1021/acsami.5b03935

[169] S. Peng, L. Li, Y.Wu, L. Jia, L.Tian, M. Srinivasan, S. Ramakrishna, Q. Yan, S. G. Mhaisalkar, Size- and shape-controlled synthesis of $ZnIn_2S_4$ nanocrystals with high photocatalytic performance, CrystEngComm 15 (2013) 1922-1930. https://doi.org/10.1039/c2ce26593a

[170] Z. Mei, S. Ouyang, D.-M. Tang, T. Kako, D. Golberg, J. Ye, An ion-exchange route for the synthesis of hierarchical $In_2S_3/ZnIn_2S_4$ bulk composite and its photocatalytic activity under visible-light irradiation, Dalton Trans. 42 (2013) 2687-2690. https://doi.org/10.1039/c2dt32271d

[171] J. Mu, B. Chen, M. Zhang, Z. Guo, P. Zhang, Z. Zhang, Y. Sun, C. Shao, Y. Liu, Enhancement of the visible-light photocatalytic activity of $In_2O_3$-$TiO_2$ nanofiber heteroarchitectures, ACS Appl. Mater. Interfaces 4 (2012) 424–430. https://doi.org/10.1021/am201499r

[172] Y. Guo, G. Zhang, J. Liu, Y. Zhang, Hierarchically structured $\alpha$-$Fe_2O_3/Bi_2WO_6$ composite for photocatalytic degradation of organic contaminants under visible light irradiation, RSC Adv. 3 (2013) 2963-2970. https://doi.org/10.1039/c2ra22741j

[173] Y. Guo, G. Zhang, H. Gan, Y. Zhang, Micro/nano-structured $CaWO_4/Bi_2WO_6$ composite: synthesis, characterization and photocatalytic properties for degradation of organic contaminants, Dalton Trans. 41 (2012) 12697-12703. https://doi.org/10.1039/c2dt31376f

[174] W. Wang, J. Wang, Z. Wang, X. Wei, L. Liu, Q. Ren, W. Gao, Y. Liang, H. Shia, p–n junction $CuO/BiVO_4$ heterogeneous nanostructures: synthesis and highly efficient visible-light photocatalytic performance, Dalton Trans. 43 (2014) 6735-6743. https://doi.org/10.1039/c3dt53613k

[175] B. Ohtani, Photocatalysis A to Z-What we know and what we don't know in ascientificsense, J. Photochem. Photobiol C: Photochem Rev. 11 (2010) 157-158. https://doi.org/10.1016/j.jphotochemrev.2011.02.001

[176] H. Tada, T. Mitsui, T. Kiyonaga, T. Akita, K. Tanaka, All-solid-state Z-scheme in CdS-Au-$TiO_2$ three-component nanojunction system, Nat. Mater. 5 (2006) 782-786. https://doi.org/10.1038/nmat1734

[177] D. Ma, J.Wu, M. Gao, Y. Xin, T. Ma, Y. Sun, Fabrication of Z-scheme g-$C_3N_4$/RGO/$Bi_2WO_6$ photocatalyst with enhanced visible-light photocatalytic

activity, Chem. Eng. J 290 (2016) 136-146.
https://doi.org/10.1016/j.cej.2016.01.031

[178] T. Arai, M. Yanagida, Y. Konishi, Y. Iwasaki, H. Sugihara, K. Sayama, Efficient complete oxidation of acetaldehyde into $CO_2$ over $CuBi_2O_4/WO_3$ composite photocatalyst under visible and UV light irradiation, J. Phys. Chem. C, Lett. 111 (2007) 7574-7577.

[179] W.K. Jo, J.Y. Lee, T.S. Natarajan, Fabrication of hierarchically structured novel redox-mediator-free $ZnIn_2S_4$ marigold flower/$Bi_2WO_6$ flower-like direct Z-scheme nanocomposite photocatalysts with superior visible light photocatalytic efficiency, Phys. Chem. Chem. Phys. 18 (2016) 1000-1016.
https://doi.org/10.1039/C5CP06176H

[180] R. Zamiri, H.A. Ahangar, D.M. Tobaldi, A. Rebelo, M.P. Seabra, M. Shabani, J.M.F. Ferreira,Fabricating and characterising ZnO–ZnS–$Ag_2S$ ternary nanostructures with efficient solar-light photocatalytic activity, Phys. Chem. Chem. Phys. 16 (2014) 22418-22425. https://doi.org/10.1039/C4CP02945C

[181] J. Li, H. Hao, Z. Zhu, Construction of g-$C_3N_4$-$WO_3$-$Bi_2WO_6$ double Z-scheme system with enhanced photoelectrochemical performance, Mater. Let. 168 (2016) 180-183. https://doi.org/10.1016/j.matlet.2016.01.058

[182] J. Li, H. Hao, J. Zhou, Z. Zhu, g-$C_3N_4$ modified flower-like $WO_3$–$Bi_2WO_6$ microspheres with enhanced photoelectrocatalytic activity, New J. Chem. 40 (2016) 9638-9647. https://doi.org/10.1039/C6NJ01749E

[183] W.K. Jo, T.S. Natarajan, Fabrication and efficient visible light photocatalytic properties of novel zinc indium sulfide ($ZnIn_2S_4$)- graphitic carbon nitride (g-$C_3N_4$)/bismuth vanadate ($BiVO_4$) nanorod-based ternary nanocomposites with enhanced charge separation via Z-scheme transfer, J. Colloid Interface Sci. 482 (2016) 58-72. https://doi.org/10.1016/j.jcis.2016.07.062

[184] S. Yin, B. Liu, P. Zhang, T. Morikawa, K.Yamanaka, T. Sato, Photocatalytic oxidation of NOx under visible LED light irradiation over nitrogen-doped titania particles with iron or platinum loading, J. Phys. Chem. C 112 (2008) 12425-12431. https://doi.org/10.1021/jp803371s

[185] Y. Sun, W. Wang, L. Zhang, S. Sun, The photocatalysis of $Bi_2MoO_6$ under the irradiation of blue LED, Mater. Res. Bulletin 48 (2013) 4357-4361.
https://doi.org/10.1016/j.materresbull.2013.07.015

[186] K. Natarajan, H.C. Bajaj, R.J. Tayade, Photocatalytic efficiency of bismuth oxyhalide (Br, Cl and I) nanoplates for RhB dye degradation under LED irradiation, J. Ind. Eng. Chem. 34 (2016) 146-156. https://doi.org/10.1016/j.jiec.2015.11.003

[187] M. Schroder, Functional metal-organic frameworks: gas storage, separation and catalysis, Springer-Verlag Berlin Heidelberg, Germany, 2010. https://doi.org/10.1007/978-3-642-14613-8

[188] C.-F. Zhang, L.-G. Qiu, F. Ke, Y-J. Zhu, Y.-P. Yuan, G.-S. Xu, X. Jiang, A novel magnetic recyclable photocatalyst based on a core–shell metal–organic framework $Fe_3O_4$@MIL-100(Fe) for the decolorization of methylene blue dye, J. Mater. Chem. A 1 (2013) 14329-14334. https://doi.org/10.1039/c3ta13030d

[189] T. Zhang, W. Lin, Metal-organic frameworks for artificial photosynthesis and photocatalysis, Chem. Soc. Rev. 43 (2014) 5982-5993. https://doi.org/10.1039/C4CS00103F

[190] C.-C. Wang, J.-R. Li, X.-L. Lv, Y.-Q. Zhang,G. Guo, Photocatalytic organic pollutants degradation in metal-organic frameworks, Energy Environ. Sci. 7 (2014) 2831-2867. https://doi.org/10.1039/C4EE01299B

[191] Y. Li, H. Xu, S. Ouyang,J. Ye, Metal-organic frameworks for photocatalysis, Phys. Chem. Chem. Phys.18 (2016)7563-7572. https://doi.org/10.1039/C5CP05885F

[192] M. Alvaro, E. Carbonell, B. Ferrer, F.X.L. Xamena, H. Garcia, Semiconductor behavior of a metal-organic framework (MOF), Chem. Eur. J. 13 (2007) 5106 - 5112. https://doi.org/10.1002/chem.200601003

[193] F.X.L. Xamena, A. Corma, H. Garcia, Applications for metal–organic frameworks (MOFs) as quantum dot semiconductors, J. Phys. Chem. C 111 (2007) 80-85. https://doi.org/10.1021/jp063600e

[194] P. Mahata, G. Madras, S. Natarajan, Novel photocatalysts for the decomposition of organic dyes based on metal-organic framework compounds, J. Phys. Chem. B 110 (2006)13759-13768. https://doi.org/10.1021/jp0622381

[195] R. Das, P. Pachfule, R. Banerjee, P. Poddar, Metal and metal oxide nanoparticle synthesis from metal organic frameworks (MOFs): finding the border of metal and metal oxides, Nanoscale 4 (2012) 591-599. https://doi.org/10.1039/C1NR10944H

[196] Y. Gu, K. Cheng, Y. Wu, Y. Wang, C. Morlay, F. Li,Metal–organic framework-templated synthesis of bifunctional N-doped $TiO_2$-carbon nanotablets via solid-

state thermolysis,ACS Sustainable Chem. Eng. 4 (2016) 6744-6753. https://doi.org/10.1021/acssuschemeng.6b01716

[197] M. Gratzel, Energy resources through photochemistry and catalysis, academic press, New York, 1983.

[198] F.E. Osterloh, Inorganic materials as catalysts for photochemical splitting of water, Chem. Mater. 20 (2008) 35-54. https://doi.org/10.1021/cm7024203

[199] A. Kudo, Y. Miseki, Heterogeneous photocatalyst materials for water splitting Chem. Soc. Rev. 38 (2009) 253-278. https://doi.org/10.1039/B800489G

[200] X. Chen, S. Shen, L. Guo, S.S. Mao, Semiconductor-based photocatalytic hydrogen generation, Chem. Rev.110 (2010)6503-6570. https://doi.org/10.1021/cr1001645

[201] J. Yu, L. Qi, M. Jaroniec, Hydrogen production by photocatalytic water splitting over Pt/$TiO_2$ nanosheets with exposed (001) facets, J. Phys. Chem. C114 (2010) 13118-13125. https://doi.org/10.1021/jp104488b

[202] M. Murdoch, G.I.N. Waterhouse, M.A. Nadeem, J.B. Metson, M.A. Keane, R.F. Howe, J. Llorca, H. Idriss, The effect of gold loading and particle size on photocatalytic hydrogen production from ethanol over Au/$TiO_2$ nanoparticles, Nat. Chem. 3 (2011) 489-492. https://doi.org/10.1038/nchem.1048

[203] S. Onsuratoom, T. Puangpetch, S. Chavadej, Comparative investigation of hydrogen production over Ag-, Ni-, and Cu-loaded mesoporous-assembled $TiO_2$-$ZrO_2$ mixed oxide nanocrystal photocatalysts, Chem. Eng. J 173 (2011) 667-675. https://doi.org/10.1016/j.cej.2011.08.016

[204] A.A. Ismail, D.W. Bahnemann, Photochemical splitting of water for hydrogen production by photocatalysis: A review, Sol. Energy Mater. Sol. Cells 128 (2014) 85-101. https://doi.org/10.1016/j.solmat.2014.04.037

[205] L. Qin, G. Si, X. Li, S.-Z. Kang, Synergetic effect of Cu–Pt bimetallic cocatalyst on $SrTiO_3$ for efficient photocatalytic hydrogen production from water, RSC Adv. 5 (2015) 102593-102598. https://doi.org/10.1039/C5RA22757G

[206] S. Sato, J.M. White, Photocatalytic water decomposition and water-gas shift reactions over NaOH-coated, platinized $TiO_2$, J. Catal. 69 (1981) 128-139. https://doi.org/10.1016/0021-9517(81)90135-4

[207] K. Yamaguti, S.Sato, Photolysis of water over metallized powdered titanium dioxide, J. Chem. Soc., Faraday Trans. 1, 81 (1985) 1237-1246. https://doi.org/10.1039/f19858101237

[208] K. Sayama, H. Arakawa, Effect of $Na_2CO_3$ addition on photocatalytic decomposition of liquid water over various semiconductor catalysis, J. Photochem. Photobiol. A: Chem. 77 (1994) 243-247. https://doi.org/10.1016/1010-6030(94)80049-9

[209] A. Kudo, Photocatalyst materials for water splitting, Catal. Surv. Asia 7 (2003) 31-38. https://doi.org/10.1023/A:1023480507710

[210] Z. Zou, H. Arakawa, Direct water splitting into $H_2$ and $O_2$ under visible light irradiation with a new series of mixed oxide semiconductor photocatalysts, J. Photochem. Photobiol. A. Chem.158 (2003) 145-162. https://doi.org/10.1016/S1010-6030(03)00029-7

[211] M. Matsumura, S. Furukawa, Y. Saho, H. Tsubomura, Cadmium sulfide photocatalyzed hydrogen production from aqueous solutions of sulfite: effect of crystal structure and preparation method of the catalyst, J. Phys. Chem. 89 (1985) 1327-1329. https://doi.org/10.1021/j100254a001

[212] J.R. Harbour, R. Wolkow, M.L. Hair, Effect of platinization on the photoproperties of cadmium sulfide pigments in dispersion. Determination by hydrogen evolution, oxygen uptake, and electron spin resonance spectroscopy, J. Phys. Chem. 85 (1981) 4026-4029. https://doi.org/10.1021/j150626a014

[213] J.F. Reber, M. Rusek, Photochemical hydrogen production with platinized suspensions of cadmium sulfide and cadmium zinc sulfide modified by silver sulphide, J. Phys. Chem. 90 (1986)824-834. https://doi.org/10.1021/j100277a024

[214] J. Yu, J. Zhang, M. Jaroniec, Preparation and enhanced visible-light photocatalytic $H_2$-production activity of CdS quantum dots-sensitized $Zn_{1-x}Cd_xS$ solid solution, Green Chem. 12 (2010) 1611-1614. https://doi.org/10.1039/c0gc00236d

[215] K. Kalyanasundaram, E. Borgarello, D. Duonghong, M. Gratzel, Cleavage of water by visible-light irradiation of colloidal CdS Solutions; inhibition of photocorrosion by $RuO_2$, Angew. Chem. Int. Ed Eng. 20 (1981) 987-988. https://doi.org/10.1002/anie.198109871

[216] K. Kobayakawa, A. Teranishi, T. Tsurumaki, Y. Sato, A. Fujishima, Photocatalytic activity of $CuInS_2$ and CuIn5S8, Electrochim. Acta 37 (1992) 465-481. https://doi.org/10.1016/0013-4686(92)87037-Z

[217] A. Kudo, K. Ueda, H. Kato, I. Mikami, Photocatalytic $O_2$ evolution under visible light irradiation on $BiVO_4$ in aqueous $AgNO_3$ solution, Catal. Lett. 53 (1998) 229-230. https://doi.org/10.1023/A:1019034728816

[218] G. Hitoki, T. Takata, J.N. Kondo, M. Hara, H. Kobayashi, K. Domen, An oxynitride, TaON, as an efficient water oxidation photocatalyst under visible light irradiation ($\lambda \leq 500$ nm), Chem. Commun. (2002) 1698-1699. https://doi.org/10.1039/B202393H

[219] K. Maeda, K. Domen, New non-oxide photocatalysts designed for overall water splitting under visible light, J. Phys. Chem. C 111 (2007) 7851-7861. https://doi.org/10.1021/jp070911w

[220] R. Abe, M. Higashi, K. Domen, Facile fabrication of an efficient oxynitride TaON photoanode for overall water splitting into $H_2$ and $O_2$ under visible light irradiation, J. Am. Chem. Soc. 132 (2010) 11828-11829. https://doi.org/10.1021/ja1016552

[221] M. Higashi, K. Domen, R. Abe, Highly stable water splitting on oxynitride TaON photoanode system under visible light irradiation, J. Am. Chem. Soc. 134 (2012) 6968-6971. https://doi.org/10.1021/ja302059g

[222] Z. Lei, W. You, M. Liu, G. Zhou, T. Takata, M. Hara, K. Domen, C. Li, Photocatalytic water reduction under visible light on a novel $ZnIn_2S_4$ catalyst synthesized by hydrothermal method, Chem. Commun. (2003) 2142-2143. https://doi.org/10.1039/b306813g

[223] B. Hinnemann, P.G. Moses, J. Bonde, K.P. Jørgensen, J.H. Nielsen, S. Horch, I. Chorkendorff, J.K. Nørskov, Biomimetic hydrogen evolution: $MoS_2$ nanoparticles as catalyst for hydrogen evolution, J. Am. Chem. Soc. 127 (2005) 5308-5309. https://doi.org/10.1021/ja0504690

[224] J. Chen, S. Shen, P. Guo, M. Wang, P. Wu, X. Wang, L. Guo, In-situ reduction synthesis of nano-sized $Cu_2O$ particles modifying g-$C_3N_4$ for enhanced photocatalytic hydrogen production, Appl. Catal. B: Environ. 152-153 (2014) 335-341. https://doi.org/10.1016/j.apcatb.2014.01.047

[225] H. Zhao, Y. Dong, P. Jiang, G. Wang, H. Miao, R. Wu, L. Kong, J. Zhang, C. Zhang, Light-assisted preparation of a ZnO/CdS nanocomposite for enhanced photocatalytic H2 evolution: An insight into importance of in situ generated ZnS, ACS Sustainable Chem. Eng. 3 (2015) 969-977. https://doi.org/10.1021/acssuschemeng.5b00102

[226] D. Zhao, C.F. Yang, Recent advances in the $TiO_2$/CdS nanocomposite used for photocatalytic hydrogen production and quantum-dot-sensitized solar cells, Renewable Sustainable Energy Rev. 54 (2016) 1048-1059. https://doi.org/10.1016/j.rser.2015.10.100

[227] Y. Xu, Y. Huang, B. Zhang, Rational design of semiconductor-based photocatalysts for advanced photocatalytic hydrogen production: the case of cadmium chalcogenides, Inorg. Chem. Front. 3 (2016) 591-615. https://doi.org/10.1039/C5QI00217F

[228] W.K. Jo, N.C.S. Selvam, Fabrication of photostable ternary CdS/$MoS_2$/MWCNTs hybridphotocatalysts with enhanced H2 generation activity, Appl. Catal. A: Gen. 525 (2016) 9-22. https://doi.org/10.1016/j.apcata.2016.06.036

[229] P.V. Kamat, Graphene-based nanoassemblies for energy conversion, J. Phys. Chem. Lett. 2 (2011) 242-251. https://doi.org/10.1021/jz101639v

[230] Q. Xiang, J. Yu, Graphene-based photocatalysts for hydrogen generation, J. Phys. Chem. Lett. 4 (2013) 753-759. https://doi.org/10.1021/jz302048d

[231] R. Abe, Development of a new system for photocatalytic water splitting into $H_2$ and $O_2$ under visible light irradiation, Bull. Chem. Soc. Jpn. 84 (2011) 1000-1030. https://doi.org/10.1246/bcsj.20110132

[232] R. Abe, K. Sayama, K. Domen, H.Arakawa, A new type of water splitting system composed of two different $TiO_2$ photocatalysts (anatase, rutile) and a $IO_3^-/I^-$ shuttle redox mediator, Chem. Phys. Letts. 344 (2001) 339-344. https://doi.org/10.1016/S0009-2614(01)00790-4

[233] H. Kato, M. Hori, R. Konta, Y. Shimodaira, A. Kudo, Construction of Z-scheme type heterogeneous photocatalysis systems for water splitting into $H_2$ and $O_2$ under visible light irradiation, Chem. Lett. 33 (2004) 1348-1349. https://doi.org/10.1246/cl.2004.1348

[234] Y. Sasaki, A. Iwase, H. Kato, A. Kudo, The effect of co-catalyst for Z-scheme photocatalysis systems with an $Fe^{3+}/Fe^{2+}$ electron mediator on overall water splitting under visible light irradiation, J. Catal. 259 (2008) 133-137. https://doi.org/10.1016/j.jcat.2008.07.017

[235] M. Higashi, R. Abe, A. Ishikawa, T. Takata, B. Ohtani, K. Domen, Z-scheme overall water splitting on modified-TaON photocatalysts under visible light ($\lambda$<500 nm), Chem. Lett. 37 (2008) 138-139. https://doi.org/10.1246/cl.2008.138

[236] X. Ma, Q. Jiang, W. Guo, M. Zheng, W. Xu, F. Ma, B. Hou, Fabrication of g-$C_3N_4$/Au/CdZnS Z-scheme photocatalyst to enhance photocatalysis performance, RSC Adv. 6 (2016) 28263-28269. https://doi.org/10.1039/C5RA27429J

[237] D. Lu, H. Wang, X. Zhao, K.K. Kondamareddy, J. Ding, C. Li, P. Fang, Highly efficient visible-light-induced photoactivity of Z-scheme g-$C_3N_4$/Ag/$MoS_2$ ternary

photocatalysts for organic pollutant degradation and production of hydrogen, ACS Sustainable Chem. Eng. 5 (2017) 1436-1445. https://doi.org/10.1021/acssuschemeng.6b02010

[238] Y. Sasaki, H. Nemoto, K. Saito, A. Kudo, Solar water splitting using powdered photocatalysts driven by Z-schematic interparticle electron transfer without an electron mediator, J. Phys. Chem. C 113 (2009)17536-17542. https://doi.org/10.1021/jp907128k

[239] G. Zhao, X. Huang, F. Fina, G. Zhang, J.T.S. Irvine, Facile structure design based on $C_3N_4$ for mediator-free Z-scheme water splitting under visible light, Catal. Sci. Technol. 5 (2015) 3416-3422. https://doi.org/10.1039/C5CY00379B

[240] F.Q. Zhou, J.C. Fan, Q.J. Xu, Y.L. Min, BiVO4 nanowires decorated with CdS nanoparticles as Z-scheme photocatalyst with enhanced $H_2$ generation, Appl. Catal. B: Environ. 201 (2017) 77-83. https://doi.org/10.1016/j.apcatb.2016.08.027

[241] J.S. Jang, S.H. Choi, D.H. Kim, J.W. Jang, K.S. Lee, J.S. Lee, Enhanced photocatalytic hydrogen production from water–methanol solution by nickel intercalated into titanate nanotube, J. Phys. Chem. C113 (2009) 8990-8996. https://doi.org/10.1021/jp900653r

[242] J. Yu, J. Ran, Facile preparation and enhanced photocatalytic $H_2$-production activity of $Cu(OH)_2$ cluster modified $TiO_2$, Energy Environ. Sci. 4 (2011) 1364-1371. https://doi.org/10.1039/c0ee00729c

[243] H. Wender, R. V. Gonçalves, C.S.B. Dias, M.J.M. Zapata, L.F. Zagonel, E. C. Mendonça, S.R. Teixeira, F. Garcia, Photocatalytic hydrogen production of $Co(OH)_2$ nanoparticle-coated $\alpha$-$Fe_2O_3$ nanorings, Nanoscale 5 (2013) 9310-9316. https://doi.org/10.1039/c3nr02195e

[244] Y.-P. Yuan, S.-W. Cao, L.-S. Yin, L. Xu, C. Xu, $NiS_2$ co-catalyst decoration on $CdLa_2S_4$ nanocrystals for efficient photocatalytic hydrogen generation under visible light irradiation, Int. J. Hydrogen Energy 38 (2013) 7218-7223. https://doi.org/10.1016/j.ijhydene.2013.03.169

[245] J. Wang, B. Li, J.Z. Chen, N. Li, J.F. Zheng, J.H. Zhao, Z.P. Zhu, Enhanced photocatalytic $H_2$-production activity of $Cd_xZn_{1-x}S$ nanocrystals by surface loading MS (M = Ni, Co, Cu) species, Appl. Surf. Sci. 259 (2012)118-123. https://doi.org/10.1016/j.apsusc.2012.07.003

[246] J. Yu, Y. Hai, B. Cheng, Enhanced photocatalytic $H_2$-production activity of $TiO_2$ by $Ni(OH)_2$ cluster modification, J. Phys. Chem. C 115 (2011) 4953-4958. https://doi.org/10.1021/jp111562d

[247] Y.-J. Yuan, H.-W. Lu, Z.-T. Yu, Z.-G. Zou, Noble-metal-free molybdenum disulfide cocatalyst for photocatalytic hydrogen production, ChemSusChem 8 (2015) 4113- 4127. https://doi.org/10.1002/cssc.201501203

[248] K. Chang, X. Hai, J. Ye, Transition metal disulfides as noble-metal-alternative co-catalysts for solar hydrogen production, Adv. Energy Mater. 6:1502555 (2016) 1-21. https://doi.org/10.1002/aenm.201502555

[249] Y. Li, G. Lu, S. Li, Photocatalytic hydrogen generation and decomposition of oxalic acid over platinized $TiO_2$, Appl. Catal. A: Gen. 214 (2001) 179-185. https://doi.org/10.1016/S0926-860X(01)00491-4

[250] Y. Li, Y. Xie, S. Peng, G. Lu, S. Li, Photocatalytic hydrogen generation in the presence of chloroacetic acids over $Pt/TiO_2$, Chemosphere 63 (2006) 1312-1318. https://doi.org/10.1016/j.chemosphere.2005.09.004

[251] J. Kim, W. Choi, Hydrogen producing water treatment through solar photocatalysis, Energy Environ. Sci. 3 (2010) 1042-1045. https://doi.org/10.1039/c003858j

[252] J. Kim, D.M. Satoca, W. Choi, Simultaneous production of hydrogen with the degradation of organic pollutants using $TiO_2$ photocatalyst modified with dual surface components, Energy Environ. Sci. 5 (2012) 7647-7656. https://doi.org/10.1039/c2ee21310a

[253] Y. Li, K. Zhang, S. Peng, G. Lu, S. Li, Photocatalytic hydrogen generation in the presence of ethanolamines over $Pt/ZnIn_2S_4$ under visible light irradiation, J Mol. Catal. A: Chem. 363-364 (2012) 354-361. https://doi.org/10.1016/j.molcata.2012.07.011

[254] W. Zhang, Y. Li, C. Wang, P. Wang, Q. Wang, Energy recovery during advanced wastewater treatment: Simultaneous estrogenic activity removal and hydrogen production through solar photocatalysis, Water Res. 47 (2013) 1480-1490. https://doi.org/10.1016/j.watres.2012.12.019

[255] J. Kim, Y. Park, H. Park, Solar hydrogen production coupled with the degradation of a dye pollutant using $TiO_2$ modified with platinum and nafion, Inter. J Photoenergy, Article ID 324859, 2014 (2014) 1-9.

[256] M. Halmann, Photoelectrochemical reduction of aqueous carbon dioxide on p-type gallium phosphide in liquid junction solar cells, Nature 275 (1978) 115-116. https://doi.org/10.1038/275115a0

[257] J.C. Hemminger, R. Carr, G.A. Somorjai, The photoassisted reaction of gaseous water and carbon dioxide adsorbed on the $SrTiO_3$ (111) crystal face to form methane, Chem. Phys. Lett.57 (1978) 100-104. https://doi.org/10.1016/0009-2614(78)80359-5

[258] K. Tennakone, Photoreduction of carbonic acid by mercury coated n-titanium dioxide, Solar Energy Mater. 10 (1984) 235-238. https://doi.org/10.1016/0165-1633(84)90063-7

[259] B.-J. Liu, T. Torimoto, H. Matsumoto, H. Yoneyama, Effect of solvents on photocatalytic reduction of carbon dioxide using $TiO_2$ nanocrystal photocatalyst embedded in $SiO_2$ matrices, J. Photochem. Photobiol. A: Chem. 108 (1997) 187-192. https://doi.org/10.1016/S1010-6030(97)00082-8

[260] I.-H. Tseng, W.-C. Chang, J.C.S. Wu, Photoreduction of $CO_2$ using sol–gel derived titania and titania-supported copper catalysts, Appl. Catal. B: Environ. 37 (2002) 37-48. https://doi.org/10.1016/S0926-3373(01)00322-8

[261] I.-H. Tseng, J.C.S Wu, H.-Y. Chou, Effects of sol–gel procedures on the photocatalysis of $Cu/TiO_2$ in $CO_2$ photoreduction, J. Catal. 221 (2004) 432-440. https://doi.org/10.1016/j.jcat.2003.09.002

[262] Y. Ku, W.-H. Lee, W.-Y. Wang, Photocatalytic reduction of carbonate in aqueous solution by $UV/TiO_2$ process, J. Mol. Catal. A: Chem. 212 (2004) 191-196. https://doi.org/10.1016/j.molcata.2003.10.047

[263] M. Anpo, H. Yamashita, Y. Ichihashi, S. Ehara, Photocatalytic reduction of $CO_2$ with $H_2O$ on various titanium oxide catalysts, J. Electroanal. Chem. 396 (1995) 21-26. https://doi.org/10.1016/0022-0728(95)04141-A

[264] H. Yamashita, Y. Fujii, Y. Ichihashi, S.G. Zhang, K. Ikeue, D.R. Park, K. Koyano, T. Tatsumi, M. Anpo, Selective formation of $CH_3OH$ in the photocatalytic reduction of $CO_2$ with $H_2O$ on titanium oxides highly dispersed within zeolites and mesoporous molecular sieves, Catal. Today 45 (1998) 221-227. https://doi.org/10.1016/S0920-5861(98)00219-3

[265] M. Anpo, H. Yamashita, K. Ikeue, Y. Fujii, S. G. Zhang, Y. Ichihashi, D.R. Park, Y. Suzuki, K. Koyano, T. Tatsumi, Photocatalytic reduction of $CO_2$ with $H_2O$ on

Ti-MCM-41 and Ti-MCM-48 mesoporous zeolite catalysts, Catal. Today 44 (1998) 327-332. https://doi.org/10.1016/S0920-5861(98)00206-5

[266] K. Ikeue, H. Yamashita, M. Anpo, Photocatalytic Reduction of $CO_2$ with $H_2O$ on titanium oxides prepared within the FSM-16 mesoporous zeolite, Chem. Lett. 28 (1999) 1135-1136. https://doi.org/10.1246/cl.1999.1135

[267] K. Ikeue, H. Yamashita, M. Anpo, T. Takewaki, Photocatalytic reduction of $CO_2$ with $H_2O$ on Ti-β zeolite photocatalysts: effect of the,hydrophobic and hydrophilic properties, J. Phys. Chem. B 105 (2001)8350-8355. https://doi.org/10.1021/jp010885g

[268] M. Anpo, H. Yamashita, Y. Ichihashi, Y. Fujii, M. Honda, Photocatalytic reduction of $CO_2$ with $H_2O$ on titanium oxides anchored within micropores of zeolites: effects of the structure of the active sites and the addition of Pt, J. Phys. Chem. B 101 (1997)2632-2636. https://doi.org/10.1021/jp962696h

[269] S. Xie, Y. Wang, Q. Zhang, W. Fan, W. Deng, Y. Wang, Photocatalytic reduction of $CO_2$ with $H_2O$: significant enhancement of the activity of Pt–$TiO_2$ in $CH_4$ formation by addition of MgO, Chem. Commun. 49 (2013) 2451-2453. https://doi.org/10.1039/c3cc00107e

[270] L. Liu, C. Zhao, H. Zhao, D. Pitts, Y. Li, Porous microspheres of MgO-patched $TiO_2$ for $CO_2$ photoreduction with $H_2O$ vapor: temperature-dependent activity and stability, Chem. Commun. 49 (2013) 3664-3666. https://doi.org/10.1039/c3cc39054c

[271] K. Hirano, K. Inoue, T. Yatsu, Photocatalysed reduction of $CO_2$ in aqueous $TiO_2$ suspension mixed with copper powder, J. Photochem. Photobiol. A: Chem. 64 (1992) 255-258. https://doi.org/10.1016/1010-6030(92)85112-8

[272] K. Kocia, K. Mateju, L. Obalovaa, S. Krejcikova, Z. Lacny, D. Placha, L. Capek, A. Hospodkova, O. Solcova, Effect of silver doping on the $TiO_2$ for photocatalytic reduction of $CO_2$, Appl. Catal. B: Environ. 96 (2010) 239-244. https://doi.org/10.1016/j.apcatb.2010.02.030

[273] Z. Zhao, J. Fan, J. Wang, R. Li, Effect of heating temperature on photocatalytic reduction of $CO_2$ by N–$TiO_2$ nanotube catalyst, Catal. Comm. 21 (2012) 32-37. https://doi.org/10.1016/j.catcom.2012.01.022

[274] Y. Izumi, Recent advances in the photocatalytic conversion of carbon dioxide to fuels with water and/or hydrogen using solar energy and beyond, Coord. Chem. Rev. 257 (2013) 171-186. https://doi.org/10.1016/j.ccr.2012.04.018

[275] W. Fan, Q. Zhang, Y. Wang, Semiconductor-based nanocomposites for photocatalytic $H_2$ production and $CO_2$ conversion, Phys. Chem. Chem. Phys.15 (2013) 2632-2649. https://doi.org/10.1039/c2cp43524a

[276] O. Ola, M.M. Maroto-Valer, Review of material design and reactor engineering on $TiO_2$ photocatalysis for $CO_2$ reduction, J. Photochem. Photobiol. C: Photochem. Rev. 24 (2015) 16-42. https://doi.org/10.1016/j.jphotochemrev.2015.06.001

[277] J. Wang, G. Ji, Y. Liu, M.A. Gondal, X. Chang, $Cu_2O/TiO_2$ heterostructure nanotube arrays prepared by an electrodeposition method exhibiting enhanced photocatalytic activity for $CO_2$ reduction to methanol, Catal. Comm. 46 (2014) 17-21. https://doi.org/10.1016/j.catcom.2013.11.011

[278] M. A. Asi, C. He, M. Su, D. Xia, L. Lin, H. Deng, Y. Xiong, R. Qiu, X. Li, Photocatalytic reduction of $CO_2$ to hydrocarbons using $AgBr/TiO_2$ nanocomposites under visible light, Catal. Today 175 (2011) 256-263. https://doi.org/10.1016/j.cattod.2011.02.055

[279] Q.D. Truong, J.-Y. Liu, C.-C. Chung, Y.-C. Ling, Photocatalytic reduction of $CO_2$ on $FeTiO_3/TiO_2$ photocatalyst, Catal. Comm. 19 (2012) 85-89. https://doi.org/10.1016/j.catcom.2011.12.025

[280] K. Tennakone, A.H. Jayatissa, S. Punchihewa, Selective photoreduction of carbon dioxide to methanol with hydrous cuprous oxide, J. Photochem. Photobiol. A 49 (1989) 369-375. https://doi.org/10.1016/1010-6030(89)87134-5

[281] Y. Matsumoto, M. Obata, J. Hombo, Photocatalytic reduction of carbon dioxide on p-type $CaFe_2O_4$powder, J. Phys. Chem. 98 (1994) 2950-2951. https://doi.org/10.1021/j100062a035

[282] Y. Liu, B. Huang, Y. Dai, X. Zhang, X. Qin, M. Jiang, M.-H. Whangbo, Selective ethanol formation from photocatalytic reduction of carbon dioxide in water with $BiVO_4$ photocatalyst, Catal. Comm. 11 (2009) 210-213. https://doi.org/10.1016/j.catcom.2009.10.010

[283] Y. Zhou, Z. Tian, Z. Zhao, Q. Liu, J. Kou, X. Chen, J. Gao, S. Yan, Z. Zou, High-yield synthesis of ultrathin and uniform $Bi_2WO_6$ square nanoplates benefitting from photocatalytic reduction of $CO_2$ into renewable hydrocarbon fuel under visible light, ACS Appl. Mater. Interfaces 3 (2011) 3594-3601. https://doi.org/10.1021/am2008147

[284] H. Cheng, B. Huang, Y. Liu, Z. Wang, X. Qin, X. Zhang, Y. Dai, An anion exchange approach to $Bi_2WO_6$ hollow microspheres with efficient visible light

photocatalytic reduction of $CO_2$ to methanol, Chem. Commun. 48 (2012) 9729-9731. https://doi.org/10.1039/c2cc35289c

[285] S. Navalon, A. Dhakshinamoorthy, M. Alvaro, H. Garcia, Photocatalytic $CO_2$ reduction using non-titanium metal oxides and sulfides, ChemSusChem 6 (2013) 562-577. https://doi.org/10.1002/cssc.201200670

[286] W. Jiang, X. Yin, F. Xin, Y. Bi, Y. Liu, X. Li, Preparation of $CdIn_2S_4$ microspheres and application for photocatalytic reduction of carbon dioxide, Appl. Surf. Sci. 288 (2014) 138-142. https://doi.org/10.1016/j.apsusc.2013.09.165

[287] Z.-Y. Wang, H.-C. Chou, J. C.S.Wu, D. P. Tsai, G. Mul, $CO_2$ photoreduction using NiO/$InTaO_4$ in optical-fiber reactor for renewable energy, Appl. Catal. A: Gen. 380 (2010) 172-177. https://doi.org/10.1016/j.apcata.2010.03.059

[288] X. Li, H. Liu, D. Luo, J. Li, Y. Huang, H. Li, Y. Fang, Y. Xu, L. Zhu, Adsorption of $CO_2$ on heterostructure $CdS(Bi_2S_3)/TiO_2$ nanotube photocatalysts and their photocatalytic activities in the reduction of $CO_2$ to methanol under visible light irradiation, Chem. Eng. J 180 (2012) 151-158. https://doi.org/10.1016/j.cej.2011.11.029

[289] X.J. Lv, W. F. Fu, C.Y. Hu, Y. Chen, W.B. Zhou, Photocatalytic reduction of $CO_2$ with $H_2O$ over a graphene-modified $NiO_x$–$Ta_2O_5$ composite photocatalyst: coupling yields of methanol and hydrogen, RSC Adv. 3 (2013) 1753-1757. https://doi.org/10.1039/c2ra21283h

[290] H. Liu, Z. Zhang, J. Meng, J. Zhang, Novel visible-light-driven $CdIn_2S_4$/mesoporous g-$C_3N_4$ hybrids for efficient photocatalytic reduction of $CO_2$ to methanol, J. Mol. Catal. A: Chem. 430 (2017) 9-19.

[291] T. Ohno, N. Murakami, T. Koyanagi, Y. Yang, Photocatalytic reduction of $CO_2$ over a hybrid photocatalyst composed of $WO_3$ and graphitic carbon nitride (g-$C_3N_4$) under visible light, J. $CO_2$ Utilization 6 (2014) 17-25. https://doi.org/10.1016/j.jcou.2014.02.002

[292] J.H. Kim, G. Magesh, H.J. Kang, M. Banu, J.H. Kim, J. Lee, J.S. Lee, Carbonate-coordinated cobalt co-catalyzed $BiVO_4/WO_3$ composite photoanode tailored for $CO_2$reduction to fuels, Nano Energy 15 (2015) 153-163. https://doi.org/10.1016/j.nanoen.2015.04.022

[293] D.-S. Lee, Y.-W. Chen, Photocatalytic reduction of carbon dioxide with water on $InVO_4$ with NiO cocatalysts, J. $CO_2$ Utilization 10 (2015) 1-6. https://doi.org/10.1016/j.jcou.2015.02.005

[294] S. Sato, T. Morikawa, S. Saeki, T. Kajino, T, Motohiro, Visible-light-induced selective $CO_2$ reduction utilizing a ruthenium complex electrocatalyst linked to a p-type nitrogen-doped $Ta_2O_5$ semiconductor, Angew. Chem. Int. Ed. 49 (2010) 5101-5105. https://doi.org/10.1002/anie.201000613

[295] S. Sato, T. Arai, T. Morikawa, K. Uemura, T.M. Suzuki, H. Tanaka, T. Kajino, Selective $CO_2$ conversion to formate conjugated with $H_2O$ oxidation utilizing semiconductor/complex hybrid photocatalysts, J. Am. Chem. Soc. 133 (2011) 15240-15243. https://doi.org/10.1021/ja204881d

[296] T. Arai, S. Sato, T. Kajino, T. Morikawa, Solar $CO_2$ reduction using $H_2O$ by a semiconductor/metal-complex hybrid photocatalyst: enhanced efficiency and demonstration of a wireless system using $SrTiO_3$ photoanodes, Energy Environ. Sci. 6 (2013) 1274-1282. https://doi.org/10.1039/c3ee24317f

[297] K. Sekizawa, K. Maeda, K. Domen, K. Koike, O. Ishitani, Artificial Z-scheme constructed with a supramolecular metal complex and semiconductor for the photocatalytic reduction of $CO_2$, J. Am. Chem. Soc. 135 (2013) 4596-4599. https://doi.org/10.1021/ja311541a

[298] W. Yu, D. Xu, T. Peng, Enhanced photocatalytic activity of g-$C_3N_4$ for selective $CO_2$ reduction to $CH_3OH$ via facile coupling of ZnO: a direct Z-scheme mechanism, J. Mater. Chem. A 3 (2015) 19936-19947. https://doi.org/10.1039/C5TA05503B

[299] J. Jin, J. Yu, D. Guo, C. Cui, W. Ho, A hierarchical Z-scheme CdS–$WO_3$ photocatalyst with enhanced $CO_2$ reduction activity, Small11 (2015)5262-5271. https://doi.org/10.1002/smll.201500926

[300] Y. He, L. Zhang, B. Teng, M. Fan, New application of Z-scheme $Ag_3PO_4$/g-$C_3N_4$ composite in converting $CO_2$ to fuel, Environ. Sci. Technol. 49 (2015) 649-656. https://doi.org/10.1021/es5046309

[301] J.-C. Wang, L. Zhang, W.-X. Fang, J. Ren, Y.-Y. Li, H.-C. Yao, J.-S. Wang, Z.-J. Li, Enhanced photoreduction $CO_2$ activity over direct Z-scheme α-$Fe_2O_3$/$Cu_2O$ heterostructures under visible light irradiation, ACS Appl. Mater. Interfaces 7 (2015) 8631-8639. https://doi.org/10.1021/acsami.5b00822

[302] J.-C. Wang, H.-C. Yao, Z.-Y. Fan, L. Zhang, J.-S. Wang, S.-Q. Zang, Z.-J. Li, Indirect Z-scheme BiOI/g-$C_3N_4$ photocatalysts with enhanced photoreduction $CO_2$ activity under visible light irradiation, ACS Appl. Mater. Interfaces 8 (2016) 3765-3775. https://doi.org/10.1021/acsami.5b09901

[303] Y. Fu, D. Sun, Y. Chen, R. Huang, Z. Ding, X. Fu, Z. Li, An amine-functionalized titanium metal–organic framework photocatalyst with visible-light-induced activity for $CO_2$ reduction, Angew. Chem. 124 (2012) 3420-3423. https://doi.org/10.1002/ange.201108357

[304] D. Sun, Y. Fu, W. Liu, L. Ye, D. Wang, L. Yang, X. Fu, Z. Li, Studies on photocatalytic $CO_2$ reduction over $NH_2$-Uio-66(Zr) and its derivatives: towards a better understanding of photocatalysis on metal–organic frameworks, Chem. Eur. J. 19 (2013) 14279-14285. https://doi.org/10.1002/chem.201301728

# Chapter 2

# Natural polymers for the removal of heavy metals

Dure Najaf Iqbal, Qudsia Kanwal, Munawar Iqbal*
Department of Chemistry, University of Lahore, 1-KM Raiwind Road Lahore, Pakistan
bosalvee@yahoo.com

**Abstract**

Heavy metals are very toxic, carcinogenic and non-biodegradable substances discharged into waste water from various chemical industries, even a small concentration of heavy metals can be lethal for the environment and public health. Advanced developments for treating of heavy metals contaminated industrial waste water involving natural polymers such as fungai, algae, microorganism, biopolymers and numerous polysaccharides have shown to be very useful.

**Keywords**

Heavy Metals, Natural Polymer, Waste Water, Polysaccharides, Biodegradable, Green Methods

## Contents

## 1. Introduction

Heavy metals are non-biodegradable metallic elements, have high molecular weight, are most persistent wastewater pollutants with density of $5 g/cm^3$[1-5]. These metals have toxic (acute, chronic or sub-chronic) [6-10], carcinogenic [11-13], mutagenic [14, 15], neurotoxic [16, 17] or teratogenic effects on human, plants and microorganisms [17-19]. Arsenic, cadmium, chromium, copper, lead, mercury, nickel, silver, and zinc are the most common toxic metals in waste water. Heavy metals have numerous agricultural, industrial, medical, domestic and commercial applications [20-23]. Their occurrence and accretion are due to certain environmental factors such as leaching of rocks [24, 25], aerial dust [26, 27], forest fires [28, 29] and flora [30, 31]. Human activities such as urbanization [32, 33] and industrialization [34, 35] have been major reasons for the entry of trace elements in water bodies. The toxicity, genotoxicity, and carcinogenic nature of heavy metals depends on the quantity, exposure limit, production rate, age, gender, heredities, and dietary status of the exposed person. Arsenic, cadmium, chromium, lead, and mercury ranked highly as toxic metals, even a small concentration of theses metallic toxicants have very lethal impact on biological tissues and environment. The International Agency for research on Cancer [36] and U.S. Environmental Protection Agency [37] declared these metals as carcinogenic.

Lead is considered a threat to human health [38]. Even lower level of exposure to lead has an influential effect on human health [39-42] and reproductive system [43, 44]. It caused a decrease in hemoglobin production [45] and mental alertness [46]. Children are a carrier of lead right from birth via blood stream of their mother in embryonic stage [47]. It is now proven that antisocial behavior and criminal attitude due to the accumulation of lead during childhood, which has lifetime effect on a child brain and intellectual capabilities. Environmental factors such as polluted dust, air, and soil are also a potential source of lead. Environmental lead is also harmful to flora, fauna, and microbes.

Another persistent cumulative poison is cadmium. A major source of kidney malfunction in human beings is the intake of cadmium by food or vegetables [48, 49]. Agriculture soil irrigated by cadmium rich water caused accumulation of cadmium in a food product. The intake of such food items by any human being results in renal tubular damage, osteomalacia, osteoporosis, cadmium pneumonitis, lung emphysema, high blood pressure and prostate cancer [50]. The International Agency for Research on Cancer (IARC) classifies cadmium in Class 1 of carcinogenic elements. Persistent exposure may result in respiratory edema and demise. Aquatic organisms absorbed Cd (II) directly from wastewater reservoirs. In plants, stomata opening, transpiration, and photosynthesis have been affected by cadmium. It causes a decrease in seed germination and lipids contents and enzymes activity. **( Table 1)** lists the sources of their toxic effects on the human health.

*Table 1. Heavy metal sources and toxicity.*

| **Heavy Metal** | **Sources** | **Toxicity** |
|---|---|---|
| Arsenic | Pesticides and metal smelters | Vomiting, abdominal pain, encephalopathy, thickening of the skin, darker skin, abdominal pain, diarrhea, heart disease, numbness, and cancer. |
| Cadmium | Fertilizers, Pesticides, electroplating, Cd Batteries and nuclear fission reactions | Cardiovascular, renal, gastrointestinal, neurological, reproductive, and respiratory disorders |
| Chromium | Mining, chemical industries, metal Piping, pesticides, electroplating, leather tanning, dyeing, and cement industries | Pneumonia, asthma, bronchitis, liver damage, a decrease in blood cells, skin ulcer |
| Lead | Pesticides, paint, mining, smoking, combustion, automobile emission | Anemia, constipation, hearing Loss, kidney damage, reduced IQ, slow development and birth defects |
| Mercury | Gold mining, coal-fired power stations, Industrial boilers and biomass combustion. | Depression, vision or hearing loss, numbness, irritability, tremors, memory problems |

Arsenic is another toxic heavy metal discharged in waste water. It binds itself with protein and damaged biological tissues. It inhibits ATP production during respiration by coagulating protein [51]. It is more dangerous in inorganic form as compared to organic arsenicals. Arsenite is more poisonous than arsenate. Arsenic compounds caused acute lethality as well as the inhibition of growth in aquatic and terrestrial biota [52].

Dental amalgam, mercury oxide batteries, electric and electronic switches, thermometers, laboratory chemical industries etc. are major sources of mercury in wastewater [53]. Methyl mercury originated from seafood badly affects human brain especially children via placenta during their birth [54]. The high toxicity of methyl mercury is well documented. Epidemiological studies revealed that infant's brain is exposed to methyl mercury through placenta via the gastrointestinal tract. However, plants are usually unresponsive to the toxic effects of mercuric compounds in waste water. Exposure to Cr(III) or Cr(VI) caused allergic reactions like irritating respiratory, stomach and blood effects [55]. Cr(VI) compounds are more toxic as compared to Cr(III) and considered as carcinogenic. There is a need to minimize human and environmental exposure to toxic heavy metals by strict regulation and developing new technologies.

## 1.1 Wastewater treatments

Numerous methodologies have been commercialized for removal of heavy metals from water sources, involving physical, chemical and biological treatments (Fig. 1). Heavy metals can be removed from sewages using adsorption with activated carbon, chemical precipitation, electrochemical treatment, ion-exchangers, membrane filtration, coagulation–flocculation, flotation precipitation with hydroxides, carbonates and sulfides are other popular techniques.

However, all the methods mentioned above become ineffective at very low metal concentration. Electrolysis and ion exchange are expensive methodology due to high capital and operational costs. Precipitation and flocculation faced a challenge because of the necessity for disposal and treatment of sludge produced during treatment. Adsorption is a well-known equilibrium separation process. For water decontamination applications, it is proven as an effective, efficient and economical process. The adsorbents are categorized as mineral, organic or biological. Activated carbons, clays, zeolites and silica gel are low-cost adsorbents of mineral origin. Agricultural wastes, biomass, and polymeric resins, microporous polymeric beads belong to the organic class of adsorbents so adsorption with activated carbon is considered more feasible and efficient treatment for removal of heavy metals from waste water. But in many cases, it's not cost effective. Consequently, low-cost sorbents derived from natural polymers are a promising alternative for reduction of water pollution [56, 57].

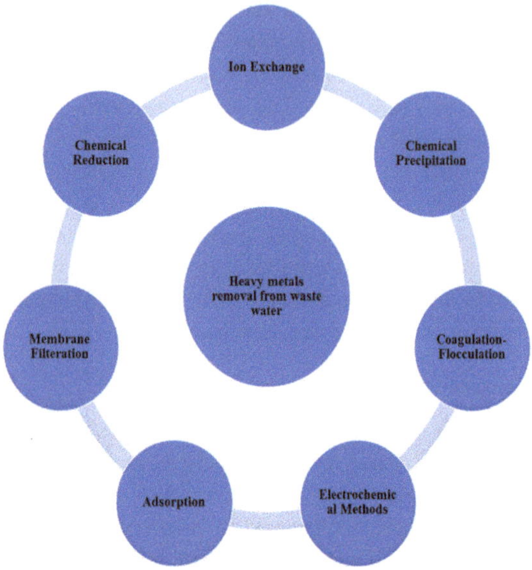


*Fig. 1. Methodologies for decontamination of waste water*

## 2. Natural polymers as biosorbents

Since the beginning of industrialization activated carbon has been the common and extensively used adsorbent for purification of waste water. Microporous structure, large surface area and chemical reactivity of carbon have made activated carbon an ideal candidate for treatments. However, activated carbon treatments are expensive and hence the need of alternative low-cost adsorbents with superior performance has always been felt. Recently natural polysaccharides have been developed as economical, more effective and environmentally friendly materials for removal of a toxic substance from contaminated water [58]. Polysaccharides based adsorbents such as cellulose [59], nitrocellulose [60], chitin[61], chitosan [62], starch [63], cyclodextrin [64] and natural gums [65] as well as their derivatives have been under consideration to purify water in an economical way. These biopolymers represent an interesting and attractive alternative as adsorbents because of their structural diversity, abundance, physio-chemical properties, stability under wide PH range, reasonable reactivity and good adsorption capacity.

Theses biopolymers have a very high swelling index due to the hydrophilic nature of monomer units associated with polymer back. Consequently, repeating anhydrous sugar units swell during adsorption process to allow the fast removal of heavy metals from the solution by diffusion. Numerous functional groups (hydroxyl, acetamido or amino functions) present on theses polymers backbone make them an ideal candidate for chemical interactions with a wide variety of molecules. Grafting of these functional groups enhanced organic solubility, polarity and hydrophilicity of natural polymer, consequently increased their adsorption capacity for heavy metals pollutants. Another advantage associated with these biopolymers is cost management so "green" and sustainable water treatments become a prerequisite. Recently, several natural polymers have been developed as economical and efficient adsorbents. The adsorption process for decontamination of water by these biopolymers is termed as biosorption. It is a mass transfer method by which heavy metals are transferred from aqueous solution and bounded by physical and chemical interactions on absorbents. Generally, a suitable adsorbent for these processes should meet following requirements.

- Efficient for wide range of environmental pollutants.
- Maximum swelling index and adsorption capacity.
- Granular and porous structural morphology.
- Biodegradable and easy regeneration ability.
- Stability under experimental conditions, temperature, pH and other variables.
- Low cost and environment benign.
- High physical strength.

The main steps involved in adsorption of pollutants on solid adsorbent are (Fig. 2)

- Metal ion transportation from the bulk solution to the external surface of the adsorbent.
- Internal mass transfer through the porous surface of adsorbents.
- Occupying of active sites of adsorbents by metal ions.

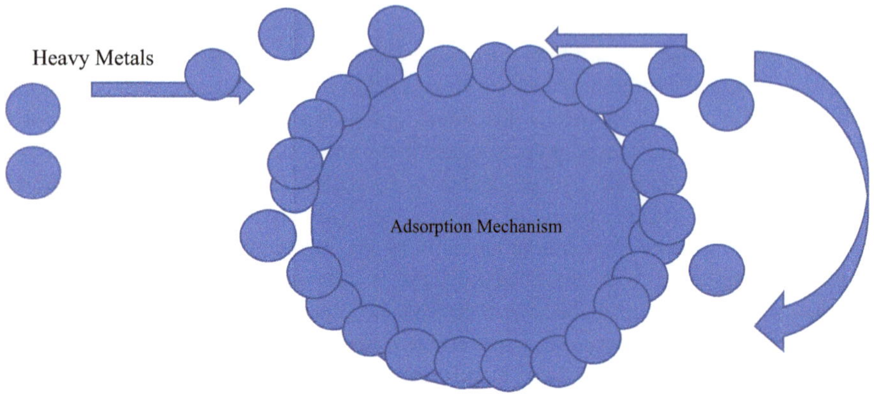


*Fig. 2. Illustrations for adsorption method for heavy metals.*

In this Chapter, numerous adsorption mechanisms have been discussed for efficient natural polymers as an adsorbent for wastewater treatment. The parameters which have been investigated for optimizing the use of adsorbent in wastewater treatment include adsorbent nature, the concentration of metal ions, temperature, time, pH of aqueous media, reaction kinetics and adsorption isotherm.

### 2.1 Cellulose as biosorbents

Most abundant biopolymer on earth is cellulose and hemicellulose (Fig. 3). Cellulose is a linear homopolymer of glucose connected by β1 4 glycosidic linkages having intramolecular and intermolecular hydrogen bonding.

Numerous derivatives of cellulose are commercially available due to structural modification by replacing free hydroxyl groups on polymer backbone by other functional groups. Cellulose was modified successfully to produce effective adsorbents using chemical and biological methods. Such as graft copolymerization, etherification and modification for corn stalk, quaternization for rice straw, and oxidation for peanut shell. These cellulosic adsorbents were characterized and then used for the removal of heavy metals from waste water.

*Fig. 3. Structure and sources of cellulose.*

Crosslinked cellulose was prepared under different synthetic conditions. Epichlorohydrin (EPI) has been considered good crosslinker for cellulose for a decade [66]. A lot of work has been done on epichlorohydrin crosslinked cellulose to form polymeric adsorbents with tunable properties. Dalbergia sissoo pods, commonly known as Indian rosewood found throughout India, Bangladesh, Pakistan and Nepal up to 900 m were assessed for sequestering of Cr (VI) from synthetic wastewater.

Cellulose based nanogels [67] have been synthesized by graft copolymerization of acrylic acid and N-isopropyl acrylamide (NIPA) onto carboxymethyl cellulose (CMC) for removal of copper and lead ions from aqueous solutions. Cellulose based adsorbents have been used for removal of $Cu^{2+}$, $Ni^{2+}$, $Zn^{2+}$, $Pb^{2+}$ and $Hg^{2+}$from waste water. Cyanoethyl cellulose [68]-based superadsorbent hydrogels were prepared and used for the adsorption of copper (II) ions from aqueous solutions. Acrylamide and acrylonitrile grafted cellulose were also examined for sorption of different metal ions as best alternative to cheap and economical sorbent materials [67, 68].

## 2.2 Chitosan as biosorbents

Chitosan is composed of randomly dispersed β-(1→4)-linked D-glucosamine (deacetylated unit) and N-acetyl-D-glucosamine (acetylated unit) chains (Fig. 4). It is produced by N-deacetylation of chitin (shrimp and other crustacean's shells) under alkaline conditions via deproteination and deacetylation mechanism. Chitosan is biocompatible, biodegradable and not-toxic hydrophilic polymer, having very reactive (-$NH_2$, -OH) groups as essential part of its backbone. Chitosan undergoes modification by various crosslinkers such as acetate, ethylene glycondiglycidyl ether, methanol, glyoxal, glycidyl chloride, glutaraldehyde, and isocyanates which enhanced the metal binding capacity of chitosan. It is a pH sensitive polymer but after substitution becomes stable over a wide pH range [69, 70].

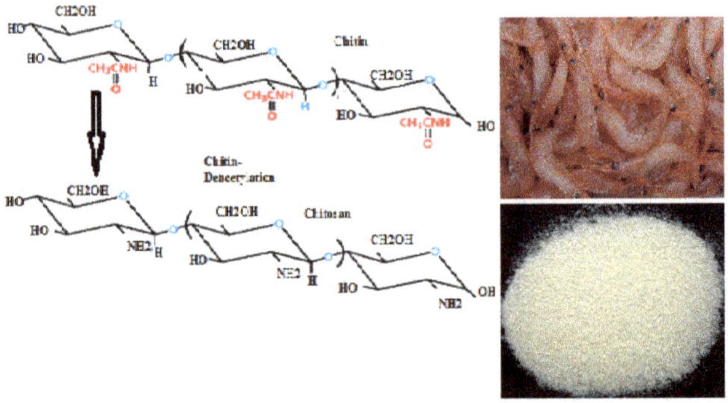


*Fig. 4. Structure of chitin and chitosan.*

Various Chitosan derivatives have been used extensively as an adsorbent. Chitosan derivatives containing nitrogen phosphorous and sulfur as hetero atoms and other derivatives such as chitosan crown ethers and chitosan ethylenediaminetetraacetic acid(EDTA)/ diethylenetriaminepentaacetic acid (DTPA) complexes. [71, 72] crosslinked via ethylene diamine have been used for removal of Cu (II) ion from wastewater.

Another approach for removal of Cu(II) ions from aqueous solution was done by pH-dependent formaldehyde crosslinked [73, 74]. Chitosan layered carbon adsorbent was developed for the removal of Cr(IV) and Cd(II) from wastewater at pH of 5.0-5.5. The reactions followed pseudo-second-order kinetics via Langmuir adsorption isotherm.

Novel chitosan composites have been fabricated for decontamination of waste water from heavy metals. Polyelectrolyte films and chitosan blends and CMC/pectin interpolymer complexes have been used for removal of Pb (II) ions from the treated waste water. Different variables studied demonstrated first order kinetic for adsorption mechanisms. [75] and Granular semi-IPN hydrogel composites were prepared for their use in removal of Cu(II) ions from the wastewater. Microporous film of polyethylene glycol cross-linked chitosan was used for Fe removal from aqueous solutions. Brominated chitosan showed enhanced Pb(II) ions adsorption capacity due to increase in the surface area and porosity. Gold ore stalk solutions having traces of cyanide have been purified by shrimp shells waste extracted chitosan. The result have shown efficient removal of Cu(II), Hg(II), Pb(II) and Zn(II).

Ethylenediaminetetraacetic acid (EDTA) ligand and chitosan-silica hybrid materials [76] were bound for creating a novel type adsorbent for removal of the Co(II), Ni(II), Cd(II), and Pb(II) ions from aqueous solutions [77].

Grafting of polyacrylonitrile on chitosan backbone has been done in the presence of ceric ammonium nitrate and the resultant modified chitosan has been used for removal of Cr(VI) and Cu(II) ions from aqueous media [78].

Thiourea modified chitosan was synthesized in two steps, used for the adsorption of Hg(II) [79]. Chitosan-thiourea and chitosan-polyacrylonitrile, chitosan-sodium alginate, mercapto-acetyl chitosan, chitosan(CRCH)-carbon tetrachloride homopolymers have been synthesized for adsorption of numerous heavy metals such as Cu(II), Cr(VI), Cd(II), Pb(II) and Ag(II) ions from wastewater. The adsorption efficacy was studied at various pH of the solution, adsorbent dosage and contact time [70].

Both alginic acid and chitosan naturally occurring polysaccharides have been the most effective adsorbents for removal of Co(II), Cu(II), and Cd(II) from wastewater [80]. The combined use of alginic acid and chitosan was expected to form a rigid matrix due to the anionic interaction between amino groups of chitosan and carboxyl groups of alginic acid and the crosslinking between the two through glutaraldehyde. The crosslinking makes the beads durable and facilitates the adsorption of Cu(II), Co(II) and Cd(II) ions under acidic conditions.

## 2.3 Starch as biosorbents

Starch is second most abundant biopolymer available naturally. It primarily consists of D-glyclopyramide homopolymerize linked together by α-1,4 and α-1,6 glycosidic linkage. Glucose diversity in starch molecules leads two types of polymers, amylose and amylopectin (Fig. 5) Amylose is an fundamentally linear polymer, whereas the amylopectin molecule is much branched and larger molecule. The structural variances among these two polymers contribute to important modifications in starch properties and functionality.

Much attention has been focused on modified starches for waste water treatments because of their ecological friendliness, easy handling, and accessibility [80].

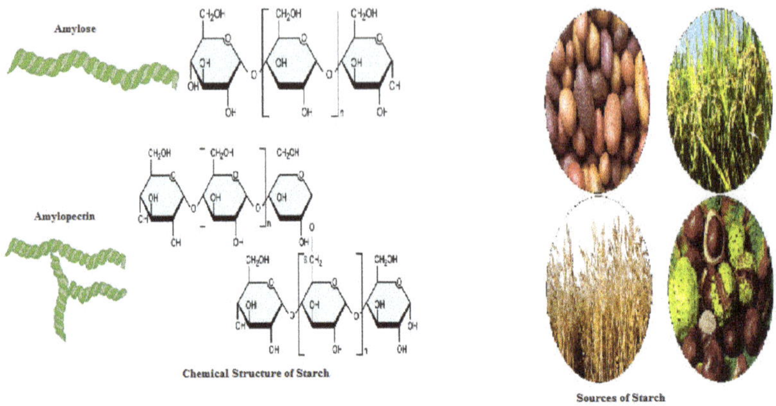


*Fig. 5. Structure of starch.*

Dithiocarbamate-modified starch (DTC) [81,82] and DTC mesoporous starch have potential for decontamination of heavy metals from waste water in the order of Cu(II)> Ni(II)> Cr(VI)> Zn(II)> Pb(II). The adsorption kinetic followed pseudo-second-order kinetics model. The adsorption kinetics of Cu(II), Cd(II), Co(II), Zn(II), Ni(II) was studied using dithiocarbamate-modified glycidyl methacrylate starch (DMGS). The adsorption of V(V) ions for removal from aqueous solutions was done by using modified starch known as dithiocarbamate- porous starch (DTCPS) [83]. The reaction kinetics was pH dependent and followed the pseudo- second order and Langmuir isotherm.

The graft copolymer of starch and acrylic acid was used as novel hydrogels and for treatments of water. Starch-g-poly- (N-methyl acrylamide-co-acrylic acid) has proved as biodegradable adsorbent for heavy metal sequence [84,85]. This synthetic graft copolymer acted as an adsorbent to remove Hg(II) from wastewater. Many grafted starch-poly (acrylic acid)/sodium humate (St-g-PAA/SH) hydrogels were used to adsorb Cu (II) from the waste aqueous solution [86]. Starch xanthate (PSX)[87], starch citrate [87] (PSC), starch-graft-polyacrylamide-co-sodium xanthate (CSAX)[88], crosslinked starches, cross-linked starch/acrylonitrile [89], cross-linked amino starch [90], hydroxyethyl starch (HES) [91], cassava starch (St-g-GMA) and many other have been reported in recent years for removal of variety of heavy metal ions to minimize water pollution.

## 2.4 Guar gum as biosorbent

Gums are important natural hydrocolloid and potential biopolymers for numerous food and non-food applications. These belongs to very important family of polysaccharides known as galactomannan and chemically consist of linear chain of D-mannose units, linked by β (1-4) glycoside linkage with α (1-6) D-galactose glycoside branching on approximately every alternate pH dependent guar gum silica nanocomposites were successfully used in elimination of Cd (II) from aqueous solution [92] as shown below in (Fig. 6).

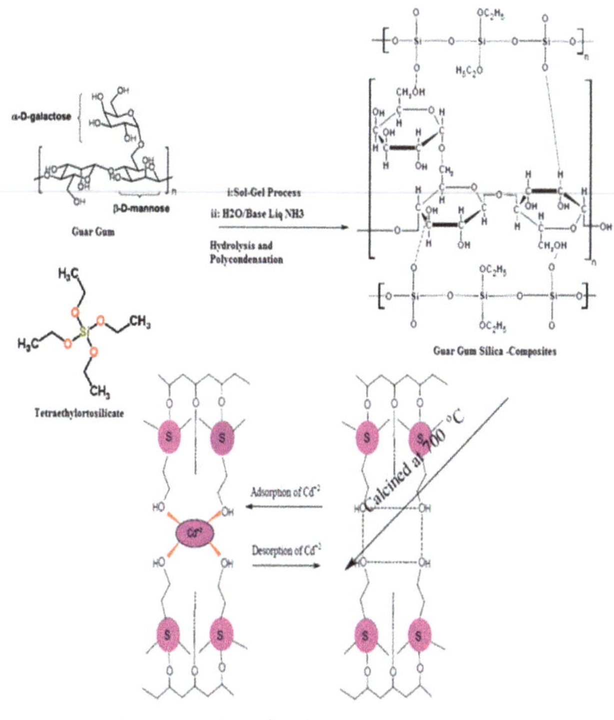


*Fig. 6. Removal of cadmium (II) by guar gum silica-nanocomposites.*

Guar gum (GG) and carbon nanotube (CNT) nanocomposite gel (MWCNT) with higher strength were prepared [93]. Iron oxide particles were then grafted on the surface of these super gels to enhance their sorption ability for decontamination of heavy metals. GG–MWCNT to formulate the magnetic GG–MWCNT–Fe3O4. Form (GG-clPDA) for the exclusion of Hg(II) ions from their aqueous solutions. Polyacrylamides and acrylic acid was polymerized with guar gum and novel hydrogels with versatile and fascinating properties were prepared by many researchers. The graft copolymer of Polyacrylamide/guar gum was then cross-linked with glutaraldehyde (GA) to obtain the sorbent material in the form of a hydrogel. The hydrogel was used for removal of chromium ion (Cr (VI)) from aqueous solution mannose units [94].

## Conclusion

There are numerous naturally occurring materials that can be used for the removal of heavy metals. Natural adsorbents such as agricultural wastes, carbohydrate derivatives, biopolymers, natural gums are most promising for removal of heavy metal due to the following considerations. They are economical, metal selective, regenerative, green and cost effective. Still, there is a need to develop novel technologies and explore the possibilities of polymers as new potential natural polymer as an adsorbents for waste water treatments.

## References

[1] Textile printing method, 1944, Google Patents.

[2] S. Khan, Q. Cao,Y.M. Zheng, Y.Z. Huang, Y.G. Zhu. Health risks of heavy metals in contaminated soils and food crops irrigated with wastewater in Beijing, China. Environmental pollution. 152 (2008) 686-692. https://doi.org/10.1016/j.envpol.2007.06.056

[3] H. A. Qdaisa, H. Moussa, Removal of heavy metals from wastewater by membrane processes: a comparative study, Desalination. 164 (2004) 105-110. https://doi.org/10.1016/S0011-9164(04)00169-9

[4] K. Kadirvelu, K. Thamaraiselvi, and C. Namasivayam, Removal of heavy metals from industrial wastewaters by adsorption onto activated carbon prepared from an agricultural solid waste, Bioresource Technology. 76 (2001) 63-65. https://doi.org/10.1016/S0960-8524(00)00072-9

[5] T.A. Kurniawan, Physico–chemical treatment techniques for wastewater laden with heavy metals, Chemical engineering journal. 118 (2006) 83-98. https://doi.org/10.1016/j.cej.2006.01.015

[6] K. E. Giller; E. Witter, S.P. Mcgrath, Toxicity of heavy metals to microorganisms and microbial processes in agricultural soils: a review, Soil Biology and Biochemistry. 30 (1998): 1389-1414. https://doi.org/10.1016/S0038-0717(97)00270-8

[7] A.M. B Påhlsson, Toxicity of heavy metals (Zn, Cu, Cd, Pb) to vascular plants, Water, Air, & Soil Pollution. 47 (1989) 287-319. https://doi.org/10.1007/BF00279329

[8] Afzali, D. Natural analcime zeolite modified with 5-Br-PADAP for the preconcentration and anodic stripping voltammetric determination of trace amount of cadmium, Analytical sciences. 21 (2005) 383-386. https://doi.org/10.2116/analsci.21.383

[9] P. Connor, Acute toxicity of heavy metals to some marine larvae, Marine Pollution Bulletin, 3 (1972) 190-192. https://doi.org/10.1016/0025-326X(72)90268-8

[10] M. Sato,and M. Kondoh, Recent studies on metallothionein: protection against toxicity of heavy metals and oxygen free radicals, The Tohoku journal of experimental medicine. 196 (2002) 9-22. https://doi.org/10.1620/tjem.196.9

[11] L. Järup, Hazards of heavy metal contamination, British medical bulletin. 68 (2003) 167-182. https://doi.org/10.1093/bmb/ldg032

[12] N. Coen, Heavy metals of relevance to human health induce genomic instability. The Journal of pathology. 195 (2001) 293-299. https://doi.org/10.1002/path.950

[13] B. Pandey, S. Suthar, and V. Singh, Accumulation and health risk of heavy metals in sugarcane irrigated with industrial effluent in some rural areas of Uttarakhand, India. Process Safety and Environmental Protection. 102 (2016) 655-666. https://doi.org/10.1016/j.psep.2016.05.024

[14] P. Wong, Mutagenicity of heavy metals, Bulletin of environmental contamination and toxicology. 40 (1988) 597-603. https://doi.org/10.1007/BF01688386

[15] H. Babich, M. Devanas, and G. Stotzky, The mediation of mutagenicity and clastogenicity of heavy metals by physicochemical factors, Environmental Research. 37 (1985) 253-286. https://doi.org/10.1016/0013-9351(85)90107-0

[16] P. Kursula, and V. Majava, A structural insight into lead neurotoxicity and calmodulin activation by heavy metals, Acta Crystallographica Section F:

Structural Biology and Crystallization Communications. 63 (2007) 653-656. https://doi.org/10.1107/S1744309107034525

[17] J. M. Matés, Roles of dioxins and heavy metals in cancer and neurological diseases using ROS-mediated mechanisms, Free Radical Biology and Medicine, 49(2010) 1328-1341. https://doi.org/10.1016/j.freeradbiomed.2010.07.028

[18] F. Calevro, Tests of toxicity and teratogenicity in biphasic vertebrates treated with heavy metals (Cr 3+, A1 3+, Cd 2+), Chemosphere. 37 (1998) 3011-3017. https://doi.org/10.1016/S0045-6535(98)00342-7

[19] S.H. Gilani, and Y. Alibhai, Teratogenicity of metals to chick embryos. Journal of Toxicology and Environmental Health, Part A Current Issues. 30 (1990) 23-31. https://doi.org/10.1080/15287399009531407

[20] A. Dąbrowski, Selective removal of the heavy metal ions from waters and industrial wastewaters by ion-exchange method, Chemosphere. 56 (2004) 91-106. https://doi.org/10.1016/j.chemosphere.2004.03.006

[21] H. Ali, E. Khan, and M.A. Sajad, Phytoremediation of heavy metals—concepts and applications, Chemosphere. 91 (2013) 869-881. https://doi.org/10.1016/j.chemosphere.2013.01.075

[22] N. Srivastava, and C. Majumder, Novel biofiltration methods for the treatment of heavy metals from industrial wastewater, Journal of hazardous materials. 151 (2008) 1-8. https://doi.org/10.1016/j.jhazmat.2007.09.101

[23] M. Algarra, Heavy metals removal from electroplating wastewater by aminopropyl-Si MCM-41, Chemosphere. 59 (2005) 779-786. https://doi.org/10.1016/j.chemosphere.2004.11.023

[24] R. A. Wuana and F.E. Okieimen, Heavy metals in contaminated soils: a review of sources, chemistry, risks and best available strategies for remediation, Isrn Ecology, 2011. https://doi.org/10.5402/2011/402647

[25] M. Saeedi, Assessment of heavy metals contamination and leaching characteristics in highway side soils, Iran, Environmental monitoring and assessment. 151 (2009) 231-241. https://doi.org/10.1007/s10661-008-0264-z

[26] K.W. Kim, Heavy metal contamination in dusts and stream sediments in the Taejon area, Korea, Journal of Geochemical Exploration. 64 (1998) 409-419. https://doi.org/10.1016/S0375-6742(98)00045-4

[27] M. Seaward, and D. Richardson, Atmospheric sources of metal pollution and effects on vegetation. Heavy metal tolerance in plants, Evolutionary aspects. (1989)75-92.

[28] J.R. McConnell and R. Edwards, Coal burning leaves toxic heavy metal legacy in the Arctic, Proceedings of the National Academy of Sciences. 105 (2008) 12140-12144. https://doi.org/10.1073/pnas.0803564105

[29] P. Nagajyoti, K. Lee, and T. Sreekanth, Heavy metals, occurrence and toxicity for plants: a review, Environmental Chemistry Letters. 8 (2010) 199-216. https://doi.org/10.1007/s10311-010-0297-8

[30] M. Arora, Heavy metal accumulation in vegetables irrigated with water from different sources, Food Chemistry. 111 (2008) 811-815. https://doi.org/10.1016/j.foodchem.2008.04.049

[31] A. Khan, Role of plants, mycorrhizae and phytochelators in heavy metal contaminated land remediation, Chemosphere. 41 (2000) 197-207. https://doi.org/10.1016/S0045-6535(99)00412-9

[32] H. Feng, A preliminary study of heavy metal contamination in Yangtze River intertidal zone due to urbanization, Marine Pollution Bulletin. 49 (2004) 910-915. https://doi.org/10.1016/j.marpolbul.2004.06.014

[33] Y. Hu, Assessing heavy metal pollution in the surface soils of a region that had undergone three decades of intense industrialization and urbanization, Environmental Science and Pollution Research. 20 (2013) 6150-6159. https://doi.org/10.1007/s11356-013-1668-z

[34] M. Nadal, M. Schuhmacher, and J. Domingo, Metal pollution of soils and vegetation in an area with petrochemical industry, Science of the total environment, 321 (2004) 59-69. https://doi.org/10.1016/j.scitotenv.2003.08.029

[35] B. Davies, Heavy metal contaminated soils in an old industrial area of Wales, Great Britain: source identification through statistical data interpretation, Water, Air, & Soil Pollution. 94 (1997) 85-98. https://doi.org/10.1007/BF02407095

[36] R.B. Hayes, The carcinogenicity of metals in humans. Cancer Causes & Control, 8 (1997) 371-385. https://doi.org/10.1023/A:1018457305212

[37] J.A. McElroy, Cadmium exposure and breast cancer risk. Journal of the National Cancer Institute, 98 (2006) 869-873. https://doi.org/10.1093/jnci/djj233

[38] L. Patrick, Lead toxicity, a review of the literature. Part I: exposure, evaluation, and treatment, Alternative medicine review. 11 (2006) 2-23.

[39] T.W. Clarkson, Metal toxicity in the central nervous system, Environmental Health Perspectives. 75 (1987) 59. https://doi.org/10.1289/ehp.877559

[40] Valciukas, J.A., Central nervous system dysfunction due to lead exposure, Science. 201 (1978) 465-467. https://doi.org/10.1126/science.663669

[41] Y. Finkelstein, M.E. Markowitz, and J.F. Rosen, Low-level lead-induced neurotoxicity in children: an update on central nervous system effects, Brain Research Reviews. 27 (1998)168-176. https://doi.org/10.1016/S0165-0173(98)00011-3

[42] A. Navas-Acien, Lead exposure and cardiovascular disease: a systematic review, Environmental Health Perspectives (2007) 472-482.

[43] S.A. Wadi and G. Ahmad, Effects of lead on the male reproductive system in mice, Journal of Toxicology and Environmental Health Part A. 56 (1999) 513-521. https://doi.org/10.1080/009841099157953

[44] E.K.R.D. Queiroz and W. Waissmann, Occupational exposure and effects on the male reproductive system, Cadernos de Saúde Pública. 22 (2006) 485-493. https://doi.org/10.1590/S0102-311X2006000300003

[45] P. Apostoli, Male reproductive toxicity of lead in animals and humans. ASCLEPIOS Study Group, Occupational and Environmental Medicine. 1998. 55 (6) 364-374. https://doi.org/10.1136/oem.55.6.364

[46] J. Domingo, Metal-induced developmental toxicity in mammals: A review. Journal of Toxicology and Environmental Health, Part A Current Issues. 42 (1994) 123-141. https://doi.org/10.1080/15287399409531868

[47] J. R. Roberts, Metal toxicity in children. Training manual on pediatric environmental health: Putting it into practice, 1999.

[48] L. Järup, Cadmium overload and toxicity. Nephrology Dialysis Transplantation, 17 (2002) 35-39. https://doi.org/10.1093/ndt/17.suppl_2.35

[49] G. Gobe, and D. Crane, Mitochondria, reactive oxygen species and cadmium toxicity in the kidney, Toxicology letters. 198 (2010) 49-55. https://doi.org/10.1016/j.toxlet.2010.04.013

[50] L. Järup, and A. Åkesson, Current status of cadmium as an environmental health problem, Toxicology and applied pharmacology. 238 (2009) 201-208. https://doi.org/10.1016/j.taap.2009.04.020

[51] R. N. Ratnaike, Acute and chronic arsenic toxicity, Postgraduate medical journal. 79 (2003) 391-396. https://doi.org/10.1136/pmj.79.933.391

[52] J. Duruibe, M. Ogwuegbu, and J. Egwurugwu, Heavy metal pollution and human biotoxic effects, International Journal of Physical Sciences. 2 (2007): p. 112-118.

[53] T. W. Clarkson, L. Magos, and G.J. Myers, The toxicology of mercury—current exposures and clinical manifestations, New England Journal of Medicine. 349 (2003) 1731-1737. https://doi.org/10.1056/NEJMra022471

[54] I.R. Ramsis, MEMS based heavy metal detector, 2012, American University in Cairo.

[55] S.B. Goldhaber, Trace element risk assessment: essentiality vs. toxicity. Regulatory toxicology and pharmacology. 38 (2003) 232-242. https://doi.org/10.1016/S0273-2300(02)00020-X

[56] A. Sonune and R. Ghate, Developments in wastewater treatment methods. Desalination, 167 (2004) 55-63. https://doi.org/10.1016/j.desal.2004.06.113

[57] G. Crini, Recent developments in polysaccharide-based materials used as adsorbents in wastewater treatment. Progress in polymer science, 30 (2005) 38-70. https://doi.org/10.1016/j.progpolymsci.2004.11.002

[58] M. J. Zohuriaan-Mehr and K. Kabiri, Superabsorbent polymer materials: a review. Iranian Polymer Journal. 17(2008) 451.

[59] D.W. O'Connell, C. Birkinshaw, and T.F. O'Dwyer, Heavy metal adsorbents prepared from the modification of cellulose: A review. Bioresource Technology, 99(2008) 6709-6724. https://doi.org/10.1016/j.biortech.2008.01.036

[60] A. Gundogdu, Biosorption of Pb (II) ions from aqueous solution by pine bark (Pinus brutia Ten.), Chemical engineering journal. 153 (2009) 62-69. https://doi.org/10.1016/j.cej.2009.06.017

[61] M. Saifuddin and P. Kumaran, Removal of heavy metal from industrial wastewater using chitosan coated oil palm shell charcoal, Electronic journal of Biotechnology. 8 (2005) 43-53.

[62] S. E. Bailey, A review of potentially low-cost sorbents for heavy metals, Water Research. 33 (1999) 2469-2479. https://doi.org/10.1016/S0043-1354(98)00475-8

[63] B. Kim and S.-T. Lim, Removal of heavy metal ions from water by cross-linked carboxymethyl corn starch, Carbohydrate polymers. 39 (1999) 217-223. https://doi.org/10.1016/S0144-8617(99)00011-9

[64] A. Badruddoza, Carboxymethyl-β-cyclodextrin conjugated magnetic nanoparticles as nano-adsorbents for removal of copper ions: synthesis and adsorption studies, Journal of hazardous materials. 185 (2011) 1177-1186. https://doi.org/10.1016/j.jhazmat.2010.10.029

[65] E. Abdel-Halim and S.S. Al-Deyab, Hydrogel from crosslinked polyacrylamide/guar gum graft copolymer for sorption of hexavalent chromium ion, Carbohydrate polymers. 86 (2011) 1306-1312. https://doi.org/10.1016/j.carbpol.2011.06.033

[66] G. Mahajan,and D. Sud, Modified agricultural waste biomass with enhanced responsive properties for metal-ion remediation: a green approach, Applied water science. 2 (2012) 299-308.

[67] S. Kamel., E. Hassan, and M. El-Sakhawy, Preparation and application of acrylonitrile grafted cyanoethyl cellulose for the removal of copper (II) ions, Journal of Applied Polymer Science. 100 (2006) 329-334. https://doi.org/10.1002/app.23317

[68] A. M. Nada, M.Y. Alkady, and H.M. Fekry, Synthesis and characterization of grafted cellulose for use in water and metal ions sorption, BioResources, 3 (2008) 46-59.

[69] Varma, A., S. Deshpande, and J. Kennedy, Metal complexation by chitosan and its derivatives: a review, Carbohydrate polymers. 55 (2004) 77-93. https://doi.org/10.1016/j.carbpol.2003.08.005

[70] W. W. Ngah, L. Teong, and M. Hanafiah, Adsorption of dyes and heavy metal ions by chitosan composites: A review, Carbohydrate polymers, 83 (2011) 1446-1456. https://doi.org/10.1016/j.carbpol.2010.11.004

[71] X.J. Hu, Adsorption of chromium (VI) by ethylenediamine-modified cross-linked magnetic chitosan resin: isotherms, kinetics and thermodynamics, Journal of hazardous materials. 185 (2011) 306-314. https://doi.org/10.1016/j.jhazmat.2010.09.034

[72] D. H. K. Reddy and S.-M. Lee, Application of magnetic chitosan composites for the removal of toxic metal and dyes from aqueous solutions, Advances in Colloid and Interface Science. 201 (2013) 68-93. https://doi.org/10.1016/j.cis.2013.10.002

[73] N. Li and R. Bai, Copper adsorption on chitosan–cellulose hydrogel beads: behaviors and mechanisms, Separation and purification technology. 42 (2005) 237-247. https://doi.org/10.1016/j.seppur.2004.08.002

[74] T. Liu, Entrapment of nanoscale zero-valent iron in chitosan beads for hexavalent chromium removal from wastewater, Journal of hazardous materials. 184 (2010) 724-730. https://doi.org/10.1016/j.jhazmat.2010.08.099

[75] W. W. Ngah, Utilization of chitosan–zeolite composite in the removal of Cu (II) from aqueous solution: adsorption, desorption and fixed bed column studies, Chemical engineering journal. 209 (2012) 46-53. https://doi.org/10.1016/j.cej.2012.07.116

[76] E. Repo, Heavy metals adsorption by novel EDTA-modified chitosan–silica hybrid materials, Journal of colloid and interface science. 358 (2011) 261-267. https://doi.org/10.1016/j.jcis.2011.02.059

[77] W. W. Ngah, C. Endud, and R. Mayanar, Removal of copper (II) ions from aqueous solution onto chitosan and cross-linked chitosan beads, Reactive and Functional Polymers. 50 (2002)181-190. https://doi.org/10.1016/S1381-5148(01)00113-4

[78] M. Yavu, An economic removal of Cu 2+ and Cr 3+ on the new adsorbents: pumice and polyacrylonitrile/pumice composite, Chemical engineering journal. 137 (2008) 453-461. https://doi.org/10.1016/j.cej.2007.04.030

[79] L. Zhou, Characteristics of equilibrium, kinetics studies for adsorption of Hg (II), Cu (II), and Ni (II) ions by thiourea-modified magnetic chitosan microspheres, Journal of hazardous materials. 161 (2009) 995-1002. https://doi.org/10.1016/j.jhazmat.2008.04.078

[80] Y. Qin, Combined use of chitosan and alginate in the treatment of wastewater, Journal of Applied Polymer Science. 104 (2007) 3581-3587. https://doi.org/10.1002/app.26006

[81] B. Xiang, Dithiocarbamate-modified starch derivatives with high heavy metal adsorption performance, Carbohydrate polymers. 136 (2016)30-37. https://doi.org/10.1016/j.carbpol.2015.08.065

[82] R. Cheng and S. Ou, Application of modified starches in wastewater treatment. 2016.

[83] J. Zhiping, Chelating Property of Modified Porous Starch for Cu~(2+)[J], Environmental Protection of Chemical Industry. 2 (2009) 023.

[84] M. Khalil, and S. Farag, Utilization of some starch derivatives in heavy metal ions removal, Journal of Applied Polymer Science. 69 (1998) 45-50.

https://doi.org/10.1002/(SICI)1097-4628(19980705)69:1<45::AID-APP6>3.0.CO;2-M

[85] V. P. Mahida and M.P. Patel, Removal of heavy metal ions from aqueous solution by superabsorbent poly (NIPAAm/DAPB/AA) amphoteric nanohydrogel, Desalination and Water Treatment. 57 (2016) 13733-13746. https://doi.org/10.1080/19443994.2015.1060171

[86] Y. Zheng, S. Hua, and A. Wang, Adsorption behavior of Cu 2+ from aqueous solutions onto starch-g-poly (acrylic acid)/sodium humate hydrogels, Desalination. 263(2016) 170-175. https://doi.org/10.1016/j.desal.2010.06.054

[87] X. Ma, Modification of porous starch for the adsorption of heavy metal ions from aqueous solution, Food Chemistry. 181 (2015) 133-139. https://doi.org/10.1016/j.foodchem.2015.02.089

[88] Q. Chang, X. Hao, and L. Duan, Synthesis of crosslinked starch-graft-polyacrylamide-co-sodium xanthate and its performances in wastewater treatment, Journal of hazardous materials.159 (2008) 548-553. https://doi.org/10.1016/j.jhazmat.2008.02.053

[89] D. W. Kang, H.R. Choi, and D.K. Kweon, Stability constants of amidoximated chitosan-g-poly (acrylonitrile) copolymer for heavy metal ions, Journal of Applied Polymer Science.73 (1999) 469-476. https://doi.org/10.1002/(SICI)1097-4628(19990725)73:4<469::AID-APP2>3.0.CO;2-I

[90] A. Dong, A novel method for amino starch preparation and its adsorption for Cu (II) and Cr (VI), Journal of hazardous materials. 181 (2010) 448-454. https://doi.org/10.1016/j.jhazmat.2010.05.031

[91] S. Çavuş et al., The competitive heavy metal removal by hydroxyethyl cellulose-g-poly (acrylic acid) copolymer and its sodium salt: The effect of copper content on the adsorption capacity, Polymer Bulletin, 57 (2006) 445-456. https://doi.org/10.1007/s00289-006-0583-6

[92] V. Singh and S.K. Singh, Cadmium (II) removal from aqueous solution using guar gum-silica nanocomposite, Advanced Materials Letters. 6 (2015) 607-615. https://doi.org/10.5185/amlett.2015.5873

[93] L. Yan, Characterization of magnetic guar gum-grafted carbon nanotubes and the adsorption of the dyes, Carbohydrate polymers. 87 (2012) 1919-1924. https://doi.org/10.1016/j.carbpol.2011.09.086

[94] T. A. Khan, M. Nazir, I. Ali, A. Kumar, Removal of Chromium (VI) from aqueous solution using guar gum–nano zinc oxide biocomposite adsorbent, Arabian Journal of Chemistry, Volume 10, Supplement 2, May 2017, Pages S2388-S2398. https://doi.org/10.1016/j.arabjc.2013.08.019

# Chapter 3

# Ion selective membrane electrodes as sensors for detection of heavy metal ions

Tauseef Ahmad Rangreez[1], Inamuddin[1,*], Rizwana Mobin[1]

Advanced Functional Materials Laboratory, Department of Applied Chemistry, Faculty of Engineering and Technology, Aligarh Muslim University, Aligarh 202002, India

inamuddin@rediffmail.com

**Abstract**

The present chapter deals with the basic information regarding pollution caused by heavy metals and their effect on our natural resources, economy and human health. It also gives an idea on the role of nano-composites in the remedial and purification processes. The principle, procedure and advantages of ion-selective membrane electrodes along with the work carried out during the last decade in the field to develop various lead and cadmium selective cation-exchangers is included in this chapter. The utility of organic-inorganic composite material based sensors and their application in the development of ion-selective membrane electrodes (ISME) for the detection of heavy metals, which render portable water unsafe for use and pose a threat to the wellbeing of man is also discussed.

**Keywords**

Composite Cation Exchanger, Ion-selective Membrane Electrodes, Heavy Metal Pollutants, Electro Chemical Sensors, Selectivity

## Contents

## 1. Introduction

The environment has been continually been neglected by the rapid, unplanned and over ambitious industrial development. Man has entered a blind race of exploitation of natural resources in the name of industrialization. The natural resources are being overburdened and as such the environmental balance is disrupted which has threatened the very existence of life on this planet [1,2]. All the essential components of environment-air, water and soil are highly polluted due to the unrestricted use and ill monitored flux of industrial and domestic discharges. Our present day environment is highly polluted and a safe, pure and supportive environment for humans is rapidly becoming at thing of the past. Pollution can be defined as any undesirable change in the physical, chemical or biological characteristics of air, land and water, that may adversely affect human life or that of other desirable species or industrial processes, living conditions and cultural assets or that may waste or deteriorate natural resources [3]. A pollutant may be a solid, liquid or gaseous substance and may be in a chemical or energy form. Some pollutants are produced naturally but within such a range that doesn't cause any alarming problem.

Today, man is collectively facing a daunting task in the form of pollution and more importantly the means and measures that need to be taken in order to control this menace and prevent further deterioration of our valuable environmental assets. Present day research focuses not only on effective treatment of the contaminants but prevention and minimizing the production of such toxic and harmful substances in the first place.

## 2. Water and its significance

Life on earth cannot be imagined without water. Water covers about 70% of our earth and is one of the most important constituents of the environment which facilitates life on earth. About 97% of available water is present in oceans, ice constitutes about 2% of the total water and only 1% of water is available to sustain life on earth as present in rivers, lakes and reservoirs constituting the hydrological cycle of earth[4](Fig. 1).

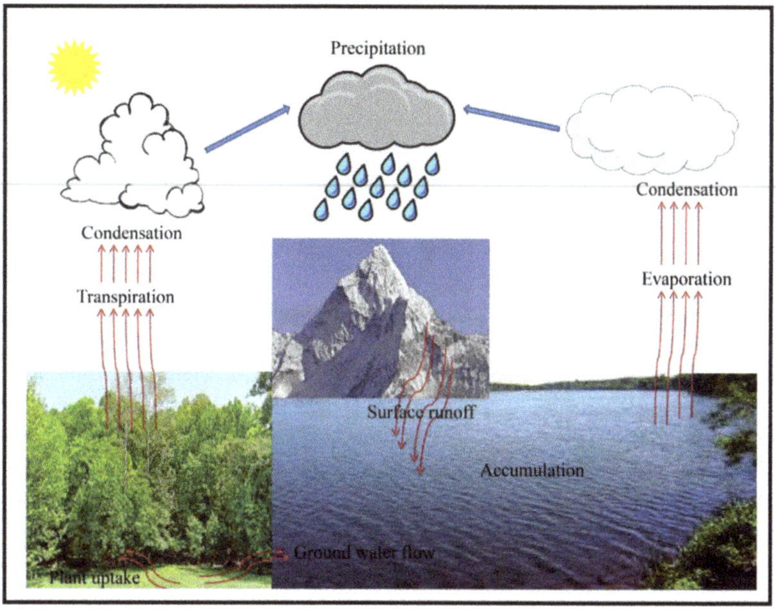


*Fig. 1: Water cycle*

Water has very wide impact on different aspects of life as food, health, economy and energy [5,6]. Water plays a leading role in photosynthesis and respiration that form the basis of life on this planet. It works as an excellent solvent in several metabolic processes in both plant and animal kingdom. About 69% of human body weight, 70% of brain and 80% of blood volume is made up of water. Water got first place in the green solvent list proposed by the royal society of chemistry as it is easily available, has no harmful effect on health and environment, is high solubilising and does not produce any type of waste. Over the last three centuries uptake of water has increased by more than 35 times. Annual utilisation and withdrawal of fresh water is about 3240 cubic km, agriculture gets the major share about 69% of this water, industry gets 23% and only 8% of this water is available for domestic purpose [7].

## 3. Water and its pollution

Entry of any toxic organic, inorganic, biological and radiological substance which changes the physical, chemical and biological properties of water is called water pollution. The Economic European Community (EEC) [8] defined pollution in its 'article one' as "the discharge by man, directly or indirectly, of substances or energy into the aquatic environment, the results of which are such as to cause potential health hazards to humans, harmful to living resources and aquatic ecosystems, damage to amenities or interferences with other legitimate uses of water. Industrial revolution, uncontrolled use of natural water resources, ever increasing human population, urbanisation and change in the lifestyle of the people have been the main cause of water pollution [9]. Deterioration in the quality of water is mainly due to industrial waste and human activity resulting in the pollution/contamination of water in all forms. Water quality parameters for domestic water supply in the form of permissible limits suggested by different organisation such as United States Public Health drinking water standards (USPH) and Indian Standard Institution (ISI) are represented in **Table 1** [10,11]. Deterioration in the quality of water affects the aquatic ecosystem, groundwater deposits and human life. Water pollution manifested as water borne diseases is the main cause of about 10-20 million human deaths annually [12]. Diarrhea causes about 6000 deaths in children every day [13,14]. An estimated 0.78 billion humans in the world don't have access to safe drinking water [15].

***Table 1:*** *Water quality parameters as per USPH and ISI*

| Parameters | ISI | USPH |
|---|---|---|
| **Essential characteristics** | | |
| Color | 15 Hazen | Colorless |
| Odour | Agreeable | Odourless |
| Taste | Agreeable | Tasteless |
| pH value | 6.5-8.5 | - |
| Turbidity | 5 | - |
| Dissolved oxygen (ppm) | - | 4.0-6.0 |
| Total dissolved solid(ppm) | 2000 | 500 |
| Total hardness (as $CaCO_3$) (ppm) | 600 | - |
| **Undesirable substances** | | |
| Aluminium (ppm) | 0.2 | - |
| Chloride (ppm) | 1000 | 250 |
| Fluoride (ppm) | 1.5 | 1.5 |
| Calcium (ppm) | 200 | 100 |
| Magnesium (ppm) | 100 | 30 |
| Nitrate (ppm) | 45 | <10 |
| Sulphate (ppm) | 400 | 250 |
| Phosphate (ppm) | - | 0.1 |
| Manganese (ppm) | 0.3 | <0.05 |

| | | |
|---|---|---|
| Iron (ppm) | 0.3 | <0.3 |
| Copper (ppm) | 1.5 | 1.0 |
| Zinc (ppm) | 15 | 5.5 |
| **Carcinogenic substances** | | |
| Cadmium (ppm) | 0.003 | 0.01 |
| Chromium (ppm) | 0.05 | 0.05 |
| Lead (ppm) | 0.01 | <0.05 |
| Arsenic (ppm) | 0.05 | 0.05 |
| Mercury (ppm) | 0.001 | 0.001 |
| Nickel (ppm) | 0.02 | - |
| Molybdenum (ppm) | 0.07 | - |
| Pesticides | - | 0.005 |

## 4. Sources of water pollution

The major sources of water pollution are waste water discharge from industries, urban and rural runoff, agricultural (increased use of chemicals, fertilizers, pesticides and surfactants), animal matter discharge into surface and ground water and municipal discharge into water bodies [16,17]. Fresh water bodies are under constant stress due to ever increasing demands of food, energy etc. [18,19] and is a leading cause of pollution of the reservoirs. These sources of water pollution are broadly categorized into two types:

**1) Point sources and 2) Non-point sources**

Point sources are single identifiable sources of pollution and include the outlets from industries, discharging wastes into water bodies while as non-point sources often referred to as 'diffusible' sources include such sources which are not localized and are spread over a large area (Fig. 2). Depending upon the nature and quality of wastes disposed by point sources, these are regulated by several environmental protection agencies. From the pollution control point of view, the non-point sources are more troublesome and problematic compared to the point sources because these contain varied types of contaminants. Paper, chemicals, petroleum and metal industries produce about 97% of water contaminants.

Water pollutants in general are grouped into the following four types: 1) organic pollutants 2) inorganic pollutants 3) suspended solid sediments and 4) radioactive materials. Among these, organic and heavy metal pollutants are of major concern. Organic pollutants are carcinogenic and heavy metals are non-biodegradable and their long exposure can lead to several serious ailments/ health hazards and even death.

## 5. Heavy metal pollution

Heavy metal pollution mainly originates from the industries, geological weathering, volcanic eruptions, forest fires, ore processing, metal coating, use of chemicals, fertilizers in agriculture, landfill waste, sewage effluents etc. Such metals are used for manufacturing of batteries, pigments, as catalyst in petroleum industries, protective coating of other metals, chemical synthesis and in metal processing plants [20,21]. The heavy metals from these sources and other anthropogenic activities make their entry into various spheres of the environment and pose a serious threat to life. The main problem arises due to their non-biodegradable and accumulative nature. Once the heavy toxic metals enter into the food chain or the food web, next is their bio-accumulation in various tissues [22]. Some heavy metals (copper, zinc, iron and trivalent chromium) are important for the human body at very low concentration levels but some of the heavy metals/metalloids are highly toxic such as mercury, cadmium, arsenic and lead. Mercury, lead and cadmium are also known as the "big three" poisonous heavy metals [23]. Due to their highly toxic nature and already well established harmful effects on human health, several heavy metals have been placed on a priority list by various environmental protection agencies. **Table2** shows the maximum permissible limit of some heavy metals/metalloids in drinking water [10,11,24-26].

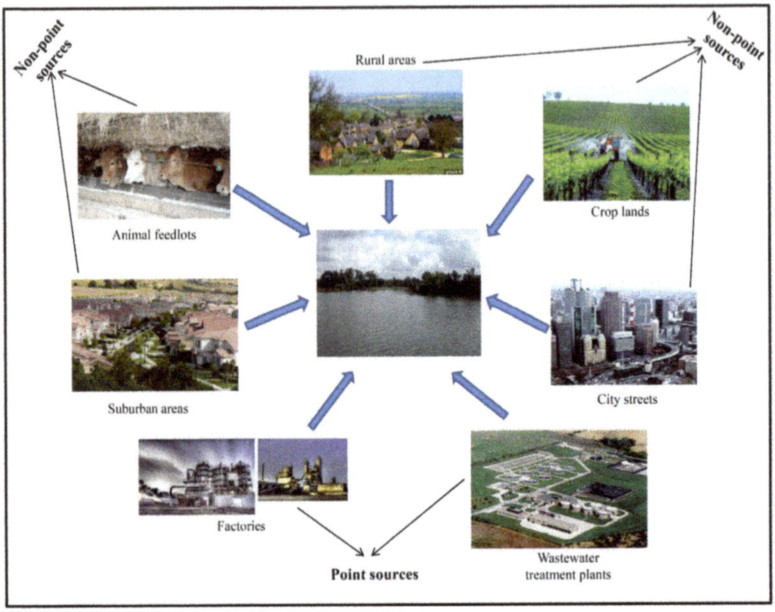


*Fig. 2: Sources of water pollution.*

## 6. Effect of heavy metal/metalloid pollution on humans

Following their route through the food chain, heavy metals accumulate in the human body as these are non-biodegradable in nature and are also not converted into harmless end products through metabolism, causing severe, carcinogenic as well as chronic health problems. Some of these metals/ metalloids are given below.

### 6.1 Cadmium

Cadmium is generally found in industrial work places and is extremely toxic in nature. It mostly gets released into the water bodies through industrial discharges, cadmium-nickel batteries, phosphate fertilizers, pigments and alloys [20] and causes stomach pain, severe vomiting, damage to kidneys, lungs, reproductive failure and destruction of red blood cells [27-29].

***Table 2:*** *Maximum permissible limits ($\mu g\ L^{-1}$) of heavy metals/metalloids in drinking water.*

| Heavy metals concentration | WHO | European Union | United States | China | Canada | ISI |
|---|---|---|---|---|---|---|
| **Cd** | 3 | 5 | 5 | 5 | 5 | 3 |
| **Cr** | 50 | 50 | 100 | 50 | 50 | 50 |
| **Cu** | - | 2000 | - | 1000 | 1000 | 1500 |
| **Pb** | - | 10 | 15 | 10 | 10 | 300 |
| **Hg** | 6 | 1 | 2 | 0.05 | 1 | 1 |
| **Ni** | - | 20 | - | - | - | 20 |
| **As** | 10 | 10 | 10 | 50 | 10 | 50 |
| **Fe** | - | - | - | - | 300 | 300 |
| **Mn** | - | - | - | - | 50 | 30 |
| **Se** | 40 | 10 | 50 | 10 | 10 | 10 |

### 6.2 Lead

Lead is one of the most prominent heavy toxic metals. It gets released into the environment by both natural as well as anthropogenic activities. The main sources include volcanic eruptions, sea aerosols, the manufacturing of lead acid batteries, metal processing plants, paints, lead pipes, solders and gasoline. Lead poisoning is also referred to as plumbism. Its exposure causes irritability, insomnia, damage to kidneys, brain, reproductive and central nervous system [21,30,31].

### 6.3 Arsenic

Arsenic exposure is related to occurrence of high blood pressure and peripheral vascular diseases [32-35]. The skin is the most badly affected part of the human body due to arsenic exposure. Dermatological hyper-pigmentation, based cell carcinoma (BCC), squamous cell carcinoma (SCC) and Bowen's disease (BD) have also been reported due to chronic arsenic exposure [36,37]. It is also related to cancer of several body organs as urinary bladder, lung and pancreas [38,39]. It gets released into the environment by both natural and anthropogenic sources [40,41].

### 6.4 Chromium

Chromium can exist as hexavalent or trivalent form in the aqueous environment. Out of these two forms Cr(VI) is highly toxic and carcinogenic as compared to Cr(III). It is used in tanning, hard chromium plating, textiles, thermal power plants and sugar mills [42-45]. Cr(VI) causes skin rashes, stomach ulcers, respiratory problems, kidney and liver damage, mutations and lung cancer.

### 6.5 Mercury

The high reactivity and relative solubility of mercury in water and other tissues make mercury a potent toxic heavy metal [46]. It is used in production of batteries, cameras, cathode tubes, calculators, dental amalgam fillings, mercury vapour lamps and barometers [47]. It gets easily accumulated in the food chain and is a precursor of several health hazards in humans viz, nausea, abdominal pain, damages kidney, gastrointestinal tract, hearing impairment, teratogenic effect [48,49]. **Table 3** represents some heavy metals/metalloids, their sources and effect on human health.

## 7. Some heavy metal disasters

Chisso Corporation released sewage into Minamata Bay in Japan containing mercury (1932-1968), resulting in its accumulation in aquatic organisms over a period of time, leading to mercury poisoning [50]. In 1956, the first case of Minamata disease appeared in Japan by consumption of mercury polluted fish.

In 1986, water used to extinguish major fire in Mexico carried ~30 tons of fungicide containing mercury into the upper Rhine and causing a large kill off of aquatic lives [51].

In 1998, Coto De Donana nature reserve (Spain) got contaminated with toxic chemicals by bursting of a dam belonging to a mine. Almost 5 million metric tons of mud containing several harmful metals (Pb, Cd) flew down Rio Guadimar. Permanent damage was caused to bird sanctuary, agriculture and fisheries [52].

***Table 3:****Sources and effects of metals/metalloids on human health.*

| Heavy metal | Sources | Effect on human health |
|---|---|---|
| **Chromium** | Cooling tower, dyes, electroplating, ink, anodizing, paints, tanning etc. | Cancer |
| **Cadmium** | Coal combustion, metal plating, water pipe, pigments, phosphate fertilizers etc. | Cardiovasculardisease, reproductive failure, kidney damage and hypertension |
| **Copper** | Pulp and paper, electrical goods, utensils, chemicals etc. | Caner (suspected) |
| **Lead** | Metal processing plants batteries, paints etc. | Affects nervous and renal systems, headache, constipation, cancer etc. |
| **Nickel** | Diesel oil, coal, steel & non-ferrous alloys, tobacco smoke | Lung cancer, respiratory Symptoms and haemorrhage |

| | | |
|---|---|---|
| | etc. | |
| **Zinc** | Galvanizing, alloys rayon, paper etc. | Cancer |
| **Mercury** | Coal combustion, electrical batteries, Chlor-alkali industry | Nerves damage, death, kidney and brain damage |
| **Iron** | Steel, dye, textiles, machinery, medicines etc. | Cancer (suspected) |
| **Cobalt** | Alloys, steel, electroplating glass, enamel etc. | Cancer |
| **Manganese** | Metal alloys, power plants, gasoline etc. | Nervous system damage |

## 8. Removal of heavy metals

The aquifers are fast depleting owing to several factors. Advanced purification technologies can release pressure of water scarcity, health and climate change. The same can be achieved by purifying and reuse of waste water, which in effect involves the development of efficient, novel, reliable and cost effective materials and methods [53]. A number of methods have been used for the determination/removal as well as treatment of heavy metals from waste water bodies. These methods are:

- Chemical and electrochemical precipitation
- Evaporation
- Membrane filtration
- Reverse osmosis

- Adsorption
- Electrochemical sensors
- Ion-selective membrane

Biological treatment in the form of activated sludge and trickling filters are not able to remove a broad array of contaminants as majority of the compounds remain soluble in the effluent [54,55]. Similarly, physico-chemical treatments such as flocculation, coagulation and lime softening are ineffective in removing various endocrine disrupting compounds (EDCs) and pharmaceutical compounds [56,57]. The chlorination provides protection against regrowth of bacteria and pathogens [58,59] but produces undesirable taste and odour [60], along with the formation of different disinfection by-products (DBPs) in portable water [61,62] while ozonisation is not an effective alternative considering its high cost. Ultraviolet and ion-exchange are considered as advance methods for the treatment process.

## 9. Nano-composites

At present the most surprising and promising field in technological developments is nanotechnology. The word "nano" comes from Greek word "*nanos*" meaning "dwarf" as in nano-science materials used are in nanometer range. Nanoparticles are embedded into metal, ceramic and polymer matrix material in nano-composites. These embedded nano-particles act as matrix reinforcement and change the physical properties of the base material. High surface to volume ratio of reinforcing material or high aspect ratio makes nano-composite different from conventional composite materials with better and interesting properties. Being environmental friendly with interesting properties, nano-composites provide a new technology and wide opportunity to all sectors of industry in order to fulfil the ever growing demands of society.

### 9.1 Classification of nano-composites

On the basis of matrix material, nano-composites are classified as:

1. Ceramic matrix nano-composites
2. Metal matrix nano-composites
3. Polymer matrix nano-composites

### 9.1.1 Ceramic matrix nano-composites

In ceramic matrix nano-composites (CMNC), the main volume of composite material is occupied with ceramic and metal is the second component. Nano-composites obtained with the combination show improved electrical, magnetic, optical, corrosion resistance and other protective properties. Pioneer work of Niihara [63] revealed the potential of CMNC made up of $Al_2O_3$/SiC system.

### 9.1.2 Metal matrix nano-composites

In this type of nano-composites, metal reinforcement is embedded into metal matrix material. Carbon nanotube (CNT) metal matrix composites are one of the important nano-composites of this class and emerge as new material with high tensile strength and electrical conductivity. Metal matrix nano-composites have been widely used in the development of structural materials, aerospace and automotive industries.

### 9.1.3 Polymer matrix nano-composites

Nano-particles such as carbon nanotubes, graphene, tungsten disulfide and molybdenum disulfide reinforced into polymer matrix enhance the compressive and flexural mechanical properties of polymeric substances. These types of nano-composites are light weight and mechanically strong, hence used in bone implantation. For enhancing the electrical conductivity, multiwalled carbon nanotubes based polymer nano-composites are used.

On the basis of nature of association between organic and inorganic components, nano-composites are classified as:

1. Inorganic-organic nano-composites
2. Organic-inorganic nano-composites

### 9.1.4 Inorganic-organic nano-composites

In this type, inorganic nano-particles are embedded into organic matrix material. Inorganic nano-particles (guest particle) are introduced into the host (organic) matrix through chemical route or by electrochemical incorporation method. Colloidal stability, better optical and catalytic properties are the main features of inorganic-organic nano composites.

### 9.1.5 Organic-inorganic nano-composites

In organic-inorganic nano-composites, organic nano-particles are embedded into the inorganic matrix material. Inorganic materials possess well defined, ordered intralamellar space that is accessible to foreign substance producing interesting lamellar organic-inorganic nano-composite materials. These lamellar nano-composites are further divided into two classes, intercalated and exfoliate.

In intercalated nano-composites, definite number of polymer chains with fixed compositional ratio is present in interlamellar space of the inorganic layer. The number of polymer chains between the inorganic layers is continuously varying in exfoliate nano-composites with a distance >100Å between layer stands.

## 9.2 Application of nano-composites

Nano-composites are used in various fields and new applications are being continuously developed. Some applications [64,65] of nano-composites are given below:

- Thin-film capacitors for computer chips
- Solid polymer electrolytes for batteries
- Automotive engine parts and fuel tanks
- Oxygen and gas barrier
- Food packaging
- Battery cathodes and ionics
- Non-linear optics
- Nano-wire sensors and others systems
- Impellers and blades

## 9.3 Nanotechnology in water purification and development of sensors/detectors

Nano-technology has the potential to control materials at nano-scale resulting in the formation of materials with specialized function [64,66]. These materials possess very high surface to volume ratio and hence can be used to detect sensitive contaminants [65,67]. Nano-technology, when applied to industrial processes also finds its application to prevent the formation of pollutants. It finds application in environmental processes at several levels viz, 1) pollution control, 2) purification (remediation) of contaminated material, 3) sensing and detection. The development and evolution of such instruments is inevitable keeping in view the rapidly changing scenario of pollutants [68,69]. The remedial process involves the removal of pollutants in water that pose a threat to human health and eco-systems. Several nano-materials have found application in water remedial processes; e.g. zeolites, carbon nanotubes (CNTs), nano-particles of zero valent iron, self assembled monolayers on mesoporous support [70]. The CNTs exhibit high adsorption capability for several heavy metals and metalloids such as $Pb^{2+}$, $Cd^{2+}$, $Cr^{3+}$, $Zn^{2+}$ and $As^{3+}$[71-74]. Nano-fibres of poly vinyl alcohol (PVA) and poly acrylonitrile (PAN) containing silver nano-particles exhibit antimicrobial activity; with PVA nano-fibres reducing ~99% bacteria in polluted water and PAN nano-fibres killing ~100% [75].

Nano-wires and tubes offer greater opportunities in the field of chemical/ biological sensors [76,77]. The development of such sensors which provide fast and accurate information about contaminants at low levels is going to boost the ability to protect the environment and human health [78].

## 10. Electrochemical sensors

The electrochemical sensors represent a rapidly progressing division of analytical chemistry [79]. These sensors essentially represent electrochemical cells employing two/three electrode system. These have attained a great significance owing to their on-site monitoring of various environmentally hazardous substances, thus addressing several environmental concerns [80]. Fig. 3 represents some of the important aspects to be considered while choosing an electrochemical sensor.

A sensor can be defined as something viz an apparatus, a device which is used to sense or detect a particular substance or species. In general, a sensor consists of an active material for sensing and a signal transducer. A chemical sensor is defined as a device which provides fast and continuous information about its immediate surrounding environment. The chemical sensors recognise the analyte and their chemical changes are manifested in the form of a signal which can be easily interpreted [79,81]. The chemical sensor senses a particular analyte and transforms the chemical interaction with the analyte into an electrical signal in the form of potential of an electrode with respect to the reference electrode. The magnitude of signal is determined both by the quality and quantity of the analyte and hence, the linear response range can be treated as the activity range of the sensor [82]. The chemical sensing involves both recognition of an analyte at a particular site and the transduction (Fig. 4). The chemical sensors find wide variety of application particularly in the field of clinical diagnostics, safety, industrial hygiene, process controls, quality controls, human comfort, critical care, emissions monitoring, automotive, safety alarms and more recently homeland security. The wide spread application of chemical sensors in these fields has economic, aesthetic and social utility

[83]. These sensors are mostly categorised as: amperometric, conductometric and potentiostatic.

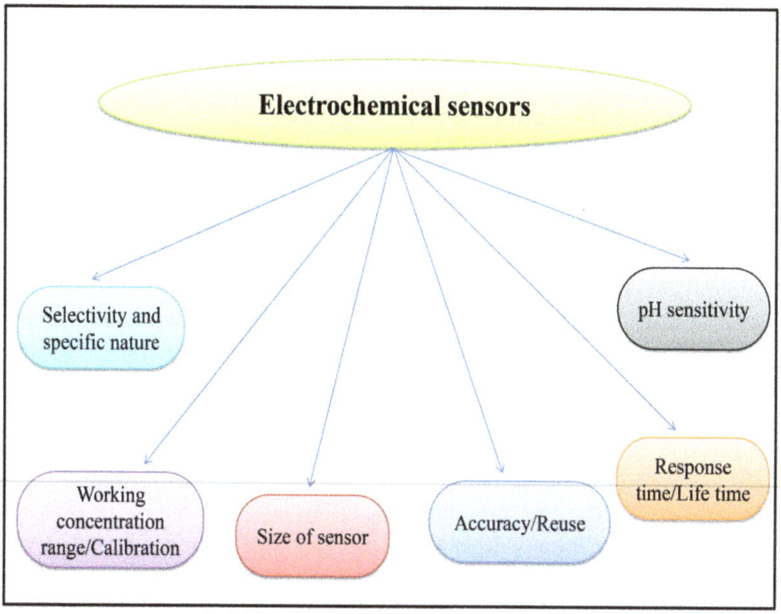


*Fig. 3: Different aspects of electrochemical sensors.*

The voltametric sensors involve the measurement of current when the potential of the cell is ramped from one preset potential value to another. Amperometric sensors on the contrary determine the current flow at a fixed applied potential by the detection of chemically active species in the process. The current is measured as a function of time and the potential of the working electrode is kept constant with respect to the reference electrode. Conductometric sensors depend on the measurement of conductivity changes of a film or bulk material on interaction with an analyte. These methods, principally non-selective, mostly find application only on development of modified surfaces along with improved instrumentation. The conducting polymers have been used as conductometric sensors e.g. pyrrole detects volatile amines and if doped with $ClO_4^-$ and tosylate, acts as $NH_3$ sensor [84].

Potentiometric sensors on interaction with a specific analyte convert the chemical information related to the analyte into a useful potential signal. The signal obtained, corresponds to the amount (logarithmic) of the species concerned. The ion-selective electrodes represent an important and widely used class of potentiometric sensors.

## 11. Ion-Selective membrane electrode

According to the International Union of Pure and Applied Chemistry (IUPAC) recommendation [85] an ion-selective electrode (ISE) is: An electrochemical sensor, based on thin films or selective membranes as recognition elements, and is an electrochemical half-cell equivalent to other half-cells of the zeroth (inert metal in a redox electrolyte), 1$^{st}$, 2$^{nd}$ and 3$^{rd}$ kinds. These devices are distinct from systems that involve redox reactions (electrodes of zeroth, 1$^{st}$, 2$^{nd}$ and 3$^{rd}$ kinds), although they often contain a 2$^{nd}$ kind electrode as the "inner" or "internal" reference electrode. The potential difference response has, as its principal component, the Gibbs energy change associated with perms-elective mass transfer (by ion-exchange, solvent extraction or some other mechanism) across a phase boundary. The ISE must be used in conjunction with a reference electrode (i.e. "outer" or "external" reference electrode) to form a complete electrochemical cell. The measured potential differences (ISE versus outer reference electrode potentials) are linearly dependent on the logarithm of the activity of a given ion in solution.

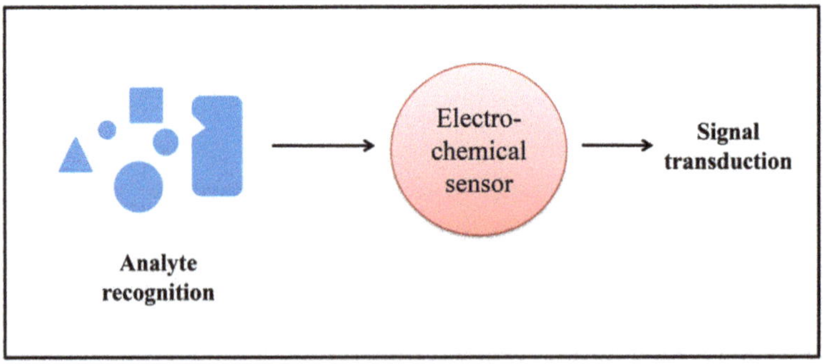


*Fig. 4: Representation of electrochemical sensing.*

Ostwald (1890) laid the foundation of work on ion-selective membrane with the discovery that for an electrolyte a membrane can be impermeable [86]. The difference of concentration at the membrane gives rise to membrane potential at the boundary. The same was later confirmed by Donan (1911) [87]. The work on collodion membrane by Michaelis and Fujita is considered as the basic work related to ion-selective membranes [88].

Juda and MC Rae [89] and Winger et al. [90] were the first to develop a stable and highly selective ion-exchange membrane with low electric resistance and this led to the use of ion-exchange membranes for industrial processes. The ion-selective membrane electrode selectively determines the activity of a particular ionic species. These electrodes mainly consist of membrane-based devices, composed of perm selective ion-conducting materials separating the sample from the inside of the electrode (Fig. 5). The potential of a working electrode is determined mainly by its environment while the reference electrode potential is kept constant by a solution containing the ion of interest at a constant activity. The potential difference (cell potential) is related to the concentration of the dissolved ion as the potential of the reference electrode remains constant. The development of a specific ion-selective membrane electrode is very typical and different strategies to produce specific electrode, selective to particular species depend primarily on the nature and composition of the membrane material. As such research in this area is wide open and has applications to an almost unlimited number of analytes, only limitation being the preparation of particular analyte selective membrane electrode based on dopant and ionophore matrix of the membrane. The ion-exchange membranes have been used in diverse fields of food, drug, biological, waste material treatment and environmental monitoring. All this has become possible due to the discovery of better selectivity, low electrical resistance, improved thermal and mechanical properties in ion-exchange membranes. The ion-selective membrane electrodes offer low cost, non-destructive, robust, portable, higher sensitive and selective procedure for the determination of heavy toxic metal ions. The harmful/hazardous effects of heavy toxic metals have necessitated the rapid, on-spot, selective determination of these ions in portable and ground water. The ion-selective electrodes can be used for the detection of both cationic and anionic ions.

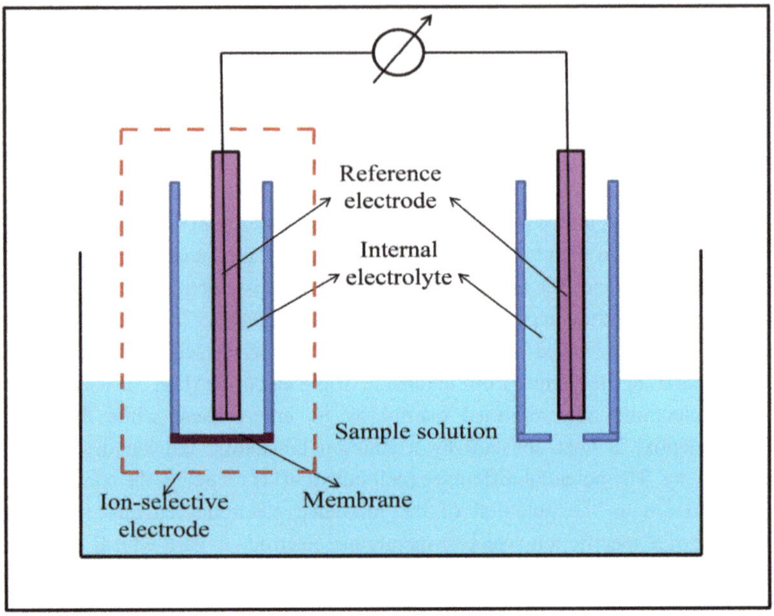


*Fig. 5: Schematic diagram of ion-selective membrane electrode.*

### 11.1 Membrane

The IUPAC recommendations [85] define membrane as: A continuous layer, usually consisting of a semi-permeable material, with controlled permeability, covering a structure, such as carbon or an inert metal, or separating two electrolyte solutions. This latter case is the most general form of an ISE. The membrane separates the internal components of the ISE from the test solution. The membrane of an ion-selective electrode is responsible for the electromotive force (emf) response and selectivity of the entire electrode.

Sollner [91] has defined membrane as: "Membrane is a phase or structure interposed between two phases or compartments which obstructs or completely prevents gross mass movement between the later, but permits passage, with varying degree of restriction of one or several species of particles from one to the other or between the two adjacent phases or compartments, which thereby acting as a physico-chemical machine transforms with various degree of efficiency according to its nature and composition of the two adjacent phases or compartment". R. Schlogl [92] described membrane as a phase, finite in space, which separates two other phases and exhibits individual resistance to the permeation of different species.

## 11.2 Physico-chemical properties of ion-selective membrane electrodes

The characteristics of an electrode can be studied by evaluating the electrode response or membrane potential, detection limit, response time, working pH range, selectivity and interference.

### 11.2.1 Electrode response or membrane potential

The use of ion-selective electrodes involves the potential determination. However, the potential of an individual electrode cannot be determined and it requires the measurement of emf of the half cell of an electrochemical cell. The two half cells are separated by the ion-selective membrane with their respective reference electrodes. A membrane potential (emf) is developed across the membrane separating the two solutions 1 and 2 with both having the same ions. This potential developed across the membrane is given by the equation (1.1)

$$E_m = \frac{RT}{Z_A F}\left[\ln\frac{[a_A]_1}{[a_A]_2} - (Z_y - Z_A)\int_1^2 t_y d\ln a_\pm\right] \quad (1.1)$$

where A = counter ion, Y = co-ion, Z = charge on ions, $t_y$ = transference number of co-ions in the membrane phase, $[a_A]_1$ and $[a_A]_2$ = activities of the counter ions in the solution 1 and 2, $a_\pm$ = mean ionic activity of the electrolyte.
For an ideally perm-selective membrane-

$$E_m = \pm\frac{RT}{Z_A F}\left[\ln\frac{[a_A]_1}{[a_A]_2}\right] \quad (1.2)$$

The above eq. (1.2) represents the Donnan potential for an ideally perm-selective membrane. The concentration of the internal or reference solution is kept constant (usually 0.1M). This solution, a standard calomel electrode (SCE) and the membrane together form an ion-selective membrane electrode. The membrane potential measurement is carried out using the following cell set up:

| Solution 2 | | | | Solution 1 |
|---|---|---|---|---|
| External saturated calomel electrode (SCE) | Test or external solution | Membrane | Internal solution | Internal saturated calomel electrode (SCE) |
| $E_L(2)$ | | | $E_L(1)$ | |

This membrane electrode is then immersed in solution 2, usually referred as external solution or test solution, having its separate external reference electrode. The emf of this cell is given by the following expression-

$$E_{cell} = E_{SCE} + E_{L(2)} + E_m + E_{L(1)} - E_{SCE} \quad (1.3)$$

where $E_{SCE}$, $E_L$ and $E_m$ refer to calomel electrode, junction and membrane potential, respectively. By combining equation (1.2) and (1.3), the following equation takes form-

$$E_{cell} = E_{SCE} - E_{SCE} + E_{L(2)} + E_{L(1)} \pm \frac{RT}{Z_A F} ln \frac{[a_A]_2}{[a_A]_1} \quad (1.4)$$

For cation-exchange membrane,

$$E_{cell} = E_{L(2)} + E_{L(1)} - \frac{RT}{Z_A F} ln[a_A]_1 + \frac{RT}{Z_A F} ln[a_A]_2 \quad (1.5)$$

The above equation reduces to

$$E_{cell} = E^0 + \frac{RT}{Z_A F} ln[a_A]_2 \quad (1.6)$$

since the internal solution concentration and the values of $E_{L(1)}$ and $E_{L(2)}$ are also almost constant. Considering the negligible values of $E_{L(1)}$ and $E_{L(2)}$ (due to salt bridge in use), the cell potential in above equation may approximately be taken as membrane potential. Hence, it is clear from equation (1.6) that the cell potential would change with the activity of the cation in external or test solution 2. At 25 °C, value of $RT/Z_AF$ comes out to be $0.059/Z_A$ volts. The membrane is said to give Nernstain response if the slope of cell potential versus log activity comes out to be $0.059/Z_A$ volts.

### 11.2.2 Detection limit

The ISEs follow the Nernstian equation within a particular concentration range of the analyte, and this region gives the working range of ISE. It is also defined as the activity range of an ion between the upper and lower detection limits of an ion-selective electrode. The ISEs exhibit a point (detection limit) at which the electrode response begins to deviate from the theoretical value and the ISE loses its specificity towards the primary ion. The presence of interfering ions in the form of impurities plays a major role in governing the detection limit of ISE. The limit of detection for an ion-selective electrode can be determined by extrapolating the linear part of the ion-selective electrode calibration curve (Fig. 6).

### 11.2.3 Response time

It is one of the essential factors of an ion selective electrode. The IUPAC recommendations describe response time as: The time which elapses between the instant when an ion-selective electrode and a reference electrode (ISE cell) are brought into contact with a sample solution (or at which the activity of the ion of interest in a solution is changed) and the first instant at which the emf/time slope ($\Delta E/\Delta t$) becomes equal to a limiting value selected on the basis of the experimental conditions and/or requirements concerning the accuracy (e.g. 0.6 mV/min.).

The expeditious measurement of the ISE represents an important criterion for membrane electrodes and has also been described viz the time necessary to attain an equilibrium value (steady potential) within ±1 mV after a ten-fold increase or decrease in the concentration of the test-solution.

### 11.2.4 Selectivity coefficients

The selectivity coefficient is one of the important factors of ion-selective electrodes, which determines the potential applicability of an electrode. The ISME yield a potential that is primarily due to the ion of interest usually called the primary ion while other interfering ions are called secondary ions. However, no ISME is exclusively specific to a single ion. The actual response of a binary mixture of primary (A) and secondary (B) ions is given by *Nikoloskii-Eisenman* equation (1.7) [93]. Detailed theory of the processes at the interference of these membranes, which generate the potential, is available elsewhere [94-96].

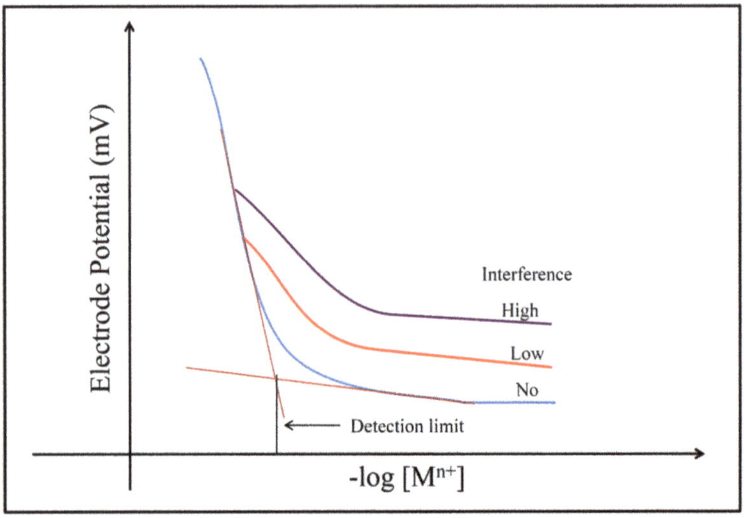


*Fig. 6: Calibration curve of ion-selective membrane electrode.*

$$E = E_0 + \frac{2.303\,RT}{z_A F} log\left[a_A + \sum K_A^{pot}(a_A)^{\frac{z_A}{z_B}}\right] \tag{1.7}$$

The lower the value of $K_{A,B}^{pot}$, the more selective is the electrode. For ideally selective electrodes, the $K_{A,B}^{pot}$ would be zero. Thus, the selectivity coefficient plays a great role while determining the selectivity of a particular ion-selective electrode. Fig. 6 shows the effect on selectivity of ISME due to presence of various interfering ions. The inconsistent values of selectivity coefficient may cause problems [97-99]. Various methods [100] have been suggested for determining the selectivity coefficient, however, it falls into two main groups, namely- (1) Separate-solution method and (2) Mixed-solution method.

#### 11.2.4.1 Separate-solution method

#### 11.2.4.1(a) Separate-solution method ($a_A$=$a_B$)

The potential of ISME is measured with two separate solutions, one containing the ion A at the activity $a_A$ (but no B), the other one containing the ion B at the same activity $a_A$=$a_B$ (but no A). If the measured values are $E_A$ and $E_B$, respectively, the value is calculated from the equation-

$$logK_{A,B}^{pot} = \frac{(E_B - E_A)Z_A F}{RTln10} + \left(1 - \frac{Z_A}{Z_B}\right) loga_A \quad (1.8)$$

**11.2.4.1 (b) Separate-solution method ($E_A$=$E_B$)**

The log of activity versus electrode potential relations of an ISE for the primary and interfering ions are obtained independently. Then, the activities that correspond to the same electrode potential value are used to determine the $K_{A,B}^{pot}$.

$$K_{A,B}^{pot} = \frac{a_A}{(a_B)^{\frac{Z_A}{Z_B}}} \quad (1.9)$$

**11.2.4.2 Mixed-solution method**

**11.2.4.2 (a) Fixed interference method**

The emf of ISE cell is measured for solutions of constant activity of the interfering ion, $a_B$ and varying activity of the primary ion, $a_A$. The emf values thus obtained are plotted against the logarithmic value of concentration of the primary ion. The intersection of the extrapolated linear portions of this plot indicates the value of $a_A$ that is to be used to calculate $K_{A,B}^{pot}$ from the following equation-

$$\boldsymbol{K_{A,B}^{pot}} = \frac{\boldsymbol{a_A}}{(\boldsymbol{a_B})^{\frac{Z_A}{Z_B}}} \quad (1.10)$$

where both $Z_A$ and $Z_B$ have the same sign (positive or negative).

**11.2.4.2 (b) Fixed primary ion method**

The emf of ISE cell is measured for solutions of constant activity of the primary ion, $a_A$ and varying activity of the interfering ion, $a_B$. The emf values obtained are plotted against the logarithmic concentration of the interfering ion. The intersection of the extrapolated linear portions of this plot indicates the value of $a_B$ that is to be used to calculate $K_{A,B}^{pot}$ from the equation (1.11)

$$\boldsymbol{K_{A,B}^{pot}} = \frac{\boldsymbol{a_A}}{(\boldsymbol{a_B})^{\frac{Z_A}{Z_B}}} \quad (1.11)$$

### 11.2.5 Life time

The life time of an ion selective membrane electrode denotes the period of time during which the slope and detection limit do not undergo any significant change. The deterioration of the plasticiser, polymer and the electro-active material is the main reason for the deviation in the response of membrane electrode. Thus, the period during which the electrode response remains reasonably constant represents the life time of ISE and it mostly varies between one to four months.

### 11.2.6 Effect of pH

The ion-selective electrodes are very specific in their response towards analyte species. These electrodes work only within a specific pH range and as such it is essential to find out the effect of pH on electrode response. In order to determine the working pH range of the ion-selective electrode, different solutions of varying pH with constant ion concentration are prepared and the electrode potential is recorded at each pH. The electrode potential is plotted against pH value and the pH over which the potential of ion-selective electrode remains constant is taken as the working pH range of the electrode.

## 11.3 Application of ion-selective electrodes

The ion-selective electrodes (ISEs) belong to the potentiometric class of chemical sensors. During the past few decades the use of ISEs has achieved new heights due to their wide spread applicability. The ISEs find application in various interdisciplinary fields as laboratory analysis, industry, process control [101], physiological measurements, environmental monitoring, critical care [102], safety [103], industrial hygiene, product quality controls [104], human comfort controls, medical health care, biochemistry lab tests, emissions monitoring, automotive, clinical diagnostics [105] and safety alarms. These sensors find such vast applications owing to a number of characteristic features as ease of preparation, low cost, on-spot determination, fast response, wide range of application, portability and need less skilled labour [106-114]. Several analytical techniques have been used for the determination of trace metal ions. These methods include atomic absorption spectroscopy (AAS) [115,116], electro thermal atomic absorption spectroscopy (ET-AAS), atomic emission spectroscopy (AES), ion chromatography [117] and inductively coupled plasma (ICP) [118,119]. However, all these methods are highly expensive, time consuming and require sophisticated equipment and well skilled labour [120-123]. Owing to these short comings, there has been a continuous need for development of other methods for easy and on-spot determination of heavy metals. The electrochemical methods have emerged as viable alternate. The ion-selective membrane electrodes have several attractive features as simple instrument, high sensitivity and selectivity, low cost and fast response [124-130].

## 12. Literature review

The contamination of water streams by heavy metal ions is a cause of great worry to mankind since their harmful effects are well established on human health. The presence of heavy metal ions is also considered as a potent threat to the wellbeing of aquatic life due to their toxic and non-biodegradable nature. A variety of approaches for the detection and removal of heavy metals in water bodies are reported in the literature viz, chemical precipitation, membrane filtration, electrochemical treatment technologies, thermal treatment adsorption, ion-exchange etc. However, all these methods are highly expensive, time consuming and require sophisticated equipment and skilled manpower. In order to overcome these limitations, there is a continuous need for development of other methods for easy and on spot determination of heavy metals. The electrochemical methods have emerged as viable alternate. The ion-selective membrane electrodes (ISME) have several attractive features as simple instrument, high sensitivity and selectivity, low cost and fast response. These ion-selective electrodes consist mainly of a perm-selective ion-conducting material and are primarily membrane-based devices, which separate the sample from the inside of the electrode. The membranes with higher selectivity bind the analyte ion, leaving interfering ions behind. Hence, different materials are used to impart ion-recognition properties to the membrane electrodes and hence higher selectivity. The development of electro-active materials for preparation of ISE is of primary importance, since this material is responsible for the recognition of analyte of primary importance. The nano-composite materials are the latest development for making electrochemical sensors for determination of heavy metals. The nanotechnology represents a promising field in technological developments. Nano-particles are embedded into metal, ceramic and polymer matrix material in nano-composite materials. These embedded nano-particles act as matrix reinforcement leading to changes in the physical properties of the base material. The higher surface to volume ratio of reinforcing material makes nano-composite different from conventional composite materials. The environmental friendly nature and the ability to modify the material at nano level, provide new opportunity to all sectors of industry in order to fulfil the growing demand of society. A vast number of ISE have been developed during last decade; as such it becomes imperative to sum-up the published work for better understanding. Although, it is very difficult to compile the work carried out in this field, an attempt has been made to arrange the work carried out during last ten years. **Table4** represents various cadmium and lead selective ion-selective electrodes and **Table5** represents some important lead and cadmium selective composite ion-exchangers.

***Table 4:*** *Cadmium and lead ion-selective membrane electrodes.*

| Ionophore/Electro-active material | Response time (s) | pH | Slope (mV/decade) | Concentration range (M) | Detection limit (M) | Ref. |
|---|---|---|---|---|---|---|
| **Cadmium** | | | | | | |
| Cetylpyridinium-tetraiodocadmate | 3 | 2.0-8.0 | -29.8 | $1.5\times10^{-6}$ -$1.0\times10^{-1}$ | $6.0\times10^{-7}$ | [131] |
| Cetylpyridinium-tetrabromocadmate | 7 | 2.0-8.0 | -25.1 | $1.0\times10^{-6}$ -$1.0\times10^{-1}$ | $8.0\times10^{-7}$ | [131] |
| Dicyclohexano-24-crown-8 | 23 | 2.0-5.4 | 30.0±1.0 | $3.0\times10^{-5}$-$1.0\times10^{-1}$ | - | [132] |
| Tetraethyl-*o*-silicate treated zeolite Y | - | - | $12.0\times10^{-3}$ | $1.0\times10^{-6}$-$15\times10^{-3}$ | - | [133] |
| N,N-[Bis(pyridin-2-yl)formylidene]butane-1,4-diamine | 10 | 2.0-8.0 | - | $7.9\times10^{-8}$-$1.0\times10^{-1}$ | $5.0\times10^{-8}$ | [134] |
| Orion-9648BN | - | - | - | $2.5\times10^{-5}$-$5.0\times10^{-4}$ | - | [135] |
| Thiacalix[4]arene | ~8 | 4.5-6.5 | 29.5 | $3.2\times10^{-6}$-$1.0\times10^{-1}$ | - | [136] |
| 1, 2-Bis(quinoline-2-carboxamido)-4-chlorobenzene | <10 | 2.4-9.0 | 30.3 ± 0.4 | $1.0\times10^{-6}$-$1.0\times10^{-1}$ | $8.0\times10^{-7}$ | [137] |
| N,N′-(4-Methyl-1,2-phenylene)diquinoline-2-carboxamide | 3-8 | 4.0-9.0 | 29.5 ± 0.5 | $1.0\times10^{-6}$-$1.0\times10^{-1}$ | $8.0\times10^{-7}$ | [138] |
| 4-Hydroxy salophen | 20 | 2.8-8.1 | 30.1 ± 1.0 | $1.0\times10^{-6}$-$1.0\times10^{-1}$ | $8.4\times10^{-7}$ | [139] |
| $[Mo_2(OAc)_2(H_2$-calix[4]arene)] | 12 | 1.0-7.0 | 30.0 ± 1.0 | $9.9\times10^{-8}$-$1.0\times10^{-1}$ | $9.8\times10^{-8}$ | [140] |
| Poly-*o*-anisidine Sn(IV) phosphate | 3-5 | - | 26.0 | $1.0\times10^{-7}$-$1.0\times10^{-1}$ | $1.0\times10^{-6}$ | [141] |

| | | | | | | |
|---|---|---|---|---|---|---|
| N,N,N’,N’-Tetrabutyl-3,6-dioxaoctanedi(thioamide) | 5-10 | 2.0-9.0 | 29.4 ± 0.3 | $1.0\times10^{-7}$-$1.0\times10^{-3}$ | $1.3\times10^{-8}$ | [142] |
| $N_1$, $N_2$-Dicyanoethyl-$N_1$,$N_2$-bis(pyridin-2-ylmethyl) benzene-1, 2-diamine | 12 | 2.0-8.0 | 29.5 | $2.5\times10^{-7}$-$1.0\times10^{-1}$ | $7.8\times10^{-8}$ | [143] |
| 1,3-Bis(2-cyanobenzene)triazene | 2 | 6.0-9.0 | 29.2 | $1.0\times10^{-5}$-$1.0\times10^{-2}$ | $8.0\times10^{-6}$ | [144] |
| 25,27-Bis(ethyl-2-(bis(2-pyridylmethyl)aminomethyl)aniline)-26,28-dihydroxy *p*-tert-butylcalix[4]arene | 10 | 6.0-9.0 | 29.4 ± 0.6 | $1.6\times10^{-6}$-$1.0\times10^{-2}$ | - | [145] |
| Poly-*o*-toluidine Ce(IV) phosphate | ~20 | 3.0-7.0 | 24.1 | $1.0\times10^{-6}$-$1.0\times10^{-1}$ | $1.0\times10^{-6}$ | [146] |
| 1-Ethyl-3-methylimidazolium chloride doped membrane | 5.0-8.0 | 3.8-8.0 | 30.3 | $1.0\times10^{-8}$-$1.0\times10^{-1}$ | $1.4\times10^{-9}$ | [147] |
| Poly(4-vinyl pyridine) | <60 | 4.0-7.5 | 27.7 ± 0.8 | $1.0\times10^{-7}$-$1.0\times10^{-1}$ | $2.51\times10^{-8}$ | [148] |
| N,N,N’,N’-Tetrabutyl-3,6-dioxaoctanedi(thiomide) | - | - | - | $1.0\times10^{-9}$-$1.0\times10^{-7}$ | $2.8\times10^{-10}$ | [149] |
| 5,5’-(5,5’-(Benzo[c][1,2,5]thiadiazole-4, 7-diyl)bis(thiophene-5,2-diyl))bis-($N_1$,$N_1$,$N_3$,$N_3$-tetraphenylbenzene-1,3-diamine) | 10 | 3.0-10.0 | 29.6 ± 0.3 | $6.3\times10^{-8}$-$1.0\times10^{-1}$ | $3.6\times10^{-8}$ | [150] |
| Ion imprinted polymer dispersed in 2- | 15 | 4.0-7.0 | 29.9 | $2.0\times10^{-7}$-$1.0\times10^{-2}$ | $1.0\times10^{-7}$ | [151] |

| | | | | | | |
|---|---|---|---|---|---|---|
| nitrophenyloctyl ether | | | | | | |
| 2,2'-Thiobis[ 4-methyl(2-amino phenoxy) phenyl ether] | 6 | 3.0-5.5 | 29.95 ± 0.1 | $3.0\times10^{-8}$-$1.0\times10^{-1}$ | $7.5\times10^{-9}$ | [152] |
| Benzil bis(carbohydrazone) | ~8 | 2.0-9.0 | 29.7 | $1.0\times10^{-8}$-$1.0\times10^{-1}$ | $3.2\times10^{-8}$ | [153] |
| Cupric oxide nanosturucture | 10 | 5.0-10.0 | 29.3 ± 0.3 | $1.0\times10^{-9}$-$1.0\times10^{-1}$ | $5.0\times10^{-10}$ | [154] |
| Modified *p*-tert butylcalix[8]arene | - | - | - | - | $1.0\times10^{-7}$ | [155] |
| Poly-*o*-toluidine Sn(IV) tungstate | 20 | 4.0-8.0 | 27.4 | $1.0\times10^{-8}$-$1.0\times10^{-1}$ | - | [156] |
| *p*-Tert-butylcalix[6]arene | 35 | 2.8-6.2 | 29.0 ± 1.0 | $9.7\times10^{-5}$-$1.0\times10^{-1}$ | - | [157] |
| 1,13-Bis(8-quinolyl)-1,4,7,10,13-pentaoxatridecane | 15 | 1.0-6.0 | 29.8 ± 0.1 | $1.0\times10^{-5}$-$1.0\times10^{-1}$ | $8.4\times10^{-6}$ | [158] |
| Polyaniline Sn(IV) silicate | 10 | 3.5-6.5 | 28.53 | $1.0\times10^{-7}$-$1.0\times10^{-1}$ | $1.0\times10^{-7}$ | [159] |
| 5-Amino-1,3,4-thiadiazole-2-thiol | 10 | 2.5-8.5 | 29.6 | $2.1\times10^{-8}$-$1.0\times10^{-1}$ | $7.58\times10^{-9}$ | [160] |
| 1-Butyl-3-methylimidazolium hexafluorophosphate | 12 | - | 30.2± 0.5 | $1.0\times10^{-5}$-$1.0\times10^{-1}$ | $1.0\times10^{-7}$ | [161] |
| Single walled carbon nanotubes Ce(IV) phosphate | 10 | - | 27.4 | $1.0\times10^{-7}$-$1.0\times10^{-1}$ | $1.0\times10^{-7}$ | [162] |
| **Lead** | | | | | | |
| N,N'-Bis(salicylidene)-2,6-pyridinediamine | 10 | 5.0-7.0 | 29.4 | $9.0\times10^{-3}$-$1.0\times10^{-3}$ | - | [163] |
| Bis(thiophenol)-4,4'-methylenediamine | 10-20 | 4.0-8.2 | 29.1 | $5.0\times10^{-6}$-$1.0\times10^{-1}$ | $3.0\times10^{-6}$ | [164] |
| Nano-sized lead oxide particles | 10 | 2.0-8.0 | 29.0± 1.0 | $2.5\times10^{-5}$-$1.0\times10^{-1}$ | $8.0\times10^{-6}$ | [165] |

| | | | | | | |
|---|---|---|---|---|---|---|
| 1-Phenyl-2-(2-quinolyl)-1,2-dioxo-2-(4-bromo)phenylhydrazone | 14 | 3.0-6.0 | 28.7 | $1.0\times10^{-6}$-$1.0\times10^{-1}$ | $6.0\times10^{-7}$ | [166] |
| Polypyrrole Th(IV) phosphate | 35 | 3.0-8.5 | 29.1 | $5.0\times10^{-6}$-$1.0\times10^{-1}$ | - | [167] |
| Potassium tetrakis(4-chlorophenyl-borate) | <15 | 3.5-6.5 | 29.7 | $1.0\times10^{-6}$-$1.0\times10^{-1}$ | $5.0\times10^{-7}$ | [168] |
| 5,11,17,23-Tetra-tert-butyl-25,26,27,28-tetrakis-(diphenylphosphinoylmethoxy)calix[4]arene | 17 | 3.5-5.0 | 28.0± 0.2 | $1.0\times10^{-5}$-$1.0\times10^{-2}$ | $1.4\times10^{-6}$ | [169] |
| 2,12-Dimethyl-7,17-diphenyltetrapyrazole | ~10 | 1.6-6.0 | 30.0 | $2.5\times10^{-6}$-$5.0\times10^{-2}$ | - | [170] |
| Tetraethyl-*o*-silicate treated zeolite Y | - | - | $12.5\times10^{-3}$ | $1.0\times10^{-6}$-$8.0\times10^{-3}$ | - | [133] |
| N,N'-Bis-thiophen-2-ylmethylene-pyridine-2,6-diamine | <5 | - | 29.59± 0.5 | $9.0\times10^{-3}$-$1.0\times10^{-3}$ | -5.74 | [171] |
| Acrylamide Zr(IV) Arsenate | 20 | 2.0-7.0 | 30.0 ± 1 | $1.0\times10^{-7}$-$1.0\times10^{-1}$ | $5.0\times10^{-6}$ | [172] |
| 3,15,21-Triaza-4,5;13,14-dibenzo-6,9,12-trioxabicycloheneicosa-1,17,19-triene-2,16-dione | ~16 | 3.7-6.5 | - | $3.0\times10^{-6}$-$1.3\times10^{-2}$ | $2.0\times10^{-6}$ | [173] |
| Polyaminoanthraquinone | 12 | 2.8-5.2 | 28.9 | $2.5\times10^{-6}$-$1.0\times10^{-1}$ | $7.76\times10^{-7}$ | [174] |
| Poly(n-butyl acrylate) | - | - | 26.6 ± 1.6 | $1.0\times10^{-9}$-$1.0\times10^{-2}$ | - | [175] |
| 5,10,15,20-Tetra-(3-bromo-4-hydroxyphenyl)porphyrin | 240 | - | - | $5.0\times10^{-6}$-$4.0\times10^{-4}$ | $2.0\times10^{-8}$ | [176] |

| | | | | | | |
|---|---|---|---|---|---|---|
| Tetra-(3-bromo-4-hydroxyphenyl)porphyrin | 120 | - | - | $5.0\times10^{-6}$-$3.0\times10^{-4}$ | $4.0\times10^{-8}$ | [176] |
| 4-Hydroxy salophen | 600 | 3.1 | - | $1.0\times10^{-7}$-$1.0\times10^{-3}$ | $8.6\times10^{-8}$ | [177] |
| Poly-*o*-anisidine Sn(IV) arsenophosphate | ~30 | 4.0-8.0 | 32.8 | $1.0\times10^{-6}$-$1.0\times10^{-1}$ | $1.0\times10^{-6}$ | [178] |
| Trans-1,2-diaminocyclohexaneN,N,N',N'-tetraacetic acid | - | 5.5-10 | - | $1.0\times10^{-6}$-$1.0\times10^{-2}$ | $8.93\times10^{-5}$ | [179] |
| 1,3-Bis(N′-benzoylthioureido)benzene | 14 | 2.2-6.0 | 31.5 ± 1.6 | $4.0\times10^{-6}$-$1.0\times10^{-2}$ | - | [180] |
| 1,3-Bis(N′-furoylthioureido)benzene | 14 | 2.2-6.0 | 30.0 ± 1.3 | $5.0\times10^{-6}$-$1.0\times10^{-2}$ | - | [180] |
| 3,7,11-Tris (2-pyridylmethyl)-3,7,11,17-tetraazabicyclo [11.3.1] heptadeca-1(17),13,15-triene (I) | 20 | 5.0-8.0 | 28.5± 0.2 | $1.0\times10^{-6}$-$1.0\times10^{-1}$ | - | [181] |
| 5,8-Bis((5-chloro-8-hydroxy-7-quinolinyl)methyl)-2,11-dithia-5,8-diaza-2,6-pyridinophane | < 300 | 5.5 | - | $3.0\times10^{-7}$-$2.5\times10^{-2}$ | - | [182] |
| 18-Crown-6 | 6 | 4.3-5.3 | 30.0 | $1.0\times10^{-6}$-$1.0\times10^{-3}$ | $5.6\times10^{-7}$ | [183] |
| Dicyclohexyl18-crown-6 | <12 | 4.6-5.2 | 30.0 | $1.0\times10^{-5}$-$1.0\times10^{-2}$ | $6.3\times10^{-6}$ | [183] |
| Dibenzo18-crown-6 | 6 | 3.1-4.9 | 28.0 | $1.0\times10^{-5}$-$1.0\times10^{-2}$ | $7.1\times10^{-6}$ | [183] |
| Poly(2-methoxy-5-(2'-ethylhexyloxy)-p-phenylene vinylene) | 10-15 | - | 29.1± 0.8 | $2.0\times10^{-9}$-$1.0\times10^{-3}$ | $1.0\times10^{-9.2}$ | [184] |
| Functionalized multi walled carbon | 25 | 2.5-6.5 | 29.5± 0.3 | $5.9\times10^{-10}$-$1.0\times10^{-2}$ | $3.2\times10^{-10}$ | [185] |

| | | | | | | |
|---|---|---|---|---|---|---|
| nanotubes | | | | | | |
| Poly(*m*-phenylenediamine) | 14 | 3.0-5.0 | 29.8 | $3.1\times10^{-6}$-$0.31\times10^{-1}$ | $6.3\times10^{-7}$ | [186] |
| 2-Carboxaldehydefuran benzoylhydrazone | 8-10 | 6.2-7.8 | - | $9.1\times10^{-7}$-$1.0\times10^{-2}$ | $7.9\times10^{-7}$ | [187] |
| Polyvinylchloride membrane based on ionic liquid | 5-7 | 3.5-7.3 | 29.8 | $1.0\times10^{-8}$-$1.0\times10^{-1}$ | $4.3\times10^{-9}$ | [187] |
| 1,1′-Thiobis(naphthalene-2,1-diyl)bis(2-aminobenzoate) | 8 | 3.5-7.0 | 29.0± 0.2 | $8.0\times10^{-8}$-$1.0\times10^{-2}$ | $5.0\times10^{-8}$ | [188] |
| Triazolo-thiadiazin derivative | 180 | 5.0-6.0 | - | $5.0\times10^{-8}$-$3.8\times10^{-4}$ | $2.2\times10^{-8}$ | [189] |
| Schiff base complex $[Co(L)_2](ClO_4)\cdot(C_3H_6O)\cdot(H_2O)]$ | 10 | 4.0-13.0 | 23.9 | $1.0\times10^{-5}$-$1.0\times10^{-2}$ | $4.6\times10^{-6}$ | [190] |
| 2-Carboxaldehydethiophene benzoylhydrazone | 8-10 | 6.2-7.8 | - | $5.9\times10^{-7}$ - $1.0\times10^{-2}$ | $3.9\times10^{-7}$ | [191] |
| Poly(vinyl chloride) membrane coated electrode | - | 3.5-7.5 | 29.1 | $1.5\times10^{-6}$-$1.3\times10^{-1}$ | $9.0\times10^{-7}$ | [192] |
| Polyaniline Ti(IV)phosphate | 10 | 2.5–6.5 | 29.48 | $1.0\times10^{-8}$-$1.0\times10^{-1}$ | $1.0\times10^{-9}$ | [193] |
| Polyaniline Ti(IV) arsenophosphate | 20 | 3.0-6.0 | 29.3 | $1.0\times10^{-6}$-$1.0\times10^{-1}$ | $1.0\times10^{-6}$ | [194] |
| Poly(methyl methacrylate)-Cerium Molybdate | 8 | 2.5-6.0 | 29.48 | $1.0\times10^{-9}$-$1.0\times10^{-1}$ | $5.0\times10^{-9}$ | [195] |
| Polyaniline tungstophosphate | ~35 | 4.0-7.0 | 29.6 | $1.0\times10^{-7}$-$1.0\times10^{-1}$ | $1.0\times10^{-7}$ | [196] |

| | | | | | | |
|---|---|---|---|---|---|---|
| Diethylsulfide-1,5-bis(2'-hydroxy-4'-nitrophenoxy)-3-thiapentane/diethylsulfide-1,5- Bis(8'-oxybenzopyridine)-3-thia pentane | 5 | 1.6-7.0 | 31.57± 0.3 | $2.0\times10^{-9}$-$1.0\times10^{-1}$ | $1.0\times10^{-9}$ | [197] |
| Nitrobenzo 18-crown-6 | 6 | 2.2-4.1 | 30.5± 0.5 | $1.0\times10^{-5}$-$1.0\times10^{-1}$ | $8.9\times10^{-6}$ | [198] |
| Zr(IV) phosphosulphosalicylate | 15 | 4.0–6.5 | 43.8 | $1.0\times10^{-5}$-$1.0\times10^{-1}$ | $4.78\times10^{-6}$ | [199] |
| N,N,N',N'-Tetrabutyl-3,6-dioxaoctanedi(thiomide) | - | - | - | $3.0\times10^{-9}$-$1.0\times10^{-7}$ | $6.6\times10^{-10}$ | [149] |
| Poly-*o*-toluidine/multi walled carbon nanotubes/Sn(IV) tungstate | 50 | - | - | $1.0\times10^{-9}$-$1.0\times10^{-1}$ | - | [200] |
| 4-(Vinylamino)pyridine-2,6-dicarboxylicacid coated on $Fe_3O_4$ nano-particles | - | - | - | - | $0.9\times10^{-9}$ | [201] |
| Carboxymethyl cellulose Sn(IV) Phosphate | 10 | 2.0-4.0 | 28.1 | $1.0\times10^{-6}$-$.01\times10^{-1}$ | $1.0\times10^{-6}$ | [202] |
| 25,26,27,28-Tetrakis(piperidinylthiocarbonylmethylene)-p-tert-butylcalix[4] arene | - | - | - | $1.0\times10^{-6}$-$1.0\times10^{-4}$ | - | [203] |
| Dithiozone | - | 5.2 | - | $1.1\times10^{-8}$-$2.4\times10^{-6}$ | $4.0\times10^{-9}$ | [204] |
| Poly(3,4-ethylenedioxythiophene)- | 11 | 3.0-6.5 | 27.4 | $1.0\times10^{-7}$-$1.0\times10^{-1}$ | $1.0\times10^{-7}$ | [205] |

| | | | | | | |
|---|---|---|---|---|---|---|
| poly(styrenesulfonate) Zr(IV) monothiophosphate | | | | | | |
| Zr(IV) phosphosulfosalicylate | 15 | 4.0-6.5 | 43.8 | $1.0\times10^{-5}$-$1.0\times10^{-1}$ | $4.78\times10^{-6}$ | [206] |

***Table 5:*** *Some important lead and cadmium selective composite ion-exchangers*

| **S. No.** | **Organic-Inorganic Composite Material** | **Application** | **References** |
|---|---|---|---|
| **1** | Polypyrrole Th(IV) phosphate | Pb(II) selective | [167] |
| **2** | Poly-*o*-toluidine Ce(IV) phosphate | Cd(II) selective | [146] |
| **3** | Polyaniline Sn(IV) phosphate | Pb(II) selective | [207] |
| **4** | Poly(methylmethacrylate) Zr(IV) phosphate | Pb(II) selective | [208] |
| **5** | Polyacrylamide Th(IV) phosphate | Pb(II) selective | [209] |
| **6** | Acrylamide aluminumtungstate | Pb(II) selective | [210] |
| **7** | Poly-*o*-anisidine Sn(IV) arsenophosphate | Pb(II) selective | [211] |
| **8** | Acrylonitrile Sn(IV) tungstate | Pb(II) selective | [212] |
| **9** | Polyacrylamide apatite | Pb(II) selective | [213] |
| **10** | Macrocyclic diamide sulphur | Cd(II) selective | [214] |
| **11** | $B_2O_3/TiO_2$ nano-composite | Cd(II) selective | [215] |
| **12** | Ethylenediamine tetra acetic acid Sn(IV) iodate | Pb(II) selective | [216] |
| **13** | Acrylamide Zr(IV) arsenate | Pb(II) selectve | [217] |

| | | | |
|---|---|---|---|
| **14** | Acrylamide stannic silicomolybdate | Pb(II) selective | [218] |
| **15** | Poly(methylmethacrylate) Ce(IV) molybdate | Pb(II) selective | [194] |
| **16** | Polyaniline Ce(IV) molybdate | Cd(II) selective | [219] |
| **17** | Poly-*o*-methoxyaniline Zr(IV) molybdate | Cd(II) selective | [220] |
| **18** | Poly-*o*-toluidine stannic molybdate | Pb(II) selective | [221] |
| **19** | Poly-*o*-anisidine Sn(IV) phosphate | Cd(II) selective | [141] |
| **20** | Polyaniline Sn(IV) molybdate | Pb(II) selective | [222] |
| **21** | Polyaniline Sn(IV) silicomolybdate | Pb(II) selective | [223] |
| **22** | Polyaniline zirconium titanium phosphate | Pb(II) selective | [224] |
| **23** | Acrylamide Sn(IV)molybdate | Cd(II) selective | [225] |
| **24** | Polyaniline Zr(IV) arsenate | Pb(II) selective | [226] |
| **25** | Poly-*o*-toluidine Zr(IV) tungstate | Pb(II) selective | [227] |
| **26** | Polyaniline Ti(IV) arsenate | Pb(II) selective | [228] |
| **27** | Polyaniline Ti(IV) phosphate | Pb(II) selective | [193] |
| **28** | Polyaniline Sn(IV) tungstoarsenate | Cd(II) selective | [229] |
| **29** | Polyaniline Sn(IV) silicate | Cd(II) selective | [230] |
| **30** | Polyvinyl alcohol zinc oxide | Cd(II) & Pb(II) selective | [231] |
| **31** | Poly-*o*-toluidine Sn(IV) tungstate | Pb(II) selective | [156] |
| **32** | Polyaniline Sn(IV) tungstomolybdate | Pb(II) selective | [232] |

| | | | |
|---|---|---|---|
| **33** | Polyaniline Ti(IV) arsenophosphate | Cd(II) selective | [233] |
| **34** | Polyaniline Sn(IV) silicophosphate | Pb(II) selective | [234] |
| **35** | Poly(3,4-ethylenedioxythiophene)-poly(styrenesulfonate)Zr(IV) phosphate | Cd(II) selective | [235] |
| **36** | Polyaniline tungstophosphate | Pb(II) selective | [196] |
| **38** | Carboxymethyl cellulose Sn(IV) phosphate | Pb(II) selective | [202] |
| **39** | Polyaniline Zr(IV) sulfosalicylate | Pb(II) selective | [236] |
| **40** | Poly(3,4-ethylenedioxythiophene)-poly(styrenesulfonate)Zr(IV) monothiophosphate | Pb(II) selective | [237] |
| **41** | Cellulose acetate Zr(IV) molybdophosphate | Cd(II) & Pb(II) selective | [238] |
| **42** | Single walled carbon nanotubes Ce(IV) phosphate | Cd(II) selective | [162] |

**Conclusion**

The water bodies are depleting rapidly owing to several factors. Only advanced purification technologies can relieve pressure of water scarcity, health and climate change by purifying and reuse of wastewater, which in effect involves the development of efficient, novel, reliable and cost effective materials and methods. The organic-inorganic nano-composites are prepared by embedding organic materials into the inorganic matrix material. Electrochemical sensors are a rapidly progressing section of analytical chemistry. These have attained a great significance because of their on-site monitoring of various environmentally hazardous substances. The ion-exchange membranes find wide spread application as food, drug, biological, waste material treatment and environmental monitoring due to their better selectivity, low electrical resistance, improved thermal and mechanical properties in ion-exchange membranes. These ISME provide low cost, non-destructive, robust, portable, higher sensitive and selective procedure for the determination of heavy toxic metal ions. The ion-exchangers and ISME listed here have been found to be selective for heavy metal ions viz, Pb (II) and Cd(II).

**References**

[1] D. Sud, G. Mahajan, M.P. Kaur, Agricultural waste material as potential adsorbent for sequestering heavy metal ions from aqueous solutions-A review, Bioresour. Technol. 99 (2008) 6017-6027. https://doi.org/10.1016/j.biortech.2007.11.064

[2] T. Deblonde, C.C. Leguille, P. Hartemann, Emerging pollutants in wastewater: A review of the literature, Int. J. Hyg. Environ. Health 214 (2011) 442-448.

[3] M.N. Obasi, Environmental pollution with particular reference to chemical pollutant in Nigeria. 1st ed., White and White Publishers Ltd., Owerri, Nigeria, 1999.

[4] H. El-Dessouky, H. Ettouney, Teaching desalination-a multidiscipline engineering science, Heat Tran. Eng. 23 (2002) 1-3.

[5] D.D. Mara,Water, sanitation and hygiene for the health of developing nations, Public Health 117 (2003) 452-456. https://doi.org/10.1016/S0033-3506(03)00143-4

[6] M. Moore, P. Gould, B.S. Keary,Global urbanization and impact on health, Int. J. Hyg. Environ. Health 206 (2003) 269-278. https://doi.org/10.1078/1438-4639-00223

[7] A.C. Sethi, Key note address, International Conference on Fresh Water, Bonn, Germany (2001).

[8] Directive 76/464/EEC, Official J. European Communities, LI29 (1976).

[9] M.A. Shannon, P.W. Bohn, M. Elimelech, J.G. Georgiadis, B.J. Marias, A.M. Mayes, Science and technology for water purification in the coming decades, Nature 452 (2008) 301-310. https://doi.org/10.1038/nature06599

[10] Drinking water specification, IS, 2nd ed. (2012).

[11] United States Public Health, Drinking water standards (2012).

[12] P. Leonard, S. Hearty, J. Brennan, L. Dunne, J. Quinn, T. Chakraborty, R.O. Kennedy, Advances in biosensors for detection of pathogens in food and water, Enzyme Microb. Tech. 32 (2003) 3-13. https://doi.org/10.1016/S0141-0229(02)00232-6

[13] N.J. Ashbolt, Microbial contamination of drinking water and disease outcomes in developing regions, Toxicology 198 (2004) 229-238. https://doi.org/10.1016/j.tox.2004.01.030

[14] G. Hutton, L. Haller, J. Bartram, Public Health and Environment, World Health Organization, Geneva (2007).

[15] WHO and UNICEF, Progress on Sanitation and Drinking Water, World Health Organization, Geneva (2013).

[16] J. Fawell, M.J. Nieuwenhuijsen, Contaminants in drinking water, Br. Med. Bull. 68 (2003) 199-208. https://doi.org/10.1093/bmb/ldg027

[17] S.R. Mozaz, M.J.L. de Alda, D. Barcelo, Monitoring of estrogens, pesticides and bisphenol A in natural waters and drinking water treatment plants by solid phase extraction-liquid chromatography-mass spectrometry, J. Chrom. 1045 (2004) 85-92. https://doi.org/10.1016/j.chroma.2004.06.040

[18] K. Mulder, N. Hagens, B. Fisher, Burning water: A comparative analysis of the energy return on water invested, Ambio 39 (2010) 30-39. https://doi.org/10.1007/s13280-009-0003-x

[19] H.C.J. Godfray, J.R. Beddington, I.R. Crute, L. Haddad, D. Lawrence, J.F. Muir, J. Pretty, S. Robinson, S.M. Thomas, C. Toulmin, Food Security: The Challenge of Feeding 9 Billion People, Science 327 (2010) 812-818. https://doi.org/10.1126/science.1185383

[20] K.S. Low, C.K. Lee,Cadmium uptake by the Moss, Calymperesdelessertii, Besch, Bioresour. Technol. 38 (1991) 1-6. https://doi.org/10.1016/0960-8524(91)90214-5

[21] D.W.O. Connell, C. Birkinshaw, T.F.O. Dwyer, Heavy metal adsorbents prepared from the modification of cellulose: A review, Bioresour. Technol. 99 (2008) 6709-6724. https://doi.org/10.1016/j.biortech.2008.01.036

[22] M.E. Argun, S. Dursun, C. Ozdemir, M. Karatas, Heavy metal adsorption by modified oak sawdust: Thermodynamics and kinetics, J. Hazard. Mater. 141 (2007) 77-85. https://doi.org/10.1016/j.jhazmat.2006.06.095

[23] G.M. Naja, B. Volesky, Treatment of metal-bearing effluents: removal and recovery in: L.K. Wang, J.P. Chen, Y.T. Hung, N.K. Shammas (Eds.), Handbook on Heavy Metals in the Environment, Taylor and Francis, Boca Raton FL USA 2009, pp. 247-291.

[24] Guidelines for Drinking Water Quality, 4th ed., World Health Organisation (2011).

[25] European Water Quality Standards (2013).

[26] Guidelines for Canadian Drinking Water Quality (2014).

[27] T. Alizadeh, M.R. Ganjali, P. Nourozi, M. Zare, M. Hoseini, A carbon paste electrode impregnated with $Cd^{2+}$ imprinted polymer as a new and high selective electrochemical sensor for determination of ultra-trace $Cd^{2+}$ in water samples, J. Electroanal. Chem. 657 (2011) 98-106. https://doi.org/10.1016/j.jelechem.2011.03.029

[28] J.O. Nriagu, J.M. Pacyna, Quantitative assessment of worldwide contamination of air, water and soils by trace metals, Nature 333 (1988) 134-139. https://doi.org/10.1038/333134a0

[29] M. Borsari, Encyclopedia of Inorganic Chemistry, 1st ed., Wiley, New York 1994.

[30] K. Kadirvelu, K. Thamaraiselvi, C. Namasivayam, Removal of heavy metals from industrial wastewaters by adsorption onto activated carbon prepared froman agricultural solid waste, Bioresour. Technol. 76 (2001) 63. https://doi.org/10.1016/S0960-8524(00)00072-9

[31] JM Pearce, Burton's line in lead poisoning, Eur. Neurol. 57 (2007) 118-119. https://doi.org/10.1159/000098100

[32] L.N. Abhyankar, M.R. Jones, E. Guallar, A.N. Acien, Arsenic exposure and hypertension: A systematic review, Environ. Health Perspect.120 (2012) 494-500. https://doi.org/10.1289/ehp.1103988

[33] C.J. Chen, H.Y. Chiou, M.H. Chiang, L.J. Lin, T.Y. Tai, Dose-response relationship between ischemic heart disease mortality and long-term arsenic exposure, Arterioscler. Thromb.Vasc. Biol. 16 (1996) 504-510. https://doi.org/10.1161/01.ATV.16.4.504

[34] Y. Chen, P.F. Litvak, G.R. Howe, J.H. Graziano, P.B. Rauf, F. Parvez, Arsenic exposure from drinking water, dietary intakes of B vitamins and folate, and risk of high blood pressure in Bangladesh: a population-based, cross-sectional study, Am. J. Epidemiol. 165 (2007) 541-552. https://doi.org/10.1093/aje/kwk037

[35] C.H. Tseng, C.K. Chong, C.J. Chen, T.Y. Tai, Dose-response relationship between peripheral vascular disease and ingested inorganic arsenic among residents in blackfoot disease endemic villages in Taiwan, Atherosclerosis 120 (1996) 125-133. https://doi.org/10.1016/0021-9150(95)05693-9

[36] N. Debendra, G. Mazumder, R. Haque, N. Ghosh, K.D. Binay, A. Santra, D. Chakraborty, A.H. Smith, Arsenic levels in drinking water and the prevalence of skin lesionsin West Bengal, India, Int. J. Epidemiology 27 (1998) 871-877. https://doi.org/10.1093/ije/27.5.871

[37] P. Ghosh, M. Banerjee, S.D. Chaudhuri, J.K. Das, N. Sarma, A. Basu, A.K. Giri, Increased chromosome aberration frequencies in the Bowen's patients compared to non-cancerous skin lesions individuals exposed to arsenic, Mutat. Res. Genet. Toxicol. Environ. Mutagen. 632 (2007) 104-110. https://doi.org/10.1016/j.mrgentox.2007.05.005

[38] F. Catterina, Y. Yan, C. Jacqueline, B. Hugo, P.L. Roxana, A. Johanna, A.H. Smith, L. Jane, S. Craig, Arsenic, tobacco smoke, and occupation: associations of multiple agents with lung and bladder cancer, Epidemiology 24 (2013) 898-905. https://doi.org/10.1097/EDE.0b013e31829e3e03

[39] J.S. Tsuji, D.D. Alexander, V. Perez, P.J. Mink, Arsenic exposure and bladder cancer: Quantitative assessment of studies in human populations to detect risks at low doses, Toxicology317 (2014) 17-30.

[40] H.M. Anawar, J. Akai, H. Sakugawa, Mobilization of arsenic from subsurface sediments by effect of bicarbonate ions in groundwater, Chemosphere 54 (2004) 753-762. https://doi.org/10.1016/j.chemosphere.2003.08.030

[41] H.T. Lin, M.C. Wang, K. Seshaiah, Mobility of adsorbed arsenic in two calcareous soils as influenced by water extract of compost, Chemosphere 71 (2008) 742-749. https://doi.org/10.1016/j.chemosphere.2007.10.022

[42] Deepali, K.K. Gangwar, Metals concentration in textile and tannery effluents, Associated Soils and Ground Water, N. Y. Sci. J. 3 (2010) 82-89.

[43] P. Sharma, V. Bihari, S.K. Agarwal, V. Verma, C.N. Kesavachandran, B.S. Pangtey, N. Mathur, K.P. Singh, M. Srivastava, S.K. Goel, Groundwater contaminated with hexavalent chromium [Cr (VI)]: A health survey and clinical examination of community inhabitants (Kanpur, India), PLOS ONE (2012) DOI: org/10.1371/journal.pone.0047877.

[44] M. Javed, N. Usmani, Assessment of heavy metal (Cu, Ni, Fe, Co, Mn, Cr, Zn) pollution in effluent dominated rivulet water and their effect on glycogen metabolism and histology of Mastacembelusarmatus, SpringerPlus 2 (2013) 390-402. https://doi.org/10.1186/2193-1801-2-390

[45] M. Javed, N. Usmani, Investigation on accumulation of toxicants and health status of freshwater fish Channapunctatus, exposed to sugar mill effluent, Int. J. Zool. Res. 3 (2013) 43-48.

[46] G.W. James, P.K. David, H.H. Gary, M.S. Anton, Handbook of Ecotoxicology, $2^{nd}$ ed., CRC Press (2002).

[47] A.A. Khan, Inamuddin, Applications of Hg(II) sensitive polyanilineSn(IV) phosphate composite cation-exchange material in determination of $Hg^{2+}$ from aqueous solutions and in making ion-selective membrane electrode, Sensor. Actuator. B Chem. 120 (2006) 10-18. https://doi.org/10.1016/j.snb.2006.01.033

[48] A.K. Jain, S.M. Sondhi, V.K. Sharma, Synthesis, characterization and Hg(II) ion selectivity of 1-(2-Nitro-4-methyl phenyl)-6-methyl-6-methoxy-1,4,5,6-tetrahydro-pyrimidine-2-(3H) thione (TPT), Electroanalysis 12 (2000) 301-305. https://doi.org/10.1002/(SICI)1521-4109(20000301)12:4<301::AID-ELAN301>3.0.CO;2-2

[49] G.E.M. Eyssen, J. Ruedy, Methyl mercury exposure in northern quebec I. neurologic findings in adults, Am. J. Epidemiol. 118 (1983) 461-469. https://doi.org/10.1093/oxfordjournals.aje.a113651

[50] M. Harada, Minamata disease: methylmercury poisoning in Japan caused by environmental pollution, Crit. Rev. Toxicol. 25 (1995) 1-24. https://doi.org/10.3109/10408449509089885

[51] W. Giger, The Rhine red, the fish dead-the 1986 Schweizerhalle disaster, a retrospect and long-term impact assessment, Environ.Sci. Pollut. Int. 16 (2009) 98-111. https://doi.org/10.1007/s11356-009-0156-y

[52] A. Blowers, S. Hinchliffe (Eds.), Environmental Responses, John Willey & Sons Ltd., Chichester, UK, 2003.

[53] M.T. Amin, A.A. Alazba, U. Manzoor, A review of removal of pollutants from water/wastewater using different types of nanomaterials, Adv. Mater. Sci. Eng. 2014 (2014), DOI:org/10.1155/2014/825910.

[54] M.R. Servos, D.T. Bennie, B.K. Burnison, A. Jurkovic, R. Mclnnis, T. Neheli, Distribution of estrogens, 17beta-estradiol andestrone, in Canadian municipal wastewater treatment plants, Sci. Total Environ. 336 (2005) 155-170. https://doi.org/10.1016/j.scitotenv.2004.05.025

[55] T. Urase, T. Kikuta, Separate estimation of adsorption and degradation of pharmaceutical substances and estrogens in the activated sludge process, Water Res. 39 (2005) 1289-1300. https://doi.org/10.1016/j.watres.2005.01.015

[56] P. Westerhoff, Y. Yoon, S. Snyder, E. Wert, Fate of endocrine-disruptor, pharmaceutical, and personal care product chemicals during simulated drinking water treatment processes, Environ. Sci. Tech. 39 (2005) 6649-6663. https://doi.org/10.1021/es0484799

[57] N. Vieno, T. Tuhkanen, L. Kronberg, Removal of pharmaceuticals in drinking water treatment: effect of chemical coagulation, Environ. Tech. 27 (2006) 183-192. https://doi.org/10.1080/09593332708618632

[58] U. Szewzyk, R. Szewzyk, W. Manz, K.H. Schleifer, Microbiological Safety of Drinking Water, Ann. Rev. Microbiol. 54 (2000) 81-127. https://doi.org/10.1146/annurev.micro.54.1.81

[59] W. Zhang, F.A. Di Giano, Comparison of bacterial regrowth in distribution systems using free chlorine and chloramine: a statistical study of causative factors, Water Res. 36 (2002) 1469-1482. https://doi.org/10.1016/S0043-1354(01)00361-X

[60] I.H. Suffet, J. Ho, D. Chou, D. Khiari, J. Mallevialle, Taste-and- odor problems observed during drinking water treatment in: I.H. Suffet, J. Mallevialle, E. Kawczynski, (Eds.), Advances in Taste-and-odor Treatment and Control, American Water Works Association, 1995.

[61] R.M. Hozalski, L. Zhang, W.A. Arnold, Reduction of haloacetic acids by Fe0: implications for treatment and fate, Environ. Sci. Tech. 35 (2001) 2258-2263. https://doi.org/10.1021/es001785b

[62] R. Sadiq, M.J. Rodriguez, Disinfection by-products (DBPs) in drinking water and predictive models for their occurrence: a review, Sci. Total Environ. 321, (2004) 21-46. https://doi.org/10.1016/j.scitotenv.2003.05.001

[63] K. Niihara, New design concept of structural ceramicsceramic nanocomposites, J. Ceram. Mc. Jpn. 99 (1991) 974-982. https://doi.org/10.2109/jcersj.99.974

[64] L.L. Beecroft, C.K. Ober, Nanocomposite Materials for Optical Applications, Chem. Mat. J. 9 (1997) 1302-1317. https://doi.org/10.1021/cm960441a

[65] G. Cao, M.F. Gaurcia, M. Aleala, L.F. burgess, T.E. Mallouk, Chiral Molecular Recognition in Intercalated Zirconium Phosphate, J. Am. Chem. Soc. 114 (1992) 7574-7575. https://doi.org/10.1021/ja00045a046

[66] M.C. Roco, S. Williams, P. Alivisatos, IWGN Workshop Report, U.S. National Science and Technology Council, Washington (1999).

[67] G.Q. Lu, X.S. Zhao, Nanoporous Materials: Science and Engineering, World Scientific, Singapore (2004). https://doi.org/10.1142/p181

[68] M.A.H. Hyder, Nanotechnology and environment: potential application and environmental implications of nanotechnology, Technical University of Hamburg-Harburg, Germany (2003).

[69] A. Indarto, J.W. Choi, H. Lee, H.K. Song, Decomposition of greenhouse gases by plasma, Environ. Chem. Lett. 6 (2008) 215-222. https://doi.org/10.1007/s10311-008-0160-3

[70] Y.C. Sharma, V. Srivastav, V.K. Singh, S.N. Kaul, C.H. Weng,Nano-adsorbents for the removal of metallic pollutants from water and wastewater, Environ. Tech. 30 (2009) 583-609. https://doi.org/10.1080/09593330902838080

[71] Y.H. Li, J. Ding, Z. Luan, Competitive adsorption of Pb2+, Cu2+ and Cd2+ ions from aqueous solutions by multiwalled carbon nanotubes, Carbon 41 (2003) 2787-2792. https://doi.org/10.1016/S0008-6223(03)00392-0

[72] Z.C. Di, J. Ding, X.J. Peng, Y.H. Li, Z.K. Luan, J. Liang, Chromium adsorption by aligned carbon nanotubes supported ceria nanoparticles, Chemosphere 62 (2006) 861-865. https://doi.org/10.1016/j.chemosphere.2004.06.044

[73] C. Lu, H. Chiu, C. Liu, Removal of Zinc(II) from aqueous solution by purified carbon nanotubes: kinetics and equilibrium studies, Ind. Eng. Chem. Res. 45 (2006) 2850-2855. https://doi.org/10.1021/ie051206h

[74] X. Peng, Z. Luan, J. Ding, Z. Di, Y. Li, B. Tian, Ceria nanoparticles supported on carbon nanotubes for the removal of arsenate from water, Mater. Lett. 59 (2005) 399-403. https://doi.org/10.1016/j.matlet.2004.05.090

[75] D.M. du Plessis, Fabrication and characterization of anti-microbial and biofouling resistant nanofibers with silver nanoparticles and immobilized enzymes for application in water filtration, University of Stellenbosch, (2011).

[76] S.K. Smart, A.I. Cassady, G.Q. Lu, D.J. Martin,The biocompatibility of carbon nanotubes, Carbon 44 (2006) 1034-1047. https://doi.org/10.1016/j.carbon.2005.10.011

[77] M. Staiano, M. Baldassarre, M. Esposito, E. Apicella, R. Vitale, V. Aurilia, S.D. Auri, New trends in bio/nanotechnology: stable proteins as advanced moleculartools for health and environment, Environ. Tech. 31 (2010) 935-942. https://doi.org/10.1080/09593331003639575

[78] T. Masciangioli, W.X. Zhang, Environmental technologies at the nanoscale, Environ. Sci. Tech. 37 (2011) 102A-108A. https://doi.org/10.1021/es0323998

[79] N.R. Stradiotto, H. Yamanaka, M.V.B. Zanoni, Electrochemical sensors: A powerful tool in analytical chemistry, J. Braz. Chem. Soc. 14 (2003) 159. https://doi.org/10.1590/S0103-50532003000200003

[80] C.M.A. Brett, Electrochemical sensors for environmental monitoring. Strategy and examples, Pure Appl. Chem. 73 (2001) 1969-1977. https://doi.org/10.1351/pac200173121969

[81] R.W. Cattrall, Chemical Sensors, Oxford Chemistry Premiers, New York, 1997.

[82] U. Yogeswaran, S.M. Chen, A review on the electrochemical sensors and biosensors composed of nanowires as sensing material, Sensors 8 (2008) 290-313. https://doi.org/10.3390/s8010290

[83] C.C. Liu, Electrochemical Sensors, in: J.D. Bronzino (Ed.), The Biomedical Engineering Handbook, 2$^{nd}$ ed., CRC Press, LLC, 2000.

[84] H. Bai, G. Shi, Gas sensors based on conducting polymers, Sensors 7 (2007) 267-307. https://doi.org/10.3390/s7030267

[85] R.P. Buck, E. Lindner, Recommendations for nomenclature of Ion-selective electrodes, Pure Appl. Chem. 66 (1994) 2527-2536. https://doi.org/10.1351/pac199466122527

[86] W. Ostwald, Elektrische Eigenschaften Halbdurchlassiger Scheidewande., J. Phys. Chem. 6 (1890) 71-82.

[87] F.G. Donan, The theory of membrane equilibrium in presence of a non-dialyzable electrolyte, Z. Elektrochem. 17 (1911) 572-581.

[88] L. Michaelis, A. Fujita, Permselectivity of biological membranes, Biochem. Z. 148 (1925) 28-37.

[89] W. Juda, W.A. McRae, Coherent ion-exchange gels and membranes, J. Am. Chem. Soc. 72 (1950) 1044-1044. https://doi.org/10.1021/ja01158a528

[90] A.G. Winger, G.W. Bodamer, R. Kunin, Some electrochemical properties of new synthetic ion exchange membranes, J. Electrochem. Soc. 100 (1953) 178-184. https://doi.org/10.1149/1.2781103

[91] K. Sollner, in: T. Shedlobsy, (Ed.) Electrochemistry in Biology and Medicines, Wiley, New York, 1955, pp. 50.

[92] R. Schlogl, Mass Transport through Membrane, Willey-VCH, Weinheim, Germany 1964

[93] B.P. Nikoloskii, Acta Physiochim. URSS 7 (1937) 597.

[94] R.P. Buck, Electroanalytical chemistry of membranes, Crit. Rev. Anal. Chem. 5 (1976) 323-420. https://doi.org/10.1080/10408347608085583

[95] D. Amman, W. Morf, P. Anker, P. Meier, E. Pret, W. Simon, Neutral carrier based ion-selective electrodes, Ibid. 5 (1983) 3-92. https://doi.org/10.1016/b978-0-08-031492-1.50005-x

[96] A.K. Covington, Ion-selective electrodes, Crit. Rev. Anal. Chem. 3 (1973) 355-406. https://doi.org/10.1080/10408347308542665

[97] M.F. Wilson, E. Haikala, P. Kivalo, An evaluation of some sodium ion-selective glass electrodes in aqueous solution: Part I. Electrode calibration characteristics and selectivity with respect to hydrogen ions, Anal. Chim. Acta 74 (1975) 395. https://doi.org/10.1016/S0003-2670(01)95790-3

[98] A. Hulanicki, A. Lewenstam, Model for treatment of selectivity coefficients for solid-state ion-selective electrodes, Anal. Chem. 53 (1981) 1401-1405. https://doi.org/10.1021/ac00232a024

[99] D.J. Clarke, D.B. Kell, J.G. Morris, A. Burns, The role of ion-selective electrodes in microbial process control, Ion Sel. Electrode Rev. 4 (1982) 75-131.

[100] Y. Umezawa (Ed.); Handbook of ion selective electrodes: selectivity coefficients, CRC Press, Boca Raton, Florida, 1990.

[101] A.K. Jain, V.K. Gupta, L.P. Singh, P. Srivastava, J.R. Raisoni, Anion recognition through novel C-thiophenecalix[4]resorcinarene: PVC based sensor for chromate ions, Talanta 65 (2005) 716-721. https://doi.org/10.1016/j.talanta.2004.07.041

[102] R. Ioana, S.V. Staden, A.A. Ratko, Enantioselective, potentiometric membrane electrodes based on cyclodextrins: Application for the determination of R-baclofen in its pharmaceutical formulation, Talanta 69 (2006) 1049-1053. https://doi.org/10.1016/j.talanta.2005.12.022

[103] M. Kaniewska, T. Sikora, R. Kataky, M. Trojanowicz, Enantioselectivity of potentiometric sensors with application of different mechanisms of chiral discrimination, J. Biochem.Biophys.Meth. 70 (2008) 1261-1267. https://doi.org/10.1016/j.jbbm.2007.09.006

[104] A. Abbaspour, A.R. Esmaeilbeig, A.A Jarrahpour, B. Khajeh, R. Kia, Aluminium(III)-selective electrode based on a newly synthesized tetradentate Schiff base, Talanta 58 (2002) 397-403. https://doi.org/10.1016/S0039-9140(02)00290-4

[105] H.Y.Aboul-Enein, X.X. Sun, C.J. Sun, Ion selective PVC membrane electrode for the determination of methacycline hydrochloride in pharmaceutical formulation, Sensors 2 (2002) 424-431. https://doi.org/10.3390/s21000424

[106] S. Anastasova, A. Radu, G. Matzeu, C. Zuliani, U. Mattinen, J. Bobacka, D. Diamond, Disposable solid-contact ion-selective electrodes for environmental monitoring of lead with ppb limit-of-detection, Electrochim. Acta 73 (2012) 93-97. https://doi.org/10.1016/j.electacta.2011.10.089

[107] M. Guzinski, G. Lisak, J. Kupis, A. Jasinski, M. Bochenska, Lead(II)-selective ionophores for ion-selective electrodes: A review, Anal. Chim. Acta 791 (2013) 1-46. https://doi.org/10.1016/j.aca.2013.04.044

[108] B. Rezaei, S. Meghdadi, F. Zarandi, A fast response cadmium-selective polymeric membrane electrode based on N,N′-(4-methyl-1,2 phenylene)diquinoline-2-carboxamide as a new neutral carrier, J. Hazard. Mater. 153 (2008) 179-186. https://doi.org/10.1016/j.jhazmat.2007.08.033

[109] X.J. Yuan, R.Y. Wang, C.B. Mao, L. Wu, C.Q. Chu, R. Yao, Z.Y. Gao, B.L. Wu, H.Y. Zhang, New Pb(II)-selective membrane electrode based on a new Schiff base

complex, Inorg. Chem. Comm. 15 (2012) 29-32. https://doi.org/10.1016/j.inoche.2011.09.031

[110] H.M. Rong, R.X. Wu, L.X. Gui, Nitrogen-bearing organic compounds as carriers for lead ion-selective electrodes with excellent response, Chin. J. Anal. Chem. 36 (2008) 1735-1741. https://doi.org/10.1016/S1872-2040(09)60011-6

[111] K.K. Tiwari, M.C. Chattopadhyaya, Heterogeneous precipitate based Cu(II)-ionselective electrode and its application in the determination of stability constant of Cu(II) complex with 4-(2-pyridylazo)-resorcinol, Indian J. Chem. 40A (2001) 619-621.

[112] A.A. Khan, Inamuddin, M.M. Alam, Determination and separation of $Pb^{2+}$ from aqueous solutions using a fibrous type organic–inorganic hybrid cation-exchange material: Polypyrrole thorium(IV) phosphate, React. Funct.Polym. 63 (2005) 119-133. https://doi.org/10.1016/j.reactfunctpolym.2005.02.001

[113] A.P. Gupta, Renuka, Studies on araldite based zirconium tungstophosphate membrane-a lead(II) ion selective electrode, Indian J. Chem. 36A (1997) 1073.

[114] A.A. Khan, A. Khan, Inamuddin, Preparation and characterization of a new organic–inorganic nano-composite poly-o-toluidine Th(IV) phosphate: Its analytical applications as cation-exchanger and in making ion-selective electrode, Talanta. 72 (2007) 699–710. https://doi.org/10.1016/j.talanta.2006.11.044

[115] P.J. Parsons, W. Slavin, A rapid Zeeman graphite furnace atomic absorption spectrometric method for the determination of lead in blood, Spectrochim.Acta B Atom.Spectros. 48 (1993) 925-939. https://doi.org/10.1016/0584-8547(93)80094-B

[116] J.E. Tahan, V.A. Granadillo, R.A. Romero, Electrothermal atomic absorption spectrometric determination of Al, Cu, Fe, Pb, V and Zn in clinical samples and in certified environmental reference materials, Anal. Chim. Acta 295 (1994) 187-197. https://doi.org/10.1016/0003-2670(94)80350-1

[117] M.F. Mousavi, M.B. Barzegar, S. Sahari, A PVC-based capric acid membrane potentiometric sensor for lead(II) ions, Sensor. Actuator. B Chem. 73 (2001) 199-204. https://doi.org/10.1016/S0925-4005(00)00698-5

[118] H.W. Liu, S.J. Jiang, S.H. Liu, Determination of cadmium, mercury and lead in seawater by electro thermal vaporization isotope dilution inductively coupled plasma mass spectrometry, Spectrochim. Acta B Atom.Spectros. 54 (1999) 1367-1375. https://doi.org/10.1016/S0584-8547(99)00081-6

[119] R.J. Bowins, R.H. McNutt, Electrothermal isotope dilution inductively coupled plasma mass spectrometry method for the determination of sub-ng-mL$^{-1}$ levels of lead in human plasma, J. Anal. Atom.Spectrom. 9 (1994) 1233-1236. https://doi.org/10.1039/JA9940901233

[120] M.N. Bui, C.A. Li, K.N. Han, X.H. Pham, G.H. Seong, Electrochemical determination of cadmium and lead on pristine single-walled carbon nanotube electrodes, Anal. Sci. 28 (2012) 699-704. https://doi.org/10.2116/analsci.28.699

[121] F. Arduini, J.Q. Calvo, A. Amine, G. Palleschi, D. Moscone, Bismuth-modified electrodes for lead detection, Trends Anal. Chem. 29 (2010) 1295-1304. https://doi.org/10.1016/j.trac.2010.08.003

[122] V.L. Dressler, D. Pozebon, A.J. Curtius, Determination of heavy metals by inductively coupled plasma mass spectrometry after on-line separation and preconcentration, Spectrochim. Acta B Atom.Spectros. 53 (1998) 1527-1539. https://doi.org/10.1016/S0584-8547(98)00180-3

[123] Z. Lu, S. Yang, Q. Yang, S. Luo, C. Liu, Y. Tang, A glassy carbon electrode modified with graphene, gold nanoparticles and chitosan for ultrasensitive determination of lead(II), MicrochimicaActa 180 (2013) 555-562. https://doi.org/10.1007/s00604-013-0959-x

[124] S. Anastasova, A. Radu, G. Matzeu, C. Zuliani, U. Mattinen, J. Bobacka, D. Diamond, Disposable solid-contact ion-selective electrodes for environmental monitoring of lead with ppb limit-of-detection, Electrochim. Acta 73 (2012) 93-97. https://doi.org/10.1016/j.electacta.2011.10.089

[125] M. Guzinski, G. Lisak, J. Kupis, A. Jasinski, M. Bochenskaa, Lead(II)-selective ionophores for ion-selective electrodes: A review, Anal. Chim.Acta, 791 (2013) 1-46. https://doi.org/10.1016/j.aca.2013.04.044

[126] M.R. Huang, X.W. Rao, X.G. Li, Nitrogen-bearing organic compounds as carriers for lead ion-selective electrodes with excellent response, Chin. J. Anal. Chem. 36 (2008) 1735-1741. https://doi.org/10.1016/S1872-2040(09)60011-6

[127] B. Rezaei, S. Meghdadi, F. Zarandi, A fast response cadmium-selective polymeric membrane electrode based on N,N'-(4-methyl-1,2-phenylene) diquinoline-2 carboxamide as a new neutral carrier, J. Hazard. Mater. 153 (2008) 179-186. https://doi.org/10.1016/j.jhazmat.2007.08.033

[128] X.J. Yuan, R.Y. Wang, C.B. Mao, L. Wu, C.Q. Chu, R. Yao, Z.Y. Gao, B.L. Wu, H.Y. Zhang, New Pb(II)-selective membrane electrode based on a new Schiff base

complex, Inorg. Chem. Comm. 15 (2012) 29-32. https://doi.org/10.1016/j.inoche.2011.09.031

[129] K.K. Tiwari, M.C. Chattopadhyaya, Heterogeneous precipitate based Cu(II) ionselective electrode and its application in the determination of stability constant of Cu(II) complex with 4-(2-pyridylazo)-resorcinol, Indian J. Chem. 40A (2001) 619.

[130] A. Mohammad, Inamuddin, S. Hussain, Synthesis and physicochemical characterization of excellent thermally stable and mercury selective organic-inorganic composite cation exchanger polyvinyl alcohol thorium(IV) phosphate, Desalin. Water Treat. 57 (2016) 13795-13806. https://doi.org/10.1080/19443994.2015.1065768

[131] M.N. Abbas, E. Zahran, Novel solid-state cadmium ion-selective electrodes based on its tetraiodoand tetrabromo-ion pairs with cetylpyridinium, J. Electroanal. Chem. 576 (2005) 205-213. https://doi.org/10.1016/j.jelechem.2004.10.017

[132] V.K. Gupta, A.K. Jain, P. Kumar, PVC-based membranes of dicyclohexano-24-crown-8 as Cd(II) selective sensor, Electrochim. Acta 52 (2006) 736-741. https://doi.org/10.1016/j.electacta.2006.06.009

[133] S. Senthilkumar, A.J. King, S.M. Holmes, R.A.W. Dryfe, R. Saraswathi, Potentiometric sensing of heavy metal ions using a novel zeolite y membrane, Electroanalysis 18 (2006) 2297-2304. https://doi.org/10.1002/elan.200603661

[134] V.K. Gupta, A.K. Singh, B. Gupta, Schiff bases as cadmium(II) selective ionophores in polymeric membrane electrodes, Anal. Chim. Acta 583 (2007) 340-348. https://doi.org/10.1016/j.aca.2006.10.039

[135] S. Tascioglu, D. Tas, Surfactant effect on determination of $Cu^{2+}$ and $Cd^{2+}$ ions by ion-selective electrodes providing evidence for the discrepancy between the point of zero charge and the isoelectric point of CdS, Colloid. Surface. Physiochem. Eng. Aspect. 302 (2007) 349-353. https://doi.org/10.1016/j.colsurfa.2007.02.058

[136] V.K. Gupta, A.K. Jain, R. Ludwig, G. Maheshwari, Electroanalytical studies on cadmium(II) selective potentiometric sensors based on t-butyl thiacalix[4]areneand thiacalix[4]arene in poly(vinyl chloride), Electrochim. Acta 53 (2008) 2362-2368. https://doi.org/10.1016/j.electacta.2007.10.001

[137] B. Rezaei, S. Meghdadi, S. Bagherpour, Cadmium selective PVC-membranes sensor based on 1, 2-bis (quinoline-2-carboxamido)-4-chlorobenzene as a neutral

carrier, IEEE Sensor. J. 8 (2008) 1469-1477. https://doi.org/10.1109/JSEN.2008.920719

[138] B. Rezaei, S. Meghdadi, R.F. Zarandi, A fast response cadmium-selective polymeric membrane electrode based on N,N′-(4-methyl-1,2-phenylene) diquinoline-2-carboxamide as a new neutral carrier, J. Hazard. Mater. 153 (2008) 179-186. https://doi.org/10.1016/j.jhazmat.2007.08.033

[139] A.A. Ensafi, S. Meghdadi, S. Sedighi, Sensitive cadmium potentiometric sensor based on 4-hydroxy salophen as a fast tool for water samples analysis, Desalination 242 (2009) 336-345. https://doi.org/10.1016/j.desal.2008.06.002

[140] S. Tyagi, H. Agarwal, S. Ikram, A polyvinylchloride-based cadmium ion-selective electrode using [$Mo_2(OAc)_2(H_2$-calix[4]arene)] as an electroactive material, Water Sci. Tech. 62 (2010) 2510-2518. https://doi.org/10.2166/wst.2010.774

[141] A.A. Khan, A. Khan, Ion-exchange studies on poly-o-anisidineSn(IV) phosphate nano composite and its application as Cd(II) ion-selective membrane electrode, Cent. Eur. J. Chem. 8 (2010) 396-408.

[142] S. Yu, F. Li, W. Qin, An all-solid-state $Cd^{2+}$-selective electrode with a low detection limit, Sensor. Actuator. B Chem. 155 (2011) 919-922. https://doi.org/10.1016/j.snb.2011.01.052

[143] J. Singh, A.K. Singh, A.K. Jain, Fabrication of novel coated graphite electrodes for the selective nano-level determination of $Cd^{2+}$ ions in biological and environmental samples, Electrochim. Acta 56 (2011) 9095-9104. https://doi.org/10.1016/j.electacta.2011.06.106

[144] M. Shamsipur, S. Sahari, M. Payehghadr, K. Alizadeh, Flow injection potentiometric determination of $Cd^{2+}$ ions using a coated graphite plasticized pvc-membrane electrode based on 1,3-bis(2-cyanobenzene)triazene, ActaChim. Slov. 58 (2011) 555-562.

[145] U. Khamjumphol, S. Watchasit, C. Suksai, W. Janrungroatsakul, S. Boonchiangma, T. Tuntulani, W. Ngeontae, New polymeric membrane cadmium(II)-selective electrodes using tripodal amine based ionophores, Anal. Chim.Acta 704 (2011) 73-86. https://doi.org/10.1016/j.aca.2011.08.005

[146] A.A. Khan, T. Akhtar, Synthesis, characterization and analytical application of nano-composite cation-exchange material, poly-o-toluidine Ce(IV) phosphate: Its application in making Cd(II) ion selective membrane electrode, Solid State Sci. 13 (2011) 559-568. https://doi.org/10.1016/j.solidstatesciences.2010.12.026

[147] C. Wardak, A comparative study of cadmium ion-selective electrodes with solid and liquid inner contact, Electroanalysis 24 (2012) 85-90. https://doi.org/10.1002/elan.201100362

[148] J.L.W. Ling, A. Khan, B. Saad, S.A. Ghani, Electro polymerized 4-vinyl pyridine on 2B pencil graphite as ionophore for cadmium (II), Talanta 88 (2012) 477-483. https://doi.org/10.1016/j.talanta.2011.11.018

[149] S.W. Jing, W.X. Wei, D.J. Wang, Z. Jun, Z.R. Ming, Q. Wei, Electrochemical sensing system for determination of heavy metals in seawater, Chin. J. Anal. Chem. 40 (2012) 670-674. https://doi.org/10.1016/S1872-2040(11)60545-8

[150] A.K. Singh, A.K. Jain, A. Upadhyay, K.R.J. Thomas, P. Singh, Electroanalytical performance of Cd(II) selective sensor based on PVC membranes of 5,5′-(5,5′-(benzo[c][1,2,5]thiadiazole-4,7-diyl)-bis(thiophene-5,2-diyl)bis-(N1,N1,N3,N3-tetraphenylbenzene-1,3-diamine), Int. J. Environ. Anal. Chem. 93 (2013) 813-827. https://doi.org/10.1080/03067319.2011.649744

[151] S.A.R. Ivari, A. Darroudi, M.H.A. Zavar, G. Zohuri, N. Ashraf, Ion imprinted polymer based potentiometric sensor for the trace determination of cadmium (II) ions, Arabian J. Chem., 10 (2013)S864-S869. https://doi.org/10.1016/j.arabjc.2012.12.021

[152] H. Bagheri, A. Afkhami, A. Shirzadmehr, H. Khoshsafar, H. Khoshsafar, H. Ghaedi, Novel potentiometric sensor for the determination of $Cd^{2+}$ based on a new nano-composite, Int. J. Environ. Anal. Chem. 93 (2013) 578-591. https://doi.org/10.1080/03067319.2011.649741

[153] S. Chandra, Deepshikha, A. Sarkar, Synthesis, spectral characterization, thermal investigation and electrochemical evaluation of benzilbis(carbohydrazone) as Cd(II) ion selective electrode, Arabian J. Chem.10 (2013) S1306-S1315. https://doi.org/10.1016/j.arabjc.2013.03.015

[154] K. Khun, Z.H. Ibupoto, M. Willander, Urea assisted synthesis of flower like CuO nanostructures and their chemical sensing application for the determination of cadmium ions, Electroanalysis 25 (2013) 1425-1432. https://doi.org/10.1002/elan.201200660

[155] C. Dernane, A. Zazoua, I. Kazane, N.J. Renault, Cadmium-sensitive electrode based on tetracetone derivatives of p-tert-butylcalix[8]arene, Mater. Sci. Eng. C 33 (2013) 3638-3643. https://doi.org/10.1016/j.msec.2013.04.049

[156] A.A. Khan, S. Shaheen, Synthesis and characterization of a novel hybrid nano composite cation exchanger poly-o-toluidine Sn(IV) tungstate: Its analytical applications as ion-selective electrode, Solid State Sci. 16 (2013) 158-167. https://doi.org/10.1016/j.solidstatesciences.2012.10.024

[157] V.K. Gupta, S. Kumar, R. Singh, L.P. Singh, S.K. Shoora, B. Sethi, Cadmium (II) ion sensing through p-tert-butyl calix[6]arene based potentiometric sensor, J. Mol. Liq. 195 (2014) 65-68. https://doi.org/10.1016/j.molliq.2014.02.001

[158] A. Ghaemi, H. Tavakkoli, T. Mombeni, Fabrication of a highly selective cadmium (II) sensor based on 1,13-bis(8-quinolyl)-1,4,7,10,13-pentaoxatridecane as a supramolecularionophore, Mater. Sci. Eng. C 38 (2014) 186-191. https://doi.org/10.1016/j.msec.2014.02.006

[159] M. Naushad, Inamuddin, T.A. Rangreez, Potentiometric determination of Cd(II) ions using PVC-based polyaniline Sn(IV) silicate composite cation-exchanger ion-selective membrane electrode, Desalin. Water Treat. 55 (2014) 463-470. https://doi.org/10.1080/19443994.2014.915389

[160] M.K. Sahani, A.K. Singh, A.K. Jain, A. Upadhyay, A. Kumar, U.P. Singh, S. Narang, Fabrication of novel coated pyrolytic graphite electrodes for the selective nano-level monitoring of Cd2+ ions in biological and environmental samples using polymeric membrane of newly synthesized macrocycle, Anal. Chim. Acta 860 (2015) 51-60. https://doi.org/10.1016/j.aca.2014.12.023

[161] C. Wardak, Solid contact cadmium ion-selective electrode based on ionic liquid and carbon nanotubes, Sensor.Actuator. B Chem. 209 (2015) 131-137. https://doi.org/10.1016/j.snb.2014.11.107

[162] Inamuddin, T.A. Rangreez, A. Khan, Synthesis of single-walled carbon nanotubes cerium(IV) phosphate composite cation exchanger: ion exchange studies and its application as ion-selective membrane electrode for determination of Cd(II) ions, Polymer Compos. (2015) DOI: 10.1002/pc.

[163] T. Jeong, H.K. Lee, D.C. Jeong, S. Jeon, A lead (II) selective PVC membrane based on a Schiff base complex of N,N'-bis(salicylidene)-2,6-pyridineamine, Talanta 65 (2005) 543-548. https://doi.org/10.1016/j.talanta.2004.07.016

[164] M.M. Ardakani, M.K. Kashani, M.S. Niasari, A.A. Ensafi, Lead ion-selective electrode prepared by sol-gel and PVC membrane techniques, Sensor.Actuator. B Chem. 107 (2005) 438-445. https://doi.org/10.1016/j.snb.2004.10.036

[165] S. Li, W. Yang, M. Chen, J. Gao, J. Kang, Y. Qi, Preparation of PbO nanoparticles by microwave irradiation and their application to Pb(II)-selective electrode based on cellulose acetate, Mater. Chem. Phys. 90 (2005) 262-269. https://doi.org/10.1016/j.matchemphys.2004.02.022

[166] H.R. Zare, M.M. Ardakani, N. Nasirizadeh, J. Safari, Lead-selective poly(vinyl chloride) membrane electrode based on 1-phenyl-2-(2-quinolyl)-1, 2-dioxo-2-(4-bromo) phenylhydrazone, Bull. Kor. Chem. Soc. 26 (2005) 51-56. https://doi.org/10.5012/bkcs.2005.26.1.051

[167] A.A. Khan, Inamuddin, M.M. Alam, Preparation, characterization and analytical applications of a new and novel electrically conducting fibrous type polymeric-inorganic composite material: polypyrrole Th(IV) phosphate used as a cation-exchanger and Pb(II) ion-selective membrane electrode, Mater. Res. Bull. 40 (2005) 289-305. https://doi.org/10.1016/j.materresbull.2004.10.014

[168] L. Chen, J. Zhang, W. Zhao, X. He, Y. Liu, Double-armed calix[4]arene amide derivatives as ionophores for lead ion-selective electrodes, J. Electroanal. Chem. 589 (2006) 106-111. https://doi.org/10.1016/j.jelechem.2006.02.001

[169] M.R. Yaftian, S. Rayati, D. Emadi, D. Matt, A coated wire-type lead(ii) ion-selective electrode based on a phosphorylated calix[4]arene derivative, Anal. Sci. 22 (2006) 1075-1078. https://doi.org/10.2116/analsci.22.1075

[170] A.K. Jain, V.K. Gupta, L.P. Singh, J.R. Raisoni, A comparative study of Pb2+ selective sensors based on derivatizedtetrapyrazole and calix[4]arene receptors, Electrochim. Acta 51 (2006) 2547-2553. https://doi.org/10.1016/j.electacta.2005.07.040

[171] H. Kim, H.K. Lee, A.Y. Choi, S. Jeon, Bull. Kor. Chem. Soc. 28 (2007) 538-542. https://doi.org/10.5012/bkcs.2007.28.4.538

[172] S.A. Nabi, A.H. Shalla, S.A. Ganai, Sorption of metal ions on acrylamidezirconium (IV) arsenate and its synthesis of PVC based lead (II) selective electrode, Separ. Sci. Tech. 43 (2008) 164-178. https://doi.org/10.1080/01496390701764882

[173] S.Y. Kazemi, M. Shamsipur, H. Sharghi, Lead-selective poly(vinyl chloride) electrodes based on some synthesized benzo-substituted macrocyclicdiamides , J. Hazard. Mater. 172 (2009) 68-73. https://doi.org/10.1016/j.jhazmat.2009.06.145

[174] X.G. Li, X.L. Ma, M.R. Huang, Lead(II) ion-selective electrode based on polyaminoanthraquinone particles with intrinsic conductivity, Talanta 78 (2009) 498-505. https://doi.org/10.1016/j.talanta.2008.11.045

[175] A. Michalska, M. Wojciechowski, E. Bulska, J. Mieczkowski, K. Maksymiuk, Poly(n-butyl acrylate) based lead (II) selective electrodes, Talanta 79 (2009) 1247-1251. https://doi.org/10.1016/j.talanta.2009.05.028

[176] S.S. Bozkurt, S. Ayata, I. Kaynak, Fluorescence-based sensor for Pb(II) using tetra-(3-bromo-4-hydroxyphenyl)porphyrin in liquid and immobilized medium, Spectrochim. Acta Mol. Biomol.Spectros. 72 (2009) 880-883. https://doi.org/10.1016/j.saa.2008.12.012

[177] A.A. Ensafi, A. Katiraei Far, S. Meghdadi, Highly selective optical-sensing film for lead(II) determination in water samples, J. Hazard. Mater. 172 (2009) 1069-1075. https://doi.org/10.1016/j.jhazmat.2009.07.112

[178] A.A. Khan, U. Habiba, A. Khan, Synthesis and characterization of organic-inorganic nanocomposite poly-o-anisidine Sn(IV) arsenophosphate: Its analytical applications as Pb(II) ion-selective membrane electrode, Int. J. Anal.Chem. (2009) DOI:10.1155/2009/659215. https://doi.org/10.1155/2009/659215

[179] A.S. Alasier, S.A. Entisar, H.A. Elbishti, Lead-sensitive membrane electrode based on chelating ion exchanger, Arabian J. Chem. 3 (2010) 89-94. https://doi.org/10.1016/j.arabjc.2010.02.004

[180] D. Wilson, M.A. Arada, S. Alegret, M.D. Valle, Lead(II) ion selective electrodes with PVC membranes based on two bis-thioureas as ionophores: 1,3-bis(N′-benzoylthioureido)benzene and 1,3-bis(N-furoylthioureido)benzene J. Hazard. Mater. 181 (2010) 140-146. https://doi.org/10.1016/j.jhazmat.2010.04.107

[181] I. Ion, A. Culetu, J. Costa, C. Luca, A.C. Ion, Polyvinyl chloride-based membranes of 3,7,11-tris (2-pyridylmethyl)-3,7,11,17-tetraazabicyclo [11.3.1] heptadeca-1(17),13,15-triene as a Pb(II)-selective sensor, Desalination 259 (2010) 38-43. https://doi.org/10.1016/j.desal.2010.04.038

[182] A.A. Zamani, N. Khorsidhi, Z. Mofidi, M.R. Yatifan, Crown ethers bearing 18c6 unit; sensory molecules for fabricating PVC membrane lead ion-selective electrodes, J. Chin. Chem. Soc. 58 (2011) 673-680. https://doi.org/10.1002/jccs.201190105

[183] M. Shamsipur, M. Sadeghi, K. Alizadeh, A. Bencini, B. Valtancoli, A. Garau, V. Lippolis, Novel fluorimetric bulk optode membrane based on 5,8-bis((5′-chloro-8′-

hydroxy-7′-quinolinyl)methyl)-2,11-dithia-5,8-diaza-2,6-pyridinophane for selective detection of lead(II) ions , Talanta 80 (2010) 2023-2033. https://doi.org/10.1016/j.talanta.2009.11.011

[184] S. Yu, F. Li, T. Yin, Y. Liu, D. Pan, W. Qin, A solid-contact Pb2+-selective electrode using poly(2-methoxy-5-(2′-ethylhexyloxy)-p-phenylenevinylene) as ion-to-electron transducer, Anal. Chim. Acta 702 (2011) 195-198. https://doi.org/10.1016/j.aca.2011.06.049

[185] J. Guo, Y. Chai, R. Yuan, Z. Song, Z. Zou, Lead (II) carbon paste electrode based on derivatized multi-walled carbon nanotubes: Application to lead content determination in environmental samples, Sensor. Actuator. B Chem. 155 (2011) 639-645. https://doi.org/10.1016/j.snb.2011.01.023

[186] M.R. Huang, X.W. Rao, X.G. Li, Y.B. Ding, Lead ion-selective electrodes based on polyphenylenediamine as unique solid ionophores, Talanta 85 (2011) 1575-1584. https://doi.org/10.1016/j.talanta.2011.06.049

[187] C. Wardak, A highly selective lead-sensitive electrode with solid contact based on ionic liquid, J. Hazard. Mater. 186 (2011)1131-1135. https://doi.org/10.1016/j.jhazmat.2010.11.103

[188] M.H. Mashhadizadeh, H. Khani, A. Shockravi, M. Sadeghpour, Determination of ultratrace levels of lead (II) in water samples using a modified carbon paste electrode based on a new podand, Mater. Sci. Eng. C 31 (2011) 1674-1680. https://doi.org/10.1016/j.msec.2011.07.021

[189] N. Aksuner, Development of a new fluorescent sensor based on a triazolo-thiadiazin derivative immobilized in polyvinyl chloride membrane for sensitive detection of lead(II) ions, Sensor. Actuator. B Chem. 157 (2011) 162-168. https://doi.org/10.1016/j.snb.2011.03.044

[190] X.J. Yuan, R.Y. Wang, C.B. Mao, L. Wu, C.Q. Chu, R. Yao, Z.Y. Gao, B.L. Wu, H.Y. Zhang, New Pb(II)-selective membrane electrode based on a new Schiff base complex, Inorg. Chem. Comm. 15 (2012) 29-32. https://doi.org/10.1016/j.inoche.2011.09.031

[191] S.M. Saadeh, H.M.A. Shawish, H.M. Dalloul, N.M. EL-Halabi, B.K. Daher, Lead(II) complexes with some SNO and ONO tridentate Schiff base ligands and their evaluation as lead(II) sensors , Mater. Sci. Eng. C 32 (2012) 619-624. https://doi.org/10.1016/j.msec.2012.01.001

[192] A. Soleymanpour, B. Shafaatian, K. Kor, A.R. Hasaninejad, Coated wire lead (II)-selective electrode based on a Schiff base ionophore for low concentration measurements, Monatsh. Chem. 143 (2012) 181-188. https://doi.org/10.1007/s00706-011-0634-z

[193] A.A. Khan, U. Baig, Electrically conductive membrane of polyaniline-titanium(IV)phosphate cation exchange nanocomposite: Applicable for detection of Pb(II) using its ion-selective electrode, J. Ind. Eng. Chem. 18 (2012) 1937-1944. https://doi.org/10.1016/j.jiec.2012.05.008

[194] R. Bushra, M. Shahadat, M.A. Khan, Inamuddin, R. Adnan, M. Rafatullah, Optimization of polyaniline supported Ti(IV) arsenophosphate composite cation exchanger based ion-selective membrane electrode for the determination of lead, Ind. Eng. Chem. Res. 53 (2014) 19387-19391. https://doi.org/10.1021/ie5034655

[195] A.A. Khan, L. Paquiza, Synthesis and characterization of in situ polymerized poly(methyl methacrylate)-cerium molybdate nanocomposite for electroanalytical application, J. Appl. Polymer Sci. 127 (2013) 3737- 3748. https://doi.org/10.1002/app.37672

[196] A. Khan, A.M. Asiri, M.A. Rub, N. Azum, A.A.P. Khan, S.B. Khan, M.M. Rahman, I. Khan, Synthesis, characterization of silver nanoparticle embedded polyanilinetungstophosphate-nanocompositecation exchanger and its application for heavy metal selective membrane, Compos. B Eng. 45 (2013) 1486-1492. https://doi.org/10.1016/j.compositesb.2012.09.023

[197] S. Singh, G. Rani, G. Singh, H. Agarwal, Comparative study of lead(II) selective poly(vinyl chloride) membrane electrodes based on podand derivatives as ionophores, Electroanalysis 25 (2013) 475-485. https://doi.org/10.1002/elan.201200404

[198] P. Heidari, S. Jalali, Z. Mofidi, A. Zamani, M.R. Yaftian, Preparation of a lead ion-selective electrode based upon crown ether nitrobenzo18-crown-6, Anal. Bioanal. Electrochem. 5 (2013) 305-315.

[199] A.A. Khan, S. Shaheen, Synthesis and characterization of a novel hybrid nano composite cation exchanger poly-o-toluidine Sn(IV) tungstate: Its analytical applications as ion-selective electrode, Solid State Sci. 16 (2013) 158-167. https://doi.org/10.1016/j.solidstatesciences.2012.10.024

[200] A.A. Khan, S. Shaheen, Preparations and characterizations of poly-o-toluidine/multiwalled carbon nanotubes/Sn(IV) tungstate composite ion exchange

thin films and their application as a Pb(II) selective electrode, RCS Adv. 4 (2014) 23456-23463. https://doi.org/10.1039/c4ra01594k

[201] O. Sayar, N.A. Torbati, H. Saravani, K. Mehrani, A. Behbahani, H. Reza, M. Zadeh, A novel magnetic ion imprinted polymer for selective adsorption of trace amounts of lead(II) ions in environment samples, J. Ind. Eng. Chem. 20 (2014) 2657-2662. https://doi.org/10.1016/j.jiec.2013.10.052

[202] Inamuddin, M. Naushad, T.A. Rangreez, Z.A. Al-Othman, Ion-selective potentiometric determination of Pb(II) ions using PVC-based carboxymethyl cellulose Sn(IV) phosphate composite membrane electrode, Desalin. Water Treat. 56 (2015) 806-813. https://doi.org/10.1080/19443994.2014.941307

[203] A. Jasinski, M. Guzinski, G. Lisak, J. Bobacka, M. Bochenska, Solid-contact lead(II) ion-selective electrodes for potentiometric determination of lead(II) in presence of high concentrations of Na(I), Cu(II), Cd(II), Zn(II), Ca(II) and Mg(II), Sensor. Actuator. B Chem. 218 (2015) 25-30. https://doi.org/10.1016/j.snb.2015.04.089

[204] K. Zargoosh, F.F. Babadi, Highly selective and sensitive optical sensor for determination of Pb2+ and Hg2+ ions based on the covalent immobilization of dithizone on agarose membrane, Spectrochim. Acta Mol. Biomol. Spectros. 137 (2015) 105-110. https://doi.org/10.1016/j.saa.2014.08.043

[205] Inamuddin, T.A. Rangreez, M. Naushad, A.A. Ahmad, Synthesis and characterisation of poly(3,4-ethylenedioxythiophene)-poly(styrenesulfonate) (PEDOT:PSS) Zr(IV) monothiophosphate composite cation exchanger: analytical application as lead ion selective membrane electrode, Int. J. Environ. Anal. Chem. 95 (2015) 312-323. https://doi.org/10.1080/03067319.2015.1016016

[206] M. Rashid, F. Khan, Lutfullah, R. Wahab, Zirconium(IV) phosphosulphosalicylate-based ion selective membrane electrode for potentiometric determination of Pb(II) ions, Arabian J. Chem. (2015), DOI: org/10.1016/j.arabjc.2014.12.013.

[207] A.A. Khan, Inamuddin, Preparation, physico-chemical characterization, analytical applications and electrical conductivity measurement studies of an 'organic–inorganic' composite cation-exchanger: PolyanilineSn(IV) phosphate, React. Funct.Polym. 66 (2006) 1649-1663. https://doi.org/10.1016/j.reactfunctpolym.2006.06.007

[208] W.A. Siddiqui, S.A. Khan, Inamuddin, Synthesis, characterization and ion-exchange properties of a new and novel 'organic–inorganic' hybrid cation-

exchanger: Poly(methyl methacrylate) Zr(IV) phosphate, Colloid. Surface. Physicochem. Eng. Aspect. 295 (2007) 193-199. https://doi.org/10.1016/j.colsurfa.2006.08.053

[209] M. Islam, R. Patel, Polyacrylamide thorium (IV) phosphate as an importantlead selective fibrousion exchanger: Synthesis, characterization and removal study, J. Hazard.Mater. 156 (2008) 509-520. https://doi.org/10.1016/j.jhazmat.2007.12.046

[210] S.A. Nabi, A.S. Ganai, A.H. Shalla, New organic-inorganic type acrylamide aluminumtungstate: preparation, characterization and analytical applications as a cation exchange material, Separ. Sci. Tech. 43 (2008) 3695-3711. https://doi.org/10.1080/01496390802212674

[211] A.A. Khan, U. Habiba, A. Khan, Synthesis and characterization of organic-inorganic nanocomposite poly-o-anisidinesn(IV) arsenophosphate: its analytical applications as Pb(II) ion-selective membrane electrode, Int. J. Anal. Chem. (2009) Article, ID659215, 10.

[212] S.A. Nabi, M. Naushad, R. Bushra, Synthesis and characterization of a new organic–inorganic Pb2+ selective composite cation exchanger acrylonitrile stannic(IV) tungstate and its analytical applications, Chem. Eng. J. 152 (2009) 80-87. https://doi.org/10.1016/j.cej.2009.03.033

[213] U. Ulusoy, R. Akkaya, Adsorptive features of polyacrylamide-apatite composite for Pb2+, UO22+ and Th4+, J. Hazard. Mater. 163 (2009) 98-108. https://doi.org/10.1016/j.jhazmat.2008.06.064

[214] M. Shamsipur, A.S. Dezaki, M. Akhond, H. Sharghi, Z. Paziraee, K. Alizadeh, Novel PVC-membrane potentiometric sensors based on a recently synthesized sulfur-containing macrocyclicdiamide for Cd2+ ion. Application to flow-injection potentiometry, J. Hazard.Mater. 172 (2009) 566-573. https://doi.org/10.1016/j.jhazmat.2009.07.003

[215] O.M. Kalfa, O. Yalcinkaya, A.R. Turker, Synthesis of nano B2O3/TiO2 composite material and its application to preconcentration and separation of cadmium, J. Hazard. Mater. 166 (2009) 455-461. https://doi.org/10.1016/j.jhazmat.2008.11.112

[216] S.A. Nabi, A.H. Shalla, EDTA-stannic (IV) iodate: preparation, characterization and its analytical applications for metal content determination in real and synthetic samples, J. Porous Mater., 16 (2009) 587-597. https://doi.org/10.1007/s10934-008-9236-5

[217] S.A. Nabi, A.H. Shalla, Synthesis, characterization and analytical application of hybrid; Acrylamide zirconium (IV) arsenate a cation exchanger, effect of dielectric constant on distribution coefficient of metal ions, J. Hazard. Mater. 163 (2009) 657-664. https://doi.org/10.1016/j.jhazmat.2008.07.011

[218] A.M. Khan, S.A. Ganai, S.A. Nabi, Synthesis of a crystalline organic-inorganic composite exchanger, acrylamide stannic silicomolybdate: Binary and quantitative separation of metal ions, Colloid. Surface. Physicochem. Eng. Aspect. 337 (2009) 141-145. https://doi.org/10.1016/j.colsurfa.2008.12.012

[219] Z. Alam, Inamuddin, S.A. Nabi, Synthesis and characterization of a thermally stable strongly acidic Cd(II) ion selective composite cation-exchanger: PolyanilineCe(IV) molybdate, Desalination 250 (2010) 515-522. https://doi.org/10.1016/j.desal.2008.09.008

[220] Inamuddin, Y.A. Ismail, Synthesis and characterization of electrically conducting poly-o-methoxyanilineZr(1V) molybdate Cd(II) selective composite cation-exchanger, Desalination 250 (2010) 523-529. https://doi.org/10.1016/j.desal.2008.06.033

[221] S.A. Nabi, R. Bushra, M. Naushad, A.M. Khan, Synthesis, characterization and analytical applications of a new composite cation exchange material poly-o-toluidine stannic molybdate for the separation of toxic metal ions, Chem. Eng. J. 165 (2010) 529-536. https://doi.org/10.1016/j.cej.2010.09.064

[222] Z.A. Al Othman, M. Naushad, A. Nilchi, Development, characterization and ion exchange thermodynamics for a new crystalline composite cation exchange material: application for the removal of Pb2+ ion from a standard sample (rompin hematite), J. Inorg. Organomet. Polymer Mater. 21 (2011) 547-559. https://doi.org/10.1007/s10904-011-9491-9

[223] S.A. Nabi, S.A. Ganai, A.M. Khan, Synthesis, characterization and ion exchange behavior of polyaniline stannic silicomolybdate, an organic–inorganic composite material: Quantitative separation of Pb2+ ions from industrial effluents, J. Inorg. Organomet.Polym. 21 (2011) 25-35. https://doi.org/10.1007/s10904-010-9423-0

[224] A.A. Khan, L. Paquiza, Characterization and ion-exchange behavior of thermally stable nano-composite polyaniline zirconium titanium phosphate: Its analytical application in separation of toxic metals, Desalination 265 (2011) 242-254. https://doi.org/10.1016/j.desal.2010.07.058

[225] M. Shahadat, A.H. Shalla, A.S. Raeissi, Synthesis, characterization, and sorption behavior of a novel composite cation exchange adsorbent, Ind. Eng. Chem. Res. 51 (2012) 15525–15529. https://doi.org/10.1021/ie3014555

[226] S.A. Nabi, R. Bushra, M. Shahadat, Removal of toxic metal ions by using compositecation-exchange material, J. Appl. Polymer Sci. 125(2012) 3438-3446. https://doi.org/10.1002/app.36325

[227] R. Bushra, M. Shahadat, A.S. Raeissi, S.A. Nabi, Development of nano-composite adsorbent for removal of heavy metals from industrial effluent and synthetic mixtures; its conducting behavior , Desalination 289 (2012)1-11. https://doi.org/10.1016/j.desal.2011.12.013

[228] M. Shahadat, S.A. Nabi, R. Bushra, A.S. Raeissi, K. Umar, M.O. Ansari,Synthesis, characterization, photolytic degradation, electrical conductivity and applications of a nanocomposite adsorbent for the treatment of pollutants, RSC Adv. 2 (2012) 7207-7220. https://doi.org/10.1039/c2ra20589k

[229] M.M. Alam, Z.A. Alothman, M. Naushad, Analytical and environmental applications of polyaniline Sn(IV) tungstoarsenate and polypyrrolepolyantimonic acid composite cation-exchangers, J. Ind. Eng. Chem. 19 (2013) 1973-1980. https://doi.org/10.1016/j.jiec.2013.03.006

[230] M. Naushad, Z.A. AL-Othman, M. Islam, Adsorption of cadmium ion using a new composite cation-exchanger polyaniline Sn(IV) silicate: kinetics, thermodynamic and isotherm studies, Int. J. Environ. Sci. Technol. 10 (2013) 567-578. https://doi.org/10.1007/s13762-013-0189-0

[231] M.M.A. Khan, Rafiuddin, Inamuddin, PVC based polyvinyl alcohol zinc oxide composite membrane: Synthesis and electrochemical characterization for heavy metal ions, J. Ind. Eng. Chem. 19 (2013) 1365-1370. https://doi.org/10.1016/j.jiec.2012.12.041

[232] R. Bushra, Mu.Naushad, R. Adnan, M.N.M. Ibrahim, M. Rafatullah, J. Ind. Eng. Chem., (2014), DOI: org/10.1016/j.jiec.2014.05.022.

[233] R. Bushra, M. Shahadat, A. Ahmad, S.A. Nabi, K.Umar, M. Oves, A.S. Raeissi, M. Muneer, Synthesis, characterization, antimicrobial activity and applications of polyanilineTi (IV) arsenophosphate adsorbent for the analysis of organic and inorganic pollutants, J. Hazard. Mater. 264 (2014) 481-489. https://doi.org/10.1016/j.jhazmat.2013.09.044

[234] M.A. Khan, A. Ahmad, K. Umar, S.A. Nabi, Synthesis, characterization, and biological applications of nanocomposites for the removal of heavy metals and dyes, Ind. Eng. Chem. Res. 54 (2015) 76-82. https://doi.org/10.1021/ie504148k

[235] A. Mohammad, Inamuddin, S. Hussain, Poly (3,4-ethylenedioxythiophene): polystyrene sulfonate (PEDOT:PSS) Zr(IV) phosphate composite cation exchanger : sol-gel synthesis and physicochemical characterization, Ionics 21 (2014) 1063-1071. https://doi.org/10.1007/s11581-014-1247-4

[236] M. Shahadat, R. Bushra, Synthesis, characterization and significant applications of PANI-Zr(IV)sulphosalicylate nanocomposite, Advances in nanotechnology 6 (2015) 181-206.

[237] T.A. Rangreez, Inamuddin, M. Naushad, H. Ali, Synthesis and characterisation of poly(3,4-ethylenedioxythiophene)-poly(styrenesulfonate) (PEDOT:PSS) Zr(IV) monothiophosphate composite cation exchanger: analytical application in the selective separation of lead metal ions, Int. J. Environ. Anal. Chem. 95 (2015) 556-568. https://doi.org/10.1080/03067319.2015.1036863

[238] T.A. Khan, Inamuddin, M. Naushad, Heavy metal ion-exchange kinetic studies over cellulose acetate Zr(IV) molybdophosphate composite cation-exchanger, Desalin. Water Treat. 53 (2015) 1675-1682. https://doi.org/10.1080/19443994.2013.838527

# Chapter 4

# A study on the potential applications of rice husk derivatives as useful adsorptive material

Mohammad Kashif Uddin[1]* and P. Fazul Rahaman[2]

[1]Basic Engineering Sciences, College of Engineering, Majmaah University, Al-Majmaah 11952, Saudi Arabia

[2]Department of Zoology, Maulana Azad National Urdu University, Gachibowli, Hyderabad - 500 032, India

mohdkashifchem@gmail.com, m.kashifuddin@mu.edu.sa

**Abstract**

Rice husk is one of the by-products of rice production, left after the burning of rice husk. It can cause environmental pollution, as its disposal is difficult. Hence its proper reuse is necessary, and because it is mainly composed of carbon and silica, it could be used in adsorption processes for removal of toxic heavy metals from water and wastewaters. Rice husk is available in ample amount. Because of its high specific surface area, it has proven to be a potential low-cost material in the applications of water treatment and building materials. This literature reviews the properties, uses, and the importance of rice husk and provides an effective collection of studies to utilize rice husk derivatives. This economically valuable agriculture waste product is a great source of silica and has many comprehensive applications.

**Keywords**

Rice Husk, Low Cost Material, Applications, Adsorption, Adsorption Capacity

## Contents

## 1. Introduction

Natural water, which has miscellaneous uses, is one of the most precious gift for the living beings to survive on earth. Any human or industrial activity that negatively affects water quality is called *water pollution.* Water pollution is a critical global problem, which needs to be evaluated constantly, and for this reason, it is important to revise water resource policies regularly at all levels. Water resources are becoming highly polluted mainly because of the worldwide expansion of industrialization, which has created huge environmental problems and spreading toxicity in living beings. Highly saline wastewater is very common nowadays because of the rapid growth of various industries like textile, coal mining, leather, chemical, pharmaceutical, petroleum, coal, machine and steel building, food processing etc. Clean freshwater is increasingly difficult to obtain and thus recycling of the wastewater is one of the treatment solutions. Many countries are spending billions of dollars on the treatment and desalination technologies to convert wastewater into drinking water.

There are many hazardous pollutants and materials, which get mixed in the water and create water problems. It is assumed that water pollution is the main cause of worldwide diseases. The presence of heavy metal ions in both surface and groundwater is a major concern because they are not easily decomposable and due to their non-biodegradable nature, these are considered as main environmental hazards. Excessive amounts of many heavy metals may cause severe toxic effects on human health. When these metal ions enter into the water bodies, they become harmful to the living organisms present in water. Heavy metals may also accumulate throughout the food chain and their specific chemical form is potentially toxic to affect the human health. The problem is the unregulated discharge of many organic and inorganic ions in municipal and industrial wastewater. Due to their severe toxicity, Environmental Protection Agency (EPA) the USA and, World Health Organisation (WHO) have set the tolerance limits of various heavy metal ions into the water as listed in Table 1 [1].

***Table 1*** *Permissible limits of various toxic heavy metals in drinking water according to various standards.*

| | **Permissible limit for drinking water (mg/L)** | | |
|---|---|---|---|
| **Metal** | **IS 10500** | **WHO** | **USEPA** |
| Nickel | 0.020 | 0.020 | 0.100 |
| Lead | 0.010 | 0.010 | 0.015 |
| Zinc | 5.000 | 3.000 | 5.000 |
| Copper | 0.050 | 2.000 | 1.300 |
| Cadmium | 0.003 | 0.003 | 0.005 |
| Mercury | 0.001 | 0.001 | 0.002 |
| Arsenic | 0.010 | 0.010 | 0.010 |
| Chromium | 0.050 | 0.050 | 0.100 |
| Manganese | 0.100 | 0.500 | 0.050 |
| Iron | 0.300 | 0.200 | 0.300 |

From many years up to now, the removal of various heavy metals from contaminated wastewater has been one of the most important research topics. Many effective conventional separation technologies methods including precipitation, separation, chemical oxidation, reverse osmosis, electrochemical treatment, emulsion, ultrafiltration, photocatalysis, ion exchange, preconcentration, evaporation, sedimentation, adsorption etc. [2] have been developed to remove heavy metals from aqueous solution .Out of these, adsorption has been widely used and is a most efficient method to eliminate heavy metals from contaminated water. This technique is superior because of its low cost, ease of operation, efficiency in treatment, high capacity, reliability, less energy consumption, and simplicity [3,4,5]. Therefore, there is an increasing interest among researchers to prepare, search and utilize an inexpensive, efficient, easily available, environmentally friendly material to serve as an adsorbent for the treatment of wastewater. The removal of heavy metals using agricultural waste materials has been extensively explored during the last decade. Natural substances, such as agricultural waste [3], pottery glaze [4], leechi fruit peel waste [8], rice straw [5], rice husk ash [6], rice hulls [7], alfalfa plants [8], green adsorbent [9], coriander seed powder [10], seeds of artimisia absinthium [11], onosmabracteatum [12], spent coffee grounds [13], P. eldarica leaves [14], activated alumina, fly ash, saw dust [15], peanut shell [16], spent mushroom [17], rice wine processing waste sludge [18], pyrolyzed olive pomace [19], coal fly ash [20], modified corn stalk [21], phoenix tree [22], defatted papaya seed [23], tea-industry waste activated carbon [24], sawdust of Jackwood [25], natural limestones [26], coconut coir and derived chars [27], Pleurotusostreatus [28], modified walnut shells [29], peat [30], have been reported to be useful in the removal of heavy metal ions from aqueous system.

This article compiles information on rice husk and highlights its potential as an effective adsorbent to remove various heavy metals, by collecting the literary research works carried out during the last 10 years (2006-2016). It is aimed to provide the knowledge of the excellent adsorptive potential of rice husk derivatives, which can be an economic utilization and superior recycle of agriculture waste to replace expensive commercial materials.

## 2. Rice husk and its derivatives

Rice, is a major source of food in the various countries of the world. Rice husk is the hard outer coating on grains of rice which protects the seed during the growing period. Around 20% of the total paddy weight is a husk, which is made of opaline silica and lignin [35].

The worldwide annual rice husk output is about 80 million tons and over 97% of the husk is generated in the developing countries [36,37]. Rice husk is an abundant agricultural waste that is generated at 545 million tons annually, which accounts about one-fifth of the annual gross rice production of the world [2,38].

*Rice Husk,* can be recycled to produce high value eco-materials, such as silicon (Si) [39], silica ($SiO_2$) [40-41], silicon carbide (SiC) [42], silicon nitride ($Si_3N_4$) and graphene (G) [43]. The chemical composition of raw rice husk has been reported to contain both organic (74%) and inorganic constituents (26%) [44]. The organic constituents include cellulose, hemicellulose, lignin, L-arabinose, methylglucuronic acid, D-galactose, proteins, and vitamins that can be removed from rich husk during the burning process [45]. The major inorganic component is $SiO_2$ (80%), and the minor inorganic constituents include alumina (3.93%), sulfur trioxide (0.78%), iron oxide (0.41%), calcium oxide (3.84%), magnesium oxide (0.25%), sodium oxide (0.67%) and, potassium oxide (1.45%) [44].

Rice husk ash (RHA) is formed by the burning or combustion process of Rice hull (RH). Rice husk ash is the by-product, which also contains some amount of carbon. Rice husk ash can be used as a soil amendment, and as an additive in cement and steel [46]. Biochar can also be produced by thermal decomposition of the rice husk, under a limited supply of oxygen ($O_2$) and at relatively low temperatures (less than 700°C). The carbonized rice husk is formed by incomplete burning and can be used as a soil amendment and, activated carbon [46] etc. However, it may cause a serious environmental and human health related problems, due to its high disposal in landfills. This problem is even worse due to its low bulk density. The typical **chemical composition and physical properties** of RHA are listed in Table 2. It can be observed from the Table that silica ($SiO_2$) is the major component of RHA [47-62].

***Table 2*** *Chemical and physical properties of rice husk ash.*

| $SiO_2$ (%) | $Al_2O_3$ (%) | $Fe_2O_3$ (%) | CaO (%) | MgO (%) | $K_2O$ (%) | $Na_2O$ (%) | LOI (%) | Specific gravity ($g/cm^3$) | References |
|---|---|---|---|---|---|---|---|---|---|
| 0.10-1.23 | - | - | - | 0.01-1.96 | 0.10-2.54 | 0.01-1.58 | 5.90 | - | [47] |
| 87.86 | 0.68 | 0.93 | 1.30 | 0.35 | 2.37 | 0.12 | - | - | [48] |
| 92.40 | 0.30 | 0.40 | 0.70 | 0.30 | 2.54 | 0.07 | 2.31 | 2.10 | [49] |
| 94.60 | 0.30 | 0.30 | 0.40 | 0.30 | 1.30 | 0.20 | 1.80 | 2.05 | [50] |
| 91.71 | 0.36 | 0.90 | 0.86 | 0.31 | 1.67 | 0.12 | 3.13 | - | [51] |
| 86.98 | 0.84 | 0.73 | 1.40 | 0.57 | - | 2.46 | 5.14 | 2.10 | [52] |
| 87.30 | 0.15 | 0.16 | 0.55 | 0.35 | 3.68 | 1.12 | 8.55 | 2.06 | [53] |
| 90.70 | 0.40 | 0.40 | 0.40 | 0.50 | 2.20 | 0.10 | 4.80 | 2.05 | [54] |
| 94.50 | 0.21 | 0.62 | 0.48 | 0.23 | - | - | - | - | [55] |
| 87.40 | 0.40 | 0.30 | 0.90 | 0.60 | 3.39 | 0.04 | - | - | [56] |
| 90.00 | 0.28 | 0.14 | 0.45 | 0.28 | 1.55 | 0.08 | - | - | [57] |
| 95.79 | 0.11 | 0.12 | 1.06 | 0.33 | 2.17 | 0.30 | - | - | [58] |
| 94.95 | 0.39 | 0.15 | 0.43 | 0.70 | 0.72 | 0.50 | 24.30 | - | [59] |
| 89.91 | 0.13 | 0.95 | 0.76 | 0.30 | 2.75 | 0.01 | 2.99 | 2.27 | [60] |
| 98.61 | 0.10 | 0.21 | 0.21 | 0.10 | 0.16 | - | - | - | [61] |
| 89.00 | 0.21 | - | 0.68 | 0.86 | 2.60 | - | 4.20 | - | [62] |

Rice husk also contains black particles, which represent carbon with incomplete combustion [63]. Liu [64] had concluded that RHA had three layered structures: inner, outer, and interface with interstitial and honeycombed pores, which were, in fact, the major cause for the high chemical activity and large specific surface area of RHA, which meant that RHA had a certain reaction activity.

## 3. Characteristics and uses of rice husk ash

RHA can be obtained from burning of rice husk in concrete and uncontrolled combustion. It is also clear from the Table 2, that RHA contains the highest content of silica, and is commonly regarded as a pozzolanic material. Pozzolans are those silica and aluminous based materials which can finely disperse, and react with calcium hydroxide when coming in contact with water to form hydrated cement [65]. RHA is a cheap alternative which is used to improve the property of cementitious products [66-68]. A lot of research work has been conducted on the use of RHA as cementitious material [69-81]. From recent publications, it can be concluded that rice husk ash has been used to reduce the cost of expensive construction materials. However continuous research to further improve the pozzolanic activity of RHA is still needed [64].

High-value applications of RHA include manufacturing of silica gels, silicon chip, synthesis of activated carbon and silica, production of lightweight construction materials and insulation, catalysts, zeolites, ingredients for lithium ion batteries, graphene, energy storage/capacitor, carbon capture, and in drug delivery vehicles were also conducted. Husk has a high calorific value and therefore can also be used as a renewable fuel [82].

On the other hand, rice husk is used as a raw material for the production of organic substances, such as xylitol, furfural, ethanol and acetic acid, and used as a cleaning or polishing agent in the metal/machine industry [83-84].

RHA is also been mixed in cement as stabilizer to improve the properties of residual soil [85]. RHA has also been used in controlling insect pests in soya beans [86].

## 4. Rice husk and rice husk ash as potential adsorbent

Advantages of using rice husk derivatives as biosorbent/ adsorbent/ sorbent are their biodegradability and good adsorption properties, which can be due to their morphology and surface functional groups. Their application has received much attention in sorption of various pollutants from aqueous media [87-92]. Table 3 summarizes the maximum adsorption capacities of rice husk derivatives towards different metal ions on the basis of various factors such as pH, contact times, temperature, dose, and initial concentration [93]. In one of the reported research works, activated carbon rice husk was used for Pb(II) metal adsorption. This study was aimed to determine the influence of changing factors like contact time, pH, and concentration on the adsorption of Pb(II)by using activated rice husk. Adsorption capacity, kinetic parameters, and thermodynamic parameters (Ea, $\Delta G$, $\Delta H$, $\Delta S$) for Pb(II) adsorption were determined. The optimum conditions for Pb(II) adsorption were at a contact time of 50 minutes, pH 5, temperature 40°C, and concentration of 300 ppm [94].

The rice husk also proved to be useful in the treatment of wastewater containing zinc and lead. A study showed that the percent adsorption of $Zn^{2+}$ and $Pb^{2+}$ ions increased with increase in dosage and contact time of rice husk. The binding process was strongly affected by pH and the optimum pH values for $Zn^{2+}$ and $Pb^{2+}$ ions were 7.0 and 9.0, respectively [95].

Zhang et al. concluded that RHA was an effective and alternative material for the removal of Fe(II) and Mn(II) ions from wastewater, because of its high biosorption capacity, cost effectiveness, and abundant availability. The correlation coefficients ($R^2$) of Langmuir and Freundlich isotherm models were 0.995 and 0.901 for Fe(II), 0.9862 and 0.8924 for Mn(II), respectively. These results concluded that Langmuir model fitted the equilibrium data better than the Freundlich isotherm model. The mean free energy

(E*a*) values evaluated in this study from D-R model, which indicated that the biosorption of Fe(II) and Mn(II) onto RHA was physical in nature. Experimental data showed that the biosorption processes of both metal ions complied with the pseudo-second-order kinetics [93].

Novel Fe–loaded chitosan carbonized rice husk beads (Fe–CCRB) were prepared by blending Fe (metal)–loaded chitosan (organic compound) and carbonized rice husk. This prepared Fe–CCRB was then used and proved to be an effective adsorbent for the removal of toxic chromium from aqueous solution [50]. Adsorption of Cr (VI) ions was highly pH dependent and the results showed that the optimum pH for the removal of Cr (VI) ions was 2.0. Equilibrium adsorption data for chromium removal were best represented by Freundlich isotherm, which further confirmed by intra particle and mass transfer diffusion studies [96].

RHA from northern Brazil was used as a bio-adsorbent for the removal of Pb(II) and Cu(II) ions from metal-containing effluents [97]. The maximum adsorption capacities obtained for lead and copper ions in the fixed bed were 0.0561 and 0.0682 mmol/g (at 20°C), respectively. The thermodynamic studies indicated that the lead adsorption process was exothermic and spontaneous, while the copper adsorption process was endothermic and spontaneous. The physical characterization of rice husks in nature (raw rice husk), rice husk ash before and after adsorption of metal ions indicated the presence of functional groups properties (carboxyl, silanol, hydroxyl, etc.), of amorphous silica and also the fibrous and longitudinal structure of the material [97].

Rice husk (a residue of the agricultural activity) was used to prepare activated carbon with high surface area for the adsorption of Pb(II) from aqueous solution. The adsorption capacity of the carbon sample was related to the porosity and the surface chemistry. Lead adsorption was dramatically enhanced after the carbon was oxidized, which indicated that adsorption was mainly dependent on coordination with surface functional groups. It was found that solution pH exhibited remarkable impact on the adsorption process and the maximum amount adsorbed was obtained at pH 5 [98].

Microwave incinerated rice husk ash (MIRHA), produced from the rice husk has been used as an adsorbent for heavy metals in used engine oil [99]. In another conducted study using rice husk as an adsorbent by batch adsorption experiments, the maximum percentage removal (99%) of Cr(VI) was observed under optimum operating conditions. A hybrid mathematical model was proposed to simulate the dynamic behavior in this study, which could also be potentially applicable to any similar other solid–liquid adsorption–diffusion system [100].

During the removal of multi-metal ions, Pb(II) showed a higher affinity for rice husk compared to Cu(II) and Cd(II) under the same conditions. The kinetic studies of sorption were well obeyed to a pseudo-second-order equation for all metals. The studies on the metal uptake as a function of pH also demonstrated that the mechanism was not limited to ion exchange, and an equation was proposed to evaluate the affinity of sorption by means of the overall exchange coefficients [5].

The adsorption behavior of rice husk ash with respect to manganese and iron has been studied by batch methods, to consider its application for water and wastewater treatment. Adsorption equilibrium time in this reported study was 120 min. The results revealed that manganese and iron were considerably adsorbed on the adsorbent. The rice husk can be an economic alternative for the removal of these metals from aqueous solutions [101].

The carbonaceous sorbent was produced from rice husk via sulphuric acid treatment, to remove Cr(VI) from aqueous solutions. The study revealed that the removal rate of Cr(VI) reduction at pH 2 followed a pseudo first-order model while the rate of sorption of total chromium followed pseudo-second-order model. The maximum sorption appeared due to Cr(VI) reduction with a rise in the solution pH, which allowed most of Cr (III) to be sorbed. Thus, sorption maxima for Cr(VI) onto a sorbent can be considered as an indication of reduction–sorption processes [37].

Ethylamine-modified biopolymer (chitosan) and carbon from biowaste (rice husk) composite beads (EAM-CCRCB) were prepared for the adsorption of hexavalent chromium ions from aqueous solution. Regeneration studies were also conducted to check the stability and activity of the adsorbent. Percentage removal increased with increased adsorbent dosage, and adsorption capacity increased with decreased dosage [102].

Novel biosorbent materials (RH-2 and RH-3) obtained from agricultural waste materials rice husks (RH-1) were successfully developed through fast and facile esterification reactions with hydroxylethylidenediphosphonic acid and nitrilotrimethylenetriphosphonic acid, respectively. The study results suggested that organomultiphosphonated rice husks could be recommended as potentially low-cost biosorbents with high efficiency for water purification and metals uptake [2].

The adsorption performances of NaOH-treated rice straw and rice husk for Cu(II) and Cd(II)were examined in a study [90]. The authors found that the percentage removal of Cu(II) by rice husk reached 97 % at pH 9 and 95 % by rice straw at pH 6. Kinetic studies for both metals revealed that the biosorption of Cu(II) and Cd(II) onto rice straw and rice husk followed pseudo-second-order kinetics. The salinity studies showed that the presence of NaCl discouraged the biosorption, as the performance of metal uptake increased in the absence of NaCl. Furthermore, the rapid uptake of heavy metal ions by rice residues was observed during the first 20 min and reached equilibrium after 30 min [90]. Another study also qualified rice straw to be an extraordinary agent for aquatic Cd(II) adsorption over wide pH range, short biosorption equilibrium time and high biosorption capacity. The adsorbed rice straw confirmed to be a high-quality bio-ore with a high Cd content, physical softness and easy metal extractability [103]. The study indicated the excellent potential of a raw rice straw in the adsorption of Cd(II), without any costly modification or pre-treatment.

Rice husk biochar is a new soil amendment, which has a potential of controlling the fate of trace elements in a soil system. The batch experiment was employed to determine the adsorption capacity of these biochars for As(V) and Cd(II). The porous structure with high alkaline and adsorption sites of these biochars can influence As and Cd mobilization by sorption and complexation mechanisms. Both biochars are alkaline in nature (>pH 9) and contain substantial amounts of N, P, K, Ca and Mg. The results of this study showed that RH biochar has the potential for use as a good sorbent for As and Cd [104]. Different materials containing carbon and silica were obtained by pyrolysis of rice husks and characterized to evaluate their adsorption properties towards some metal ions [92]. It was found that the pyrolysis temperature influenced the phase composition and the structural characteristics, nature, and the amount of finely dispersed fluids on their surfaces. It was concluded that the prepared four composite materials could be used for the individual and simultaneous removal of Fe(III), Pb(II), Cu(II), and Cr(III) ions from contaminated aqueous solutions.

The preparation of nanoparticles based on treated low-value agricultural by-product rice husk (TARH) and their use as an adsorbent for the removal of Pb(II) ion have been reported. Kinetic data following the pseudo-second-order kinetic model which suggested that the rate-limiting step might be chemical sorption rather than diffusion [105]. The modified rice husks were characterized in terms of morphology (scanning electron microscopy), functional groups (Fourier transform infrared spectroscopy), surface charge (pHpzc), and elemental composition to indicate the importance of treatments that led to significant changes to their surface chemistry. The potential application of these rice husk based adsorbents towards Hg(II) containing wastewaters was demonstrated [106]. A study was conducted to analyze the biosorption of manganese(II) from aqueous solutions by RHA in a continuously packed column reactor. The results showed that the equilibrium adsorption ($mgg^{-1}$) has increased after increasing the influent Mn(II) concentration, but decreased when increasing the flow rate and the bed height [107].

The technical feasibility of rice husk ash for adsorptive removal of Cr(VI) and Cu(II) from aqueous solutions was examined in a packed bed. The capacity of the bed to adsorb metal ions was found to increase with an increase in bed height, and decrease in flow rate, as well as metal ion concentration [108]. The results of another study indicated that the maximum removal of Cr(III) and Cu(II) by rice husk occurred at pH 5–6 and removal rate increased with increasing pH from 1 to 6. It was concluded, that the removal efficiency was enhanced by increasing wastewater initial concentration in the first percentage of adsorption and then decreased due to saturation of rice husk particles. The rate of reaction was fast and metal ions were removed with 15–20 min of the reaction [109].

The efficiency of rice husk to remove Hg(II) from river waters spiked with realistic environmental concentrations was evaluated. The remediation process with rice husk biowaste was extremely efficient in river waters spiked with lower levels of Hg(II). It was also able to eliminate the toxicity to the exposed organism's algae (*Pseudokirchneriella subcapitata* and *Brachionus calyciflorus)* and ensured the total survival of Daphnia magna species [110].

The rice hull ash, composed $SiO_2$(94%),could be recovered and purified by depolymerization reaction yielding an organo-silicic gel. A reported paper showed that this silica can be used to fix $Cd^{2+}$ from aqueous solution. The competition with $Ca^{2+}$ ions in the solutions was also studied in order to evaluate the selectivity of the $Cd^{2+}$ fixation. It was found that the rice hull ash has a higher capacity to fix $Cd^{2+}$ than the rice hull derivatives [111].

Characterization of the fresh rice waste adsorbents as well as of Cr(VI) loaded adsorbents was carried out by FTIR studies. Results showed that the different functional groups (surface hydroxyl, alkene, aromatic nitro, carboxylate anion and silicon oxide etc.) were responsible for the adsorption process. The developed adsorption method using rice husk silica powder was highly effective for removing Cr(VI) in the tannery industrial wastewater, which proved the applicability of adsorbent as an economical and worthwhile alternative to other conventional methods [112].

The adsorption of As(III) onto RH was favourable and influenced by temperature. The thermodynamic study showed the spontaneous nature of the sorption of As(III) onto RH [113]. A study was conducted to determine the selected physicochemical properties of two biochars, one commercially produced from rice husks and the other from oil palm empty fruit bunches. The adsorption capacities were also evaluated for Zn, Cu, and Pb by a batch equilibrium method. There was no significant difference between the carbon contents of biochars formed from empty fruit bunches (EFBB) and rice husks (RHB) [114].

A simple one-step synthetic approach using rice husk has been developed to prepare magnetic $Fe_3O_4$ loaded porous carbons composite (MRH) for removal of As(V). In view of the combined advantages of rice husk carbons and $Fe_3O_4$ nanoparticles, the synthesized MRH composites showed excellent adsorption efficiency for aqueous As(V) [115].

A commercial resin (Lewatit TP 214), and rice husk (agricultural waste) were prepared from synthetic dilute gold chloride solutions with 100-ppm gold ($Au^{3+}$) and then used to investigate gold uptake. The study also provided a comparison between these two sorbents. It was found that before the rice husks were activated, the gold uptake (%) was poor in comparison with Lewatit TP 214, but when the husk became activated at 1000°C under an argon atmosphere, the gold uptake (%) increased four times. The study concluded that the gold uptake (%) increased with increasing sorbent dosage, contact time, and temperature [116].

The synchronous removal of $Pb^{2+}$ and $Cd^{2+}$ ions from aqueous solutions by colloidal particles made from RHA in aqueous sodium dodecyl sulfate (SDS) solution was investigated. The RHA-SDS combination showed higher adsorption capacity for $Cd^{2+}$ ion compared to $Pb^{2+}$ ion under the experimental conditions [117].

RHA was chemically modified using acetic acid to increase the level of –COOH functional groups on the surface. Adsorption studies using $Cu^{2+}$ ions as the target pollutant showed 99% removal efficiency after modification, as compared to 95% removal efficiency before modification in 3 hour reaction period. The increment in sorption efficiency might be attributed to the changes in the specific surface area, –COOH groups and pHzpc [118]. Both unmodified and modified forms of RHA were found effective for the removal of $Cu^{2+}$ from aqueous solution.

The adsorption capability and adsorption rate of rice husk ash were considerably higher and faster for Zn(II) ions than for Se(IV) ions. Zn(II) adsorption was found faster in reaching equilibrium within ≃1 h while Se(IV) adsorption was slower in reaching equilibrium within ≃100 h [119]. The kinetics and thermodynamics of water adsorption onto rice husks ash filled polypropene composites during soaking were studied at different temperatures, quantities, and nature of fillers. The equilibrium uptake values of water adsorption were found to increase with respect to the amounts of the fillers in the composites [120].

Rice hulls have been used as an adsorbent to remove copper ions. The functional groups such as hydroxyl, carboxyl, and amino groups play an important role in the removal of copper ions. The optimal pH for copper ion uptake was found to be 4 [121]. Shells of lentil (LS), wheat (WS) and rice (RS) have appeared to be promising adsorbents for the removal of Cu(II) from aqueous solution. The process using LS, WS, and RS for the removal and recovery of heavy metal was potentially more economical than current process technology [122].

Removal of some heavy metal ions from wastewater was conducted by preparing a copolymer of iron and aluminum after impregnated with active silica derived from rice husk ash. This study highlighted the sodium silicate (activated silica) as one of coagulant modifiers which had many features like very fast hydrolysis, promoted rapid settling rates, extremely efficient in cold water, gave low soluble residual aluminum, consumed less alkalinity, very little effect on pH, increased the stability of coagulant and led to a denser compacted sludge [123]. The increase in the mass of adsorbent (rice husk ash) led to increasing in lead(II)ion adsorption due to increase in a number of adsorption sites. Maximum uptake was obtained at an adsorbent dose of 5 g/L, which was considered as optimum adsorbent dosage level under specified conditions [124]. Adsorption experiments were carried out using waste rice straw of several kinds as a biosorbent to adsorb Cu(II), Zn(II), Cd(II) and Hg(II) ions from aqueous solutions at room temperature. Based on the experimental data and the Freundlich model, the adsorption order on the rice straw was Cd(II) > Cu(II) > Zn(II) > Hg(II). This adsorption process reached the equilibrium before 1.5 h, with maximum adsorptions at pH 5.0 [125].

A study suggested that polymer grafted rice husk/sawdust could be effectively used as adsorbents for removal of $Cd^{2+}$ ions from an aqueous medium. The values of $R_{L,}$ which is a dimensionless constant referred to as separation factor or equilibrium, have been found to be between 0 and 1, suggested the favorable adsorption [126]. The results of another study revealed that the hexavalent chromium was considerably adsorbed on rice husk and it could be an economical method for the removal of hexavalent chromium from an aqueous system. The removal of chromium was dependent on the physicochemical characteristics of the adsorbent, adsorbate concentration and other studied process parameters [127].

The adsorption characteristics of $Cu^{2+}$ on raw rice husk (RRH) and expanding rice husk (ERH) were studied in a fixed bed column with varying parameters of inlet concentration, flow rate, and bed depth. The adsorption of $Cu^{2+}$ in different column design parameters was found depended strongly on bed depth, flow rate, and initial concentration. ERH showed significantly higher adsorption capacity of high efficiency towards Cu (II). Ion exchange and hydrogen bonding were the major mechanisms of copper retention by ERH [128].

The chemical improvement by partial alkali digestion in the autoclave has enhanced the surface area and facilitated the transport of metal ions to the binding sites onto biomatrix. The newly developed biomatrix obtained by partial alkali digestion of rice husk could be used to remove as many as eight different metal ions (Ni, Zn, Cd, Mn, Co, Cu, Hg and Pb) in single and mixed metal solutions effectively [129]. It was found that the adsorption capacity and adsorption rate constant were increased and the minimum column bed depth required was reduced when the rice husk was treated with phosphate when compared with that of raw rice husk. From the column studies for individual metal ions, it was observed that the increase in the metal uptake capacity with the increase of bed height was due to the increase in the contact time [130].

The removal of Cr(VI) ions from aqueous solutions via adsorption onto black rice husk ash was studied from kinetics point of view under different conditions. The most favorable conditions for removing Cr(VI) from aqueous solutions via adsorption onto black rice husk ash (BRHA) were found to be pH 2, adsorbent dosage level of 15 $gL^{-1}$ in the concentration range 25–200 mg $L^{-1}$, temperature interval 10–30°C and contact time 30–120min [131].

White rice husk ash (WRHA) was obtained via thermal degradation of rice husks in air, which was used to remove Mn(II) ions from aqueous solutions. The obtained results showed that the removal of Mn(II) ions from aqueous solutions using WRHA was via chemisorption process, where the OH-groups of $SiO_2$were mainly responsible for effective removal of Mn(II) ions from water solutions [132].

Polyaniline/rice husk composites were prepared by coating the rice husk substrate with aniline in 1M HCl solution using the chemical oxidative polymerization method. The adsorption of arsenic ions from aqueous solution on polyaniline/rice husk (PAn/RH) nanocomposite was investigated. The experimental data were better described by pseudo-second- order model as evident from the correlation coefficient ($R^2$) [133].

A novel magnetic porous sorbent from agricultural waste rice husk was successfully synthesized through a simple carbon-thermal method. The results illustrated that rice husk-derived magnetic carbonaceous materials were potential adsorbents for Cr(VI) pollution treatment. Fan et al. in the same study also provided a suitable method for the effective conversion of biomass waste, which might solve the problem of waste disposal and widen the applications of the materials [134].

The adsorption of Ni(II) on a new porous carbonaceous adsorbent, pre-treated rice husk particles, was investigated in both batch and column experiments. The maximum adsorption capacity (7.6 mg/g) estimated by the mass transport model was better in accordance with the results of batch experiments. The work indicated the ability of the proposed model to predict column behavior using batch results [135].

A work was conducted to describe the sorption of metal ions using montmorillonite-rice husk composite as a novel adsorbent. The adsorption capacity of the new adsorbent was compared with the unmodified, alkaline and acid treated montmorillonite adsorbents. This study showed that the addition of rice husk to a montmorillonite as a cheaper adsorbent recorded a higher adsorption for metal ions as compared to the more expensive acid, alkaline and unmodified forms. This new adsorbent favorably recorded an increase in surface area and porosity [136].

Composites based on polyaniline (PANi) and rice husk (RH) were prepared by different methods [137]. PANi–RH composites proved to be useful adsorbents for removing lead(II) and cadmium(II) ions from solution by adsorption following Langmuir isotherm model. However, the Freundlich and Temkin isotherm models could be used only for lead(II), not for cadmium ion adsorption. The PANi–RH composite prepared by soaking method was better to remove lead(II) ion, but not for cadmium(II) ion in comparison with that prepared by the chemical method.

Adsorbents carbonized rice husk (CRH) and activated rice husk (ARH) prepared from rice husks were investigated as viable materials for treatment of Pb, Cd, Cu, and Zn-containing industrial wastewater at controlled pH. The results obtained from the batch experiments revealed a relative ability of the rice husk in removing some heavy metals at pH 7. The CRH adsorption capacity decreased in the order of Cu>Pb>Zn>Cd in batch adsorption whereas the adsorption capacity in case of small-scale column tests was in the order of Zn>Cu>Pb>Cd. However, ARH adsorption capacity performance was similar to CRH [138].

Simple thermal treatment was applied on rice husk to investigate its effect on sorption potential towards removal of Zn(II) ions from aqueous solution. The mechanisms of Zn(II) ions sorption were proposed based on equilibrium data obtained from the study. The sorption process was favorable as indicated by the separation factor ($R_L$) value between 0 and 1 [139].

Rice husk (RH), rice husk carbon (RHC), and peat were used to adsorb Cr, Fe, Mn, Ni, Zn, and Pb from aqueous solutions. The results suggested that all three low-cost adsorbents could be used to remove Cr(III), Mn(II), Fe(II), Ni(II), Zn(II), and Pb(II) from aqueous solutions, with limitations inherent to each material. The used adsorbents (excluding peat) were advantageous for this application because they were by products of agricultural activity and, unlike peat and mineral carbon, they did not need to be extracted by mining [34].

Kinetic and isotherm studies revealed that MIRH (microwave irradiated rice husk), an inexpensive material, could be used as a potential adsorbent for the removal of Ni(II) ions from aqueous solutions. The adsorption data fitted the Freundlich isotherm model reasonably well and the maximum adsorption capacity obtained using the Langmuir model was comparable with both the experimental data and the qe values obtained from the pseudo-second-order kinetic model [140].

The fixed-bed column treatment system developed by Z. Asifet al. was a most suitable domestic approach to remove arsenic from local areas, because of its simplicity, easy operation, and handling. The proposed method was effective for a wide range of arsenic concentrations (i.e., 15–70 ppb), usually observed in contaminated groundwater of Asian countries. The study also highlighted the recycling potential of RHA that can be used as cement/insulation panels and roof sheets [141].

Three kinds of agricultural by-products, wheat stem, corncob, and rice husk, were tested as biosorbents for cadmium removal from aqueous solution. The initial sorption rate, as well as the maximum sorption capacity (qmax) calculated from the Langmuir isotherm equation, showed the following tendency: rice husk>wheat stem>corncob [142]. Here, rice husk was again proved to be the best natural adsorbent for the removal of heavy metals from aqueous solutions. In yet another study, the technical feasibility of rice husk ash for adsorptive removal of Cr (VI) and Cu(II) was examined in packed bed from aqueous solutions. The capacity of the bed to adsorb metal ions was found to increase with an increase in bed height and a decrease in flow rate, as well as metal ion concentration. The Yoon–Nelson model and Thomas model were applicable for Cr(VI) removal, and for Cu(II)Thomas model was suited [108].

Studies on sorption of Fe(III), Cu(II), Cd(II), and Pb(II) ions sorption from individual solutions by rice husk-based sorbents indicated that the sorption capacity of the sorbents depended on their preparation procedure. The highest sorption capacity with respect to Cu(II), Cd(II), and Pb(II) ions was exhibited by the sorbent prepared via alkaline hydrolysis of the rice husk, and with respect to Fe(III) ions, that prepared via heat treatment of the residue from acid hydrolysis of the husk, which was essentially amorphous silicon dioxide. The sorbent prepared by alkaline hydrolysis of the rice husk could be recommended for separation of lead and copper ions from cadmium ions in mixed solutions [143].

In a recent report by Saadat et al. suggests that rice husk could be used effectively as an adsorbent for pre-treatment for tannery wastewater [135]. They found that rice husk could be employed as a better adsorbent for preliminary treatment of tannery wastewater. The pollutant parameters of tannery wastewater like turbidity, BOD, COD were reduced to the satisfactory level.

Sugashini and Begum investigated the removal of heavy metal ions using Fe-loaded chitosan carbonized rice husk beads (Fe-CCRB) from synthetic and industrial effluents using fixed bed column. The Fe-CCRB adsorbent was used to treat the electroplating effluent and it was found to be more efficient in adsorption of anionic pollutant e.g. Cr(VI) ions when compared to cationic pollutants Ni(II) and Zn(II). Fe-CCRB adsorbent was used to treat chrome plating effluent for adsorption of Cr(VI) ions [96].

The magnetic mesoporous silica of magMCM-41 with large surface area (695 $m^2\ g^{-1}$) and high magnetization (10.79 emu $g^{-1}$) was synthesized using extracted amorphous silica from rice husk. The synthesized materials were applied for adsorption of Cd(II) ions from aqueous solution in batch operation systems. A highly selective adsorbent was obtained by grafting 3- aminopropyltrimethoxysilane on the pores of the magMCM-41, in which the adsorption capacity of Cd(II) ions increased from 41.8 to 86 mg $g^{-1}$, under the same conditions. The best responses for Cd(II) adsorption capacity and removal efficiency were 493.21 mg $g^{-1}$ and 60.25%, respectively, which were achieved at pH of 5.05, sorbent dosage of 0.1 g $L^{-1}$ and Cd(II) concentration of 150 mg $L^{-1}$ [144].

The elemental mercury ($Hg^0$) capture of potassium permanganate ($KMnO_4$)-modified calcium-based with rice husk ash sorbent (CaO/RHA) was tested to evaluate the effectiveness of $Hg^0$ adsorption. The sorbent physical properties (total surface area and pore size distribution) were noted to affect mercury sorptive efficiency [145]. The dilute (0.013 M) solutions of NaOH, KOH, and $Ca(OH)_2$ in the mercerization of rice husk resulted in the increase of the removal of Cd(II) ions in aqueous solutions as compared to that by unmodified RH. Thus, this study revealed that after mercerization using dilute alkaline solution or by using the 2-batch treatment approach, RH could be better utilized as an efficient, low-cost biosorbent for the removal of Cd(II) ions from aqueous solutions [146].

A new approach adopted for the removal of Mo(VI) (approximately 99.50 %) from Mo-containing surface water by using the modified rice husk and corn straw [147]. Reutilization of both CS and RH gels was possible up to six adsorption-elution cycles in column mode operation with no damage of gels, packed in the column.

A laboratory-scale study has explored the use of RHA as an admixture to stabilize and solidify the waste sludge from a chrome tanning process [148]. The small amount of rice husk (0.25 and 0.50 g $L^{-1}$) was able to reduce the Hg(II) levels by about 80% for an initial concentration of 0.05 $mgL^{-1}$and 90% for 0.50 $mgL^{-1}$. The biosorbent was reused further for cleaning treatments, maintaining the efficiency and high performance [149].

***Table 3*** *Maximum adsorption capacities of rice husk derivatives towards various metal ions.*

| **Rice husk derivatives as an adsorbent** | **Targeted Metal ion** | **Experimental conditions** | **Adsorption capacity, qm (mg/g)** | **References** |
|---|---|---|---|---|
| Rice husk ash | Fe(II) | Contact time: 60 min, Concentration: 2-40 mg/L, pH: 5, Temperature: 30°C, Dose: 0.6 g | 6.21 | [93] |
| Rice husk ash | Mn(II) | Contact time: 60 min, Concentration: 2-40 mg/L, pH: 6, Temperature: 30°C, Dose: 0.6 g | 3.02 | [93] |
| Activated carbon developed from rice husk | Pb(II) | Contact time: 0-300 min, Concentration: 50-500 mg/L, pH: 5, Temperature: 30°C | 263.00 | [98] |
| Rice husk ash | Mn(II) | Contact time: 300 min, Concentration: 10-50 mg/L, pH: 3-6, Temperature: 30°C, Dose: 0.1 g | 3.17 | [93] |
| Rice husk ash | Fe(II) | Contact time: 300 min, Concentration: 10-50 mg/L, pH: 3-6, Temperature: 30°C, Dose: 0.5 g | 66.66 | [93] |
| Dry carbonaceous sorbent produced from rice husk | Cr(VI) | Concentration: 100-500 mg/L, pH: 5-6, Temperature: 25°C, Dose: 0.1 g | 278.00 | [37] |
| Ethylamine-modified chitosan carbonized rice husk composite beads | Cr(VI) | Contact time: 150 min, Concentration: 100-500 mg/L, pH: 2, Temperature: 70°C, Dose: 0.1 g | 88.40 | [102] |
| Rice straw | Cu (II) | Contact time: 1440 min, Concentration: 5-60 mg/L, pH: 3, Dose: 0.2 g | 9.94 | [90] |
| Rice husk | Cu (II) | Contact time: 1440 min, Concentration: 5-60 mg/L, pH: 3, Dose: 0.2 g | 5.92 | [90] |
| Rice straw | Cd (II) | Contact time: 1440 min, Concentration: 0.5-8.0 mg/L, pH: 3, Dose: 0.2 g | 1.23 | [90] |

| | | | | |
|---|---|---|---|---|
| Rice husk | Cd (II) | Contact time: 1440 min, Concentration: 0.5-8.0 mg/L, pH: 3, Dose: 0.2 g | 0.80 | [90] |
| Unmodified rice straw | Cd (II) | Contact time: 180 min, Concentration: 50 mg/L, pH: 2-6, Temperature: 25°C, Dose: 1.0 g | 13.90 | [103] |
| Rice husk biochar | As(V) | Contact time: 960 min, Concentration: 0-200 mg/L, pH: 2-8, Temperature: 25°C, Dose: 2.0 g | 0.35 | [104] |
| Rice husk biochar | Cd(II) | Contact time: 960 min, Concentration: 0-200 mg/L, pH: 2-8, Temperature: 25°C, Dose: 2.0 g | 3.19 | [104] |
| Pyrolyzed rice husk | Fe(III) | Contact time: 300 min, Concentration: 50 mg/L, pH: 2.3, Dose: 0.2 g | 21.73 | [92] |
| Pyrolyzed rice husk | Pb(II) | Contact time: 300 min, Concentration: 50 mg/L, pH: 3.6, Dose: 0.2 g | 10.41 | [92] |
| Pyrolyzed rice husk | Cu(II) | Contact time: 300 min, Concentration: 50 mg/L, pH: 3.6, Dose: 0.2 g | 4.60 | [92] |
| Pyrolyzed rice husk | Cr(III) | Contact time: 300 min, Concentration: 50 mg/L, pH: 3.6, Dose: 0.2 g | 3.47 | [92] |
| Rice husk ash | Cr(VI) | Contact time: 240 min, Concentration: 10-300 mg/L, pH: 3, Temperature: 30°C, Dose: 10 g | 25.64 | [19] |
| Low-cost nanoparticles sorbent from modified rice husk (CNR@TARH) | Pb(II) | Contact time: 90 min, Concentration: 300 mg/L, pH: 5, Temperature: 30°C, Dose: 0.1 g | 93.45 | [105] |
| Cross linked chitosan carbonized rice husk (CCACR) | Cr(VI) | Contact time: 120 min, Concentration: 255 mg/L, pH: 2, Temperature: 30°C, Dose: 0.44 g | 55.40 | [102] |
| Sulfur-functionalized rice husk | Hg(II) | Contact time: 2880 min, Concentration: 25-200 mg/L, pH: 5, Temperature: 30°C, Dose: 0.02 g | 89.00 | [106] |
| Organosilane-grafted rice husk | Hg(II) | Contact time: 2880 min, Concentration: 25-200 mg/L, pH: 5, Temperature: 30°C, Dose: 0.02 g | 118.00 | [106] |
| Rice husk ash | Mn(II) | Concentration: 5-20 mg/L, pH: 6, Temperature: 30°C | 7.57 | [107] |

| | | | | |
|---|---|---|---|---|
| Rice husk | Cr(III) | Contact time: 30 min, Concentration: 100-800 mg/L, pH: 5-6, Temperature: 25°C, Dose: 5.00 g | 30.00 | [109] |
| Rice husk | Cu(II) | Contact time: 30 min, Concentration: 100-800 mg/L, pH: 5-6, Temperature: 25°C, Dose: 4.00 g | 22.50 | [109] |
| Rice straw | Cu(II) | Contact time: 30 min, Concentration: 5-60 mg/L, pH: 3, Temperature: 25°C, Dose: 0.2 g | 9.94 | [90] |
| Rice straw | Cd(II) | Contact time: 30 min, Concentration: 5-60 mg/L, pH: 3, Temperature: 25°C, Dose: 0.2 g | 1.23 | [90] |
| Rice husk | Cu(II) | Contact time: 30 min, Concentration: 5-60 mg/L, pH: 3, Temperature: 25°C, Dose: 0.2 g | 5.92 | [90] |
| Rice husk | Cd(II) | Contact time: 30 min, Concentration: 5-60 mg/L, pH: 3, Temperature: 25°C, Dose: 0.2 g | 1.35 | [90] |
| Rice husk ash | Cr(VI) | Contact time: 180 min, Concentration: 10 mg/L, pH: 2, Temperature: 25°C, Dose: 40 g | 0.49 | [150] |
| Modified rice husk ash | Cr(VI) | Contact time: 180 min, Concentration: 10 mg/L, pH: 2, Temperature: 25°C, Dose: 40 g | 0.84 | [150] |
| Rice husk silica | Cr(VI) | Contact time: 150 min, Concentration: 292 mg/L, pH: 4, Dose: 15 g | 78.12 | [151] |
| Rice husk | As(III) | Contact time: 5 min, Concentration: 20-1000 mg/L, pH: 6, Temperature: 10-45°C, Dose: 6 g | 171.10 | [113] |
| Rice husk | Zn(II) | Contact time: 1440 min, Concentration: 4-400 mg/L, pH: 6, Temperature: 25°C, Dose: 0.2 g | 34.30 | [114] |
| Rice husk | Cu(II) | Contact time: 1440 min, Concentration: 4-400 mg/L, pH: 6, Temperature: 25°C, Dose: 0.2 g | 37.50 | [114] |
| Rice husk | Pb(II) | Contact time: 1440 min, Concentration: 4-400 mg/L, pH: 6, Temperature: 25°C, Dose: 0.2 g | 43.90 | [114] |
| Carbons developed from rice husk | As(V) | Contact time: 60 min, Concentration: 1-10 mg/L, pH: 7, Temperature: 25°C, Dose: 2 g | 4.33 | [115] |
| Activated rice husk | Au(III) | Contact time: 240 min, Concentration: 1-10 mg/L, pH: 2, Temperature: 60°C, Dose: 0.005 g | 93.30 | [116] |
| Rice husk | Pb(II) | Contact time: 1440 min, Concentration: 10-100 mg/L, pH: 5.5, Temperature: 25°C, Dose: 0.005 g | 101.10 | [152] |

| | | | | |
|---|---|---|---|---|
| Rice husk | Cd(II) | Contact time: 1440 min, Concentration: 10-100 mg/L, pH: 5.5, Temperature: 25°C, Dose: 0.005 g | 103.10 | [152] |
| Rice husk ash | Cu(II) | Contact time: 180 min, Concentration: 100 mg/L, pH: 6.3, Temperature: 25°C, Dose: 0.25 g | 1.22 | [118] |
| Modified rice husk ash | Cu(II) | Contact time: 180 min, Concentration: 100 mg/L, pH: 6.3, Temperature: 25°C, Dose: 0.25 g | 3.83 | [118] |
| Rice husk ash | Zn(II) | Contact time: 5-180 min, Concentration: 10-100 mg/L, pH: 6, Temperature: 25°C, Dose: 0.25 g | 9.58 | [119] |
| Rice husk ash | Se(IV) | Contact time: 4.8-15000 min, Concentration: 10-100 mg/L, pH: 2, Temperature: 25°C, Dose: 0.25 g | 2.00 | [119] |
| Tartaric acid modified rice husk | Ni(II) | Contact time: 45 min, Concentration: 60 mg/L, pH: 4, Temperature: 25°C, Dose: 0.3 g | 55.50 | [153] |
| Tartaric acid modified rice husk | Cd(II) | Contact time: 45 min, Concentration: 60 mg/L, pH: 6, Temperature: 25°C, Dose: 0.3 g | 45.50 | [153] |
| Rice husk | Cu(II) | Contact time: 180 min, Concentration: 25-500 mg/L, pH: 6, Temperature: 40°C, Dose: 1.0 g | 16.07 | [122] |
| Copolymer derived from rice husk ash | Pb(II) | Contact time: 5 min, Concentration: 70 mg/L, pH: 7-8.5, Temperature: 27°C, Dose: 0.07 g | 416.00 | [123] |
| Copolymer derived from rice husk ash | Fe(II) | Contact time: 5 min, Concentration: 28 mg/L, pH: 7-8.5, Temperature: 27°C, Dose: 0.02 g | 222.00 | [123] |
| Copolymer derived from rice husk ash | Mn(II) | Contact time: 5 min, Concentration: 28 mg/L, pH: 7-8.5, Temperature: 27°C, Dose: 0.02 g | 158.00 | [123] |
| Copolymer derived from rice husk ash | As(V) | Contact time: 5 min, Concentration: 14 mg/L, pH: 7-8.5, Temperature: 27°C, Dose: 0.01 g | 146.00 | [123] |
| Rice husk ash | Pb(II) | Contact time: 60 min, Concentration: 10 mg/L, pH: 5, Temperature: 30-50°C, Dose: 5 g | 91.74 | [124] |
| Grafted rice husk | Cd(II) | Contact time: 180 min, Concentration: 30-300 mg/L, pH: 9, Temperature: 30°C, Dose: 2.5 g | 0.889 | [126] |
| Pre-boiled rice husk (BRH) | Cr(VI) | Contact time: 180 min, Concentration: 100 mg/L, pH: 2, Temperature: 25°C, Dose: 20 g | 8.50 | [127] |
| Formaldehyde treated rice husk (FRH) | Cr(VI) | Contact time: 180 min, Concentration: 100 mg/L, pH: 2, Temperature: 27°C, Dose: 20 g | 10.40 | [127] |

| | | | | |
|---|---|---|---|---|
| Polyaniline/rice husk nanocomposite | As(V) | Contact time: 5-60 min, Concentration: 100 mg/L, pH: 10, Temperature: 30°C, Dose: 0.12-1.00 g | 34.48 | [133] |
| Rice husk derived magnetic sorbents | Cr(VI) | Contact time: 90 min, Concentration: 50-300 mg/L, pH: 2, Temperature: 30-50°C, Dose: 0.02 g | 157.70 | [134] |
| Pre-treated rice husk | Ni(II) | Contact time: 1440 min, Concentration: 17-88 mg/L, pH: 2, Temperature: 25°C, Dose: 15 g | 8.00 | [135] |
| Rice husk ash | Cd(II) | Contact time: 120 min, Concentration: 3-100 mg/L, pH: 5, Temperature: 30°C, Dose: 7.50 g | 35.84 | [154] |
| Thermally treated rice husk | Zn(II) | Contact time: 180 min, Concentration: 30 mg/L, pH: 6, Temperature: 30°C, Dose: 2.00 g | 6.55 | [139] |
| Rice husk carbon | Cr(VI) | Contact time: 1-1440 min, Concentration: 24.80 mg/L, pH: 1.70, Temperature: 25°C, Dose: 0.30 g | 37.20 | [34] |
| Rice husk carbon | Mn(II) | Contact time: 1-1440 min, Concentration: 24.80 mg/L, pH: 1.70, Temperature: 25°C, Dose: 0.30 g | 23.90 | [34] |
| Rice husk carbon | Fe(II) | Contact time: 1-1440 min, Concentration: 24.80 mg/L, pH: 1.70, Temperature: 25°C, Dose: 0.30 g | 29.60 | [34] |
| Rice husk carbon | Ni(II) | Contact time: 1-1440 min, Concentration: 50 mg/L, pH: 1.70, Temperature: 25°C, Dose: 0.30 g | 9.62 | [34] |
| Rice husk carbon | Zn(II) | Contact time: 1-1440 min, Concentration: 24.80 mg/L, pH: 1.70, Temperature: 25°C, Dose: 0.30 g | 8.21 | [34] |
| Rice husk carbon | Pb(II) | Contact time: 1-1440 min, Concentration: 24.80 mg/L, pH: 1.70, Temperature: 25°C, Dose: 0.30 g | 7.96 | [34] |
| Microwave irradiated rice husk | Ni(II) | Contact time: 210 min, Concentration: 6 mg/L, Temperature: 35°C, Dose: 2.8 g | 1.17 | [140] |
| Untreated rice husk | U(VI) | Contact time: 5-720 min, Concentration: 10-100 mg/L, Temperature: 30°C, pH: 4, Dose: 0.05 g | 45.24 | [155] |
| Treated rice husk | U(VI) | Contact time: 5-720 min, Concentration: 10-100 mg/L, Temperature: 30°C, pH: 4, Dose: 0.05 g | 47.16 | [155] |
| Immobilized rice husk | U(VI) | Contact time: 5-720 min, Concentration: 10-100 mg/L, Temperature: 30°C, pH: 4, Dose: 0.05 g | 40.00 | [155] |
| Pyrolyzed rice husk | Fe(III) | Contact time: 300 min, Concentration: 50-500 mg/L, pH: 2.3, Dose: 0.2 g | 21.73 | [92] |

| | | | | |
|---|---|---|---|---|
| Pyrolyzed rice husk | Pb(II) | Contact time: 300 min, Concentration: 50-500 mg/L, pH: 3.6, Dose: 0.2 g | 10.41 | [92] |
| Pyrolyzed rice husk | Cr(III) | Contact time: 300 min, Concentration: 50-200 mg/L, pH: 3.6, Dose: 0.2 g | 4.60 | [92] |
| Pyrolyzed rice husk | Cu(II) | Contact time: 300 min, Concentration: 50-200 mg/L, pH: 3.6, Dose: 0.2 g | 3.47 | [92] |
| Rice husk-based sorbents | Fe(III) | Contact time: 1440 min, Concentration: 5-300 mg/L, Temperature: 30°C, pH: 3.00-4.35 | 4.66 | [143] |
| Rice husk-based sorbents | Cu(II) | Contact time: 1440 min, Concentration: 50-200 mg/L, pH: 5.55-5.73 | 6.76 | [143] |
| Rice husk-based sorbents | Cd(II) | Contact time: 1440 min, Concentration: 50-200 mg/L, pH: 5.95-6.02 | 7.09 | [143] |
| Rice husk-based sorbents | Pb(II) | Contact time: 1440 min, Concentration: 50-200 mg/L, pH: 5.65-6.10 | 20.14 | [143] |
| Magnetic mesoporous silica of magMCM-41 with rice husk silica | Cd(II) | Contact time: 120 min, Concentration: 150 mg/L, Temperature: 25°C, pH: 5.05, Dose: 0.1 g | 493.21 | [144] |
| $Ca(OH)_2$ -modified rice husk | Cd(II) | Contact time: 30 min, Concentration: 5 mg/L, Temperature: 25°C, pH: 6, Dose: 0.2 g | 10.46 | [146] |

## Conclusions

The successfully reported studies of rice husk derivatives clarify their advantages and applications in the development of environmental science. Both raw and modified rice husk forms have been highly effective in the removal of heavy metals after tested in batch experiments by changing the parameters such as pH, sorbent dose, temperature and initial concentration etc.

## References

[1] M. K. Uddin, A review on the adsorption of heavy metals by clay minerals, with special focus on the past decade, Chem. Eng. J. 308 (2017) 438-462. https://doi.org/10.1016/j.cej.2016.09.029.

[2] M. Xu, P. Yin, X. Liu, Q. Tang, R. Qu, Q. Xu, Utilization of rice husks modified by organomultiphosphonic acids as low-cost biosorbents for enhanced adsorption of heavy metal ions, Bioresour. Technol. 149 (2013) 420–424. https://doi.org/10.1016/j.biortech.2013.09.075.

[3] M.A. Khan, M.K. Uddin, R. Bushra, A. Ahmad, S.A. Nabi, Synthesis and characterization of polyaniline Zr(IV) molybdophosphate for the adsorption of phenol from aqueous solution, React. Kinet. Mech. Catal. 113 (2014) 499-517. https://doi.org/10.1007/s11144-014-0751-x.

[4] R.A.K. Rao, M. Kashifuddin, Adsorption studies of Cd ( II ) on ball clay : Comparison with other natural clays, Arab. J. Chem. 9 (2016) S1233–S1241. https://doi.org/10.1016/j.arabjc.2012.01.010.

[5] M.G.Guiso, G.Alberti, G.Emma, M.Pesavento, Pb ( II ), Cu ( II ) and Cd ( II ) Removal through Untreated Rice Husk ; Thermodynamics and Kinetics, Anal. Sci, 28 (2012) 993-999.

[6] W.C. Li, F.Y. Law, Y.H.M. Chan, Biosorption studies on copper ( II ) and cadmium ( II ) using pretreated rice straw and rice husk, 24 (2017)8903–8915. https://doi.org/10.1007/s11356-015-5081-7.

[7] R.A.K. Rao, M. Kashifuddin, Pottery glaze—An excellent adsorbent for the removal of Cu(II) from aqueous solution, Chin.J.Geochem. 31 (2012) 136–146. https://doi.org/10.1007/s11631-012-0560-8.

[8] R.A.K. Rao, F. Rehman, M. Kashifuddin, Removal of Cr(VI) from electroplating wastewater using fruit peel of Leechi (Litchi chinensis) Desal. Water Treat.49 (2012) 136-146. https://doi.org/10.1080/19443994.2012.708211.

[9] D. Augusto, M. Zaia, R. Ventura, A. Alberto, Use of rice straw as biosorbent for removal of Cu ( II ), Zn ( II ), Cd ( II ) and Hg ( II ) ions in industrial effluents, J. Hazard. Mater. 166 (2009) 383–388. https://doi.org/10.1016/j.jhazmat.2008.11.074.

[10] T. Kumar, A. Kumar, S. Mandal, S. Kumar, The sorption of lead ( II ) ions on rice husk ash, J. Hazard. Mater. 163 (2009) 1254–1264. https://doi.org/10.1016/j.jhazmat.2008.07.119.

[11] C. Jeon, Removal of copper ion using rice hulls, J. Ind. Eng. Chem. 17 (2011) 517–520. https://doi.org/10.1016/j.jiec.2010.10.020.

[12] Z. Elouear, F. Bouhamed, N. Boujelben, J. Bouzid, Application of sheep manure and potassium fertilizer to contaminated soil and its effect on zinc , cadmium and lead accumulation by alfalfa plants, Sustain. Environ. Res. (2016) 5–9. https://doi.org/10.1016/j.serj.2016.04.004.

[13] R.M. Ali, H.A. Hamad, M.M. Hussein, G.F. Malash, Potential of using green adsorbent of heavy metal removal from aqueous solutions : Adsorption kinetics , isotherm , thermodynamic , mechanism and economic analysis, Ecol. Engg. 91 (2016) 317–332.

[14] R.A.K. Rao, M. Kashifuddin, Adsorption Properties of Coriander Seed Powder (Coriandrum sativum): Extraction and Pre-concentration of Pb(II), Cu(II) and Zn(II) Ions from Aqueous Solution, Adsorpt. Sci. Technol. 30 (2012) 127-146.

[15] R. A.K. Rao, S. Ikram, M.K. Uddin, Removal of Cr(VI) from aqueous solution on seeds of Artimisia absinthium (novel plant material) Desalin. Water Treat. 54 (2015) 3358–3371. https://doi.org/10.1080/19443994.2014.908147.

[16] R.A.K. Rao, S. Ikram, M.K. Uddin, Removal of Cd(II) from aqueous solution by exploring the biosorption characteristics of gaozaban (Onosma bracteatum)J. Environ. Chem. Eng. 2 (2014) 1155–1164.

[17] R. Lavecchia, F. Medici, M.S. Patterer, A. Zuorro, Lead Removal from Water by Adsorption on Spent Coffee Grounds, 47 (2016) 295–300. https://doi.org/10.3303/CET1647050.

[18] S. Asgarzadeh, R. Rostamian, E. Faez, A. Maleki, Biosorption of Pb ( II ), Cu ( II ), and Ni ( II ) ions onto novel lowcost Peldarica leaves-based biosorbent :

isotherm , kinetics , and operational parameters investigation, 57 (2016)14544-14551. https://doi.org/10.1080/19443994.2015.1067831.

[19] A.K. Bhattacharya, T.K. Naiya, S.N. Mandal, S.K. Das, Adsorption, kinetics and equilibrium studies on removal of Cr(VI) from aqueous solutions using different low-cost adsorbents, Chem. Eng. J. 137 (2008) 529–541. https://doi.org/10.1016/j.cej.2007.05.021.

[20] A. Witek-krowiak, R.G. Szafran, S. Modelski, Biosorption of heavy metals from aqueous solutions onto peanut shell as a low-cost biosorbent, Desal.. 265 (2011) 126–134. https://doi.org/10.1016/j.desal.2010.07.042.

[21] X. Jing, Y. Cao, X. Zhang, D. Wang, X. Wu, H. Xu, Biosorption of Cr ( VI ) from simulated wastewater using a cationic surfactant modi fi ed spent mushroom, Desal.. 269 (2011) 120–127. https://doi.org/10.1016/j.desal.2010.10.050.

[22] C. Chung, Y. Shen, Y. Min, M. Kuang, C. San, C. Yi, Y. An, Biosorption of chromium , copper and zinc on rice wine processing waste sludge in fixed bed, Desal. 267 (2011) 20–24. https://doi.org/10.1016/j.desal.2010.08.040.

[23] A.H. El-sheikh, M.M.A. Hilal, J.A. Sweileh, Bioresource Technology Bio-separation , speciation and determination of chromium in water using partially pyrolyzed olive pomace sorbent, Bioresour. Technol. 102 (2011) 5749–5756. https://doi.org/10.1016/j.biortech.2011.03.021.

[24] A.D. Papandreou, C.J. Stournaras, D. Panias, I. Paspaliaris, Adsorption of Pb ( II ), Zn ( II ) and Cr ( III ) on coal fly ash porous pellets, Miner. Eng. 24 (2011) 1495–1501. https://doi.org/10.1016/j.mineng.2011.07.016.

[25] S. Chen, Q. Yue, B. Gao, Q. Li, X. Xu, K. Fu, Adsorption of hexavalent chromium from aqueous solution by modified corn stalk : A fixed-bed column study, Bioresour. Technol. (2011). https://doi.org/10.1016/j.biortech.2011.11.110.

[26] S. Liang, N. Ye, Y. Hu, Y. Shi, W. Zhang, W. Yu, X. Wu, Lead adsorption from aqueous solutions by a granular adsorbent prepared from phoenix tree, RSC Adv. 10 (2016) 25393–25400. https://doi.org/10.1039/C6RA03258C.

[27] Z.N. Garba, I. Bello, A. Galadima, A.Y. Lawal, Optimization of adsorption conditions using central composite design for the removal of copper ( II ) and lead ( II ) by defatted papaya seed, Karbala Int. J. Mod. Sci. 2 (2016) 20–28. https://doi.org/10.1016/j.kijoms.2015.12.002.

[28] C. Duran, D. Ozdes, A. Gundogdu, M. Imamoglu, H. Basri, Tea-industry waste activated carbon , as a novel adsorbent , for separation , preconcentration and

speciation of chromium, Anal. Chim. Acta. 688 (2011) 75–83. https://doi.org/10.1016/j.aca.2010.12.029.

[29] T.S.Anurandhan, .P.G. Radhakrishnan, Thermodynamics of chromium ( III ) adsorption onto a cation exchanger derived from saw dust of Jack wood, Env. Chem. Lett. (2011) 121–125. https://doi.org/10.1007/s10311-009-0255-5.

[30] A. Sdiri, T. Higashi, F. Jamoussi, S. Bouaziz, Effects of impurities on the removal of heavy metals by natural limestones in aqueous systems, J. Environ. Manage. 93 (2012) 245–253. https://doi.org/10.1016/j.jenvman.2011.08.002.

[31] Y. Shen, S. Wang, Y. Tzou, Y. Yan, W. Kuan, Removal of hexavalent Cr by coconut coir and derived chars – The effect of surface functionality, Bioresour. Technol. 104 (2012) 165–172. https://doi.org/10.1016/j.biortech.2011.10.096.

[32] A. Javaid, R. Bajwa, U. Shafique, J. Anwar, Removal of heavy metals by adsorption on Pleurotus ostreatus, Biomass and Bioenergy. 35 (2011) 1675–1682. https://doi.org/10.1016/j.biombioe.2010.12.035.

[33] T. Altun, E. Pehlivan, Removal of Cr ( VI ) from aqueous solutions by modified walnut shells, Food Chem. 132 (2012) 693–700. https://doi.org/10.1016/j.foodchem.2011.10.099.

[34] R.M. Lattuada, M.C.R. Peralba, J.H.Z. Dos Santos, A.G. Fisch, Peat , Rice Husk and Rice Husk Carbon as Low- Cost Adsorbents for Metals from Acidic Aqueous Solutions, Sep. Sci. Tech. 49 (2014) 101- 111. https://doi.org/10.1080/01496395.2013.815476.

[35] S. Noor Syuhadah, H. Rohasliney, Rice Husk as biosorbent: A Review, Heal. Environ. J. 3 (2012) 89–95.

[36] L. Armesto, A. Bahillo, K. Veijonen, A. Cabanillas, J. Otero, Combustion behaviour of rice husk in a bubbling fluidised bed, Biomass Bioenergy. 23 (2002) 171–179.

[37] E.I. El-Shafey, Behaviour of Reduction–Sorption of Chromium (VI) from an Aqueous Solution on a Modified Sorbent from Rice Husk, Water, Air, Soil Poll. 163 (2005) 81–102.

[38] K. Kenes, O. Yerdos, M. Zulkhair, D. Yerlan, Study on the effectiveness of thermally treated rice husks for petroleum adsorption, J. Non. Cryst. Solids. 358 (2012) 2964–2969. https://doi.org/10.1016/j.jnoncrysol.2012.07.017.

[39] D.S. Jung, M.-H. Ryou, Y.J. Sung, S. Bin Park, J.W. Choi, Recycling rice husks for high-capacity lithium battery anodes., Proc. Natl. Acad. Sci. U. S. A. 110 (2013) 12229–34. https://doi.org/10.1073/pnas.1305025110.

[40] T.H. Liou, C.C. Yang, Synthesis and surface characteristics of nanosilica produced from alkali-extracted rice husk ash, Mater. Sci. Eng. B 176 (2011) 521–529. https://doi.org/10.1016/j.mseb.2011.01.007.

[41] V. Della, I. Kuhn, D. Hotza, Rice husk ash as an elemente source for active silicaproduction, Mater. Lett. 57 (2002) 818–821. https://doi.org/10.1016/S0167-577X(02)00879-0.

[42] R. V Krishnarao, M.M. Godkhindi, Distribution of Silica in Rice Husks and its effect on the Formation of Silicon Carbide, Ceram. Int. 18 (1992) 243–249. https://doi.org/10.1016/0272-8842(92)90102-J.

[43] H. Muramatsu, Y.A. Kim, K.S. Yang, R. Cruz-Silva, I. Toda, T. Yamada, M. Terrones, M. Endo, T. Hayashi, H. Saitoh, Rice husk-derived graphene with nano-sized domains and clean edges, Small. 10 (2014) 2766–2770. https://doi.org/10.1002/smll.201400017.

[44] S. Sankar, S.K. Sharma, N. Kaur, B. Lee, D.Y. Kim, S. Lee, H. Jung, Biogenerated silica nanoparticles synthesized from sticky, red, and brown rice husk ashes by a chemical method, Ceram. Int. 42 (2016) 4875–4885. https://doi.org/10.1016/j.ceramint.2015.11.172.

[45] N. Yalçin, V. Sevinç, Studies on silica obtained from rice husk, Ceram. Int. 27 (2001) 219–224. https://doi.org/10.1016/S0272-8842(00)00068-7.

[46] http://rkb.irri.org/step-by-step-production/postharvest/rice-by-products/rice-husk.

[47] S. Apoorv, R. Khalid, Physical and Chemical Properties of Rice Husk Ash and Ground Granulated Blast Furnace Slag- A Review, Glob. J. Eng. Sci. Res. 1 (2014).

[48] A.N. Givi, S.A. Rashid, F.N.A. Aziz, M.A.M. Salleh, Assessment of the effects of rice husk ash particle size on strength, water permeability and workability of binary blended concrete, Constr. Build. Mater. 24 (2010) 2145–2150. https://doi.org/10.1016/j.conbuildmat.2010.04.045.

[49] Q. Feng, H. Yamamichi, M. Shoya, S. Sugita, Study on the pozzolanic properties of rice husk ash by hydrochloric acid pretreatment, Cem. Concr. Res. 34 (2004) 521–526. https://doi.org/10.1016/j.cemconres.2003.09.005.

[50] M. Nehdi, J. Duquette, A. El Damatty, Performance of rice husk ash produced using a new technology as a mineral admixture in concrete, Cem. Concr. Res. 33 (2003) 1203–1210. https://doi.org/10.1016/S0008-8846(03)00038-3.

[51] W. Xu, T.Y. Lo, S.A. Memon, Microstructure and reactivity of rich husk ash, Constr. Build. Mater. 29 (2012) 541–547. https://doi.org/10.1016/j.conbuildmat.2011.11.005.

[52] D.D. Bui, J. Hu, P. Stroeven, Particle size effect on the strength of rice husk ash blended gap-graded Portland cement concrete, Cem. Concr. Compos. 27 (2005) 357–366. https://doi.org/10.1016/j.cemconcomp.2004.05.002.

[53] M.H. Zhang, R. Lastra, V.M. Malhotra, Rice-husk ash paste and concrete: Some aspects of hydration and the microstructure of the interfacial zone between the aggregate and paste, Cem. Concr. Res. 26 (1996) 963–977. https://doi.org/10.1016/0008-8846(96)00061-0.

[54] M. Reddy, D.V., Alvarez, Marine Durability Characteristics of Rice Husk Ash-Modified Reinforced Concrete, in: Fourth LACCEI Int. Lat. Am. Caribb. Conf. Eng. Technol., Mayagüez, Puerto Rico, 2006.

[55] G.R. Rao, A.R.K. Sastry, P.K. Rohatgi, Nature and reactivity of silica available in rice husk and its ashes, Bull. Mater. Sci. 12 (1989) 469–479. https://doi.org/10.1007/BF02744917.

[56] V.T.A. Van, C. Rößler, D.D. Bui, H.M. Ludwig, Rice husk ash as both pozzolanic admixture and internal curing agent in ultra-high performance concrete, Cem. Concr. Compos. 53 (2014) 270–278. https://doi.org/10.1016/j.cemconcomp.2014.07.015.

[57] A.L.G. Gastaldini, G.C. Isaia, A.P. Saciloto, F. Missau, T.F. Hoppe, Influence of curing time on the chloride penetration resistance of concrete containing rice husk ash: A technical and economical feasibility study, Cem. Concr. Compos. 32 (2010) 783–793. https://doi.org/10.1016/j.cemconcomp.2010.08.001.

[58] P.M. Yeletsky, V.A. Yakovlev, M.S. Mel'gunov, V.N. Parmon, Synthesis of mesoporous carbons by leaching out natural silica templates of rice husk, Microporous Mesoporous Mater. 121 (2009) 34–40. https://doi.org/10.1016/j.micromeso.2008.12.025.

[59] U.R. Lakshmi, V.C. Srivastava, I.D. Mall, D.H. Lataye, Rice husk ash as an effective adsorbent: Evaluation of adsorptive characteristics for Indigo Carmine

dye, J. Environ. Manage. 90 (2009) 710–720. https://doi.org/10.1016/j.jenvman.2008.01.002.

[60] M. Jamil, M.N.N. Khan, M.R. Karim, A.B.M.A. Kaish, M.F.M. Zain, Physical and chemical contributions of Rice Husk Ash on the properties of mortar, Constr. Build. Mater. 128 (2016) 185–198. https://doi.org/10.1016/j.conbuildmat.2016.10.029.

[61] Q. Feng, Q. Lin, F. Gong, S. Sugita, M. Shoya, Adsorption of lead and mercury by rice husk ash, J. Colloid Interface Sci. 278 (2004) 1–8. https://doi.org/10.1016/j.jcis.2004.05.030.

[62] K.Y. Foo, B.H. Hameed, Utilization of rice husk ash as novel adsorbent: A judicious recycling of the colloidal agricultural waste, Adv. Colloid Interface Sci. 152 (2009) 39–47. https://doi.org/10.1016/j.cis.2009.09.005.

[63] R. V Krishnarao, J. Subrahmanyam, T.J. Kumar, Studies on the Formation of Black Particles in Rice Husk Silcia Ash, J. Eur. Ceram. Soc. 21 (2001) 99–104.

[64] X. Liu, X. Chen, L. Yang, H. Chen, Y. Tian, Z. Wang, A review on recent advances in the comprehensive application of rice husk ash, Res. Chem. Intermed. 42 (2016) 893–913. https://doi.org/10.1007/s11164-015-2061-y.

[65] K. Xu, Q. Sun, Y. Guo, Y. Zhang, S. Dong, Preparation of super-hydrophobic white carbon black from nano-rice husk ash, Res. Chem. Intermed. 40 (2014) 1965–1973. https://doi.org/10.1007/s11164-013-1094-3.

[66] P. Chindaprasirt, S. Rukzon, V. Sirivivatnanon, Resistance to chloride penetration of blended Portland cement mortar containing palm oil fuel ash, rice husk ash and fly ash, Constr. Build. Mater. 22 (2008) 932–938. https://doi.org/10.1016/j.conbuildmat.2006.12.001.

[67] W. Tangchirapat, R. Buranasing, C. Jaturapitakkul, P. Chindaprasirt, Influence of rice husk-bark ash on mechanical properties of concrete containing high amount of recycled aggregates, Constr. Build. Mater. 22 (2008) 1812–1819. https://doi.org/10.1016/j.conbuildmat.2007.05.004.

[68] A.L.G. Gastaldini, G.C. Isaia, T.F. Hoppe, F. Missau, A.P. Saciloto, Influence of the use of rice husk ash on the electrical resistivity of concrete: A technical and economic feasibility study, Constr. Build. Mater. 23 (2009) 3411–3419. https://doi.org/10.1016/j.conbuildmat.2009.06.039.

[69] M.R. Karim, M.F.M. Zain, M. Jamil, F.C. Lai, M.N. Islam, Use of wastes in construction industries as an energy saving approach, Energy Procedia. 12 (2011) 915–919. https://doi.org/10.1016/j.egypro.2011.10.120.

[70] A. Nazari, Predicting the total specific pore volume of geopolymers produced from waste ashes by gene expression programming, Neural Comput. Appl. 15 (2012) 1–7. https://doi.org/10.1007/s00521-012-1135-7.

[71] U.. Hassan, M.. Maharaz, The effects of Rice Husk Ash on Porcelain composition, Sci. World J. 10 (2015) 7–16.

[72] B.M. Miyandehi, A. Feizbakhsh, M.A. Yazdi, Q. feng Liu, J. Yang, P. Alipour, Performance and properties of mortar mixed with nano-CuO and rice husk ash, Cem. Concr. Compos. 74 (2016) 225–235. https://doi.org/10.1016/j.cemconcomp.2016.10.006.

[73] S.M.S. Kazmi, S. Abbas, M.A. Saleem, M.J. Munir, A. Khitab, Manufacturing of sustainable clay bricks: Utilization of waste sugarcane bagasse and rice husk ashes, Constr. Build. Mater. 120 (2016) 29–41. https://doi.org/10.1016/j.conbuildmat.2016.05.084.

[74] R. Siddique, K. Singh, Kunal, M. Singh, V. Corinaldesi, A. Rajor, Properties of bacterial rice husk ash concrete, Constr. Build. Mater. 121 (2016) 112–119. https://doi.org/10.1016/j.conbuildmat.2016.05.146.

[75] W. Yang, Y. Xue, S. Wu, Y. Xiao, M. Zhou, Performance investigation and environmental application of basic oxygen furnace slag ??? Rice husk ash based composite cementitious materials, Constr. Build. Mater. 123 (2016) 493–500. https://doi.org/10.1016/j.conbuildmat.2016.07.051.

[76] P. Sturm, G.J.G. Gluth, H.J.H. Brouwers, H.C. Kühne, Synthesizing one-part geopolymers from rice husk ash, Constr. Build. Mater. 124 (2016) 961–966. https://doi.org/10.1016/j.conbuildmat.2016.08.017.

[77] A. Mohammad, Inamuddin, S. Hussain, Poly (3,4-ethylenedioxythiophene): polystyrene sulfonate (PEDOT:PSS) Zr(IV) phosphate composite cation exchanger : sol-gel synthesis and physicochemical characterization, Ionics (Kiel). 21 (2015) 1063–1071.

[78] J. Alex, J. Dhanalakshmi, B. Ambedkar, Experimental investigation on rice husk ash as cement replacement on concrete production, Constr. Build. Mater. 127 (2016) 353–362. https://doi.org/10.1016/j.conbuildmat.2016.09.150.

[79] Inamuddin, Y.A. Ismail, Synthesis and characterization of electrically conducting poly-o-methoxyaniline Zr(1V) molybdate Cd(II) selective composite cation-exchanger, Desalination. 250 (2010) 523–529.

[80] D.T. Ambas, G. Mikhael, C. Dela Rosa, L.J.A. Esquejo, J.D. Gil, M.L. Magtalas, C.J.D. Rubinas, Effect of Rice Husk Ash as Cement Replacement in the Compressive Strength of Hydraulic Cement Mortar Cube, IEEE Explor (2015) 8–10.

[81] A.A. Ramezanianpour, M. Mahdikhani, Gh. Ahmadibeni, The effect of rice husk ash on mechanical properties and durability of sustainabe concretes, Int. J. Civ. Eng. 7 (2009) 83–91.

[82] http://www.pioneerstar.com/Pages/About_Products.aspx?Type=1&ID=25.

[83] K. Mohanta, D. Kumar, O. Parkash, Properties and Industrial Applications of Rice husk : A review, Int. J. Emerg. Technol. Adv. Eng. 2 (2012) 86–90.

[84] N. Priyantha, A.N. Navaratne, T. Kulasooriya, Adsorption of Heavy Metal Ions on Rice Husk : Isotherm Modeling and Error Analysis, Int. J. Earth Sci. Eng. 8 (2015) 346–352.

[85] A.S.M. E.A. Basha , R. Hashim , H.B. Mahmud, Stabilization of residual soil with rice husk ash and cement, J. Tek. Sipil. 5 (2004) 1–84.

[86] http://www.agnet.org/library.php?func=view&id=20110712165046.

[87] J.M. Dias, M.C.M. Alvim-Ferraz, M.F. Almeida, J. Rivera-Utrilla, M. Sánchez-Polo, Waste materials for activated carbon preparation and its use in aqueous-phase treatment: A review, J. Environ. Manage. 85 (2007) 833–846. https://doi.org/10.1016/j.jenvman.2007.07.031.

[88] O. Ioannidou, A. Zabaniotou, Agricultural residues as precursors for activated carbon production-A review, Renew. Sustain. Energy Rev. 11 (2007) 1966–2005. https://doi.org/10.1016/j.rser.2006.03.013.

[89] W.S. Wan-Ngah, M.A.K.M. Hanafiah, Removal of heavy metal ions from wastewater by chemically modified plant wastes as adsorbents: A review, Bioresour. Technol. 99 (2008) 3935–3948. https://doi.org/10.1016/j.biortech.2007.06.011.

[90] W.C. Li, F.Y. Law, Y.H.M. Chan, Biosorption studies on copper (II) and cadmium (II) using pretreated rice straw and rice husk, Environ. Sci. Pollut. Res. (2015) 1–3. https://doi.org/10.1007/s11356-015-5081-7.

[91] D. Angelova, I. Uzunov, S. Uzunova, A. Gigova, L. Minchev, Kinetics of oil and oil products adsorption by carbonized rice husks, Chem. Eng. J. 172 (2011) 306–311. https://doi.org/10.1016/j.cej.2011.05.114.

[92] P. Vassileva, A. Detcheva, I. Uzunov, S. Uzunova, Removal of Metal Ions From Aqueous Solutions Using Pyrolyzed Rice Husks: Adsorption Kinetics and Equilibria, Chem. Eng. Commun. 200 (2013) 1578–1599. https://doi.org/10.1080/00986445.2012.755519.

[93] Y. Zhang, J. Zhao, Z. Jiang, D. Shan, Y. Lu, Biosorption of Fe (II) and Mn (II) Ions from Aqueous Solution by Rice Husk Ash, Biomed Res. Int. 2014 (2014) 1–10.

[94] Santi, I. Raya, M. Zakir, The Adsorption of Pb ( II ) Ions on Activated Carbon from Rice Husk , Irradiated by Ultrasonic Waves : Kinetic and Thermodynamics Studies, J. Nat. Sci. Res. 4 (2014) 18–24.

[95] A. Elham, T. Hossein, H. Mahnoosh, Removal of Zn(II) and Pb (II) ions Using Rice Husk in Food Industrial Wastewater, J. Appl. Sci. Environ. Manag. 14 (2010) 159–162. https://doi.org/http://dx.doi.org/10.4314/jasem.v14i4.63306.

[96] S. Sugashini, K.M.M.S. Begum, Adsorption and desorption studies on the performance of Fe-loaded chitosan carbonized rice husk for metal ion removal, Desalin. Water Treat. 51 (2013) 7764–7774. https://doi.org/10.1080/19443994.2013.793947.

[97] M.G.A. Vieira, A.F.D.A. Neto, M.G.C. da-Silva, C.N. Carneiro, A.A.M. Filho, Adsorption of Lead and Copper Ions From Aqueous Effluents on Rice Husk Ash in a Dynamic System, Brazilian J. Chem. Eng. 31 (2014) 519–529. https://doi.org/10.1590/0104-6632.20140312s00002103.

[98] A.M. Youssef, A.I. Ahmed, M.I. Amin, U.A. El-Banna, Adsorption of lead by activated carbon developed from rice husk, Desalin. Water Treat. 54 (2014) 1–14. https://doi.org/10.1080/19443994.2014.896289.

[99] A.I. Habib, S.R.M. Kutty, N. Shafiq, M.F. Nuruddin, Adsorption of Metals (Zn, Ca and B) in Used Engine Oil by Using Microwave Incinerated Rice Husk Ash (MIRHA), in: InCIEC (2014) 1065-1073. https://link.springer.com/chapter/10.1007/978-981-287-290-6_94.

[100] D. Sarkar, S.K. Das, A. Bandyopadhyay, Analysis of Bio-Sorption of Cr(VI) onto Raw Rice Husk by a Hybrid Theoretical Model Using Results of Batch Experiments, Adsorpt. Sci. Technol. 31 (2013) 747–766. https://doi.org/10.1260/0263-6174.31.8.747.

[101] F.A. Adekola, D.S.S. Hodonou, H.I. Adegoke, Thermodynamic and kinetic studies of biosorption of iron and manganese from aqueous medium using rice husk ash, Appl. Water Sci. 6 (2016) 319–330.

[102] S. Sugashini, K.M.M.S. Begum, Studies on the Performance of Ethylamine-Modified Chitosan Carbonized Rice Husk Composite Beads for Adsorption of Metal Ion, Bioremediat. J. 17 (2013) 97–106. https://doi.org/http://dx.doi.org/10.1080/10889868.2013.786016.

[103] Y. Ding, D. Jing, H. Gong, L. Zhou, X. Yang, Biosorption of aquatic cadmium(II) by unmodified rice straw, Bioresour. Technol. 114 (2012) 20–25. https://doi.org/10.1016/j.biortech.2012.01.110.

[104] N.A. Sari, C.F. Ishak, R.A. Bakar, Characterization of oil palm empty fruit bunch and rice husk biochars and their potential to adsorb arsenic and cadmium, Am. J. Agric. Biol. Sci. 9 (2014) 450–456. https://doi.org/10.3844/ajabssp.2014.450.456.

[105] A. Masoumi, K. Hemmati, M. Ghaemy, Low-cost nanoparticles sorbent from modified rice husk and a copolymer for efficient removal of Pb(II) and crystal violet from water, Chemosphere. 146 (2016) 253–262. https://doi.org/10.1016/j.chemosphere.2015.12.017.

[106] S.T. Song, N. Saman, K. Johari, H. Mat, Surface chemistry modifications of rice husk toward enhancement of Hg(II) adsorption from aqueous solution, Clean Technol. Environ. Policy. 16 (2014) 1747–1755. https://doi.org/10.1007/s10098-014-0803-y.

[107] N. Li, J. Zhao, Z. Jiang, B. Cao, Y. Lu, D. Shan, Y. Zhang, Continuous manganese(II) ions removal from aqueous solutions using rice husk ash-packed column reactor, Desalin. Water Treat. 3994 (2016) 1–9. https://doi.org/10.1080/19443994.2015.1128362.

[108] S. Sarkar, S.K. Das, Removal of Cr(VI) and Cu(II) ions from aqueous solution by rice husk ash—column studies, Desalin. Water Treat. 57 (2016) 20340–20349. https://doi.org/10.1080/19443994.2015.1107754.

[109] S. Sobhanardakani, H. Parvizimosaed, E. Olyaie, Heavy metals removal from wastewaters using organic solid waste-rice husk, Environ. Sci. Pollut. Res. 20 (2013) 5265–5271. https://doi.org/10.1007/s11356-013-1516-1.

[110] L.S. Rocha, I. Lopes, C.B. Lopes, B. Henriques, A.M.V.M. Soares, A.C. Duarte, E. Pereira, Efficiency of a cleanup technology to remove mercury from natural waters by means of rice husk biowaste: Ecotoxicological and chemical approach,

Environ. Sci. Pollut. Res. 21 (2014) 8146–8156. https://doi.org/10.1007/s11356-014-2753-7.

[111] N.A. Sánchez-Flores, M. Solache, M.T. Olguín, J.J. Fripiat, G. Pacheco-Malagón, J.M. Saniger, S. Bulbulian, Selectivity of the Cd2+/Ca2+ exchange on modified rice hull silica, Environ. Technol. 30 (2009) 269–275. https://doi.org/10.1080/09593330802573290.

[112] T.K. Naiya, B. Singha, S.K. Das, FTIR Study for the Cr ( VI ) Removal from Aqueous Solution Using Rice Waste, Int. Proc. Chem. Biol. Environ. Eng. 10 (2011) 114–119.

[113] S. Kamsonlian, S. Suresh, V. Ramanaiah, C.B. Majumder, S. Chand, A. Kumar, Biosorptive behaviour of mango leaf powder and rice husk for arsenic(III) from aqueous solutions, Int. J. Environ. Sci. Technol. 9 (2012) 565–578. https://doi.org/10.1007/s13762-012-0054-6.

[114] A.W. Samsuri, F. Sadegh-Zadeh, B.J. Seh-Bardan, Characterization of biochars produced from oil palm and rice husks and their adsorption capacities for heavy metals, Int. J. Environ. Sci. Technol. 11 (2014) 967–976. https://doi.org/10.1007/s13762-013-0291-3.

[115] S. Luo, M.N. Shen, F. Wang, Q.R. Zeng, J.H. Shao, J.D. Gu, Synthesis of Fe3O4-loaded porous carbons developed from rice husk for removal of arsenate from aqueous solution, Int. J. Environ. Sci. Technol. 13 (2016) 1137–1148. https://doi.org/10.1007/s13762-016-0955-x.

[116] S. Aktas, M.H. Morcali, Gold uptake from dilute chloride solutions by a Lewatit TP 214 and activated rice husk, Int. J. Miner. Process. 101 (2011) 63–70. https://doi.org/10.1016/j.minpro.2011.07.007.

[117] O.J. Achadu, O.O. Ayejuyo, F.E. Ako, C.L. Dalla, O.O. Olaoye, Synchronous Adsorption of Cadmium and Lead ions from Aqueous Media by Rice Husk Ash and Sodium Dodecyl Sulfate Combination, Int. J. Mod. Anal. Sep. Sci. 3 (2014) 20–39.

[118] P. Taylor, G.C. Pathiraja, D.K. De Silva, L. Dhanapala, Desalination and Water Treatment Investigating the surface characteristics of chemically modified and unmodified rice husk ash ; bottom- up approach for adsorptive removal of water contaminants, Desalin. Water Treat. 3994 (2015) 37–41. https://doi.org/10.1080/19443994.2014.883133.

[119] A.G. El-Said, N.A. Badawy, A.Y. Abdel-Aal, S.E. Garamon, Optimization parameters for adsorption and desorption of Zn(II) and Se(IV) using rice husk ash: Kinetics and equilibrium, Ionics. 17 (2011) 263–270. https://doi.org/10.1007/s11581-010-0505-3.

[120] L. Vlaev, S. Turmanova, A. Dimitrova, Kinetics and thermodynamics of water adsorption onto rice husks ash filled polypropene composites during soaking, J. Polym. Res. 16 (2009) 151–164. https://doi.org/10.1007/s10965-008-9213-3.

[121] C. Jeon, Removal of copper ion using rice hulls, J. Ind. Eng. Chem. 17 (2011) 517–520. https://doi.org/10.1016/j.jiec.2010.10.020.

[122] H. Aydin, Y. Bulut, Ç. Yerlikaya, Removal of copper (II) from aqueous solution by adsorption onto low-cost adsorbents, J. Environ. Manage. 87 (2008) 37–45. https://doi.org/10.1016/j.jenvman.2007.01.005.

[123] S.A. Abo-El-Enein, M.A. Eissa, A.A. Diafullah, M.A. Rizk, F.M. Mohamed, Removal of some heavy metals ions from wastewater by copolymer of iron and aluminum impregnated with active silica derived from rice husk ash, J. Hazard. Mater. 172 (2009) 574–579. https://doi.org/10.1016/j.jhazmat.2009.07.036.

[124] T.K. Naiya, A.K. Bhattacharya, S. Mandal, S.K. Das, The sorption of lead(II) ions on rice husk ash, J. Hazard. Mater. 163 (2009) 1254–1264. https://doi.org/10.1016/j.jhazmat.2008.07.119.

[125] C.G. Rocha, D.A.M. Zaia, R.V. da S. Alfaya, A.A. da S. Alfaya, Use of rice straw as biosorbent for removal of Cu(II), Zn(II), Cd(II) and Hg(II) ions in industrial effluents, J. Hazard. Mater. 166 (2009) 383–388. https://doi.org/10.1016/j.jhazmat.2008.11.074.

[126] N. Sharma, K. Kaur, S. Kaur, Kinetic and equilibrium studies on the removal of Cd2+ ions from water using polyacrylamide grafted rice (Oryza sativa) husk and (Tectona grandis) saw dust, J. Hazard. Mater. 163 (2009) 1338–1344. https://doi.org/10.1016/j.jhazmat.2008.07.135.

[127] M. Bansal, U. Garg, D. Singh, V.K. Garg, Removal of Cr(VI) from aqueous solutions using pre-consumer processing agricultural waste: A case study of rice husk, J. Hazard. Mater. 162 (2009) 312–320. https://doi.org/10.1016/j.jhazmat.2008.05.037.

[128] X. Luo, Z. Deng, X. Lin, C. Zhang, Fixed-bed column study for Cu2+ removal from solution using expanding rice husk, J. Hazard. Mater. 187 (2011) 182–189. https://doi.org/10.1016/j.jhazmat.2011.01.019.

[129] K.K. Krishnani, X. Meng, C. Christodoulatos, V.M. Boddu, Biosorption mechanism of nine different heavy metals onto biomatrix from rice husk, J. Hazard. Mater. 153 (2008) 1222–1234. https://doi.org/10.1016/j.jhazmat.2007.09.113.

[130] S. Mohan, G. Sreelakshmi, Fixed bed column study for heavy metal removal using phosphate treated rice husk, J. Hazard. Mater. 153 (2008) 75–82. https://doi.org/10.1016/j.jhazmat.2007.08.021.

[131] V.G. Georgieva, M.P. Tavlieva, S.D. Genieva, L.T. Vlaev, Adsorption kinetics of Cr(VI) ions from aqueous solutions onto black rice husk ash, J. Mol. Liq. 208 (2015) 219–226. https://doi.org/10.1016/j.molliq.2015.04.047.

[132] M.P. Tavlieva, S.D. Genieva, V.G. Georgieva, L.T. Vlaev, Thermodynamics and kinetics of the removal of manganese(II) ions from aqueous solutions by white rice husk ash, J. Mol. Liq. 211 (2015) 938–947. https://doi.org/10.1016/j.molliq.2015.08.015.

[133] M.S. Lashkenari, B. Davodi, H. Eisazadeh, Removal of arsenic from aqueous solution using polyaniline/rice husk nanocomposite, Korean J. Chem. Eng. 28 (2011) 1532–1538. https://doi.org/10.1007/s11814-011-0014-8.

[134] Y. Fan, R. Yang, Z. Lei, N. Liu, J. Lv, S. Zhai, B. Zhai, L. Wang, Removal of Cr(VI) from aqueous solution by rice husk derived magnetic sorbents, Korean J. Chem. Eng. 33 (2016) 1416–1424. https://doi.org/10.1007/s11814-015-0248-y.

[135] S. Saadat, A.A. Hekmatzadeh, A.K. Jashni, Mathematical modeling of the Ni(II) removal from aqueous solutions onto pre-treated rice husk in fixed-bed columns: a comparison, Desalin. Water Treat. 57 (2016) 16907–16918. https://doi.org/10.1080/19443994.2015.1087877.

[136] K.G. Akpomie, F.A. Dawodu, Montmorillonite-rice husk composite for heavy metal sequestration from binary aqua media: a novel adsorbent, Trans. R. Soc. South Africa. 70 (2015) 83–88. https://doi.org/10.1080/0035919X.2014.984259.

[137] T.T. Pham, T.T.T. Mai, M.Q. Bui, T.X. Mai, H.Y. Tran, T.B. Phan, Nanostructured polyaniline rice husk composite as adsorption materials synthesized by different methods, Adv. Nat. Sci. Nanosci. Nanotechnol. 5 (2014) 15010. https://doi.org/10.1088/2043-6262/5/1/015010.

[138] I. Nhapi, N. Banadda, R. Murenzi, C.B. Sekomo, U.G. Wali, Removal of Heavy Metals from Industrial Wastewater Using Rice Husks, Open Environ. Eng. J. 4 (2011) 170–180. https://doi.org/10.2174/1874829501104010170.

[139] A.F.A. Tarmizi, K.K. Ong, W.M.Z.W. Yunus, A. Fitrianto, M.L. Japit, A.G. Hussin, C.C. Teoh, Sorption Isotherm Model of Zinc ( II ) onto Thermally Treated Rice Husk, Orien. J. Chem. 32 (2016).

[140] M.G. Pillai, I. Regupathi, M.H. Kalavathy, T. Murugesan, L.R. Miranda, Optimization and analysis of nickel adsorption on microwave irradiated rice husk using response surface methodology (RSM), J. Chem. Technol. Biotechnol. 84 (2009) 291–301. https://doi.org/10.1002/jctb.2038.

[141] Z. Asif, Z. Chen, Removal of arsenic from drinking water using rice husk, Appl. Water Sci. (2015). https://doi.org/10.1007/s13201-015-0323-x.

[142] G. Tan, H. Yuan, Y. Liu, D. Xiao, Removal of Cadmium from Aqueous Solution using Wheat Stem, Corncob, and Rice Husk, Sep. Sci. Technol. 46 (2011) 2049–2055. https://doi.org/10.1080/01496395.2011.573835.

[143] I. V. Sheveleva, A.N. Kholomeidik, A.V. Voit, N.P. Morgun, L.A.. Zemnukhova, Removal of metal ions with rice husk-based sorbents, Russ. J. Appl. Chem. 82 (2009) 1840–1844. https://doi.org/10.1134/S1070427209100188.

[144] S. Kamari, F. Ghorbani, Synthesis of magMCM-41 with rice husk silica as cadmium sorbent from aqueous solutions: parameters' optimization by response surface methodology, Environ. Technol. 0 (2016) 1–18. https://doi.org/10.1080/09593330.2016.1237557.

[145] Z.C. Zhao PF, Guo X, The Removal of Elemental Mercury by Potassium Permanganate-modified Rice Husk Ash Sorbents, Energy Sources Part A Recover. Util. Environ. Eff. 34 (2012) 994–1003. https://doi.org/10.1080/15567031003735295.

[146] M.R.H. M. Haris, N.A. A. Wahab, C.W. Reng, B. Azahari, K. Sathasivam, The sorption of cadmium(II) ions on mercerized rice husk and activated carbon, Turkish J. Chem. 35 (2011) 939–950. https://doi.org/10.3906/kim-1103-62.

[147] W. Shan, Z. Zhao, D. Fang, Z. Lou, J. Xu, S. Yue, B.K. Biswas, Y. Xiong, Investigation on the selective adsorption of Mo(VI) by using modified rice husk and corn straw, Waste and Biomass Valorization. 4 (2013) 385–393. https://doi.org/10.1007/s12649-012-9149-z.

[148] P. Fongsatitkul, P. Elefsiniotis, D. Kitkaew, C. Rungsipanodorn, Use of rice husk ash as an admixture to remove chromium from a tannery waste, Water. Air. Soil Pollut. 220 (2011) 81–88. https://doi.org/10.1007/s11270-010-0736-y.

[149] L.S. Rocha, C.B. Lopes, J.A. Borges, A.C. Duarte, E. Pereira, Valuation of Unmodified Rice Husk Waste as an Eco-Friendly Sorbent to Remove Mercury: A Study Using Environmental Realistic Concentrations, Water. Air. Soil Pollut. 224 (2013) 1559. https://doi.org/10.1007/s11270-013-1599-9.

[150] J. Wongjunda, P. Saueprasearsit, Biosorption of Chromium (VI) Using Rice Husk Ash and Modified Rice Husk Ash, Env. Res. J. 4 (2010) 244-250.

[151] D. Sivakumar, Hexavalent chromium removal in a tannery industry wastewater using rice husk silica, Global J. Environ. Sci. Manage. 1 (2015) 27-40.

[152] M. Ajmal, R.A.K. Rao, S. Anwar, J. Ahmad, R. Ahmad, Adsorption studies on rice husk: removal and recovery of Cd(II) from wastewater, Biores. Tech. 2 (2003) 147-149.

[153] S. Sobhanardakani, R. Zandipak, Adsorption of Ni (II) and Cd (II) from Aqueous Solutions Using Modified Rice Husk, Iran J Health Sci 3 (2015) 1-9.

[154] T.K. Naiya, A.K. Bhattacharya, S.K. Das, Adsorptive removal of Cd(II) ions from aqueous solutions by rice husk ash, Env. Prog. Sust. Ener. 28 (2009) 535-546.

[155] A. Kausar, H.N. Bhatti, R.A. Sarfraz, M. Shahid, Prediction of optimum equilibrium and kinetic models for U(VI) sorption onto rice husk: comparison of linear and nonlinear regression methods, Desal. Water Treat. 52 (2014) 1495-1503.

# Chapter 5

# Natural husks as potential adsorbents for uptake of heavy metals

Prashant Pandey and Vipin Kumar Saini

School of Environment and Natural Resources, Doon University, Dehradun, Uttarakhand 248001, India

**Abstract**

The naturally available materials are the sustainable choice for the adsorptive removal of various pollutants from wastewater. The disposal and management of agricultural wastes have been the concern of socio-economic and environmental importance. Hence, agricultural wastes especially husks (originated from rice, peanuts, coconut, black gram, etc.) have been frequently used as anatural source for activated carbon. Studies confirm that husks based adsorbents (either modified or unmodified) have reasonable adsorption potential for uptake of metal ions from wastewater. However, there is still a need to find out the practical utility of such developed adsorbents on a commercial scale, leading to better materials for adsorptive removal applications. In this context, this chapter is intended to review the role of husks as adsorbent development, particularly for heavy metals removal.

**Keywords**

Acticated Carbon, Heavy Metals, Wastewater, Adsorption, Adsorbent, Husks

## Contents

## 1. Introduction

Minimization of water pollution is one of the most challenging environmental problems at present [1-5]. The rapid and unmanaged industrialization has resulted in the generation of various water and wastewater pollutants such as heavy metals, dyes, phenols (including other organic compounds), inorganic anions, and pesticides (which are toxic to many living life forms and organisms) [6-9]. Among these, incessant discharge of heavy metals beyond the tolerable limit attracted researchers around the world. However heavy metals are considered any natural occurring metallic elements with high atomic weight and density at least 5 times greater than water. Heavy metals are considered as trace elements due to their presence in trace concentrations (ppb range to less than 10ppm) in several environmental matrices [10, 11]. Some of the heavy metals such as arsenic (As), cadmium (Cd), chromium (Cr), copper (Cu), nickel (Ni), lead (Pb), and zinc (Zn) have been considered most hazardous because of high solubility in the water [12]. These metals enter into the food chain and get accumulated in the human body, resulting in affecting human cellular organelles and components such as cell membrane, mitochondria, lysosome, endoplasmic reticulum, nuclei, and some enzymes (involved in metabolism, detoxification, and damage repair) [13].

Therefore, a large number of water purification and recycling technologies such as chemical oxidation-reduction, reverse osmosis, coagulation, flocculation, precipitation, adsorption, ion exchange, electrodeposition and filtration have been applied to mitigate this problem [14-17]. But the economic and technical limitations restrict their use for wider applicability [16]. However, among them, adsorption based technologies offer a simple, inexpensive, effective and universal alternative [18]. The adsorbent is a critical part of all adsorption based technologies. A lot of adsorbents such as activated carbon, zeolites, clay, metal oxides, and bio-adsorbents have been studied for heavy metals removal from water and waste water [19, 20]. Activated carbon has been the most preferred due to its low cost, good sorption capacity, naturally availability of raw source, and regenerating property [21-24].

Heavy metals can be defined as any metals and metalloid elements having a density between 3.5 to 7 g/cm$^3$ along atomic weights in between 63.5 to 200.6 [25-27]. Some of the elements which are characterized under this category are mercury (Hg), cadmium

(Cd), arsenic (As), chromium (Cr), thallium (Tl), zinc (Zn), nickel (Ni), copper (Cu) and lead (Pb) etc. These are natural components of the earth's crust and can be generated as pollutant *via,* natural or anthropogenic sources. These are non-biodegradable and up to some extent enter our bodies *via* food, drinking water, and air, which results in bioaccumulation. As trace elements, some heavy metals such as copper and selenium are essential to maintaining the metabolism of the human body but at higher concentrations, they can lead to serious disease [28, 29]. Heavy metal poisoning could result, for instance, from drinking-water contamination, high ambient air concentrations near emission sources, or intake *via* the food chain. The mining wastes, landfill leaches, municipal wastewater, urban runoff and industrial wastewaters are the major sources of heavy metals in the environment. Till date, about 10% of discharged waste water from industries is treated and residual wastewater is released into water sources such as rivers, seas, and diffused into groundwater [30]. The presence of heavy metals in water and wastewater is a global issue although their concentration varies from place to place. These metals have been extensively studied for their effects on human health, and the reports are regularly reviewed by several national and international organizations such as the World Health Organisation (WHO), Central Pollution Control Board (CPCB) and others. Table 1 lists sources and toxicological effects of some of the heavy metals. Both acute and chronic heavy metal exposure directly or indirectly may damage human's central nervous function, the cardiovascular and gastrointestinal (GI) systems, lungs, kidneys, liver, endocrine glands, and bones [29, 31]. Most industrial processes pose less demanding requirements on the quality of freshwater and therefore criteria are usually developed for raw water in relation to its use as a source of water for drinking-water supply, agriculture, and recreation, or as a habitat for biological communities. Standards may also be developed, in relation to the functioning of aquatic ecosystems in general. The protection and maintenance of these usually impose different requirements on water quality and, therefore, the associated water quality criteria are often different for each use.

*Table 1: Standards for heavy metals from wastewater by central pollution control board (CPCB), India.*

| S. No. | Heavy metal | Industrial sources | Health Effects | Industrial Effluent discharge (ppm) | | |
|---|---|---|---|---|---|---|
| | | | | Surface water | Public sewers | Marine coastal |
| 1 | Arsenic | Wood processing industry | Skin manifestations, visceral cancers, vascular disease | 0.2 | 0.2 | 0.2 |
| 2 | Chromium | Electroplating and metal surface treatment processes, wood processing industry, petroleum refining | A headache, diarrhea, nausea, vomiting, carcinogenic | 2 | 2 | 2 |
| 3 | Copper | Wood processing industry | Liver damage, Wilson disease, insomnia | 3 | 3 | 3 |
| 4 | Cadmium | Electroplating and metal surface treatment processes | Kidney damage, renal disorder, human carcinogen | 2 | 1 | 2 |
| 5 | Nickel | Printed circuit board (PCB)manufacturing, petroleum refining | Dermatitis, nausea, chronic asthma, coughing, human carcinogen | 3 | 3 | 5 |
| 6 | Zinc | Electroplating and metal surface treatment processes | Depression, lethargy, neurological signs and increased thirst | 5 | 15 | 15 |

| | | | | | | |
|---|---|---|---|---|---|---|
| 7 | Lead | Printed circuit board (PCB)manufacturing | Damage the fetal brain, diseases of the kidneys, circulatory system, and nervous system | 0.1 | 1 | 2 |
| 8 | Mercury | Coal combustion, smelting, incineration, cremation | Rheumatoid arthritis, and diseases of the kidneys, circulatory system, and nervous system | 0.01 | 0.01 | 0.01 |
| 9 | Vanadium | Synthesis of sulphuric acid used as steel additive | Increased level may cause paralysis | 0.2 | 0.2 | 0.2 |
| 10 | Iron | Alloy making industries, fungicides, dye industries | Increased pulse rate, congestion, of blood vessels, hypertension | 3 | 3 | 3 |

For the development of environmental sustainable adsorption processes, the researchers are focusing on locally available, inexpensive and sustainable adsorbents as agricultural wastes, industrial by-products, and natural raw materials etc. The natural products used as adsorbents have been crab shell, egg shell, husks, bark *etc.* [32-37]. These materials are not only considered, as low cost but also require little processing, besides being abundant in nature. Husk is particularly important due to its abundant availability as agricultural waste and favourable physiochemical properties [38]. Husks are the outer covering or membrane of some fruits or seeds and are usually considered worthless, which make them a good source of activated carbon. For example, world harvest about 500 million ton of rice per year and its husk remains as an unused waste material. Besides, dumping it some small-scale industries uses it as burning material. In India, its annual generation is about 18-22 million tons [16]. Hence, such agricultural waste or by-products can be a good source of raw materials for water and wastewater treatment processes. Fig. 1 hereby shows thedevelopment of adsorbents from natural agricultural wastes for removal of heavy metals.

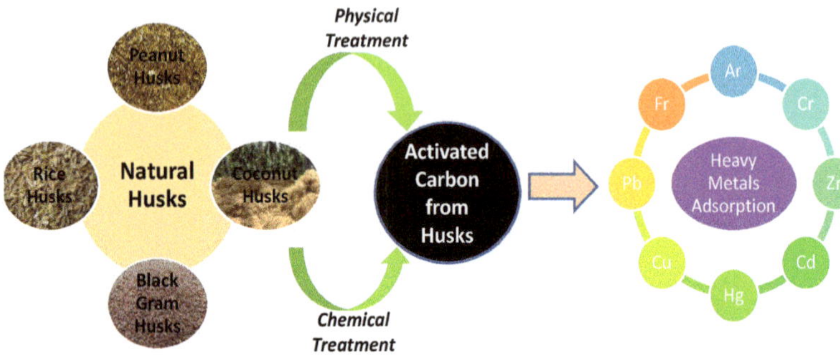


*Fig. 1: Development of activated carbon from husks for heavy metals removal.*

## 2. Mechanism of adsorption for heavy metals on husks

For adsorption of metal ions, it is very important to understand the underlying mechanism of interaction between adsorbent and adsorbate for the selection of proper adsorbent. Adsorption is the phenomenon that involves two phases one is a solid phase (adsorbent) and another one is a liquid phase (solvent) containing dissolved species to be adsorbed (hereby-heavy metals). The presence of functional groups on the surfaces of adsorbent and pores facilitate suitable conditions for the adsorbent to attract metal ions and bound by are latively complex process which is affected by several mechanisms involving physical adsorption, chemisorption, complexation, ion exchange and chelation etc. [39, 40]. Lignin and cellulose are the major constituents of husks whereas; hemicellulose, lipids, proteins, simple sugars, starch, water, hydrocarbons, and ash are minor components. Husks also constitute of some more compounds containing a variety of functional groups (carboxyl, hydroxyl, sulphydryl, amideand amine etc.) which operate in the binding process [11, 41]. Cellulose is acrystalline-homopolymer of glucose with $\beta 1 \rightarrow 4$ glycosidic linkage and intra-molecular and intermolecular hydrogen bonds [42]. Hemicellulose is a heteropolymer of main xylose with $\beta 1 \rightarrow 4$ glycosidic linkages with other substances of acetyl feruloyl and glycouronyl groups [43]. Lignins are three-dimensional polymer of aromatic compounds covalently linked with xylans in hardwoods and galactoglucomannans in softwoods [44]. Several functional groups are present in husks including acetamido, carbonyl, phenolic, amido, amino, sulphydryl, and carboxyl etc. [45, 46]. All these groups show an affinity towards metal complexation. Some of the adsorbents from husks are non-selective and bind to a wide range of heavy metals with no specific priority, whereas others are specific for certain types of metals depending upon their chemical composition [47] and the charge on heavy metal.

## 3. Application of various husks for heavy metal adsorption

### 3.1 Rice husks

India produces 104,800 million tons rice per year, which is about 20% of the total world rice production. At milling of paddy, about 78 % of its total weight is received as rice, broken rice, bran and rest 22% as husk [48]. Hence, large production of husks causes serious disposal problem even if it is biodegradable in nature. Therefore, attention has been focused on utilization of husks (waste material) that could solve both disposal problem and utilization as less-expensive material in waste-water treatment. From last three decades researchers intensively focused on preparation of effective adsorbents from rice husks for removal of heavy metals like arsenic (As), gold (Au), cadmium (Cd), cobalt (Co), chromium (Cr), copper(Cu), mercury (Hg), nickel (Ni), lead (Pb) and zinc (Zn). This effort has been possible due to useful physio-chemical properties of rice husks that fulfil the criteria to bind metal ions. Rice husks constitutes organic matter (24.3%, cellulose 34.4% and lignin 19.2%) and inorganic substances (carbon 37.05 wt.%, hydrogen 8.80 wt.%, nitrogen 11.06 wt.%, silicon 9.01 wt.% and oxygen 35.03 wt.%) and ash 18.85% [49-51]. The surface of husk contains silanol groups in form of silicon dioxide structure (Si-O-Si-OH) that are similar to silanol group of silicic acid. The presence of polar groups on the surface provides considerable cation exchange capacity to the adsorbent. Their functional groups along with other components facilitate rice husks to adsorb pollutants like metal ions [16]. The FTIR spectra of rice husks having several absorption peaks illustrate the complex nature as an adsorbent. The absorption peak around 3660.6 $cm^{-1}$ indicates the existence of free and intermolecular bonded hydroxyl groups. The peak observed at 3076.2 $cm^{-1}$ can be assigned to stretching vibration of the C-H group. The peak around 1664.4 $cm^{-1}$ corresponds to the C=O stretching that may be attributed to the lignin aromatic groups [52]. The additional peaks at 848.6 and 750.2 $cm^{-1}$ can be assigned for bending modes of aromatic compounds [53]. Roy *et al.*, [54], used rice husks for the removal of As, Cd, Cr, Pb, and Sr. The rice husks showed a greater affinity towards Pb than Sr. It was reported that at lower pH of the medium, cationic metals can be desorbed from the adsorbent. Unmodified rice husks showed a low affinity towards metal ions. Thus, surface modification fulfils all the criteria for adsorption of heavy metals [54]. Munaf and Zein [55] used, modified the rice husk modified with 1% HCl as an adsorbent for the removal of Cr, Zn, Cu, and Cd ions. Under optimised experimental conditions, Cr showed a maximum affinity for an adsorbent [55]. Wong *et al.*, [56], used tartaric acid to modify rice husk for removal of Cu and Pb from waste water. The presence of carboxylic group on the surface of adsorbent enhances their affinity for these ions and shows maximum adsorption capacity of 29 and 108 mg/g for Cu and Pb respectively [56]. Many studies have also been performed to compare the adsorption efficiencies between unmodified and modified rice husks. Tarley *et al.*, [57], used unmodified and NaOH modified adsorbents for on-line preconcentration of Cd (II) and Pb(II) [57]. Mohan and Shrilakshami, [58], compared the performance of raw rice husk and phosphate treated rice husk in the removal of lead, copper, zinc and manganese through fixed bed column study. They observed that adsorption capacity increases with the increase of contact time as well as bed height. For removal of all metal ions, minimum column height reduced three times for phosphate treated rice husks as compared to raw one [58]. Bansal *et al.*, [59], used pre-boiled and formaldehyde treated rice husks for Cr (VI) removal from aqueous solution. Formaldehyde-treated rice husk showed more affinity towards Cr (VI) than boiled rice husk [59]. M. G. A.Vieira *et al.*, [60], used natural rice husks and calcined rice husks at 500°C for 1 h for Ni (II), Cd (II), Zn (II), Pb (II) and Cu (II) removal. Calcination of results in the increase of surface area and promotes the adsorption of lead and copper metal ions [60]. Table 2 lists some adsorbents derived from rice husks for heavy metals removal from wastewater.

*Table 2: Maximum adsorption capacities of some adsorbent derived from rice husks.*

| S. No. | Rice husk | Heavy metal | Adsorption capacity (mg/g) | Reference |
|---|---|---|---|---|
| 1 | Pulverized rice husk | As (IV) | 615.1 | [54] |
| 2 | HCl modified rice husk | Cr | 0.27 | [55] |
| | | Zn | 0.17 | |
| | | Cu | 0.20 | |
| | | Cd | 0.16 | |
| 3 | Tartaric acid modified rice husks | Cu | 29.0 | [56] |
| | | Pb | 108.0 | |
| 4a | rice husk | Cd | 4.0 | [57] |
| | | Pb | 9.45 | |
| 4b | NaOH modified rice husk | Cd | 7.40 | |
| | | Pb | 21.55 | |
| 5 | Rice husk based carbon | Cr (VI) | 48.31 | [59] |
| 6a | rice husk | Ni | 0.0 | [60] |
| | | Cd | 1.98 | |
| | | Zn | 0.91 | |
| | | Pb | 6.41 | |
| | | Cu | 2.38 | |
| 6b | calcined rice husk(500°C) | Ni | 0.0 | |
| | | Cd | 0.45 | |

| | | | | |
|---|---|---|---|---|
| | | Zn<br>Pb<br>Cu | 1.10<br>6.69<br>3.25 | |
| 7 | Iron oxide coated rice husks | As (V) | 2.5 | [61] |
| 8 | Iron and Al copolymer impregnated with silica (PAlFeClSi) | Pb<br>Fe<br>Mn<br>As | 416<br>222<br>158<br>146 | [62] |
| 9 | Fe-coated rice husk | As (III)<br>As (V) | 30.7<br>16.0 | [63] |
| 10 | Hexadecylpyridinium bromide modified rice husk | Cr (VI) | 1.40 | [64] |

### 3.2 Peanut husks

Global peanut (groundnut) production is about 29 million metric tons per year. India producing about 5-8 million tons peanuts every year, is the second largest producer of peanut after China. Peanuts or groundnut kernels are approximately 70% of weight in shells. The rest 30% of total production is peanut hulls, which is an agricultural waste [65]. The vast quantity of waste is generally dumped in landfills. Therefore, its utilization as a source of activated carbons provides not only a good solution to its disposal problem but also an application in wastewater treatment process. Its macro porous nature, resistance to temperature, pH, mechanical stress and prolonged submersion in water proves it an effective alternative to binding metal ions. It constitutes primarily of cellulose (35.7%), hemicelluloses (18.7%) and lignin (30.2%). Lignin is a polymer of phenyl propene-like structures and the functional groups include methoxyl($-OCH_3$), hydroxyl (–OH), and aldehydes [11, 66]. The binding of metal ions in lignin is due to the interaction between these groups on the lignin molecules and the charged metal ions in aqueous solution. Reports showed, that in an aqueous solution of $pH > 3.5$, the functional groups of peanut husk acquire a negative charge and are able to adsorb cationic species through electrostatic attraction [67]. The presence of functional groups has been assessed by IR absorption bands. The band in the region 1000–1250 $cm^{-1}$ occur with the broadest, the strongest overlapping bands involving the stretching vibrations of the C-O bonds in peanut husks carbon after adsorption of heavy metals. The 1360–1380 $cm^{-1}$ bands may be attributed to the aromatic CH and carboxyl- carbonate structures. The 1700–1760 $cm^{-1}$ regions together with that at about 2400– 2450 $cm^{-1}$ are characteristic of the carboxylic acid groups [66].

According to Husaini, *et al.*, [68], the peanut husk releases hydrogen ions in solution as it binds metal ions and acts as an acidic ion exchanger, causing a net decrease in the pH of the solution [68]. Periasamy and Namasivayam prepared activated carbon from peanut hulls, by treating them with bicarbonates and used them to adsorb Ni (II) from aqueous solution. Their comparative study with granular activated carbon showed that the prepared adsorbent from peanut husks had 36 times higher adsorption capacity. The adsorbent from peanut husks showed 53.6 mg/g adsorption capacity at pH 6.5 and with the optimum dosage of 100 mg/100 mL for Ni (II) ions [69]. Qin Li *et al.*, [67], used modified the peanut husk for the removal of Pb(II), Cr(III) and Cu(II) ions. They used the sulphuric acid for surface modification and to reduce the organic pigments in order to increase the affinity towards Pb(II) (29.14 mg/g), Cr (III) (7.67 mg/g), Cu(II) (10.15 mg/g). To explain the adsorption mechanism, they used three different kinds of kinetic models (intraarticular diffusion model, Lagergren-first-order, and second-order equations) [67]. Salam *et al.*, [70], prepared modified peanut husks using 0.1 N HCl to study the adsorption behaviors for $Cu^{2+}$ and $Zn^{2+}$ ions [70]. Ugwekar and Lakhawat, [71], treated the peanut husks with 1 N phosphoric acid and used it to adsorb Cu ions from aqueous solution [71].Witek-Krowiak *et al.*,[72], showed the adsorption of Cu (II) and Cr(III) ions from aqueous solutions by peanut husks (raw sample only washed with water) with adsorption capacities of 25.39 mg/g for $Cu^{2+}$ and 27.86 mg/g for $Cr^{3+}$ [72]. Al-Othman *et al.*, [73], prepared activated carbon from peanut shell by chemical activation with 20% KOH at 1:1 ratio. The two samples (oxidized and unoxidized) were prepared. The unoxidized sample was prepared in the presence of nitrogen atmosphere which was then heated in air at a desired temperature to get oxidized activated carbon and then both the samples were tested for the adsorption of Cr (VI). The oxidized carbon with the higher surface area, better pore volume showed higher adsorption capacity than the unoxidized carbon [73]. Table 3 summarizessome of theadsorbents derived from peanut husks and hulls.

*Table 3: Maximum adsorbent capacities of some adsorbents derived from peanut husks.*

| S. No. | Husks | Heavy metal | Adsorption capacity (mg/g) | Reference |
|---|---|---|---|---|
| 1 | Peanut husk carbon (PHC) | Pb<br>Zn<br>Ni<br>Cu | 31.05<br>8.50<br>7.33<br>12.36 | [66] |
| 2 | Formalin and $H_2SO_4$ treated peanut husk | Pb<br>Cr<br>Cu | 29.14<br>7.67<br>10.15 | [67] |

| | | | | |
|---|---|---|---|---|
| 3 | Peanut husk carbon (PHC) | Ni | 53.6 | [69] |
| 4 | HCl treated peanut husk(charcoal) | Cu<br>Zn | 0.35<br>0.37 | [70] |
| 5 | Peanut husk | Cu<br>Cr | 25.39<br>27.86 | [72] |
| 6a | Peanut Husk Carbon (unoxidized) | Cr | 14.31 | [73] |
| 6b | Peanut husk carbon (oxidized) | Cr | 16.26 | |
| 7 | Peanut husks | Ni | 53.6 | [74] |
| 8 | Activated carbon from peanut husks | Cu | 65.6 | [75] |
| 9 | Sulphuric acid modified peanut husk | Pb<br>Cr<br>Cu | 29.14<br>7.67<br>10.15 | [76] |
| 10 | Peanut hulls | Cu | 21.25 | [77] |

### 3.3 Coconut husks

India is the third largest producer of coconut, cultivating about 10.9 million tons of coconut every year from its vast area of 2 million hectares. The coconut fruit yields 40 % coconut husks (Coconut mesocarp) containing 30% fiber, with dust making up the rest, which is agricultural waste from the local coconut industry and often causes serious disposal problems for local environment [78]. Chemically the coconut husks consist of cellulose, lignin, pyroligneous acid, gas, charcoal, tar, tannin, and potassium. Coconut husk and shells are an attractive source of activated carbon. The major advantage of using coconut biomass is that coconut is a permanent crop and available round the year so there is a constant supply. In this context, coconut husk may be one of the low-cost adsorbents that can serve intensively for the removal of various heavy metals and metalloids such as Pb, Cd, Zn, Ni, Cu, Cr, Hg and As. In coconut husk several functional groups such as aromatic groups, carbonyl and phenolic groups are present and all may be involved in metal binding [79]. The unprocessed husks exhibit lower adsorption or affinity towards heavy metals, therefore several treatments and modification methods have been developed to overcome this drawback. The adsorption through coconut husks for Cr (III) ions has been studied using radiotracer and batch techniques by Ahmad *et al.,* [80]. In their study maximum adsorption was achieved within 30 min with the maximum Langmuir sorption capacity of 18.25 μmol/g [80]. Manju *et al.,* [81] prepared an adsorbent with copper impregnation onto coconut husk which was employed for As (III) removal. The maximum adsorption capacity was observed at pH 12.0. The adsorption capacity increases with temperature from 30 to 60 °C, which was found to be 146.30 to 158.65 mg/g respectively. The experimental adsorption data fitted reasonably well to the Langmuir isotherm. Desorption studies revealed that spent adsorbent could be regenerated and reused by 30% $H_2O_2$ in 0.5 M $HNO_3$ [81].

Hasany and Ahmed, 2006 [80], explored the coconut husk's applicability for the removal of Cd(II), Cr(III) and Hg(II) ions from aqueous solutions. The adsorption data were analyzed using different sorption isotherms. The values of the Freundlich constants, 1/n and Cm, were 0.92 and 52.6 mmol/g; 0.85 and 56.0 mmol/g; and 0.88 and 6.84 mmol/g for Cd(II) Cr(III) and Hg(II) ions, respectively. The sorption process was found to be endothermic and spontaneous in nature with weak bond formation between the metal ions and coconut husk. Among various ions tested, the only borate was found to reduce Cr (III) whereas Hg(II) and Cd(II) sorption decreased in the presence of thiosulfate and ascorbate ions, respectively. A combination of ion exchange and surface complexation of metal ions was proposed as the possible mechanism for the sorption of Hg(II), Cd(II) and Cr(III) onto coconut husk[82]. Apaolaza and Guerrero [83, used coconut husks and mixed with composted sewage sludge containing Cd, Pb, Zn and nitrate. Their approach was to minimize the environmental risks of using sewage sludges [83]. Sreedhar *et al.*,[84], modified the coconut husk using polysulphide for Hg (II) removal from the aqueous phase. The parameters such as its concentration, agitation time, pH, sorbent dose, ionic strength and temperature affecting Hg (II) adsorption were studied to estimate optimum conditions. Maximum adsorption capacity was observed in the pH range 5.5-10.0. The adsorption isotherm was also affected by temperature since the adsorption capacity was increased by raising the temperature from 30 to 60°C. Lagergren equation and Langmuir and Freundlich isotherm models have been used to determine rate constants of adsorption and capacity of treated coconut husk [84]. Agbozu and Emoruwa [85], prepared adsorbent from coconut husks by treating with 0.1 MHCl and investigated their affinity to adsorb Cu, Pb, Fe, Cr and Cd from aqueous solution. They observed that the percentage removal of metals increases with increasing in adsorbent dose (Cr>Cu>Pb>Fe>Cd). The initial concentration of metal ions and pH showed significant effect on adsorption capacity [85]. Malik *et al.,* [86], investigated the adsorption capacity of unmodified coconut husks for the removal of heavy metal ions ($Pb^{2+}$, $Cu^{2+}$, $Ni^{2+}$ and $Zn^{2+}$) from industrial wastewater. Their results showed that maximum uptake occur at 443.0 mg/g (88.6%) for Cu, 404.5 mg/g (80.9%), for Ni 362.2 mg/g (72.4%) for $Pb^{2+}$ and 338.0 mg/g (67.6%) for $Zn^{2+}$. These adsorption capacities were found to be sensitive to the amount of adsorbent, heavy metal ion concentration, pH, temperature and contact time. The theoretical studies using quantum provide the rich electron donation sites of coconut

Husk [86]. Table 4 summarizes the adsorbents obtained from coconut husks for heavy metals removal fromwastewater.

*Table 4: Maximum adsorbent capacities of some adsorbents derived from coconut husks.*

| S. No. | Coconut husks | Heavy metal | Adsorption capacity[mg/g] | References |
|---|---|---|---|---|
| 1 | Carbonized coir pith | Cu | 39.70 | [78] |
| 2 | Coconut husk | Cr | 0.35 | [80] |
| 3 | Coconut husk carbon impregnated with $Cu^{2+}$ | As | 158.65 | [81] |
| 4 | Coconut Husk | Cd<br>Cr<br>Hg | 5912.2<br>2912.2<br>1374.8 | [82] |
| 5 | Polysulphide treated coconut husk | Hg | 53.74 | [84] |
| 6 | HCl treated coconut husk | Cu<br>Cd<br>Fe<br>Cr<br>Pb | $1.0\times10^{-4}$<br>$1.0\times10^{-4}$<br>$0.9\times10^{-4}$<br>$3.1\times10^{-4}$<br>$3.9\times10^{-4}$ | [85] |
| 7 | Unmodified coconut husks | Cu<br>Ni<br>Pb<br>Zn | 443.0<br>404.5<br>362.2<br>338.0 | [86] |

### 3.4 Black gram husks

The husk of black gram, which is also called Bengal gram or common gram, is another agricultural waste. The husks of a black gram are generated as a waste of no utility during the seed-splitting milling process. The percentage proximate composition of husk on dry weight basis revealed moisture (15.5%), protein (3.32%), fat (0.7%), crude fiber (51.9%), andash (4.93%). It's crude fibre constitutes about 69% cellulose, 11.98% hemicellulose, and 19.14% lignin [87]. Saeed and Iqbal used this husk for the removal of cadmium at 10 ppm concentration level. They used several desorbing agents for adsorbent recovery and HCl was found most appropriate with 99.89% of desorption. During desorption study, adsorption–desorption cycles in batch operations showed declining trend.The decline was only 9.71% when compensated for husk loss, which is comparable to 10.45% decline during six cycles in fixed bed column bioreactor in which sample (husk) loss was only 5.98% [87]. In 2005, Saeed and co-workers [88] again used this husk as an adsorbent for Pb, Zn, Cu, and Ni removal. The maximum amount of heavy metals adsorbed at equilibrium was 49.97, 33.81, 25.73 and 19.56 mg/g for Pb, Zn, Cu, and Ni, respectively [88]. Ramachandra *et al.,* [89], investigated the adsorption of Cr (VI) using Bengal gram husks and the maximum adsorption capacity was found as 91.64 mg/g [89]. Ahalya *et al.,*[90], shows their applicability for the adsorption of Fe (III) with adsorption capacity of 72.16 mg/g. They also carried out desorption studies at different concentrations of hydrochloric acid showing the quantitative recovery of the metal ion from black gram husks [90]. Table 5 concluded the adsorbents from black gram husks for removal of heavy metals from wastewater.

*Table 5: Maximum adsorbent capacities of some adsorbents derived from black gram husks.*

| S .No. | Black gram husks | Heavy metal | Adsorption capacity (mg/g) | Reference |
|---|---|---|---|---|
| 1 | Black gram husk | Cd | 38.76 | [87] |
| 2 | Black gram husk | Pb<br>Cd<br>Zn<br>Cu<br>Ni | 49.97<br>39.99<br>33.81<br>25.73<br>19.56 | [88] |
| 3 | Black gram husk | Cr | 91.64 | [89] |
| 4 | Black gram husk | Fe | 72.16 | [90] |
| 5 | Black gram husk | Zn | 33.81 | [91] |

## 4. Conclusion

Adsorption with husks showed significant contribution for the removal of heavy metals from wastewater. The results of the studies carried out for heavy metal ion adsorption onto agricultural waste husks indicated the potential of adsorbents for wastewater treatment applications. Natural husks have been investigated as a replacement strategy for existing conventional systems. The applicability of these low-cost adsorbents is recommended since they are relatively cheap, easily and naturally available, and renewable and show high affinity for heavy metals. The process and mechanism of interaction require further investigation in the direction of modeling, regeneration of adsorbent and recovery of metal ions and immobilization of the waste material for enhanced efficiency and recovery. The applications of husk at commercial level are limited to selected husk (for instance coconut husk). More studies are required to improve the adsorption capacity as well as selectivity of these materials for heavy metal ions. Further, all the studies are limited to the powdered form of adsorbent, which is difficult to work with due to its handling and associated health risks. Therefore, developments of structured adsorbents out of natural husks are needed to increase their industrial usefulness.

## References


[1] M.A. Shannon, P.W. Bohn, M. Elimelech, J.G. Georgiadis, B.J. Mari-as, A.M. Mayes, Science and technology for water purification in the coming decades, Nature 452 (2008)301-310. https://doi.org/10.1038/nature06599

[2] P. Anastas,N. Eghbali, Green chemistry: principles and practice. Chem. Soc. Rev. 39 (2010)301-312. https://doi.org/10.1039/B918763B

[3] [3]R.P. Schwarzenbach, B.I. Escher, K.Fenner, T.B. Hofstetter, C.A. Johnson, U. Gunten, B. Wehrli. The challenge of micropollutants in aquatic systems. Science 313 (2006)1072-1077.

[4] R. Gilpin, The challenge of global capitalism: The world economy in the 21st century, Princeton University Press Princeton, New Jersy, 2002.

[5] V. Chibber, Locked in place: State-building and late industrialization in India,Princeton University Press, New Jersy, 2003.

[6] V. Gupta, P.J.M. Carrott, M.M.L. Ribeiro, Suhas. Low-cost adsorbents: growing approach to wastewater treatment—a review.Environ. Sci. Technol. 39 (2009)783-842. https://doi.org/10.1080/10643380801977610

[7] S.K. Agarwal, Water pollution. Vol. 2. 2005: APH Publishing.

[8] S. Das, R. Raj, N. Mangwani, H.R. Dash, J. Chakraborty, Heavy metals and hydrocarbons: adverse effects and mechanism of toxicity in: S. Das (Ed.) Microbial biodegradation and bioremediation, Elsevier, Amsterdam, The Netherlands, 2014,pp. 23-54. https://doi.org/10.1016/B978-0-12-800021-2.00002-9

[9] I. Mohmood, C. Batista L. Isabel, L. Iqbal A.C. Duarte, E. Pereira. Nanoscale materials and their use in water contaminants removal—a review.Environ. Sci. Pollut. Res. 20 (2013)1239-1260. https://doi.org/10.1007/s11356-012-1415-x

[10] S. Babel, T.A. Kurniawan. Low-cost adsorbents for heavy metals uptake from contaminated water: a review. J. Hazard. Mater.97 (2003)219-243. https://doi.org/10.1016/S0304-3894(02)00263-7

[11] S.E. Bailey, T.J. Olinb, R.M.Brickab, D.D. Adriana. A review of potentially low-cost sorbents for heavy metals. Water Res. 33 (1999)2469-2479. https://doi.org/10.1016/S0043-1354(98)00475-8

[12] S.S. Ahluwalia, D. Goyal. Microbial and plant derived biomass for removal of heavy metals from wastewater, Bioresour. Technol. 98 (2007) 2243-57. https://doi.org/10.1016/j.biortech.2005.12.006

[13] L. Nash. Water quality and health. Water in Crisis: A guide to the World's fresh water resources, Oxford University Press, New York, 1993,pp. 25-39.

[14] R.K. Ibrahim, M. Hayyanl, M. AlSaadiAdeeb, A. Hayyan, S. Ibrahim. Environmental application of nanotechnology: air, soil, and water, Environ. Sci. and Poll. Res. 23 (2016)13754-13788. https://doi.org/10.1007/s11356-016-6457-z

[15] S. Srivastava, P. Goyal, Novel biomaterials: Decontamination of toxic metals from wastewater, Springer-Verlag Berlin and Heidelberg GmbH and Co., 2010. https://doi.org/10.1007/978-3-642-11329-1

[16] M. Ahmaruzzaman, V.K. Gupta. Rice husk and its ash as low-cost adsorbents in water and wastewater treatment. Ind. Eng. Chem. Res. 50 (2011)13589-13613. https://doi.org/10.1021/ie201477c

[17] F. Fu, Q. Wang. Removal of heavy metal ions from wastewaters: a review, J. Environ. Manage. 92 (2011)407-418. https://doi.org/10.1016/j.jenvman.2010.11.011

[18] W.W. Ngah, L. Teong, M. Hanafiah, Adsorption of dyes and heavy metal ions by chitosan composites: A review, Carbohydr. Polym.83 (2011)1446-1456. https://doi.org/10.1016/j.carbpol.2010.11.004

[19] J. Aguado, J.M. Arsuaga, A. Arecncibia, M. Lindo, V. Gascon, Aqueous heavy metals removal by adsorption on amine-functionalized mesoporous silic, J. Hazard. Mater. 163 (2009) 213-21. https://doi.org/10.1016/j.jhazmat.2008.06.080

[20] A. Shahbazi, H. Younesi, A. Badiei, Functionalized SBA-15 mesoporous silica by melamine-based dendrimer amines for adsorptive characteristics of Pb (II), Cu (II) and Cd (II) heavy metal ions in batch and fixed bed column, Chem. Eng.J.168 (2011)505-518. https://doi.org/10.1016/j.cej.2010.11.053

[21] G. Crini, Non-conventional low-cost adsorbents for dye removal: a review, Bioresour. Technol. 97 (2006) 1061-1085. https://doi.org/10.1016/j.biortech.2005.05.001

[22] K. Kadirvelu, M. Kavipriya, C. Karthika, M. Radhika, N. Vennilamani, S. Pattabhi, Utilization of various agricultural wastes for activated carbon preparation and application for the removal of dyes and metal ions from aqueous solutions.

Bioresour. Technol. 87 (2003)129-132. https://doi.org/10.1016/S0960-8524(02)00201-8

[23] M. Saifuddin, P.Kumaran, Removal of heavy metal from industrial wastewater using chitosan coated oil palm shell charcoal, Electronic journal of Biotechnology, 8 (2005) 43-53.

[24] M. Kobya, E. Damirbas, E. Senturk, M. Ince, Adsorption of heavy metal ions from aqueous solutions by activated carbon prepared from apricot stone, Bioresour. Technol. 96 (2005)1518-1521. https://doi.org/10.1016/j.biortech.2004.12.005

[25] B. Volesky, Biosorption of heavy metals. CRC press, Boca Raton, Florida, 1990.

[26] N. Ahalya, T. Ramachandra, R. Kanamadi, Biosorption of heavy metal, Res. J. Chem. Environ. 7 (2003) 71-79.

[27] J. Wang, S. Zheng, Y. Shao, J. Liu, Z. Xu, D. Zhu, Amino-functionalized Fe 3 O 4@ SiO 2 core–shell magnetic nanomaterial as a novel adsorbent for aqueous heavy metals removal, J. Colloid Interface Sci. 349 (2010) 293-299. https://doi.org/10.1016/j.jcis.2010.05.010

[28] S. Martin, W. Griswold, Human health effects of heavy metals. Environ. Sci. Technol. 15 (2009) 1-6.

[29] W. Vries, J.E. Groenenberg, S. Lofts, E. Tipping, Critical loads of heavy metals for soils, in: B.J. Alloway (Ed.), Heavy metals in soils, Springer, Dordrecht, Netherlands,2013,pp. 211-237. https://doi.org/10.1007/978-94-007-4470-7_8

[30] I. Abdelfattah, A.A. Ismail, F. Sayed, K.M. Aboelghait, Biosorption of heavy metals ions in real industrial wastewater using peanut husk as efficient and cost effective adsorbent, Envi. Nanotechnol., Moniti & Mang. 6 (2016) 176-183.

[31] M. Jaishankar, T. Tseten, N. Anbalagan, B.B. Mathew, Toxicity, mechanism and health effects of some heavy metals. Inter. Toxicol. 7 (2014) 60-72. https://doi.org/10.2478/intox-2014-0009

[32] K. Vijayaraghavan, R. Balasubramanian, Is biosorption suitable for decontamination of metal-bearing wastewaters? A critical review on the state-of-the-art of biosorption processes and future directions, J. Environ. Manage. 160 (2015) 283-296. https://doi.org/10.1016/j.jenvman.2015.06.030

[33] V.K. Gupta, I. Ali, Environmental water: advances in treatment, remediation and recycling, Elsevier: Amsterdam, The Netherlands, 2012.

[34] M. Arami, N.Y. Limaee, N.M. Mahmoodi, Investigation on the adsorption capability of egg shell membrane towards model textile dyes, Chemosphere. 65 (2006) 1999-2008. https://doi.org/10.1016/j.chemosphere.2006.06.074

[35] H.K. An, B. Park, D. Kim, Crab shell for the removal of heavy metals from aqueous solution, Water Res. 35 (2001) 3551-3556. https://doi.org/10.1016/S0043-1354(01)00099-9

[36] I. Tan, A.L. Ahmad, B. Hameed, Adsorption of basic dye on high-surface-area activated carbon prepared from coconut husk: Equilibrium, kinetic and thermodynamic studie, J. Hazard. Mate. 154 (2008) 337-346. https://doi.org/10.1016/j.jhazmat.2007.10.031

[37] P. Patnukao, A. Kongsuwan, P. Pavasant, Batch studies of adsorption of copper and lead on activated carbon from Eucalyptus camaldulensis Dehn. Bark, J. Environ. Sci. 20 (2008) 1028-1034. https://doi.org/10.1016/S1001-0742(08)62145-2

[38] N. Yalçın, V. Sevinc, Studies of the surface area and porosity of activated carbons prepared from rice husks, Carbon. 38 (2000) 1943-1945. https://doi.org/10.1016/S0008-6223(00)00029-4

[39] J. G.Torresdey, G. D. L. Rosa, J. P.Videa, Use of phytofiltration technologies in the removal of heavy metals: a review, Pure and Applied Chemistry. 76 (2004) 801-813.

[40] B. Volesky, Detoxification of metal-bearing effluents: biosorption for the next century, Hydrometallurgy. 59 (2001) 203-216. https://doi.org/10.1016/S0304-386X(00)00160-2

[41] A. Hashem, R.A. Akasha, A. Ghith, D.A. Hussei, Adsorbent based on agricultural wastes for heavy metal and dye removal: A review. Energy Educ. Sci. Technol. 19 (2007).

[42] A. Demirbas, Biomass resources for energy and chemical industry, Energy Edu Sci Technol. 5 (2000) 21-45.

[43] U.K. Garg, M.P. Kaur, V.K. Garg, D. Sud, Removal of Nickel (II) from aqueous solution by adsorption on agricultural waste biomass using a response surface methodological approach, Bioresour. Technol. 99 (2008) 1325-1331. https://doi.org/10.1016/j.biortech.2007.02.011

[44] U.K. Garg, M.P. Kaur, V.K. Garg, D. Sud, Removal of hexavalent chromium from aqueous solution by agricultural waste biomass, J. Hazard. Mater. 140 (2007) 60-68. https://doi.org/10.1016/j.jhazmat.2006.06.056

[45] T. Beveridge, R. Murray, Sites of metal deposition in the cell wall of Bacillus subtilis, J.Bacteriol. 141 (1980) 876-887.

[46] V.K. Gupta, I. Ali, Utilisation of bagasse fly ash (a sugar industry waste) for the removal of copper and zinc from wastewater, Sep. Purif. Technol. 18 (2000) 131-140. https://doi.org/10.1016/S1383-5866(99)00058-1

[47] D. Sud, G. Mahajan, M. Kaur, Agricultural waste material as potential adsorbent for sequestering heavy metal ions from aqueous solutions–A review, Bioresour. Technol. 99 (2008) 6017-6027. https://doi.org/10.1016/j.biortech.2007.11.064

[48] R. Milind, D. Julie, J. Snehal, Comparative adsorption studies on activated rice husk and rice husk ash by using methylene blue as dye. in International congress on environmental research at BITS Pilani, Goa,2009.

[49]. T.G. Chuah, A. Jumasiah, I. Azni, S. Katayon, S.Y.T. Choong, Rice husk as a potentially low-cost biosorbent for heavy metal and dye removal: an overview, Desalination. 175 (2005) 305-316. https://doi.org/10.1016/j.desal.2004.10.014

[50] K.K. Krishnani, X. Meng, C. Christodoulatos, Biosorption mechanism of nine different heavy metals onto biomatrix from rice husk. J. Hazard. Mater. 153 (2008) 1222-1234. https://doi.org/10.1016/j.jhazmat.2007.09.113

[51] U. Kumar, M. Bandyopadhyay, Sorption of cadmium from aqueous solution using pretreated rice husk, Bioresour. Technol. 97 (2006) 104-109. https://doi.org/10.1016/j.biortech.2005.02.027

[52] M. Bansal, U. Garg, D. Singh, V.K. Garg, Removal of Cr (VI) from aqueous solutions using pre-consumer processing agricultural waste: A case study of rice husk, J. Hazard. Mater. 162 (2009) 312-320. https://doi.org/10.1016/j.jhazmat.2008.05.037

[53] W. Nakbanpote, B.A. Goodman, P. Thiravetyan, Copper adsorption on rice husk derived materials studied by EPR and FTIR. Colloids and Surfaces A: Physicochemical and Engineering Aspects, 304 (2007) 7-13. https://doi.org/10.1016/j.colsurfa.2007.04.013

[54] D. Roy, P.N. Greenlaw, B.S. Shane, Adsorption of heavy metals by green algae and ground rice hulls, J. Environ. Sci. Health, Part A. 28 (1993) 37-50.

[55] E. Munaf, R. Zein, The use of rice husk for removal of toxic metals from waste water, Environ. Technol. 18 (1997) 359-362. https://doi.org/10.1080/09593331808616549

[56] K.K. Wong, C.K. Lee, K.S. Low, M.J. Haron, Removal of Cu and Pb by tartaric acid modified rice husk from aqueous solutions, Chemosphere. 50 (2003) 23-28. https://doi.org/10.1016/S0045-6535(02)00598-2

[57] C.R.T. Tarley, S.L.C. Ferreira, M.A.Z. Arruda, Use of modified rice husks as a natural solid adsorbent of trace metals: characterisation and development of an on-line preconcentration system for cadmium and lead determination by FAAS, Microchem. J. 77 (2004) 163-175. https://doi.org/10.1016/j.microc.2004.02.019

[58] S. Mohan, G. Sreelakshmi, Fixed bed column study for heavy metal removal using phosphate treated rice husk,J. Hazard. Mater. 153 (2008) 75-82. https://doi.org/10.1016/j.jhazmat.2007.08.021

[59] M. Bansal, D. Singh, V. Garg, A comparative study for the removal of hexavalent chromium from aqueous solution by agriculture wastes' carbons, J. Hazard. Mater. 171 (2009)83-92. https://doi.org/10.1016/j.jhazmat.2009.05.124

[60] M.G.A. Vieira, A.F. Neto, M.G. Silva, Characterization and use of in natura and calcined rice husks for biosorption of heavy metals ions from aqueous effluents, Brazilian Journal of Chemical Engineering, 29 (2012) 619-634. https://doi.org/10.1590/S0104-66322012000300019

[61] E. Pehlivan, T.H. Tran, W.K.I. Ouédraogo, C. Schmidt, Removal of As(V) from aqueous solutions by iron coated rice husk, Fuel Process. Technol. 106 (2013) 511-517. https://doi.org/10.1016/j.fuproc.2012.09.021

[62] S.A. Abo-El-Enein, M.A. Eissa, A.A. Diafullah, Removal of some heavy metals ions from wastewater by copolymer of iron and aluminum impregnated with active silica derived from rice husk ash, J. Hazard. Mater. 172 (2009) 574-579. https://doi.org/10.1016/j.jhazmat.2009.07.036

[63] A.W. Samsuri, F. Sadegh-Zadeh, B.J. Seh-Bardan, Adsorption of As (III) and As (V) by Fe coated biochars and biochars produced from empty fruit bunch and rice husk, Journal of Environmental Chemical Engineering. 1 (2013) 981-988. https://doi.org/10.1016/j.jece.2013.08.009

[64] I.O. Ali, T.M. Salama, M.A. Lateef, H.A. Gumaa, Chemical modification of rice husk by quaternized hexadecylpyridinium for removal of chromate oxyanions

from aqueous solution,Environmental Technology & Innovation. 4 (2015) 110-122. https://doi.org/10.1016/j.eti.2015.05.003

[65] I. Safarik, M. Safarikova, Magnetic fluid modified peanut husks as an adsorbent for organic dyes removal, Physics Procedia, 9 (2010) 274-278. https://doi.org/10.1016/j.phpro.2010.11.061

[66] S. Ricordel, S. Taha, I. Cisse, G. Dorange, Heavy metals removal by adsorption onto peanut husks carbon: characterization, kinetic study and modelling, Sep. Purif. Technol. 24 (2001) 389-401. https://doi.org/10.1016/S1383-5866(01)00139-3

[67] Q. Li, J. Zhai, W. Zhang, M. Wang, J. Zhou, Kinetic studies of adsorption of Pb (II), Cr (III) and Cu (II) from aqueous solution by sawdust and modified peanut husk, J. Hazard. Mater. 141 (2007) 163-167. https://doi.org/10.1016/j.jhazmat.2006.06.109

[68] S. Husaini, J. Zaidi, M. Akram, Evaluation of toxic metals in the industrial effluents and their segregation through peanut husk fence for pollution abatement, J. Radioanal. Nucl. Chem. 289 (2011) 203. https://doi.org/10.1007/s10967-011-1062-6

[69] K. Periasamy, C. Namasivayam, Removal of nickel (II) from aqueous solution and nickel plating industry wastewater using an agricultural waste: peanut hulls, Waste Manage. (Oxford) 15 (1995) 63-68. https://doi.org/10.1016/0956-053X(94)00071-S

[70] O.E.A. Salam, N.A. Reiad, M.M. ElShafei, A study of the removal characteristics of heavy metals from wastewater by low-cost adsorbents, Journal of Advanced Research. 2 (2011) 297-303. https://doi.org/10.1016/j.jare.2011.01.008

[71] R. Ugwekar, G. Lakhawat, Recovery of heavy metal by adsorption using peanut hull., Journal of Advanced Engineering Technology. 3 (2012)39-43.

[72] A. Witek-Krowiak, R.G. Szafran, S. Modelski, Biosorption of heavy metals from aqueous solutions onto peanut shell as a low-cost biosorbent, Desalination. 256 (2011) 126-134. https://doi.org/10.1016/j.desal.2010.07.042

[73] Z.A. Al-Othman, R. Ali, M. Naushad, Hexavalent chromium removal from aqueous medium by activated carbon prepared from peanut shell: adsorption kinetics, equilibrium and thermodynamic studies,Chem. Eng. J. 184 (2012) 238-247. https://doi.org/10.1016/j.cej.2012.01.048

[74] C. Namasivayam, K. Periasamy, Bicarbonate-treated peanut hull carbon for mercury (II) removal from aqueous solution, Water Res. 24 (1993) 1663-1668. https://doi.org/10.1016/0043-1354(93)90130-A

[75] K. Periasamy, C. Namasivayam, Removal of copper (II) by adsorption onto peanut hull carbon from water and copper plating industry wastewater, Chemosphere. 32 (1996) 769-789. https://doi.org/10.1016/0045-6535(95)00332-0

[76] P. Brown, I.A. Jefcoat, D. Parrish, S. Gill, Evaluation of the adsorptive capacity of peanut hull pellets for heavy metals in solution, Advances in Environmental Research. 4 (2000) 19-29. https://doi.org/10.1016/S1093-0191(00)00004-6

[77] C.S. Zhu, L.P. Wang, W.B. Chen, Removal of Cu (II) from aqueous solution by agricultural by-product: peanut hull, J. Hazard. Mater. 168 (2009) 739-746. https://doi.org/10.1016/j.jhazmat.2009.02.085

[78] C. Namasivayam, K. Kadirvelu, Agricultural solid wastes for the removal of heavy metals: adsorption of Cu (II)by coirpith carbon, Chemosphere. 34 (1997) 377-399. https://doi.org/10.1016/S0045-6535(96)00384-0

[79] A. Bhatnagar, V.J.P. Vilar, C.M.S. Botelho, Coconut-based biosorbents for water treatment—a review of the recent literature, Adv. Colloid Interface Sci. 160 (2010) 1-15. https://doi.org/10.1016/j.cis.2010.06.011

[80] R. Ahmad, S.M. Hasany, M.H. Chaudhary, Adsorption Characteristics of Cr(III)Ions onto Coconut Husk from Aqueous Solution,Adsorpt. Sci. Technol. 23 (2005) 467-478. https://doi.org/10.1260/026361705774859929

[81] G. Manju, C. Raji, T. Anirudhan, Evaluation of coconut husk carbon for the removal of arsenic from water,Water Res. 32 (1998) 3062-3070. https://doi.org/10.1016/S0043-1354(98)00068-2

[82] S.M. Hasany, R. Ahmad, The potential of cost-effective coconut husk for the removal of toxic metal ions for environmental protection,J. Environ. Manage.81 (2006) 286-295. https://doi.org/10.1016/j.jenvman.2006.01.009

[83] L. Hernandez-Apaolaza, F. Guerrero, Comparison between pine bark and coconut husk sorption capacity of metals and nitrate when mixed with sewage sludge, Bioresour. Technol. 99 (2008) 1544-1548. https://doi.org/10.1016/j.biortech.2007.04.032

[84] M. Sreedhar, A. Madhukumar, T. Anirudhan, Evaluation of an adsorbent prepared by treating coconut husk with polysulphide for the removal of mercury from wastewater, Ind. J. Eng. Mater. Sci. 6 (1999) 279-285.

[85] I. Agbozu, F. Emoruwa, Batch adsorption of heavy metals (Cu, Pb, Fe, Cr and Cd) from aqueous solutions using coconut husk, African Journal of Environmental Science and Technology. 8 (2014) 239-246. https://doi.org/10.5897/AJEST2013.1577

[86] R. Malik, S. Dahiya, An experimental and quantum chemical study of removal of utmostly quantified heavy metals in wastewater using coconut husk: A novel approach to mechanism, Int. J. Biol. Macromol. 98 (2017) 139-149. https://doi.org/10.1016/j.ijbiomac.2017.01.100

[87] A. Saeed, M. Iqbal, Bioremoval of cadmium from aqueous solution by black gram husk (Cicerarientinum), Water Res. 37 (2003) 3472-3480. https://doi.org/10.1016/S0043-1354(03)00175-1

[88] A. Saeed, M. Iqbal, M.W. Akhtar, Removal and recovery of lead (II) from single and multimetal (Cd, Cu, Ni, Zn) solutions by crop milling waste (black gram husk), J. Hazard. Mater. 117 (2005) 65-73. https://doi.org/10.1016/j.jhazmat.2004.09.008

[89] N. Ahalya, R. Kanamadi, T. Ramachandra, Biosorption of chromium (VI) from aqueous solutions by the husk of Bengal gram (Cicer arientinum). Electronic J. Biotechnol. 8 (2005). https://doi.org/10.2225/vol8-issue3-fulltext-10

[90] N. Ahalya, R.D. Kanamadi, T.V. Ramachandra, Biosorption of iron (III) from aqueous solutions using the husk of Cicer arientinum, Ind. J. Chem. Technol. 13 (2006) 122-127.

[91] J. Kumar, C. Balomajumder, P. Mondal, Application of Agro-Based Biomasses for Zinc Removal from Wastewater–A Review. Clean–Soil, Air, Water, 39 (2011) 641-652. https://doi.org/10.1002/clen.201000100

# Chapter 6

# Removal of heavy metals using graphene composites

Mukesh Kumar[1,*] Prashant Singh[2] and Manish Kumar Gautam[3]

[1]Sri Aurobindo College, Department of Chemistry, University of Delhi, Delhi-110017, India

[2]Atma Ram Sanatan Dharma College, Department of Chemistry, University of Delhi, Delhi-110021, India

[3]Dyal Singh College, Department of Chemistry, University of Delhi, Delhi-110003, India

*mknanocomposites@gmail.com

**Abstract**

The presence of heavy metal ions in water is a serious risk to the human health and other living forms on earth because the indiscriminate disposal of these non-biodegradable pollutants worldwide. Nanotechnology-based developed current nano-adsorbent are the demand of the current research for the effective removal of highly toxic heavy metal ions from the wastewaters. Graphene and graphene oxide based nano-adsorbent play a major role in the effective removal of heavy metal ions because the unique nature of graphene such as extremely high electronic mobility, large surface area to volume ratio, weight, and preponderance of exposed edge planes to enhance charge storage which are responsible for their higher absorption capacity when compared with other materials. Hence, the applicability of graphene and its composites as an adsorbent for the removal of heavy metal ions is summarized in this chapter.

**Keywords**

Graphene, Graphene Oxide, Nanoadsorbent, Surface Adsorption, Heavy Metal Ions, Grafting, Magnetite, Functionalization

## Contents

## 1. Introduction

Water is the most important natural occurring resource in the world which is essential for all types of living forms. The water crisis is a most serious problem globally due to current issues such as increasing population, pollution and global warming. Wastewaters from developing industries such as metallurgical, chemical manufacturing, papermaking, battery manufacturing and mining industries create a huge amount of environmental pollutants in the form of many kinds of toxic or poisonous heavy metals such as chromium (Cr), cobalt (Co),nickel (Ni), copper (Cu), zinc (Zn), thallium (Tl), arsenic (As), cadmium (Cd), mercury (Hg), lead (Pb) etc. These metals tend to accumulate in a living organism. Most of the heavy metals are common pollutants for potable and wastewaters. These heavy metals pose a toxicity threat to human health and other animals even at low concentration. For example, lead is extremely toxic to human health and causes toxicity to the nervous system, kidneys and reproductive system and excess exposure may lead to irreversible brain damage and encephalopathy disease symptoms. Some important severe toxic effects of the heavy metals present in potable water are summarized in Table 1. Water containing excess heavy metal ion concentration is a serious risk to the public health and as well as other living systems on earth. The presence of heavy metal ions in drinking water is a serious threat to the human health, owing to their high non-biodegradability, accrual in living organisms which cause various diseases, genetic disorders, and lethal ecological effects. Other harmful effects of heavy metals include several serious and long-lasting disorders, such as renal damage, emphysema, hypertension, testicular atrophy, and skeletal abnormality in fetuses [1]. Removal of heavy metals has become an especially challenging issue due to their high toxicity and potential carcinogenicity. The permissible values of these heavy metals in potable water are supposed to be near or less than 50ppb for Cr(VI) and 10 ppb for Pb(II) according to WHO [2].

*Table 1.Standard heavy metal concentration in potable water and their toxic effects on the human health [2].*

| Metal Name | Standards Concentration of Heavy Metals in Potable Water | Some Toxic Effects of Excess's Concentration on Human Health |
|---|---|---|
| Lead(Pb) | 0.10 mg/ L | Metabolic Poison, Aquatic Fauna and Livestock, Hypertension, Tiredness, Brain Damage, Irritation |
| Mercury(Hg) | 0.001 mg/ L | Mutagenic Effects, Disturbs the Cholesterol, Deformed Babies Birth |
| Arsenic(As) | 10.00 mg/L | Melanosis, Keratosis, Hyperpigmentation in Humans, Genotoxicity, Immunotoxic, Modulation of Co-Receptor Expression |
| Cadmium(Cd) | 0.005 mg/ L | Kidneys Damage, Bronchitis, Emphysema, Anemia, Acute Effects in Children |
| Copper(Cu) | 0.010 mg/ L | Kidney Damage, Aquatic Fauna, Phytotoxic, Mucosal Irritation, Diarrhea, Central Nervous System Damage, Causes Nose, Eyes, and Mouth Irritation |
| Chromium(Cr) | 0.05mg/L | Carcinogenic, Lever Damage, Lung Carcinoma, Pulmonary Congestion, Skin Irritation, Vomiting |
| Zinc(Zn) | 0.10 mg/ L | Health Damaging, Phytotoxic, Anemia, Lack of Muscular Coordination, Abdominal Pain, Vomiting in Children, Stomach Cramps, Skin Irritations |
| Nickel(Ni) | 0.10 mg/ L | DNA Damage, Eczema of Hands, High Phytotoxicity, Damaging Fauna |

Because of the tight regulatory systems, nowadays removal of the environmental pollutants is the prime concern. Therefore, it is necessary to remove all toxic heavy metals from wastewater before releasing it into the environment and hence the development of efficient and economical materials and technologies for the effective removal of water pollutants is necessary. There are several traditional techniques for the removal of heavy metals from water including precipitation, reverse osmosis, electrochemical, ion exchange, membrane filtration, coagulation, irradiation, and adsorption etc. [3]. Due to the simple operation, low cost and high efficiency, adsorption technology is regarded as one of the most promising technology for the effective removal of the heavy metal pollutants from wastewaters. However, adsorption technology is still having several challenges, such as selective recovery and reuse of valuable pollutants and adsorbents. The development of nanoscience and nanotechnology has drawn much attention due to their remarkable potential for the effective removal of wastewater pollutants when compared with traditional materials. Nanotechnology is the engineering tool of the functional system at the molecular scale which offers the developments of new products and processes for the heavy metal removal from wastewaters. Carbon based materials are being extensively used for the removal of heavy metals due to their favorable properties such as the large surface area to volume ratio and light weight which are responsible for their higher absorption capacity when compared with other materials. Carbon nanotube [4], activated carbon, graphene, graphene oxide (GO) etc. have drawn much attention for the removal of heavy metals as adsorbents. Functionalized carbon nanotubes represented better adsorption for the heavy metal removal when compared to non-functionalized and activated carbon materials. All the carbon-based nanomaterial's including carbon nanotubes, fullerene, graphene and graphene derivatives have shown remarkable adsorbent capacities.

Earlier research showed that carbon nanotubes could be promising adsorbent materials for environmental protection regardless of their high cost and limited availability at present and hence large-scale applications of carbon nanotubes are limited. Therefore, low cost, readily available and effective adsorbents are demandable for the removal of heavy metals from wastewater. Graphene and its derivatives may be ideal substitute materials for wastewater treatment.

Graphene is the other type of non-classical carbon material adsorbent, which is a flat, single-atom-thick sheet of $sp^2$-hybridized carbon atoms densely packed into a honeycomb two-dimensional lattice arrangement. It has been considered for a wide variety of realizing real time applications. Graphene is considered the mother of all the carbon allotropes due to its extraordinarily excellent mechanical, thermal, and electrical properties [5] which are key factors for its use in various applications, including the electronic devices [6] photonics devices [7] biomedicine [8] and contamination purification [9, 10]. Graphene has been considered a promising material for the absorption of vide variety of pollutants in aqueous system due to its large specific surface area. Wet chemical oxidation is one of the most effective method for the production of graphene from graphite where graphene oxide (GO) is intermediate with highly negative charge density and possesses various reactive functional groups such as hydroxyl, carboxyl, and epoxy groups. Hence GO can be used as a great potential scavenger for the effective removal of pollutants, such as cationic pollutants, through electro static interactions. GO is soluble in polar as well as non-polar solvents due to the presence of oxygen functionality which makes GO hydrophilic and graphene domains which make GO hydrophobic. Relative high surface area, hydroxyl, carboxyl, and epoxy functional groups present on the surface of graphene and better conductivity are essential for a better adsorption of several heavy metals through aqueous solution [1]. Nair et al. observed that a few layered thick GO membrane can completely block the passage of pollutants in the form of liquids and gasses under a dry state while facilitating the permeation of water vapor [11]. On the other hand, disadvantages of graphene layers aggregation can be partially prevented through composites formation. Additionally, the composites formation may impart enhanced removal of heavy metals from the wastewaters due to the synergistic effect of the materials used. Hence removal of heavy metals through graphene and its composites is summarized in this chapter which may be helpful for the purification of potable and safe drinking water as well as removal of heavy metals from wastewaters and cleanup the environmental problems.

Generally, graphene and graphene oxide have proved to be promising adsorbent candidates for wastewater treatment or water purification. Filtration has been considered as a promising separation and purification technology due to its easy operation, energy saving property, and high efficiency. Historically, water purification technologies have utilized high surface area carbon materials in the form of activated carbon for decolorization and heavy metal ion retention. Compared to these materials, GO is produced by room-temperature soft chemistry principles and is likely to be cost-efficient. A report on magnetite-graphene hybrid materials for magnetically controlled separation of As from water published recently [12]. The possibility of harnessing this readily available and inexpensive material has been relatively unexplored. In this chapter, we wish to report the surface modification approaches and post-synthesis assembly steps, which will enable exploitation of GO as a novel material for low-cost water purification processes

## 2. Removal of lead

Lead is one of the most useful heavy metals with the worldwide applications spectrum. However, lead present in potable and wastewater is a big threat for human beings as well as the ecosystem. Toxic effect of lead may lead to anemia, coma, renal failure, insomnia, headache, dizziness, irritability, cancer, weakness of muscles, hallucination, renal damages and subtle effects on metabolism and intelligence to the human health. Therefore, it becomes essential to remove lead from portable and wastewaters.

Ethylenediaminetetraacetic acid (EDTA) and GO adduct as shown in Fig. 1 was prepared by the reaction of N-(trimethoxysilylpropyl) ethylenediaminetetraacetic acid (EDTA-silane) with the hydroxyl functionality present on the surface of the graphene layers [13]. It was observed that the EDTA-GO shows enhanced adsorption for Pb(II) from the solution when compared with the EDTA-RGO and GO samples as shown in Fig. 1. The superior performance of the EDTA-GO was correlated with the chelating character of the EDTA and higher functionality.

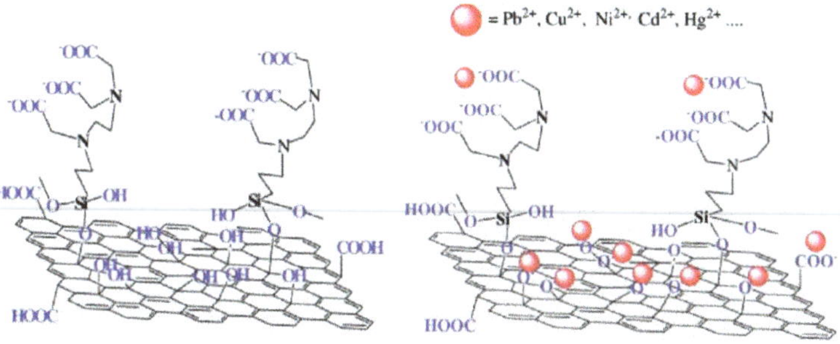


*Fig. 1. Adsorption of heavy metal ions by EDTA-GO adsorbent [13].*

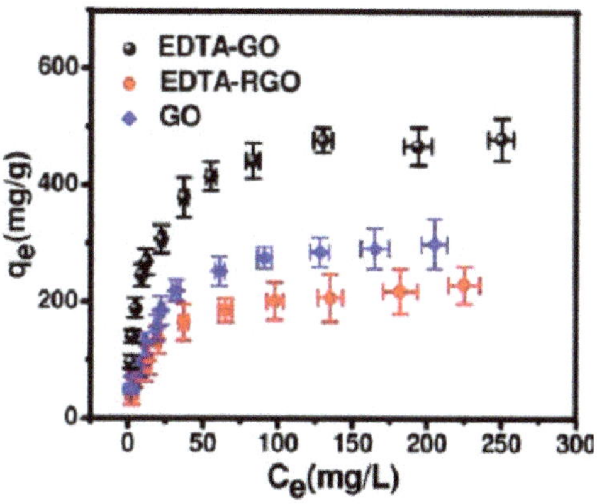


*Fig. 2. Adsorption isotherms of Pb(II) for EDTA-GO, EDTA-RGO and GO at pH 6.8 [13].*

The maximum adsorption capacities of EDTA-GO and GO for Pb(II) were found to be 479±46 and 328±39 mg/g, respectively shown in Fig. 2 which are greater than that of oxidized carbon nanotube and activated carbon [13]. It was suggested that two adsorption processes were mainly responsible for the enhanced removal of Pb(II) with EDTA-GO viz. ion-exchange reaction process between Pb(II) and carboxylic and hydroxyl functional groups and the chelate formation between Pb(II) and EDTA formation on the surface of GO.

Facile magnetic ternary composites of $Fe_3O_4$, GO and chitosan was prepared for the effective removal of metal ions as shown in scheme 1 [14]. Magnetic fluid was coated on the chitosan to improve the separation speed through improved surface area and reduce the required dosage for the better adsorption of heavy metals.

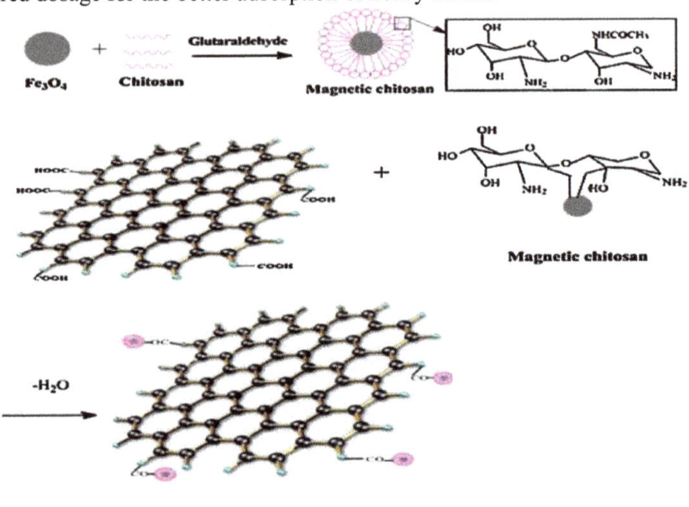


***Scheme** 1. Schematic representation of magnetic chitosan/graphene oxide (MC-GO) formation 39.57 emu/g with saturation magnetization value[14].*

The adsorption amount and rate were calculated based on the concentration difference solution before and after adsorption process using the following equation.

$$Q = \frac{(C_O - C_e)V}{W}, \quad E = \frac{(C_O - C_e)}{C_O} \times 100\ \%$$

Where, $C_0$ and $C_e$ are the initial and equilibrium concentrations of the metal ion in milligrams per liter, respectively, Vis the volume of metal ion solution in liters, and W is the amount of the adsorbent used, in grams.

The adsorption capacity of the prepared MC-GO was determined at different range pH from 1.0 to 7.0 values (Fig. 3) as it is sensitive to pH and usually do not occur at low pH.

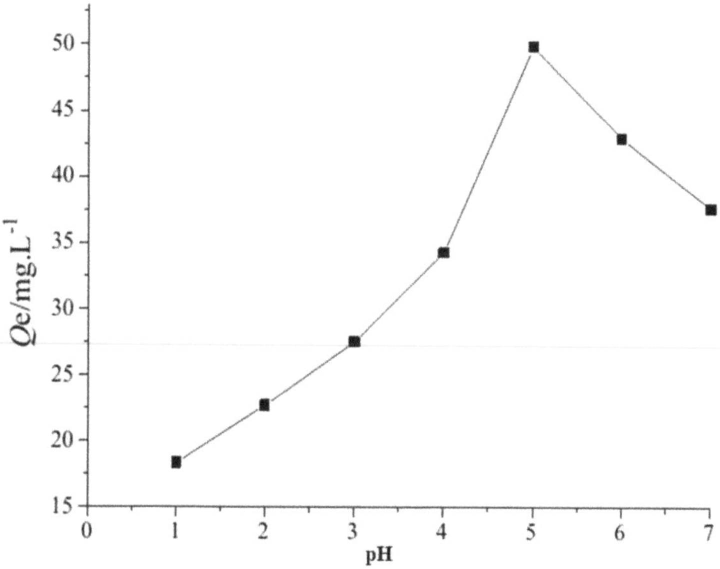


*Fig. 3. Absorption capacity measurement of MC-GO at different pH for Pb(II) adsorption[14].*

The maximum adsorption capacity of the prepared MC-GO was found to be 50.23 mg/g at pH 5 which decreases at higher pH because, at a lower pH, the amino functional groups on the surface of MC-GO could be easily protonated. In acidic media, the slight uptake decrease was attributed due to the protonation of the lone pair of the nitrogen atom of an amino group that may hinder the complex formation as shown below by the following reaction.

$$\mathbf{R{-}NH_2 + H^+ \longrightarrow R{-}NH_3^+}$$

$$R{-}NH_3^+ + Pb^{2+} \longrightarrow R{-}NH_3^+ \cdots Pb^{2+}$$

*Fig. 4 represents the effect of time on the absorption capacity of MCGO for Pb(II).*

Maximum absorption was obtained within the first 40 minutes and no significant charge was observed after one hour. The initial sorption of 60 minutes time could be attributed with the large surface area, the sufficient active sites exposure and the high surface reactivity of the MCGO which found to be decreasing later due to decrease of the bonding sites on the surface of an absorbent and the aggregation between particulates maters.

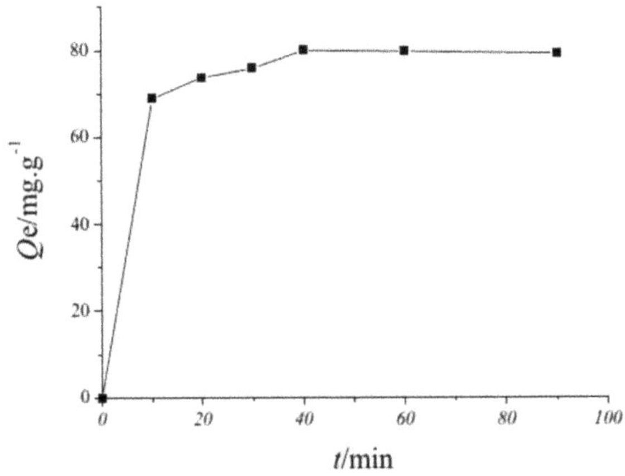


*Fig. 4. Effects of contact time on adsorption of Pb(II) on removal efficiency by MCGO [14].*

The maximum absorption capacity of MC-GO was further determined from the Langmuir isotherm and found to be 82.10 mg/ g which was quite in order with the calculated value for the monolayer adsorption. For the possibility of regeneration and reusability of MC-GO as an adsorbent, desorption experiments were performed at different pH and found that the solutions with pH 1 and 2 could efficiently remove the adsorbed metal ions and the desorption efficiency was 90.3 and 90.1%, respectively. Poly3-aminopropyltriethoxysilane (PAS) oligomers were three-dimensionally cross-linked with

GO nanosheets to get high-performance adsorbent. The three-dimensional cross linking prevents nanosheets aggregation provides easy accessibility for foreign molecules and introduces a lot of amino functional groups [15]. The adsorption performance of PAS-GO, 3-aminopropyltriethoxysilane functionalized GO and GO was investigated by removing Pb(II) ions from aqueous solution.

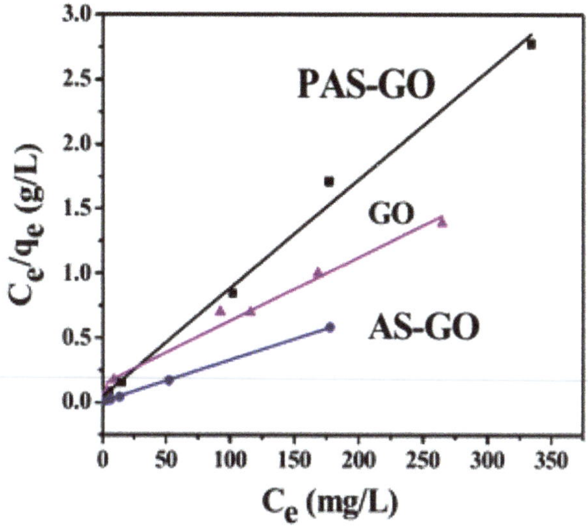


*Fig. 5. Langmuir isotherms for Pb(II) adsorption in aqueous solution for PAS-GO, AS-GO, and GO at 303 K [15].*

Langmuir adsorption isotherms represent the maximum adsorption capacities for PAS-GO in comparison with 3-aminopropyltriethoxysilane functionalized GO and pure GO samples.

The highest adsorption (312.5 mg/g) capacity of PAS-GO has been attributed to the following factors:

(1) PAS chains in PAS-GO adsorbent provided more functionality/ binding sites for binding metal ions.

(2) Multi-arm oligomer three-dimensional functionality prevented GO nanosheets from aggregation which may provide easier access to binding sites for foreigner molecules.

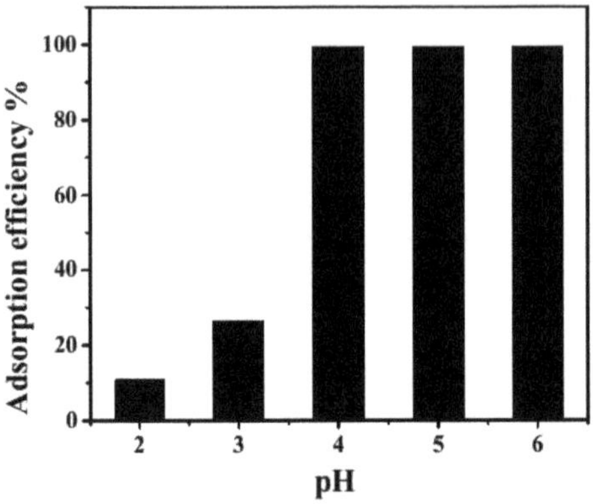


*Fig. 6. PAS-GO adsorption capacity at different pH for Pb(II) using 1g/L adsorbent at 303 K for 7h[15].*

It was observe that at pH 4.0, the effect of protonation became weak and the adsorption efficiency of PAS-GO reached to 100% than the values obtained at lower pH values of 2.0 and 3.0. When pH was allowed to further increasing to near 6.0, the adsorption performance of PAS-GO still remains at nearly 100%. Concluding that PAS-GO may be the best adsorbent for Pb(II) ions removals in an acidic medium from pH 4.0–7.0 because in basic medium these ions are precipitated. One step synthesis of polyacrylamide grafted graphene (PAM-g-graphene) composites has been reported by Xe et al. [16]. X-ray irradiation with acrylamidemonomers in aqueous at room temperature through in situ radical polymerization as shown in scheme 2. Which led to the individual sheets GO layers exfoliation into polyacrylamide. The resulted PAM-g-graphene show 24.2% degree of grafting of PAM on the surface of graphene with the thickness of 2.59 nm and exhibits superior adsorption capacity for Pb(II) ions.

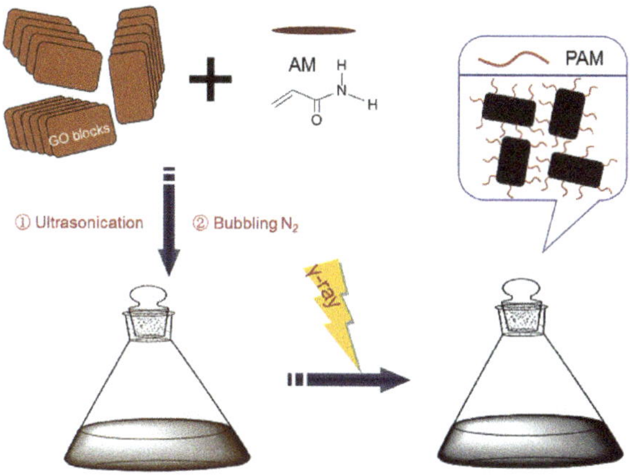


*Scheme 2. PAM-Graphene composites preparation through in-situ free radical polymerization by γ-ray irradiation[16].*

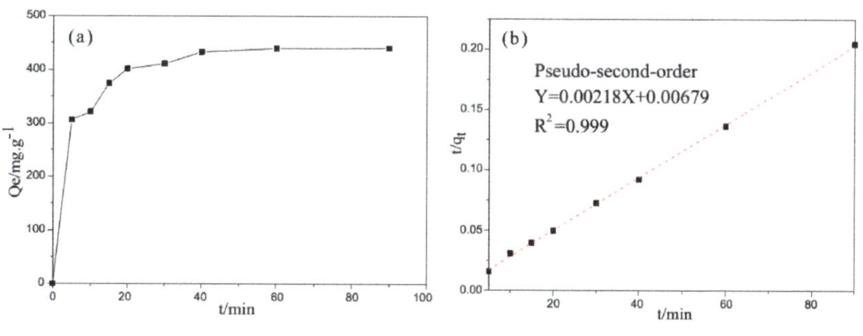


*Fig. 7. (a) Adsorption efficiency of PAM-GO with time for Pb(II) removal and (b) pseudo-second-order kinetics[16].*

Fig. 7(a) represents the adsorption efficiency of PAM-GO with contact time for Pb(II) concentration of 45 mg $L^{-1}$at pH 6. PAM-GO adsorption capacity sharply increases within the first 20 minutes and reaches equilibrium within 30 minutes. The short period of time adsorption capacity suggests that PAM-GO has higher adsorption efficiency for industrial applications. Fig. 7(b) represents the pseudo-second-order kinetic model parameters for the adsorption process of Pb(II) on PAM-GO. The experimental value (440.4 mg/ g) was found similar as the value obtained calculated (458.7 mg/ g) from the pseudo-second-order kinetic model. The correlation coefficient value ($R^2$= 0.999) obtained from the pseudo-second-order kinetic model for PAM-GO adsorbent of Pb(II) removal suggests that the adsorption process may be controlled by chemical adsorption involving valence forces through sharing or exchange of electrons between sorbent and sorbate. Comparative Langmuir and Freundlich isotherm models for GO and PAM-GO, to explore the adsorption mechanism have been represented in Fig. 8.The adsorption of Pb(II) for GO and PAM-graphene is well fitted to the Langmuir isotherm model with the higher $R^2$ value indicating the specific homogeneous adsorption within GO and monolayer adsorption for PAM-GO. The theoretical adsorption capacity, determined from the Langmuir isotherm PAM-GO for Pb(II) was found to be 819.67 mg/g, which is 20 times higher than that of graphene nanosheets.

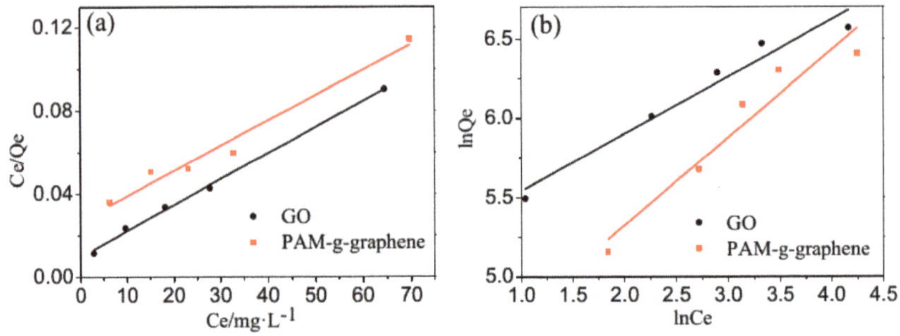


*Fig. 8. (a) The Langmuir isotherm; (b) The Freundlich isotherm for Go and PAM-GO [16].*

Hydrated manganese oxide and GO (HMO@GO) sorbents with excellent settling ability (<2 min)were fabricated through *in-situ* growing nanosized hydrated manganese oxide (HMO) on GO nanosheets surface [17]. Fig. 9 represents adsorption behavior of HMO-Go composites and GO over pH range 2.5-6.7. It was observed that as the pH increases, sorption capacity of HMO-GO and GO also increases because deprotonation increases with the increase in pH. The leaching of Mn was found to be negligible from HMO@GO during the experiment over the pH effect.

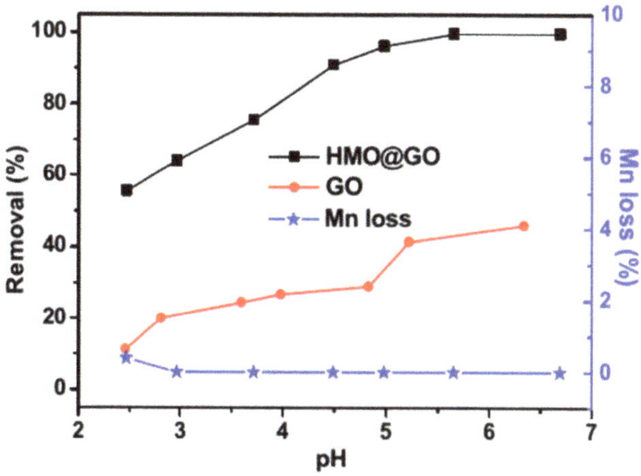


*Fig. 9. pH effect on adsorption of HMO@GO and GO for Pb(II) [17].*

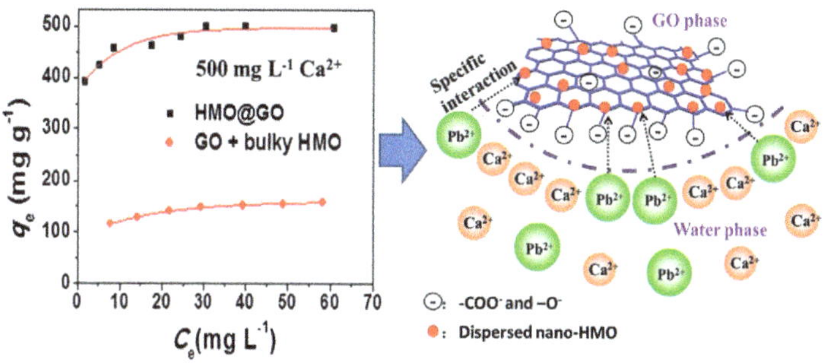


*Fig. 10. Comparative synergistic adsorption capacities of HMO@GO and GO with bulky HMO in the presence of Ca(II) and Pb(II) [17].*

Fig. 10 represents the absorption behavior and capacities of nanosize HMO@GO and GO with bulky HMO. The adsorption capacity of nanosize HMO@GO was found to be four times better than the GO with bulky HMO due to the synergistic effect of nanosize HMO deposition onto the surface of GO layers having large surface area.

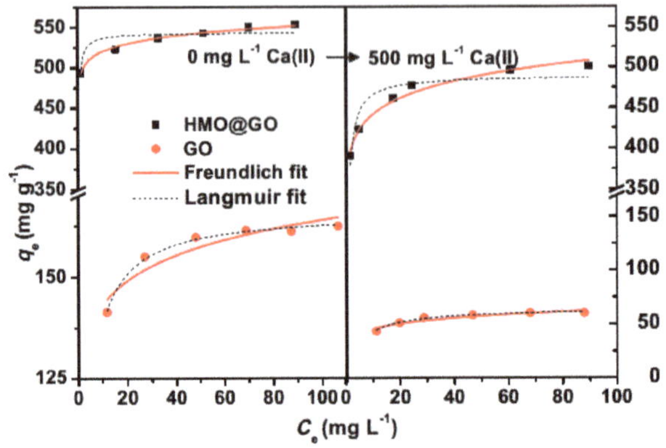


*Fig. 11. Sorption isotherms of Pb(II) onto HMO@GO and GO surfaces in the absence and presence of Ca(II) [17].*

Sorption isotherms of Pb(II) onto HMO@GO and GO at pH 5 were investigated which is based on the Freundlich and the Langmuir model as shown in Fig.11. The higher related coefficients of the Freundlich models than the Langmuir model for HMO@GO suggests a heterogeneous chemisorption process through inner-sphere complexation, whereas the opposite results were obtained for GO which suggests a monolayer adsorption on the surface through simple ion exchange. Experimentally obtained sorption capacity of Pb(II) For HMO@GO was found to be 553.6 mg/g which was higher than any other Mn-based adsorbents.

## 3. Removal of cadmium

Cadmium is found in ores with other elements and it is highly carcinogenic (U.S. Environmental Protection Agency) when exposed to human health. Cd is widely used in industrial processes (electroplating, metal finishing in nickel-cadmium batteries fabrication, television sets, ceramics, photography etc.). It is also used as an anticorrosive agent, stabilizer in PVC, color pigment, and a neutron absorber. Chronic exposure to cadmium results in kidney dysfunction, liver and blood damage, bone degeneration and high levels of exposure will result in carcinogenicity or even death.

Iron oxide and GO magnetic composites prepared and were used as an adsorbent for the effective removal of Cd(II) ions [18]. Magnetic GO adsorbent was prepared by dispersing GO in water followed by the addition of ferrous ammonium sulfate, ammonium ferric sulfate, and ammonia under oxygen-free condition. Two different sorption isotherms i.e. Langmuir isotherm for monolayer and Freundlich isotherms for multilayer sorption have been shown in Fig. 12.

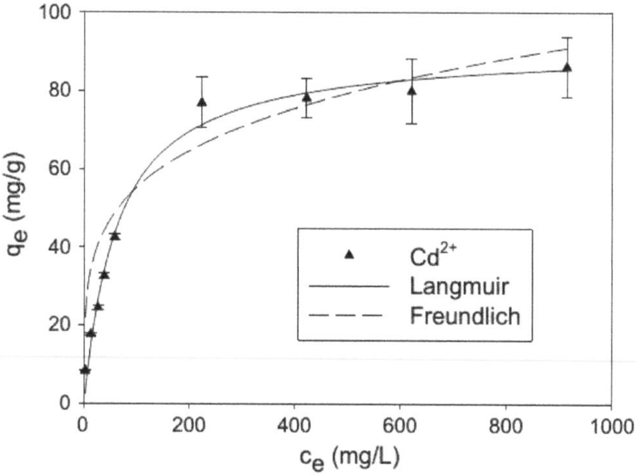


*Fig. 12. M-GO Sorption isotherms for Cd(II) at pH = 6, time 24h, temperature 25 ℃ [18].*

Fig. 12 shows that the sorption isotherms data of Cd(II) are better fitted with Langmuir model in the mono-component system. The sorption capacity of adsorbate on M-GO was 91.29 mg/g. Desorption and reusability behavior were studied using HCl solution and ethylene glycol as eluants for Cd(II) and results showed that adsorption capacities of Cd(II) decreased with increasing regeneration cycle numbers. After the first and fourth regeneration cycle, the removal efficiencies of M-GO were found to be 67.55 and 33.78%, respectively indicating the good reusability of M-GO.

Using $FeCl_3$ and GO as precursors, magnetic graphene composites were synthesized by one-step method and grafted thiol groups using (3-mercaptopropyl)trimethoxysilane (MPTES) [19, 20]. Resulted thiol-functionalized magnetic graphene composite (M-$Fe_3O_4$-GO) was used as an adsorbent for the removal of Cd(II) from aqueous solution and effect of pH, contact time and initial solution concentration of adsorption were studied. The amount of adsorbed Cd(II) by M-$Fe_3O_4$-GO adsorbent was calculated using the following equation:

$$q_e = \frac{(C_O - C_e)\ X\ V}{m}$$

Where, $C_0$ and $C_e$ are the initial and equilibrium concentrations of Cd(II) ions (mg/ L), respectively, V is the volume of the solution(mL), and mis the mass of M-$Fe_3O_4$-GO (mg). Fig. 13 represents the adsorption effect with the increasing concentration of Cd(II) which showed that with the increase of concentration, the adsorption of metal ion increases and the double adsorption (5 mg/ L) was obtained at 25 mg/ L. Results were found to be similar with previous studies which indicated that as the concentration of the solution increase, driving forces becomes dominant for or the mass transfer and subsequent for better adsorption on the nanostructures.

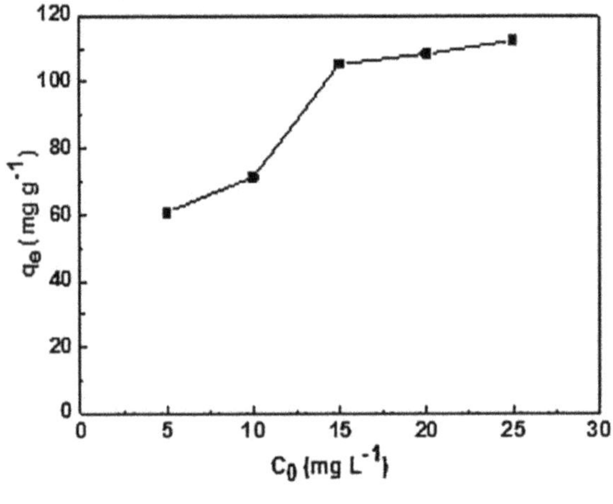


*Fig. 13.The adsorption capacity of M-$Fe_3O_4$-GO for Cd(II) in aqueous solution [19].*

The maximum adsorption capacity of M-$Fe_3O_4$-GO for Cd(II) ions obtained from the Langmuir isotherm was found 125 mg/g which was almost much higher than the value reported for magnetic adsorbents except the value reported by Wang et al. [21]. Reported single-step synthetic method was found to be more feasible in terms of economical and large-scale production. The high efficiency of the thiol-functionalized adsorbent for the removal of Cd(II) may be attributed to the synergistic effects of all the three components viz. thiol groups, $Fe_3O_4$ nanoparticles and high surface area graphene nano layers (scheme 3). This functional groups were assumed to be ion exchanged with proton, and combine Cd(II) ions on the surface of the nanoparticles.

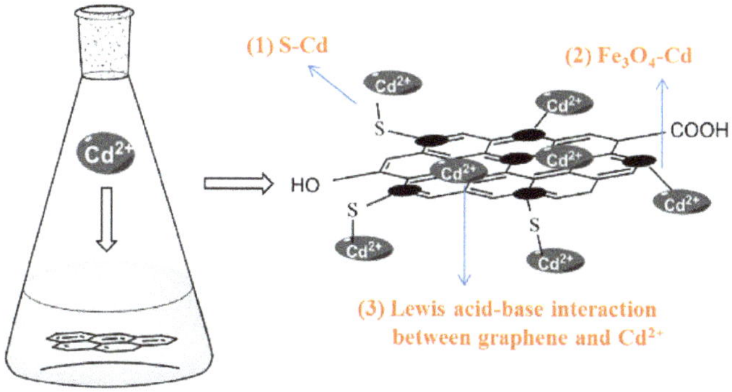


*Scheme 3. Adsorption of Cd(II) on M-$Fe_3O_4$-GO [19].*

## 4. Removal of mercury

The natural major sources of Hg may be a volcanic eruption, weathering of rocks and soils, whereas anthropogenic Hg comes from its extensive use in industrial applications, mining processing batteries, and mercury vapor lamps. The toxicity of Hg has been recognized worldwide, such as in Minamata Bay of Japan. Mentally disturbed and physically deformed babies were born to mothers who were exposed to toxic Hg due to consumption of contaminated fish. Mercury is a neurotoxin that can cause damage to the central nervous system of the living organisms. High concentrations of Hg may cause impairment of pulmonary and kidney function, chest pain, and dyspnea. Methyl mercury $(CH_3Hg)^+$ is more toxic than any other compound of Hg.

Single-step, eco-friendly strategic preparation of sulfur nanoparticle/reduced graphene oxide nanohybrid (SRGO) was carried out for the fast and efficient Hg(II) removal by Thakur et al. [22]. Hg(II) removal efficiency of the prepared SRGO was studied with the effect of exposure time at neutral pH under ambient conditions, at different pH.

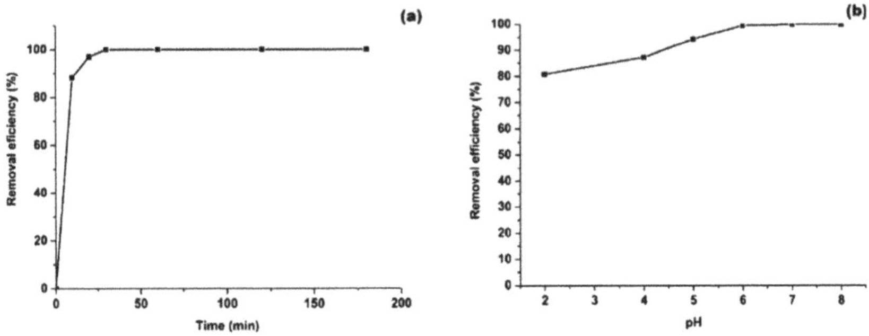


*Fig. 14. Hg(II) removal efficiency of SRGO adsorbent (a) with a variation of time and (b) at different pH [22].*

Rapid adsorption of Hg(II) ions (15 minutes) and equilibrium time (30 minutes) (Fig. 14 a) might be attributed due to the strong interactions between Hg(II) ions and the sulfur nanoparticle present at the surface of graphene nanosheets. Maximum adsorption efficiency of SRGO for the Hg(II) removal was at pH less than 6 pH (Fig. 14 b) because Hg exist in $Hg(OH)_2$ and $Hg(OH)^+$ as Hg(II) between 3 to 6 pH. The small decrease of Hg(II) ions adsorption observed in the acidic medium may be correlated with the low mobility higher hydrated species and protonation of the surface functional groups of SRGO adsorbent. Coulombic repulsions between Hg(II) (aq) species and $H^+$ or $H_3O^+$ions may be the other factor of hindered adsorption of the SRGO. At more than 6 pH, availability of the more functional may result in high adsorption due to deprotonation of the surface functionality and consequently the maximum adsorption of Hg(II) at higher pH values. Selective adsorption for Hg(II) was also performed by studying the mixed adsorbent of Hg(II), Cd(II), Zn(II), and Cu(II) aqueous solution. Nano-adsorbent, SRGO showed almost identical Hg(II) ions while other ions showed only slight/negligible adsorption. Selective adsorption may be correlated due to the presence of sulfur nanoparticles on the surface of SRGO which was suitable for Hg(II) ions than other ions.

A three-dimensional, self-assembled, graphene-diatom silica composite aerogel was prepared through simple synthetic approach as shown in Fig. 15 and adsorption performance for the removal of Hg(II) from water was studied as a function of time, pH, and solute concentration under optimized conditions [23].

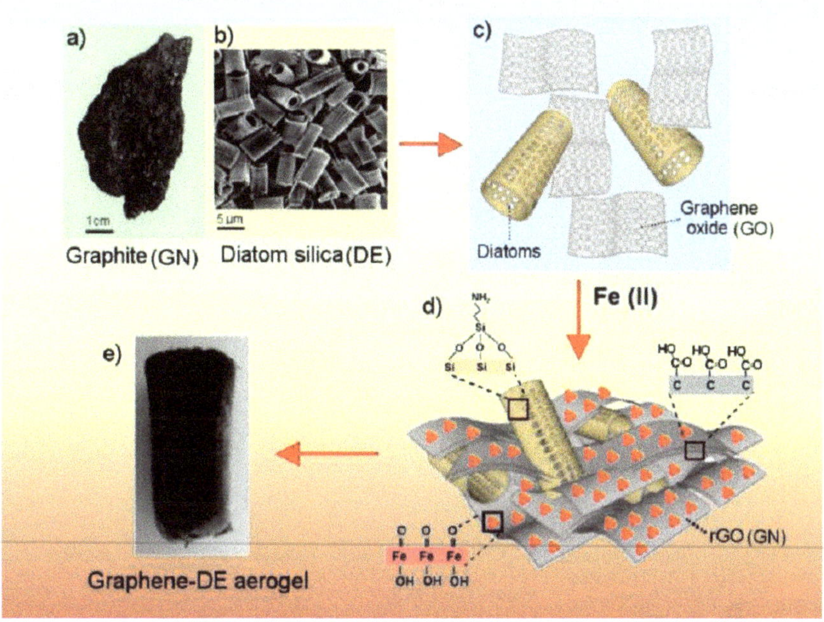


*Fig. 15. Systematic representation of three-dimensional, self-assembled, graphene-diatom silica composites aerogel preparation [24].*

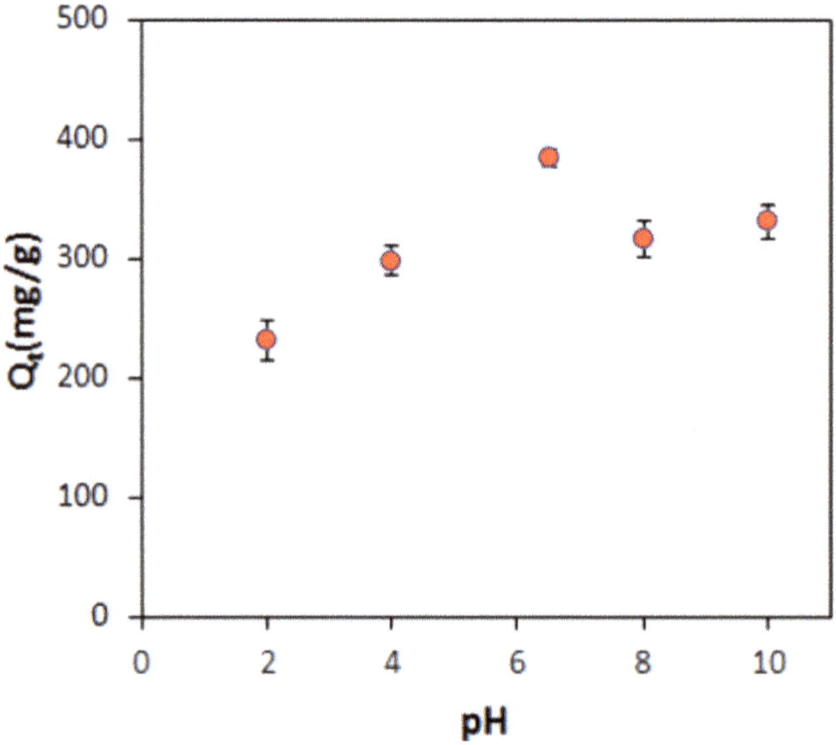


*Fig. 16.Graphene-diatom silica composites aerogel adsorption capacity for Hg(II) in the pH range of 2−10 [24].*

Fig. 16 represents the effect of pH on the adsorption capacity of adsorbent using 200 mg/L Hg(II) ions concentration. Increased adsorption capacity was observed from 2 to 6.5 pH which was then slightly decreased above 6.5 pH which may be due to physical trapping and binding with available amino and oxygen functional groups on the adsorbent surface. In addition, zeta potential below 6 pH, show that adsorbent is positively charged which results in lower adsorption of positively charge Hg(II) ions. As the pH increases, the surface of the adsorbent becomes more negative and consequently higher adsorption of Hg(II) ion on the surface.

## 5. Removal of chromium

Cr exists in several oxidation states but only two trivalent and hexavalent forms are stable enough to occur in the environment. Cr(III) exists in natural waters in hydrolyzed $Cr(H_2O)_4OH_2^+$ form and is an essential micronutrient in the human body and combines with enzymes to form sugar, protein, and fat. Cr(III), is an essential element for the functioning of living organisms, a high concentration of it can be dangerous to living organisms as it can co-ordinate with organic compounds which in turn leads to the inhibition of some metallo enzyme systems. Cr(VI), found as $CrO_4^{2-}$, $HCrO_4^-$, or $Cr_2O_7^{2-}$ depending on the pH of the medium, is more toxic than Cr(III), and is known to be carcinogenic and mutagenic and induced dermatitis. Cr is widely used in a number of commercial products, including dyes, paints pigments and salts for leather tanning, electroplating, water cooling and textile [24]. Cr(VI) is one of the most toxic pollutants worldwide as a carcinogenic and mutagenic agent present in wastewater and needs complete removal from wastewater before disposal. Cr(VI) affects human physiology, accumulates in the food chain and causes severe health problems ranging from simple skin irritation to lung carcinoma, pulmonary congestion, vomiting and liver damage etc.

Polypyrrole GO nanocomposites (PPy-GO NC) were prepared for the effective removal of Cr(VI) ions from the aqueous solution using batch and packed-bed column modes [25]. Zeta potential of the prepared PPy-GO NC was observed 6.2 indicating that at higher than 6.2 pH the surface of the adsorbent is positively charged which is suitable for anion adsorption while at lower pH the surface charge is negative, which may be beneficial for cationic adsorption. In addition, the increased surface area of the PPy-GO NC compared to PPyhomopolymer, suggests its enhanced performance in removing Cr(VI) from aqueous solutions.

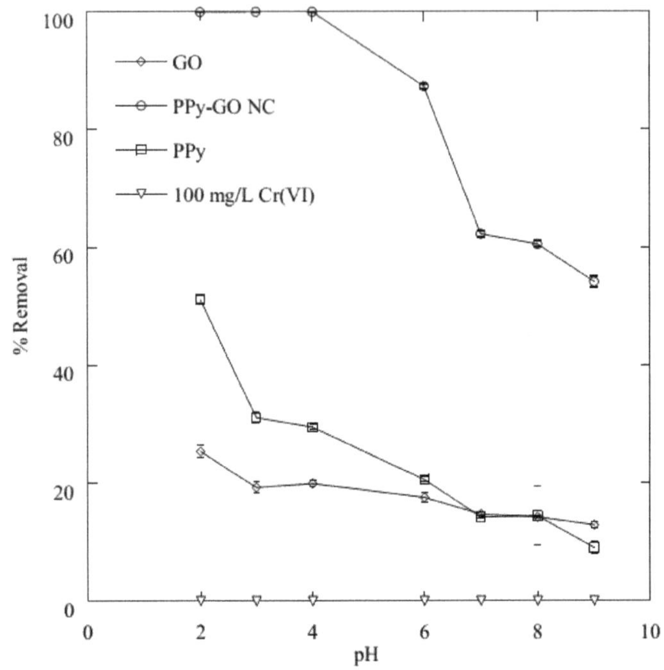


*Fig. 17. Effect of pH on Cr(VI) adsorption [26].*

Fig. 17 reveals the adsorption behavior of Cr(VI) onto GO, PPyhomopolymer and PPy–GO. Results show 99.98 percent Cr(VI) removal by PPy–GO NC which decreases with an increase in pH from 2 to 9. In comparison to the GO and PPyhomopolymer, the enhanced adsorption capacity of PPy–GO NC may be attributed to increased surface area and the presence of highly reactive exposed surface sites. Higher efficiency for Cr(VI) removal at lower pH can be attributed to the fact that below pH 6.2, the surface of PPy–GO NC is positively charged due to the dissociation of doped $Cl^+$ ions with simultaneous protonation of the nitrogen atom of the PPy matrix in the presence of $H^+$ ions [26]. Therefore, at pH less than 6, the $HCrO4^-$ions (major Cr(VI) ion), are replaced by the doped $Cl^+$ ions as well as attracted electrostatically by the protonated surface sites of the PPy–GO NC adsorbent. Moreover, at lower pH, the removal of Cr(VI) was associated with the reduction-oxidation reactions of Cr(VI) via electrons of the PPypolymer while a decrease in Cr(VI) ions removal efficiency at higher pH values may be due to the deprotonation of the nitrogen atoms of the PPy and competitive interactions between hydroxyl ions and $CrO4^{2-}$ ions present in the reaction solution.

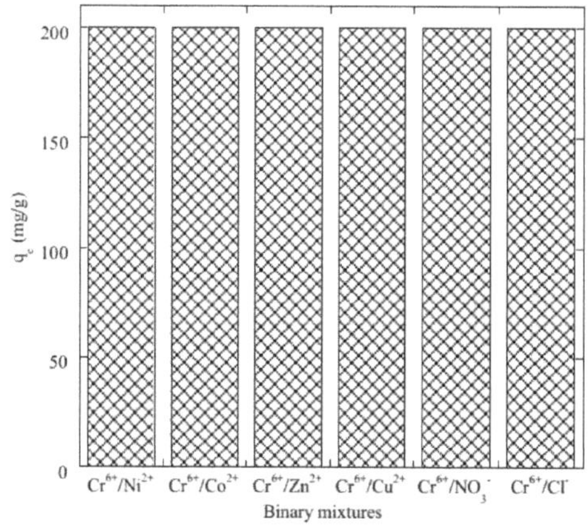


*Fig. 18. Specific PPy–GO NC adsorption study for Cr(VI) in presence of $Ni^{2+}$, $Co^{2+}$, $Zn^{2+}$, $Cu^{2+}$, $NO_3^-$, and $Cl^-$ ions. Temp: 25 C, pH: 2.0 [25].*

Removal of the particular heavy metal ions become important when more than a single ion are present in a solution. Most of the discharges from the industries contain a mixture of ions which adversely affect the water quality and environment. Therefore a comparative behavior of PPy-GO-NC for Cr(VI) in present of other metal ions was studied as shown in Fig. 18. At equilibrium, the concentrations of coexisting ions remained unchanged which indicates the selective adsorption of PPy-GO-NC.

A novel green synthetic approach was developed by Dubey et al. [27] for the preparation of graphene sand composite (GSC) as a novel adsorbent for the removal of Cr(VI) ions from aqueous solutions as shown in Fig. 19. GSC, as prepared, was used as an adsorbent for the removal of Cr(VI) by batch adsorption technique for different contact times, pH and adsorbent doses etc.

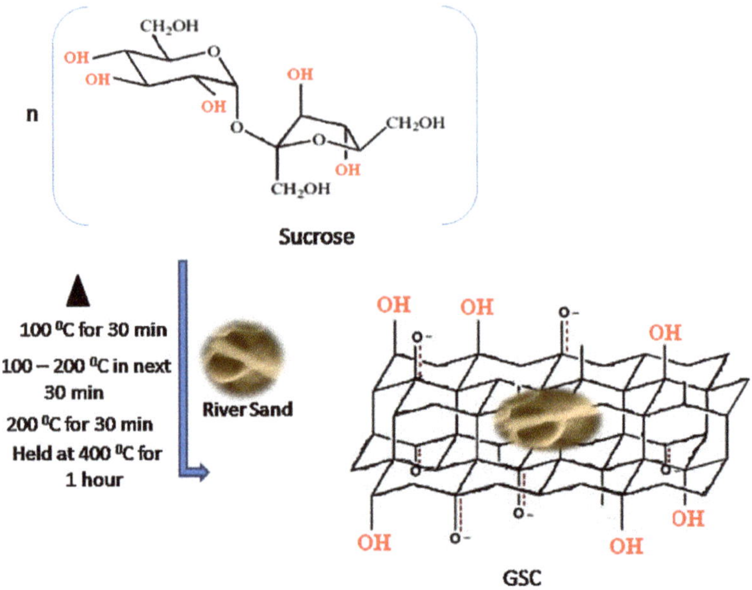


*Fig. 19. Schematic presentation of the GSC preparation [28].*

Cr occurs in different oxidation states depending on the pH of the medium. The following equilibria can exist in an aqueous solution and hence removal efficiencies of GSC for Cr(VI) were studied in the 1.5 to 11 pH range.

$$H_2CrO_4 \leftrightarrow H^+ + HCrO_4^-$$

$$HCrO_4 \leftrightarrow H^+ + CrO_4^{2-}$$

$$2HCrO_4^- \leftrightarrow Cr_2O_7^{2-} + H_2O$$

Adsorption efficiency of the GSC for the removal of Cr(VI) at 1.5 pH was 93 % indicating the maximum adsorption in acidic medium because several forms of chromium ions are found in an acidic medium such as chromate ($CrO_4^{2-}$), dichromate ($Cr_2O_7^{2-}$) and hydrogen chromate ($HCrO_4^-$). As per the literature, it has been suggested that when pH is lower than 6.8, the dominant species are $HCrO^{4-}$ ions while above 6.8, only $CrO_4^{2-}$ ions are stable [28]. The strong adsorption in the acidic medium may be attributed to increased $H^+$ ions concentration on the surface of GSC which offers strong electrostatic attraction between $H^+$ and $HCrO_4^-$ ions. Whereas at the increased pH, the Cr(VI) adsorption showed exponential decrease which may be attributed to positively charged surface in a basic medium as well as to the dual competition of both the anions chromate ($CrO_4^{2-}$) and hydroxide ($OH^-$) ions for adsorption on the surface of the GSC adsorbent. Further, zeta potentials of GSC at pH 2.2 and 10.1 was found to be 244 and 225 mV respectively which support the logic that more positive the zeta potential is, the greater will be the removal of chromium.

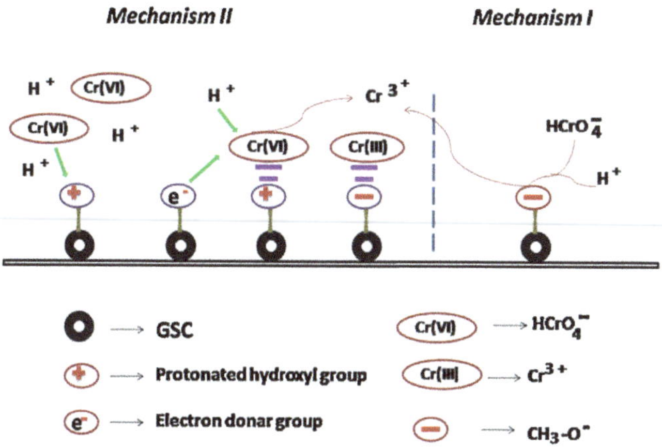


*Fig. 19. Cr(III) and Cr(VI) adsorption mechanism on GSC adsorbent surface [27].*

Based on the XPS, EDX and FT-IR studies, two adsorption mechanism were suggested for the removal of Cr(III) and Cr(VI) by GSC adsorbent as shown in Fig. 19. Binding energy obtained from the XPS results suggests that both Cr(III) and Cr(VI) co-exist on the surface of adsorbent surface and highly toxic Cr(VI) ions adsorbed onto the GSC get partially reduced to relatively less toxic Cr(III). The EDX results also confirmed the majority of Cr(III) adsorbed onto the GSC and small amount of Cr(VI). Based on the above results, two adsorption mechanisms were proposed as shown in Fig. 19. Mechanism I (direct reduction) suggests that the partial reduction of Cr(VI) to Cr(III) due to the presence of lower oxidation potential groups(electron donating groups such as -

OH, C=O, O-$CH_3$) present on the surface of GSC adsorbent in an acidic medium from 0.5 to 1.5 pH because reduction of Cr(VI) to Cr(III) requires a lot of protons and electrons. Whereas mechanism II (indirect reduction) involves the following three steps;

1. Binding of anionic Cr(VI) species to the positively-charged groups present on the GSC adsorbent surface.
2. The reduction of Cr(VI) to Cr(III) by the adjacent electron donor groups (as -OH, C=O, O-$CH_3$) present on the surface of GSC adsorbent.
3. Strong coordinate-covalent bond formation of the reduced Cr(III) with the GSC surface functional groups such as carbonyl (C=O) and methoxy groups (O-$CH_3$).

*In-situ* synthesis of sulfuric acid doped diaminopyrimidine polymers onto the surface of GO via mutual oxidation-reduction technique has been reported by Dinda et al. [29] for the removal of toxic Cr(VI) from aqueous solution as shown in Fig. 20.

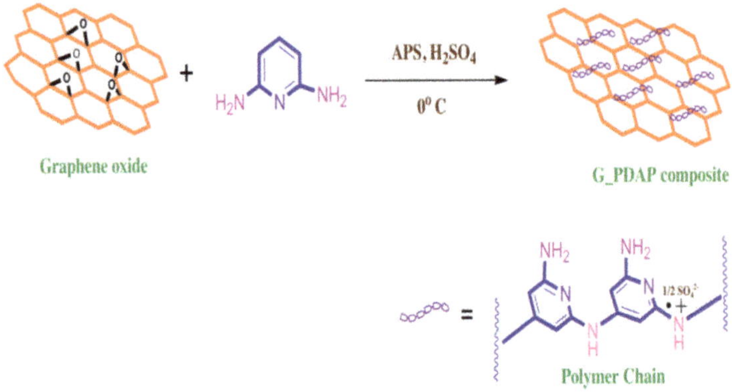


*Fig. 20. In-situ oxidative polymerization of 2, 6-diaminopyridine on GO surface [29].*

Total Cr ions removal efficiency of G-PDAP composite as an adsorbent has been presented in Fig. 21 which indicates that the adsorption efficiency decreases moderately with increase in pH from 1 to 9. Better adsorption efficiency of an adsorbent was observed in an acidic medium i.e. ~71, 59 and 46% at pH 5, 7 and 9, respectively. At lower pH values, Cr(VI) mainly exists as $HCrO_4^-$ which efficiently adsorbed by the adsorbent surface having protonated amine and imine functional groups of polymer chain on GO surface which is in turn then reduced to Cr(III) via redox reaction as shown below. The faint greenish color solution, as shown in Fig. 22 is obtained due to the dissolution of a trace amount of Cr(III) which turn the solution greenish and the remaining part get adsorbed by the adsorbent and is removed from the solution. Decreasing efficiency at higher pH may be correlated with the decreasing interactions between $HCrO_4^-$ and protonated G-PDAP as $OH^-$ competes with $HCrO_4^-$ ions.

$$Cr_2O_7^{2-} + H^+ \longrightarrow 2HCrO_4^- + H_2O$$

$$HCrO_4^- + 7H^+ + 3e^- \longrightarrow Cr(III) + 4H_2O, (E_0 = +1.35V)$$

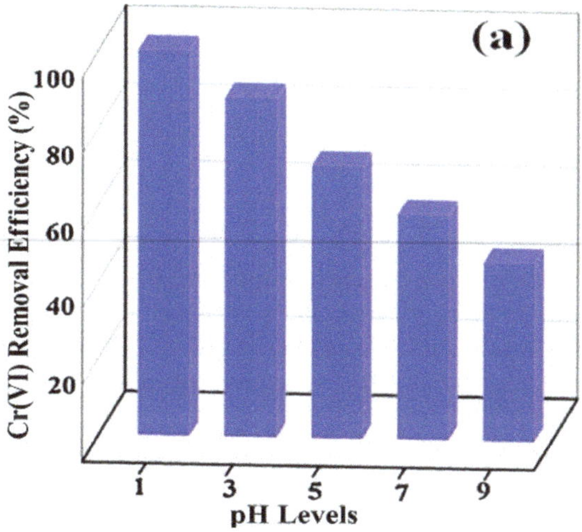


*Fig. 21. Total Cr, Cr(III) and Cr(VI) removal efficiency of G-PDAP composite at 1-9 pH [29].*

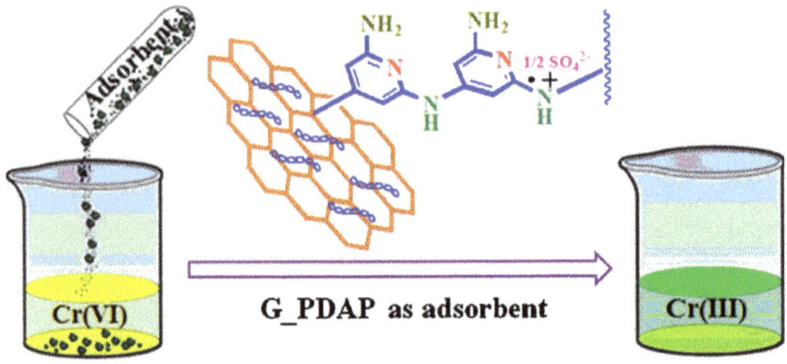


*Fig. 22. The green color of the solution due to a trace amount of Cr(III) [29].*

## 6. Removal of arsenic

Arsenic(As) is a naturally occurring element in the deposits of earth's crust worldwide and its higher concentration in groundwater is found in countries such as India, Bangladesh, Taiwan, Brazil, and Chile etc. In most of the groundwater, arsenic exists as arsenate, As(V), and arsenite, As(III) viz. monomethyl arsenic acid (MMA), dimethyl arsenic acid (DMA)etc. Among all the arsenic species, As(III) is about 60 times more toxic than arsenate [30]. It enters the environment as a pollutant through natural weathering of rocks and anthropogenic activities, mining and smelting processes, pesticide uses and coal combustion. Arsenic exists in powdery amorphous and crystalline forms in the ore. Higher concentration of arsenic creates severe health hazards to human beings and animals.

Magnetite-reduced GO (M-RGO) superparamagnetic composites with a magnetite particle size of ~10 nm were prepared by Chandra et al. which could be separated by applying external magnetic field [31]. Due to the presence of RGO, the magnetic intensities were lower than the bulk $Fe_3O_4$ and the small size of $Fe_3O_4$ nanoparticles. At room temperature, M-RGO composites exhibited a superparamagnetic state with small remnant magnetization and coercivity which are desirable to get strong magnetic signals at small applied magnetic fields. As shown in Scheme 4, prepared M-RGO has been used for the removal of arsenic species from the solution at different pH and temperature.

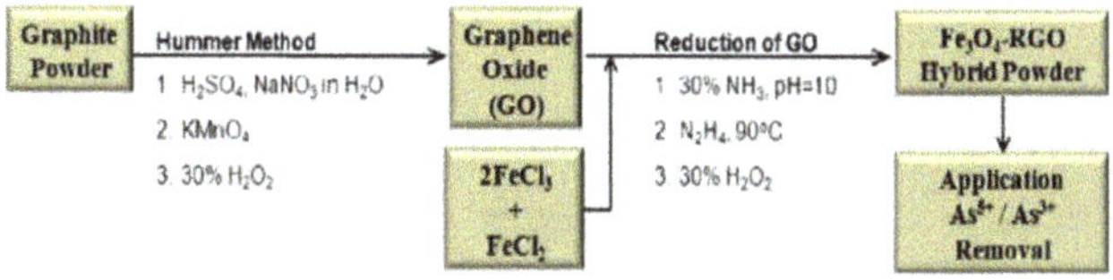


*Scheme 4. $Fe_3O_4$-RGO superparamagnetic composites preparation for As removal [31].*

Fig. 23 exhibits the maximum adsorption capacity M-RGO adsorbent for As(III) and As(V). Removal of As(III) with M-RGO was found higher than As(V) which increased with the content of nano $Fe_3O_4$in accordance with the previously reported results [32]. The removal of As(III) and As(V) ions from the wastewaters is affected by the electrostatic attractions of the ions with the adsorbent surface. Hence the removal of As(III) and As(V) was carried out at different pH levels because electrostatic attractions are pH dependable. As(V) is found in the form of negative ionic species as $H_2AsO_3^-$, under wide pH range, whereas, As(III) is in a non-ionic form ($H_3AsO_4$) [33]. Adsorptions of As(III) and As(V) were correlated with the point of zero charges ($pH_{PZC}$) of the adsorbent (M-RGO) at different pH. It was observed that when $pH_{PZC}$ of an adsorbent was higher than the pH of the case study, the surface of the adsorbent was positively charged and responsible for adsorption of As(V) anions on the surface. At the higher pH than $pH_{PZC}$ of an adsorbent, the surface was negatively charged which results in low adsorption of the pollutants on the adsorbent surface.

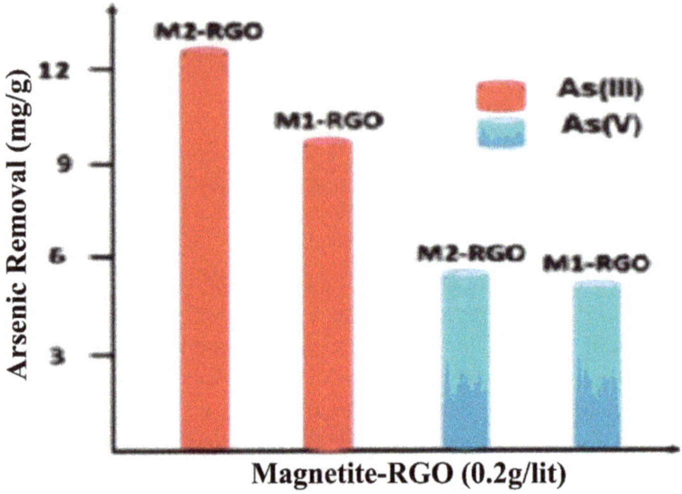


*Fig. 23. The maximum adsorption capacity of $Fe_3O_4$-RGO (M-RGO) composites for arsenic removal from water [32].*

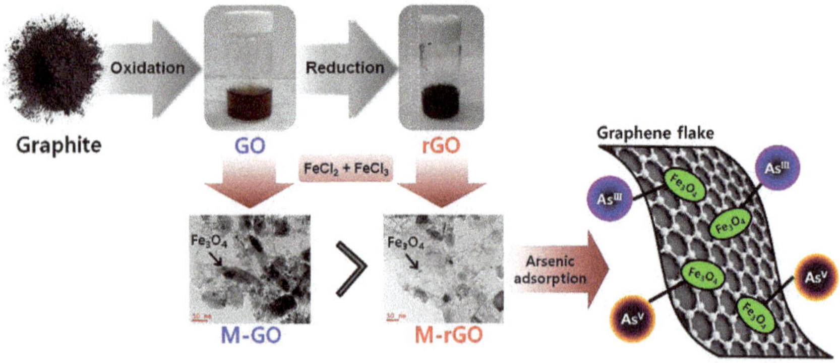


*Scheme 5. Systematic representation of M-GO and M-rGO adsorbents for arsenic removal [12].*

Systematic representation for the preparation of GO and reduced GO with magnetite ($Fe_3O_4$) viz. M-GO and M-rGO adsorbent for arsenic removal were reported by Yoon et al. [12] as shown in Scheme 5.

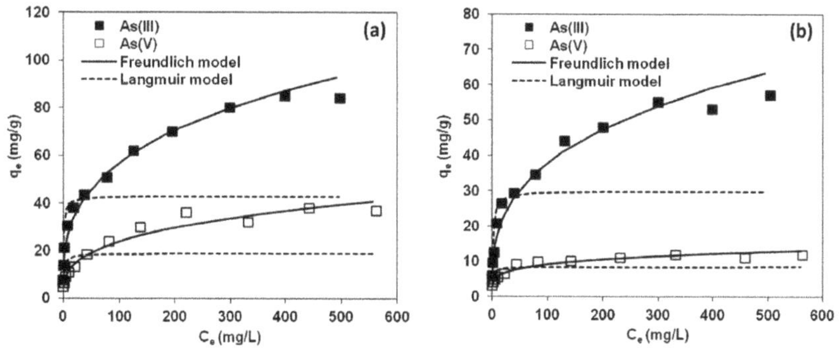


*Fig. 24. Freundlich and Langmuir adsorption isotherms of As(III) and As(V) onto (a) M-GO and (b) M-rGO adsorbents at pH 7 [12].*

Fig. 24 demonstrates the Freundlich and Langmuir adsorption (monolayer) isotherms for the adsorption of As(III) and As(V) on the surface of M-GO and M-rGO adsorbents. The content of magnetite nanoparticles loaded was found more on the surface of M-GO (630 mg/g) when compared with the (457 mg/g) because more functional groups of M-GO were able to retain the higher content of nanomagnetites. It was observed that the content of nanomagnetites was the significant factor for the effective removal of As(III) and As(V) with nano adsorbent and extra energy used in the reduction process could be saved.

The adsorption capacity of M-GO adsorbent was found to be 85 and 38 mg/g for As(III) and As(V), respectively whereas adsorption capacity of M-rGO adsorbent was found to be 57 and12mg/g for As(III) and As(V), respectively. Freundlich adsorption isotherms were found to be more linear with good correlation coefficients in comparison to the Langmuir isotherms suggesting that arsenic is adsorbed in the form of heterogeneity on the surface of adsorbents. The reason for higher adsorption capacity of an M-GO adsorbent was the higher functionality present on the oxidized form of graphite viz. GO including carboxylic, hydroxyl and other oxygen functionalities.

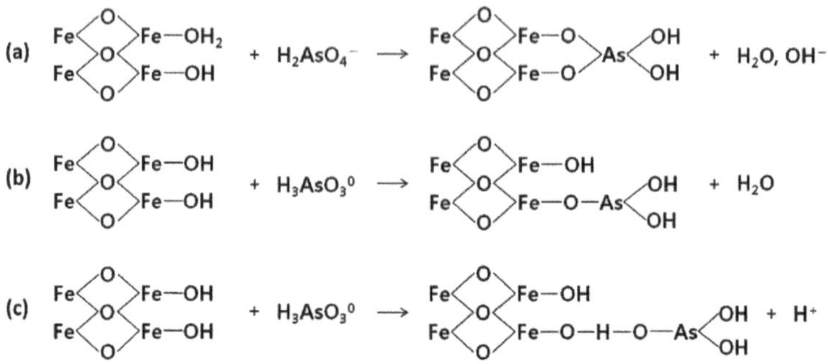


*Scheme 6. Possible As(III) and As(V) adsorption mechanism on $Fe_3O_4$ magnetite (a) bidentate binuclear–bridging complex formation with As(V), (b) monodentate complex formation with As(III), (c) outer-sphere complex formation with As(III) [12].*

Scheme 6 represent that As(III) and As(V) were adsorbed onto the iron oxide containing adsorbent by an inner-sphere ligand-exchange mechanism, in which the arsenic oxyanion exchange takes place with surface-OH or $-OH_2$ functional groups that are directly coordinated with $Fe^{3+}$at the surface of their oxide. Scheme 6 (a) shows that As(V) anion was absorbed with the two adjacent $Fe^{3+}$cations and the complex formed, is called bidentate binuclear–bridging complex while As(III) was adsorbed by iron oxide, forming both bidentate binuclear–bridging complexes and monodentate complexes as shown in scheme 6 (b). Additionally As(III) might be adsorbed due to monodentate complexes or an outer-sphere complex formation in which the ligand is not coordinated directly to the structural cation but instead is bound to the surface-OH or $-OH_2$, probably by means of a hydrogen bond as shown in scheme 6(c). Hence, the adsorption of As(III) was found more on the surface $Fe_3O_4$ loaded graphene adsorbent. Results obtained from pH study also confirmed that the adsorption of As(III) was favorably affected by the complex formation on the surface of an adsorbent, while As(V) was adsorbed by electrostatic interactions between the adsorbent and adsorbate.

## Conclusions

Nanoadsorbents based on nanomaterials such as graphene which is considered the mother of all forms of carbon allotropes because of its unique nature play a major role in the adsorption and consequently separation of heavy metal ions from the wastewaters. Graphene has extremely high electronic mobility, high surface area to volume ratio, weight, and the preponderance of exposed edge planes to enhance charge storage which is responsible for their higher absorption capacity when compared with other materials. Hence this chapter includes the detailed study of the recently developed novel nano-adsorbent based on the graphene, graphene oxide and its composites for the effective removal of heavy metal ions (Pb, Cd, Hg, Cr and As) from the wastewaters.

## References


[1] G. Zhao, J. Li, X. Ren, C. Chen, X. Wang, Few-layered graphene oxide nanosheets as superior sorbents for heavy metal ion pollution management, Environ. Sci. Technol., 45(2011) 10454-10462. https://doi.org/10.1021/es203439v

[2] G. Gollavelli, C.-C. Chang, Y.-C. Ling, Facile synthesis of smart magnetic graphene for safe drinking water: heavy metal removal and disinfection control, ACS Sust. Chem. & Eng., 1(2013) 462-472. https://doi.org/10.1021/sc300112z

[3] X. Wang, Y. Guo, L. Yang, M. Han, J. Zhao, X. Cheng, Nanomaterials as sorbents to remove heavy metal ions in wastewater treatment, Environ. & Anal. Toxicol., 2(2012) 1-5. https://doi.org/10.4172/2161-0525.1000154

[4] C. Chen, X. Wang, Adsorption of Ni(II) from Aqueous solution using oxidized multiwall carbon nanotubes, Ind. Eng. Chem. Res., 45(2006) 9144-9149. https://doi.org/10.1021/ie060791z

[5] Y. Y. Liang, D. Q. Wu, X. L. Feng, K. M€ullen, Dispersion of graphene sheets in organic solvent supported by ionic interactions, Adv. Mater., 21(2009) 1679-1683. https://doi.org/10.1002/adma.200803160

[6] X. Li, X. Wang, L. Zhang, S. Lee, H. Dai, Chemically derived, ultrasmooth graphene nanoribbon semiconductors, Science, 319(2008) 1229-1232. https://doi.org/10.1126/science.1150878

[7] T. Mueller, F. N. Xia, P. Avouris, Graphene photodetectors for high-speed optical communications, Nat. Photonics, 4(2010) 297–301. https://doi.org/10.1038/nphoton.2010.40

[8] L. Zhang, J. Xia, Q. Zhao, L. Liu, Z. Zhang, Functional graphene oxide as a nanocarrier for controlled loading and targeted delivery of mixed anticancer drugs, Small, 6(2010) 537–544. https://doi.org/10.1002/smll.200901680

[9] K. C. Kemp, H. Seema, M. Saleh, N. H. Le, K. Mahesh, V. Chandra, K. S. Kim, Environmental applications using graphene composites: Water remediation and

gas adsorption, Nanoscale, 5(2013) 3149–3171. https://doi.org/10.1039/c3nr33708a

[10] G. Zhao, J. Li, X. Ren, C. Chen, X. Wang, Few-layered graphene oxide nanosheets as superior sorbents for heavy metal ion pollution management, Environ. Sci. Technol., 45(2011) 10454–10462. https://doi.org/10.1021/es203439v

[11] R. R. Nair, H. A. Wu, P. N. Jayaram, I. V. Grigorieva, A. K. Geim, Unimpeded permeation of water through helium-leak-tight graphene-based membranes, Science, 335(2012) 442–444. https://doi.org/10.1126/science.1211694

[12] Y. Yoon, W. K. Park, T.-M. Hwang, D. H. Yoon, W. S. Yang, J.-W. Kang, Comparative evaluation of magnetite–graphene oxide and magnetite-reduced graphene oxide composite for As(III) and As(V) removal, J. Hazard. Mater., 304(2016) 196-204. https://doi.org/10.1016/j.jhazmat.2015.10.053

[13] C. J. Madadrang, H. Y. Kim, G. Gao, N. Wang, J. Zhu, H. Feng, M. Gorring, M. L. Kasner, S. Hou, Adsorption Behavior of EDTA-Graphene Oxide for Pb(II) Removal, ACS Appl. Mater. & Interf., 4(2012) 1186-1193. https://doi.org/10.1021/am201645g

[14] L. Fan, C. Luo, M. Sun, X. Li, H. Qiu, Highly selective adsorption of lead ions by water-dispersible magnetic chitosan/graphene oxide composites, Coll. and Surfac. B: Biointerf., 103(2013) 523-529. https://doi.org/10.1016/j.colsurfb.2012.11.006

[15] S. Luo, X. Xu, G. Zhou, C. Liu, Y. Tang, Y. Liu, Amino siloxane oligomer-linked graphene oxide as an efficient adsorbent for removal of Pb(II) from wastewater, J. Hazard. Mater., 274(2014) 145-155. https://doi.org/10.1016/j.jhazmat.2014.03.062

[16] Z. Xu, Y. Zhang, X. Qian, J. Shi, L. Chen, B. Li, J. Niu, L. Liu, One step synthesis of polyacrylamide functionalized graphene and its application in Pb(II) removal, Appl. Surf. Sci., 316(2014) 308-314. https://doi.org/10.1016/j.apsusc.2014.07.155

[17] S. Wan, F. He, J. Wu, W. Wan, Y. Gu, B. Gao, Rapid and highly selective removal of lead from water using graphene oxide-hydrated manganese oxide nanocomposites, J. Hazard. Mater., 314(2016) 32-40. https://doi.org/10.1016/j.jhazmat.2016.04.014

[18] J.-H. Deng, X.-R. Zhang, G.-M. Zeng, J.-L. Gong, Q.-Y. Niu, J. Liang, Simultaneous removal of Cd(II) and ionic dyes from aqueous solution using magnetic graphene oxide nanocomposite as an adsorbent, Chem. Eng. J., 226(2013) 189-200. https://doi.org/10.1016/j.cej.2013.04.045

[19] J. Liu, H. Du, S. Yuan, W. He, Z. Liu, Synthesis of thiol-functionalized magnetic graphene as adsorbent for Cd(II) removal from aqueous systems, J. Environ. Chem. Eng., 3(2015) 617-621. https://doi.org/10.1016/j.jece.2015.01.016

[20] H. He, C. Gao, Supraparamagnetic, conductive, and processable multifunctional graphene nanosheets coated with high-density Fe3O4 nanoparticles ACS Appl. Mater. Interfaces, 2(2010) 3201-3210. https://doi.org/10.1021/am100673g

[21] Y. Wang, S. Liang, B. Chen, F. Guo, S. Yu, Y. Tang, Synergistic removal of pb(ii), cd(ii) and humic acid by fe3o4@mesoporous silica-graphene oxide composites, PLOS One, 8(2013) e65634. https://doi.org/10.1371/journal.pone.0065634

[22] S. Thakur, G. Das, P. K. Raul, N. Karak, Green one-step approach to prepare sulfur/reduced graphene oxide nanohybrid for effective mercury ions removal, J. Phys. Chem. C, 117(2013) 7636-7642. https://doi.org/10.1021/jp400221k

[23] S. Kabiri, D. N. H. Tran, S. Azari, D. Losic, Graphene-diatom silica aerogels for efficient removal of mercury ions from water, ACS Appl. Mat. & Interf., 7(2015) 11815-11823. https://doi.org/10.1021/acsami.5b01159

[24] S. A. Katz, H. Salem, The Biological and environmental chemistry of chromium, VCH, New York, . (1994)

[25] K. Z. Setshedi, M. Bhaumik, M. S. Onyango, A. Maity, High-performance towards Cr(VI) removal using multi-active sites of polypyrrole–graphene oxide nanocomposites: Batch and column studies, Chem. Eng. J., 262(2015) 921-931. https://doi.org/10.1016/j.cej.2014.10.034

[26] X. Zhang, R. Bai, Surface electric properties of polypyrrole in aqueous solutions, Langmuir, 19(2003) 10703-10709. https://doi.org/10.1021/la034893v

[27] R. Dubey, J. Bajpai, A. K. Bajpai, Green synthesis of graphene sand composite (GSC) as novel adsorbent for efficient removal of Cr (VI) ions from aqueous solution, J. Wat. Proc. Eng., 5(2015) 83-94. https://doi.org/10.1016/j.jwpe.2015.01.004

[28] J. Zhu, S. Wei, H. Gu, B. Rapole, Q. Wang, Z. Luo, N. Haloolaarachchige, D. P. Young, Z. Guo, One-pot synthesis of magnetic graphene nanocomposites decorated with core@double-shell nanoparticles for fast chromium removal, Environ. Sci. Technol. , 46(2012) 977-985. https://doi.org/10.1021/es2014133

[29] D. Dinda, S. Kumar Saha, Sulfuric acid doped poly diaminopyridine/graphene composite to remove high concentration of toxic Cr(VI), J. Hazard. Mater., 291(2015) 93-101. https://doi.org/10.1016/j.jhazmat.2015.02.065

[30] J. F. Ferguson, J. Gavis, A review of the arsenic cycle in natural waters, Water Res., 6(1972) 1259-1274. https://doi.org/10.1016/0043-1354(72)90052-8

[31] V. Chandra, J. Park, Y. Chun, J. W. Lee, I.-C. Hwang, K. S. Kim, Water-dispersible magnetite-reduced graphene oxide composites for arsenic removal, ACS Nano, 4(2010) 3979-3986. https://doi.org/10.1021/nn1008897

[32] Q. Zhang, B. Pan, W. Zhang, B. Pan, Q. Zhang, H. Ren, Arsenate removal from aqueous media by nanosized hydrated ferric oxide (HFO)-loaded polymeric sorbents: Effect of HFO loadings, Ind. Eng. Chem. Res., 47(2008) 3957-396. https://doi.org/10.1021/ie800275k

[33] T.-F. Lin, J.-K. Wu, Adsorption of arsenite and arsenate within activated alumina grains: Equilibrium and Kinetics, Water Res., 35(2001) 2049-2057. https://doi.org/10.1016/S0043-1354(00)00467-X

# Chapter 7

# Magnetic mollusk shell-$Fe_3O_4$ composite powder used as seeding adsorbent to purify Zn(II) and Pb(II) contaminated wastewater

Mokgadi BOPAPE[1*], Jianwei REN[2], Myalowenkosi SABELA[3], Anthony MULIWA[1], Maurice Stephen ONYANGO[1]

[1]Department of Chemical, Metallurgical and Materials Engineering, Tshwane University of Technology, Private Bag X680, Pretoria, 0001, South Africa

[2]Material Science and Manufacturing, Council for Scientific and Industrial Research, Pretoria 0001, South Africa

[3]Department of Chemistry, Durban University of Technology, P.O Box 1334, Durban 4000, South Africa

[*]*bopapemf@tut.co.za*

**Abstract**

Biosorption of toxic heavy metal ions using inexpensive non-living biomass from aqueous solutions has been recognized as a promising technique for the decontamination of industrial effluents containing heavy metals. This chapter reports on the preparation and characterization of partially converted mollusk shell-$Fe_3O_4$ powder with sufficient magnetic saturation values. At first stage, biosorption kinetics, equilibrium and thermodynamics of Zn(II) and Pb(II) as typical aqueous heavy metal ions, onto mollusk shell-$Fe_3O_4$ composite powder were investigated in batch mode. Thereafter, the mollusk shell-$Fe_3O_4$ composite powder was explored as a seeding adsorbent to treat Zn(II) contaminated wastewater, and a lab-scale magnetically assisted water treatment system applying magnetic trap was evaluated to recover the mollusk shell-$Fe_3O_4$ powder from fluid state. In the adsorption and magnetic separation systems, Zn(II) removal and magnetic separation efficiency were used as key performance indicators. Overall equilibrium uptake capacity was 312.5 mg $g^{-1}$ at 298 K and 400 mg $g^{-1}$ at 318 K while $\Delta H^{o}_{ads}$ and $\Delta S^{o}_{ads}$ were determined to be -24.46 kJ $mol^{-1}$ and 103.33 J $mol^{-1}$ K respectively.

**Keywords**

Mollusk Shell-$Fe_3O_4$ Powder, Wastewater, Adsorption Kinetics, Equilibrium, Thermodynamics

## Contents

## 1 Introduction

One of the important applications of adsorption processes is the decontamination of toxic heavy metals from industrial effluents, using abundant, low-cost natural materials, and industrial waste [1]. Biosorption of toxic heavy metal ions from aqueous solution using inexpensive non-living biomass has emerged strongly as a promising technique for the decontamination of such industrial effluents. This is because it is simple, analogous operation to conventional ion exchange technology, efficient and makes use of cheap materials like biomass and waste bio-products [2-7]. Literature reports indicate that the total cost of treating electroplating wastewater using partially converted mollusk shells waste would be less than half the cost of conventional precipitation treatment [8]. In South Africa, for instance, tons of mollusk shells wastes are piled along the long coastline. The shells contain chitosan which is known to have a variety of functional groups such as large numbers of amino ($-NH_2$), carboxyl (-COOH) and hydroxyl (-OH) groups which make it an effective ion-exchanger [7, 9]. Among the various biosorbents found in literature, chitosan has been reported to have high affinity for metal ions, such as arsenic [10], lead, cadmium, copper [11], chromium [12], cobalt [13], manganese, zinc [14], and nickel [15] in aqueous solutions. The nitrogen in the amino group of the chitosan molecule acts as an electron donor and is presumably responsible for selective chelation with transition metal ions [16-19]. Moreover, it facilitates modifications of the polymer, thus enhancing adsorption properties.

Nanostructured adsorbents especially those containing nanosized $Fe_3O_4$ particles are considered as potential sorbents for remediation of heavy metal polluted waters. The incorporation of $Fe_3O_4$ particles not only improves the surface area of the resultant adsorbent, but also facilitates easy separation under external magnetic fields [20-22]. The adsorption characteristics of magnetic particles can also be improved by surface modifications such as physical coating and covalent binding to enable specific metal complexation [23-26]. For example, amino-functionalized $Fe_3O_4$@$SiO_2$ core-shell magnetic nano-materials were achieved by surface functionalization of $Fe_3O_4$@$SiO_2$ nano-particles using (3-aminopropyl) trimethoxysilane (APTMS) as the silylation agent, and the resulting material exhibited high adsorption affinity for Cu(II), Pb(II) and Cd(II) ions in aqueous solutions [23]. Mono-disperse $Fe_3O_4$@silica core-shell microspheres were prepared by a two-step method, and used for removal of heavy metal ions such as Hg(II), Pb(II) from industrial wastewater [24]. Similarly, Liu et al., 2008 prepared humic -acid-coated $Fe_3O_4$ nano-particles by co-precipitation method and found the sorbent to effectively remove Hg(II), Pb(II), Cd(II), and Cu(II) from water [25]. Recently, magnetic chitosan/magnetite nano-composite powder was proposed for the removal of Pb(II) and Ni(II) at pH 4-6 [26].

In view of the aforementioned reasons, the present study examined the adsorption and separation behaviour of partially converted mollusk shell-$Fe_3O_4$ powder towards Zn(II) and Pb(II) ions in aqueous solutions. The physicochemical properties of the sorbent were determined by SEM, FT-IR, and VSM characterisation techniques. At the first instance, biosorption kinetics, equilibrium and thermodynamics of Zn(II) and Pb(II) ions onto mollusk shell-$Fe_3O_4$ composite powder were investigated in batch mode.

## 2. Materials and methods

### 2.1 Materials

Zinc nitrate hexahydrate ($Zn(NO_3)_2.6H_2O$) and lead nitrate $Pb(NO_3)_2$) were purchased from Saarchem (Pty) Ltd., South Africa for use as sources of Zn(II) and Pb(II) ions solutions. All glasswares were cleaned using nitric acid solution and rinsed thoroughly with deionised water prior to use. Stock solutions of both metal ions were prepared. The pH adjustments were done using 0.1 M NaOH and 0.1 M $HNO_3$ solutions. All other chemicals including NaOH, $FeCl_3.6H_2O$, $FeCl_2.4H_2O$, HCl and KCl that were used in the experiment were of analytical reagent grade (AR) and were used as received from Sigma Aldrich suppliers in South Africa.

### 2.2 Methods

#### 2.2.1 Preparation of partially converted mollusk shell@$Fe_3O_4$ powder

The mollusk shells were collected from the west coast of Cape Town in South Africa. The shells were first rinsed with water to remove excess dirt and sand. Once the shells had dried at 120 °C overnight, they were manually crushed in a plastic bag and screened through a set of sieves to get the geometrical size of 75-150 μm. Demineralization was achieved by soaking the shell particles in 5 wt% NaOH and 5 wt% HCl solution at room temperature for 6h, respectively. The demineralized shells were subsequently soaked in 50 wt% NaOH for 6 hrs at 90 °C to induce deacetylation. After this, the solution mixture was vacuum filtered and rinsed thoroughly with deionised water until pH of the filtrate was neutral and a white powder was obtained and then air-dried.

The mollusk shell@$Fe_3O_4$ powder was synthesised by a co-precipitation method. A solution containing 0.8 M $FeCl_3.6H_2O$ and 0.4 M $FeCl_2.4H_2O$ was poured in 150 mL of 0.5 M HCl solution under nitrogen gas bubbling. After attaining homogeneity, the resultant mixture was added drop-wise into a 250 mL mixture containing 10 M NaOH solution and 20 g of mollusk shell particles (75-150 μm) in a three-necked flask while bubbling nitrogen gas. A black colour change was observed as the temperature of the reaction mixture reached to about 65 °C and by further heating the temperature was maintained at 95 °C for 1hr. Afterwards, the black mixture was cooled to room temperature, and repeatedly washed with deionized water until neutral pH was achieved. The solids were then separated magnetically by an external hand-held magnet. Finally, the solid was vacuum-dried for 12 hrs and the dry material was named as magnetic mollusk shell-$Fe_3O_4$ powder for further use.

### 2.2.2 Adsorbent characterization

The X-ray diffraction (XRD) patterns of mollusk shell, magnetite ($Fe_3O_4$) and mollusk shell@$Fe_3O_4$ particles were obtained at room temperature by PANalytical X'Pert Pro powder diffractometer with Pixcel detector using Ni-filtered Cu K-alpha radiation (0.154 nm) in the range of $2\theta$ = 1–90°, and a scanning rate of 0.02 $s^{-1}$. The morphology and composition of adsorbent media were characterized by SEM/EDX using a JEOL JSM-7600F Field Emission Scanning Emission Microscope (FeSEM), running at 2 kV accelerating voltage. The micrograph was scanned in different magnifications for clearer morphology. Fourier transform infrared spectroscopy (FT-IR) spectra were recorded with Perkin Elmer FTIR spectrometer, using NUJOL CP as a background, in the region of 400–4000 $cm^{-1}$ and with a resolution of 4 $cm^{-1}$. Magnetic measurement was performed using a SQUIT Magnetometer-VSM, Model 7 TeslaXL (Quantum Design, San Diego). The sample material was pressed into pellets for containment into the magnetometer sample environment. Magnetization was recorded as a function of field (-2.0 to +2.0 T) at 300 K temperature. The specific surface area of the particles was determined with nitrogen adsorption BET (Brunauer-Emmett-Teller) measurements performed with a Micromeritics instrument Tristar II 3020 V1.02. To determine the point-zero of charge ($pH_{pzc}$) of mollusk shell@$Fe_3O_4$ powder, 0.1 M KCl solutions were used with an initial pH adjusted between 2.0 - 12.0 using either 0.1M NaOH or HCl solutions. Then 0.05 g of mollusk shell@$Fe_3O_4$ powder was added into each solution and samples were placed in a thermostatic shaker for a period of 24 h to equilibrate. Thereafter, the mixtures were filtered and the final pH of each filtrate was measured and recorded [27, 28]. Elemental analysis for both ions was performed by inductively coupled plasma-atomic emission spectroscopy (ICP-AES) using a Jobin-Yvon JY70 Plus Spectrometer at a plasma power of 1400 W.

## 2.3 Zn(II) and Pb(II) batch biosorption experiments

### 2.3.1 Effect of initial solution pH on metal removal

The interaction between Zn(II) and Pb(II) with functional groups present in the sorbent was studied by varying the initial solution pH between pH 2–8, by 1 M NaOH or HCl. Accordingly, bio-sorption was conducted by adding 0.05 g of the mollusk shell@$Fe_3O_4$ composite powder to 50 mL of 200 mg/L Zn(II) and Pb(II) solutions in plastic sampling bottles respectively. The sample bottles were contacted for 24 h in a thermostatic shaker. After this period, the suspensions were filtered and the residual Zn(II) and Pb(II) concentrations were measured by inductively coupled plasma (ICP-AES) at a plasma power of 1400 W, coolant and nebulizer flow of 12 and 1 L $min^{-1}$ respectively.

### 2.3.2 Effect of contact time

Batch bio-sorption experiments were conducted by contacting 1.0 g of biosorbent in a 1.0 L batch reactor with initial Zn(II) concentrations of 51.55 ppm and 95.53 ppm and Pb(II) concentrations of 63.25 ppm and 102.66 ppm respectively at pH 5.5. The reactor was stirred with an overhead stirrer at a speed of 250 rpm at room temperature. At different time intervals, 10 mL of samples were collected and the residual concentration of Zn(II) and Pb(II) ion solutions were analysed by ICP-AES. The equilibrium bio-sorption capacities of the mollusk shell@$Fe_3O_4$ composite powder for Zn(II) and Pb(II) were calculated using Eq. (1) as follows:

$$q_t = \frac{(C_o - C_t)xV}{m} \tag{1}$$

where $C_o$ (mg/L) is the initial concentration, $C_t$ (mg/L) the concentration at any time, $q_t$ (mg/g) the amount of Zn(II) and Pb(II) adsorbed at any time, $m$ (g) the bio-sorbent mass and $V$ (L), the solution volume.

### 2.3.3 Effect of initial metal concentration and temperature

The sorption equilibrium data using partially converted mollusk shell@$Fe_3O_4$ composite powder (PCMS-$Fe_3O_4$) were determined by varying the initial metal solutions concentrations (100-500 mg/L) and temperature (298, 308 and 318 K). The bio-sorbent mass, pH and solution volume were fixed at 0.05 g, 5.5 and 50 mL, respectively. The sample bottles were contacted for 24 h in a thermostatic shaker. After this period, the suspensions were filtered and the concentrations of Zn(II) and Pb(II) left in the supernatant solution were determined by ICP-OES. The equilibrium sorption capacity was determined by Eq. (2) as:

$$q_e = \frac{(C_o - C_e)xV}{m} \tag{2}$$

where $q_e$ and $C_e$ are the amounts of Zn(II) and Pb(II) adsorbed (mg/g) and the equilibrium concentration (mg/L), respectively, while V is the solution volume (L).

## 3. Results and discussion

### 3.1 Bio-sorbent characterization

#### 3.1.1 X-ray diffraction (XRD) analysis

The XRD patterns of partially converted mollusk shell (spectra a), partially converted mollusk shell@$Fe_3O_4$ composite powder (spectra b) and $Fe_3O_4$ (spectra c) obtained from the experiment are shown in Fig. 1. In spectra a and b, the diffraction peak at $2\theta=18.14^o$-$20.14^o$ is the typical fingerprint of semi-crystalline chitosan [29]. All the diffraction peaks and relative intensities in spectra c match well with those for spinel $Fe_3O_4$ structure in JCPDS file (PDF No. 65-3107) [25]. The XRD pattern further shows new peaks on the partially converted mollusk shell@$Fe_3O_4$ which corresponds to intensities of the partially converted mollusk shell (specta b). By comparing patterns of spectra (b) and (c), it is evident that $Fe_3O_4$ was successfully coated onto the composite and the interaction of iron (Fe) with the mollusk shell. The average $Fe_3O_4$ particle size was approximately 10 nm calculated by the Debye-Scherrer equation ($d = k\lambda / \beta\cos\theta$), where d is the average particle size (nm) of the sample, $\lambda$ is the X-ray wavelength (nm), $k$ is the constant, $\theta$ and $\beta$ are the Bragg angle (in radians) and the excess line broadening (in radians), respectively. The average $Fe_3O_4$ particle size was calculated using the most intense peak (311) with its full width at half maximum.

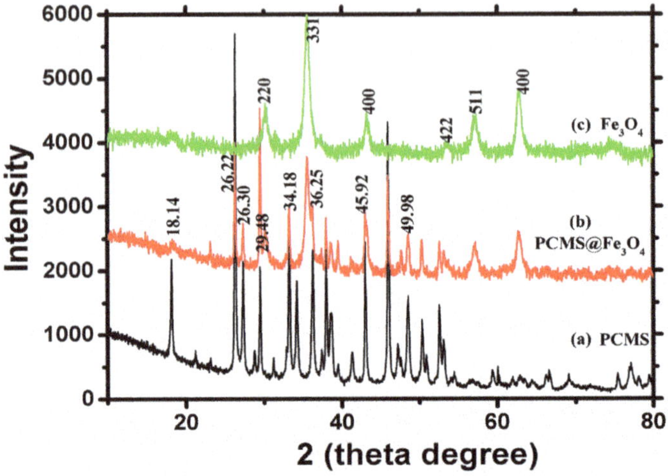


*Fig. 1. XRD patterns for $Fe_3O_4$, PCMS@$Fe_3O_4$ composite powder and PCMS.*

### 3.1.2 Scanning electron microscopy (SEM) analysis and BET

The morphological characteristics of partially converted mollusk shell@$Fe_3O_4$ composite powder were determined using a scanning electron microscopy (SEM). As observed in SEM images of Fig. 2, two distinct differences are clear; the surface of the bio-sorbent in Fig. 2(a) appears to be smooth and non-porous while that of Fig. 2(b) appears to be irregular and porous. This can be attributed to the coating of iron (Fe) on the surface of the sorbent, thus increasing the contact area and facilitating pore diffusion during sorption. The average size of mollusk particles was 75-150 µm. After coating $Fe_3O_4$ nano-particles onto the surface of the composite, the specific surface area of the sorbent measured by BET method increased from 2.95 to 36.33 $m^2/g$. Hence it can be concluded that the mollusk shell-$Fe_3O_4$ composite powder has ample morphology for ion sorption.

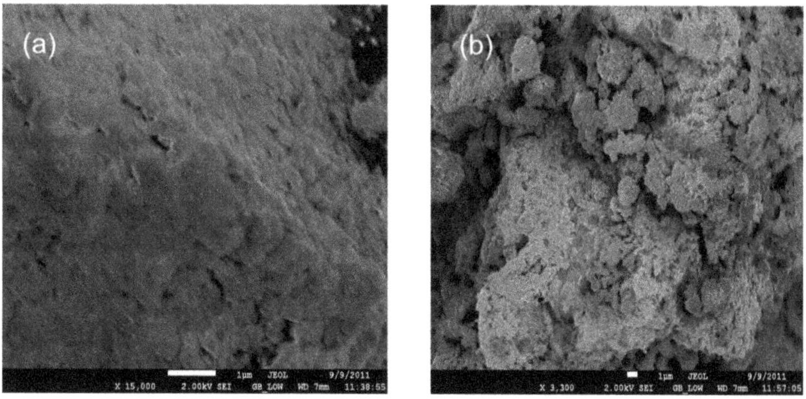


*Fig. 2. SEM micrograph of (a) pure PCMS and (b) PCMS@$Fe_3O_4$ composite.*

### 3.1.3 Fourier transform infrared (FTIR)

Fig. 3 shows the FTIR spectra of raw mollusk shell (curve a), deacetylated mollusk shell (curve b), partially converted mollusk shell@$Fe_3O_4$ (curve c), and partially converted mollusk shell@$Fe_3O_4$ after Zn(II) bio-sorption (curve d) while (curve e) shows the composite powder after Pb(II) bio-sorption. The peaks around wavenumbers 712, (2875-2920), (1680-1480), (1400-1660) and (3750-3000) $cm^{-1}$ can be attributed to the Fe-O group, C-H bending, CO bending, $NH_2$ bending vibration and OH stretching group respectively [30-33]. The -$NH_2$ group of the composite powder has a main characteristic peak in the region of 3750-3000 $cm^{-1}$, which appears to be masked by a peak due to the OH group and also a characteristic bio-sorption peak at 1640.15 $cm^{-1}$ due to its amide group. The presence of magnetite is indicated in the region of 750-500 $cm^{-1}$. The FTIR spectra indicated that both chitosan and $Fe_3O_4$ phases are present in magnetic chitosan nanoparticles and the $Fe_3O_4$ magnetic nanoparticles were coated on the chitosan [34]. After adsorption/bio-sorption, the main characteristic peak for both metals increased in intensity. This can be accredited to the interaction of the $NH_2$ group with Zn(II) and Pb(II) metal ions and further depicts the mechanism of bio-sorption onto the mollusk shell.

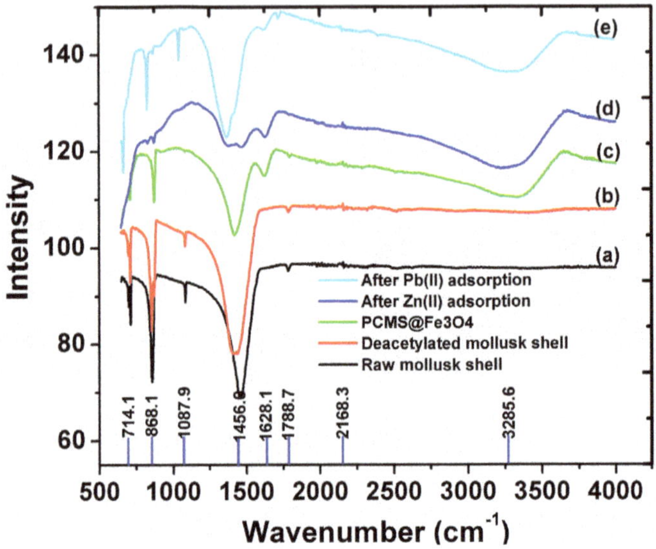


*Fig. 3. FTIR spectra of raw mollusk shell, deacetylated mollusk shell, PCMS@$Fe_3O_4$ composite, Zn(II) and Pb(II) after adsorption.*

### 3.1.3 Magnetic properties measurements

Magnetic measurements were conducted on a VSM to verify whether the synthesized composite powder could be recovered from the aqueous phase by magnetic separation processes. As can be seen in Fig. 4a, the composite did not exhibit hysteresis loop and remanence magnetization. Further, the saturation magnetization ($M_s$) of the composite was 25.3 emug$^{-1}$. These observations suggested that partially converted mollusk shell powder was super-paramagnetic [34] and had sufficient magnetism to facilitate its separation with a conventional magnet separation device. The minimum acceptable literature value for magnetic separation is 16.3 emug$^{-1}$ [19, 35]. Similar findings were reported by Tran et al. for the application of chitosan/magnetite beads onto aqueous Pb(II) and Ni(II) ions. Fig. 4b demonstrates the mollusk shell-$Fe_3O_4$ powder attracted by an external magnet.

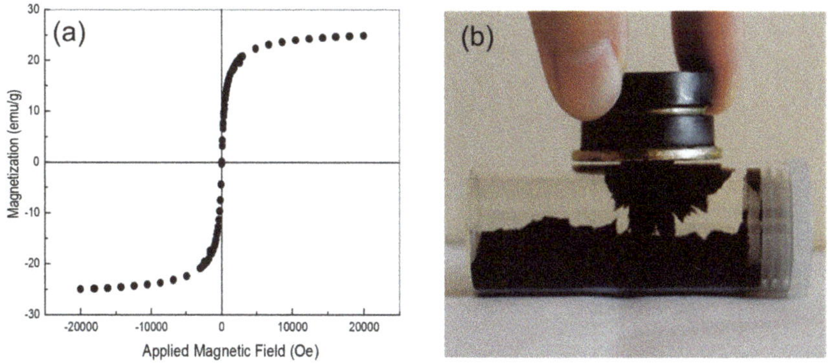


*Fig. 4. (a) Magnetization curve for PCMS-$Fe_3O_4$ composite and (b) External attraction of PCMS@$Fe_3O_4$ by a hand-held magnet.*

### 3.1.4 Point zero of charge

The pH value of an adsorbent in solution when the net surface charge is zero is defined as the point zero of charge ($pH_{PZC}$). It has a huge impact in adsorption processes because multivalent cation adsorption occurs effectively at pH below $pH_{PZC}$ [28]. The variation of the zeta potential of partially converted mollusk shell-$Fe_3O_4$ composite with respect to pH was studied to determine the surface charge on the composite. Fig. 5 shows the $pH_{PZC}$ results obtained for the partially converted mollusk shell- $Fe_3O_4$ composite powder. The point zero of charge was around 7.1 and over a wide range of pH, the composite powder exhibits a net positive zeta potential which might be due to the addition of the acetamido and amino groups protons onto partially converted mollusk shell@$Fe_3O_4$ powder. Above the $pH_{PZC}$, the composite powder demonstrated a net negative zeta potential due to protonation and attachment of $OH^-$ ions.

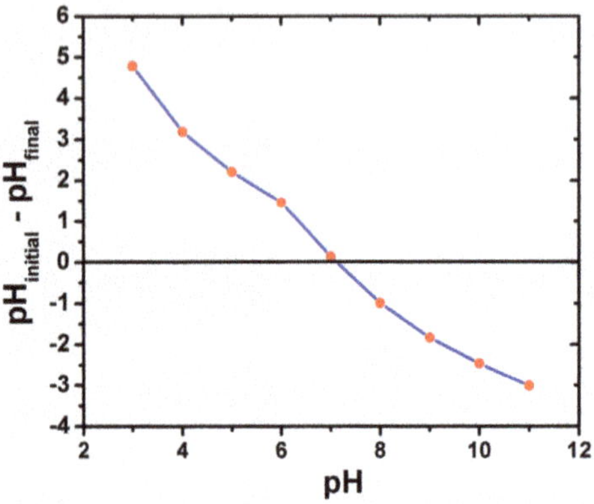


*Fig. 5. Determination of $pH_{pzc}$ on PCMS-$Fe_3O_4$ composite powder.*

## 3.2 Bio-sorption studies

### 3.2.1 Effect of solution pH on metal removal

Fig. 6. shows the effect of solution pH on the removal of Zn(II) and Pb(II). It was observed that when pH was increasing, the sorption capacity also increased, and the maximum bio-sorption efficiency attained was 90% at pH 3-6. This can be attributed to the metal chemistry in solution where the available free lone pairs of electrons onto nitrogen atoms are suitable for coordination with Pb(II) ions [39]. At pH 2, a slight decrease in the adsorption capacity might be due to the excess addition of hydrogen ($H^+$) that limits complex formation. When the initial pH of the solution was higher than 5.5, precipitation of Pb(II) ions resulted due to high concentration of hydroxyl anions in the solution. Further observations showed that adsorption of Zn(II) was mainly dependent on the proton concentration. The higher the proton concentration, the lower is the removal efficiency [36]. Further, the dependence of metal adsorption capacity on pH is related not only to the density of functional surface groups of the bio-sorbent and the chemistry of metal ions in solution [37, 38] but also on the degree of ionization. Therefore, all subsequent experiments were carried out at pH 5.5. Furthermore, it was observed that from pH 1.6 to 5, amine groups are mostly protonated and thus available for anion sorption, although the degree of protonation decreases as pH increases. At pH 6, the percentage of protonated amine groups decreases to 76% whereas at pH 8 the composite is almost completely de-protonated and thus unable to sorb metal anions [39].

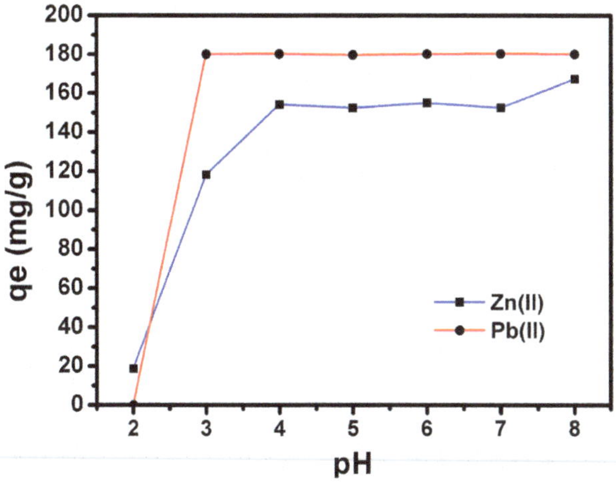


*Fig. 6. Effect of initial solution pH on Pb(II) and Zn(II) adsorption. (Adsorbent dosage: 0.05 g; initial concentration: 200 mg/L; T: 25 °C).*

### 3.2.2 Adsorption kinetics and modelling

Bio-sorption kinetics plays an important role in selection and designing of reactor systems, as well as operations. Heavy metal bio-sorption is metabolism-independent and thus, typically occurs rapidly, in particular for uptake of cationic metal ions [7]. Kinetics experiments were performed using mollusk shell@$Fe_3O_4$ powder concentration of 1.0 g/L, solution pH at 5.5 and initial Zn(II) concentration of 51.55 and 95.53 mg/L respectively at 25 °C. As shown in Fig. 7, Zn(II) adsorption occurred in two steps, the first step involving rapid metal adsorption within the first 20 min and afterwards, adsorption continuing for a longer time period until equilibrium was reached. Equilibrium was reached between 2-3 h. In the succeeding experiments, a contact time of 2 hrs was chosen to ensure equilibrium was attained. Further, the experimental data were modelled onto two kinetic models; pseudo-first order and pseudo-second order models [40, 41]. The adsorption of Pb(II) metal ion onto partially converted mollusk@$Fe_3O_4$ composite as a function of contact time was determined and results are presented in Fig. 7. The experiment was carried out at initial concentrations of 63.25 mg/L and 102.66 mg/L at 25 °C. Bio-sorption was rapid in the first stages (10-30 min) and then slowed down considerably as the reaction approached equilibrium. Specifically, up to 25.44 mg/g and 63.46 mg/g for Pb(II) uptake was recorded in 10 minutes contact time for the 63.25 mg/L and 102.66 mg/L initial concentrations, respectively. Such a rapid uptake shows that the active sites on the partially converted mollusk-$Fe_3O_4$ composite were readily accessible for Pb(II) adsorption. The absence of the internal diffusion resistance and high surface area of the partially converted mollusk@$Fe_3O_4$ composite further explains the rapid uptake as observed. As time elapsed, evidently after the first hour to the second hour, uptake of Pb(II) increased from 27.80 to 27.81 mg/g and 66.83 to 67.03 mg/g for 63.26 and 102.77 mg/L initial concentrations, respectively.

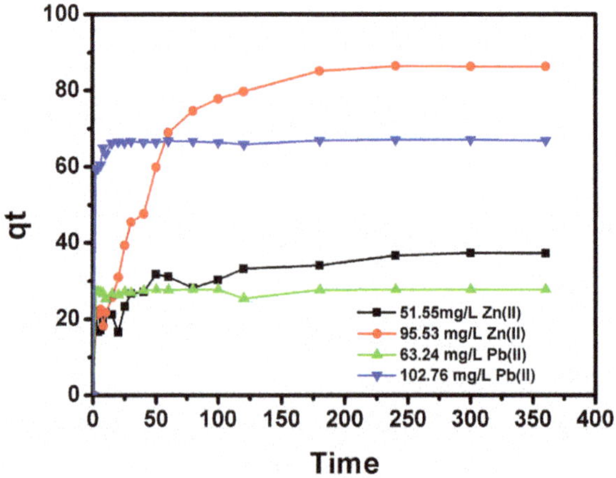


*Fig. 7. Adsorption kinetics of Zn(II) and Pb(II) ions onto mollusk shell-$Fe_3O_4$ powder. Adsorbent dosage: 1.0 g $L^{-1}$; stirring rate: 250 rpm; pH: 5.5; T: 25 °C.*

The pseudo-first order rate equation (Lagergren equation) is described by the following expression [42].

$$\frac{d_q}{d_t} = k_1(q_e - q_t) \tag{3}$$

where ($q_e$) is the quantity of the metal ions adsorbed at equilibrium/g (mg/g); (q) amount of metal ions adsorbed any time (mg/g) and ($k_1$) is the rate constant ($min^{-1}$). Eq (3) induces to Eq. 4 after integrating and applying boundary conditions, t = 0 and $q_t = 0$ to t = t and $q_t = q_t$:

$$\log(q_e - q_t) = \log q_e - \frac{k_1 t}{2.303} \tag{4}$$

The pseudo second order rate equation is described as follows:

$$\frac{t}{q_t} = \frac{1}{k_2.q_e^2} + \frac{t}{q_e} \tag{5}$$

and the initial sorption rate, $h_0$ (mg/g. min) can be defined as:

$$h_0 = k_2.q_e^2 (t \to 0) \tag{6}$$

For the model to be valid and satisfied, a linear plot of $t/q$ against $t$ in Eq. 5 should give a linear relationship and thereafter, $q_e$ and $k_e$ constants determined.

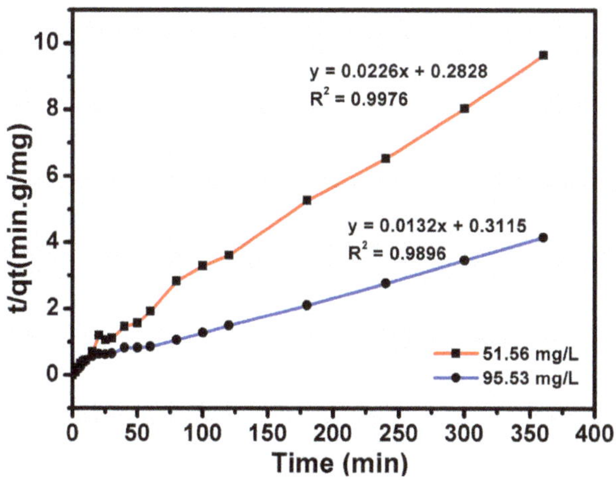


*Fig. 8. Pseudo second order plots of Zn(II) ions onto mollusk shell-$Fe_3O_4$ powder; Adsorbent dosage: 1.0 g $L^{-1}$; stirring rate: 250 rpm; pH: 5.5; T: 25 °C.*

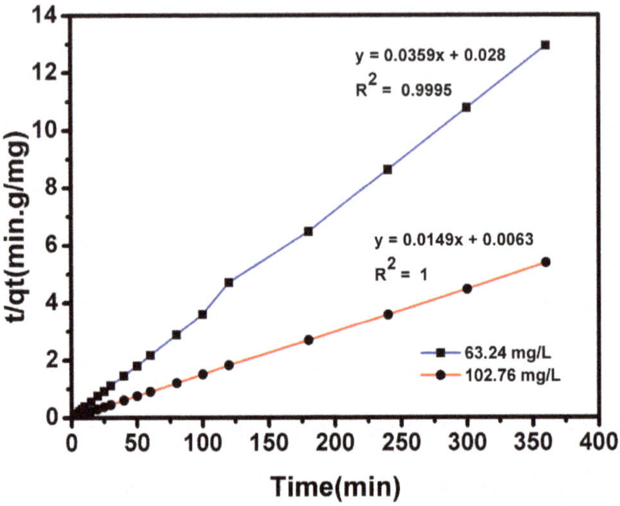


*Fig. 9. Pseudo second order plots on Pb(II) ions onto mollusk shell @$Fe_3O_4$ powder. Adsorbent dosage: 1.0 g $L^{-1}$; stirring rate: 250 rpm; pH: 5.5; T: 25 °C*

The plot fitted very well to the experimental data and thus sorption was described as a pseudo-second order rate equation for both metal ions. Additional data are presented in Table 1. The $R^2$ values are the same for all initial concentrations of Zn(II) ions with a value of 0.99. A linear plot of the model is presented in Fig.8 and thereafter rate constants $q_e$ and $k_2$ are determined from the slope of Eq 5. Nonetheless, the reaction is not likely to be expressed as a first order process if the equilibrium uptake of metal ions doesn't equal the intercept, even if the plot has a high correlation coefficient with the experimental data [42]. From our study, the intercept was found to be 80.1 and this was not equal to the equilibrium uptake therefore confirming that the process did not follow the first order mechanism. The $R^2$ values were 0.95 and 0.99 respectively. Table 1 shows the data obtained with the rate constants calculated from the slope of Eq. 5. The experimental data in Fig. 9 matched well the model data as reflected by the extremely high correlation coefficients at both concentrations ($R^2$=0.99). The findings indicate that the rate limiting step might be due to chemical sorption which involved exchange of electrons between Pb(II) molecules and partially converted shell-$Fe_3O_4$ composite. Moreover, from the experimental results reported in Table 1, the adsorption capacity $q_e$ depended on the initial concentration. The pseudo-first order model was also modelled but did not give good correlation coefficients. Thus the latter is not discussed.

*Table 1. Summary of kinetic parameters on Zn(II) and Pb(II) adsorption.*

| | | | Pseudo-1st order | | | Pseudo-2nd order | | |
|---|---|---|---|---|---|---|---|---|
| **Metal** | **Conc (ppm)** | **$q_e$ (exp)** | **$R^2$** | **$k_1$ (min-1)** | **qe (mg/g)** | **$R^2$** | **$k_2$ (g/mg.min)** | **$q_e$ (mg/g)** |
| Zn(II) | 50 | 51.55 | 0.89 | 0.013 | 20.07 | 0.999 | 0.00214 | 38.02 |
| | 100 | 95.53 | 0.95 | 0.022 | 80.11 | 0.999 | 0.0004 | 93.46 |
| Pb(II) | 50 | 63.26 | 0.05 | 0.005 | 2.29 | 0.999 | 0.05 | 27.86 |
| | 100 | 102.77 | 0.05 | 0.003 | 1.5 | 0.999 | 0.0002 | 67.15 |

### 3.2.3 Sorption isotherms

The correct relationship needs to be established for batch equilibrium experiments as such results help in modelling and designing adsorption systems [26]. Adsorption isotherms provide necessary data for understanding the mechanism of adsorption processes, surface properties and the interaction between the sorbate and sorbent. The Langmuir and Freundlich [43] isotherm linear equations were applied to depict the equilibrium sorption relationship of Zn(II) and Pb(II), and are presented by the following expressions:

$$\frac{c_e}{q_e} = \frac{1}{q_m . k_L} + \frac{c_e}{q_m} \quad (7)$$

$$\log q_e = \log K_F + \frac{1}{n} \log c_e \quad (8)$$

where $q_m$ (mg/g) is the maximum adsorption capacity, $k_L$ (L/mg) is the adsorption equilibrium constant. A plot of $C_e/q_e$ vs. $C_e$ should yield a straight line if Langmuir isotherm is obeyed; $q_m$ and $k_L$ are calculated from the slope and intercept, respectively. $K_F$ is Freundlich parameter relating adsorption capacity and $1/n$, a parameter related to adsorption intensity. The slope ranges between 0 and 1, however the process becomes more heterogeneous as the slope approaches a value close to zero [44].

Furthermore, the essential characteristics of the Langmuir isotherm can further described by a separation factor, which is defined as:

$$R_L = \frac{1}{(1+bC_e)} \tag{9}$$

The value of $R_L$ shows the nature of adsorption and shape of the isotherm. When the value of $R_L$ is within a range of 0-1, adsorption is favourable.

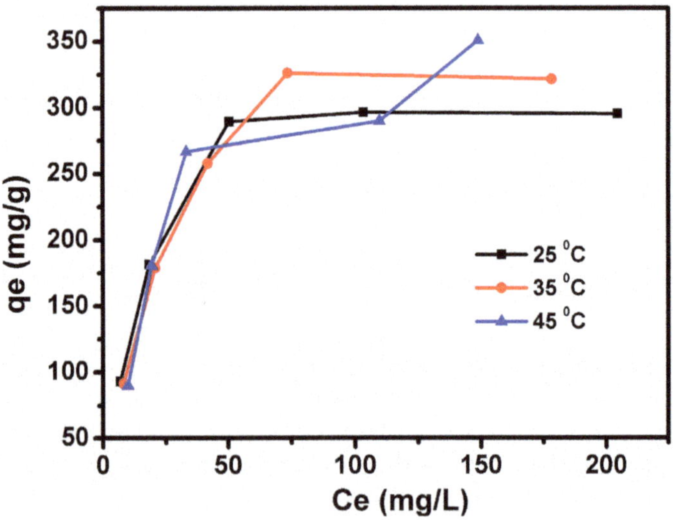


*Fig. 10. Adsorption isotherms for Zn(II) adsorption onto partially converted mollusk shell-$Fe_3O_4$ composite.*

Fig. 10 presents the adsorption isotherms of Zn(II) at different temperatures. As observed, the adsorption curves indicated a slight increase in the adsorption capacity of Zn(II) metal uptake as temperature increased. The Langmuir sorption model was chosen to estimate the maximum uptake capacity and affinity for Zinc ions. From the Langmuir fit summarized in Table 2, correlation coefficients ($R^2$) obtained were 0.99 at 298K and 308K respectively, where at 318K, $R^2$ was 0.98, thus showing a better fit as compared to the Freundlich model (not shown). This indicated that adsorption occurred at specific homogeneous sites within the adsorbent and formed a monolayer cover of Zn(II) at the surface of the sorbent. The sorption parameters are presented in Table 2. The maximum adsorption capacity ($q_m$) seemed to increase with an increase in temperature. Specifically, the maximum uptake capacity was 312.5 mg/g at 298 K and 400 mg/g at 318 K.

*Table 2. Langmuir and Freundlich isotherms parameters for Zn(II) and Pb (II) adsorption.*

| Analyte | Langmuir | | | | Freundlich | | |
|---|---|---|---|---|---|---|---|
| | T (K) | $R^2$ | b (L/mg) | $q_m$ (mg/g) | $R^2$ | $K_f$ | 1/n |
| Zn | 298 | 0.99 | 0.071 | 312.5 | 0.89 | 57.91 | 0.34 |
| | 308 | 0.99 | 0.054 | 357.1 | 0.87 | 45.72 | 0.42 |
| | 318 | 0.98 | 0.038 | 400 | 0.83 | 44.17 | 0.47 |
| Pb | 298 | 0.99 | 0.109 | 769.23 | 0.66 | 210 | 0.255 |
| | 308 | 0.99 | 0.10 | 769.23 | 0.59 | 214 | 0.239 |
| | 318 | 0.99 | 0.076 | 909.09 | 0.99 | 100 | 0.457 |

The interactions between partially converted mollusk shell-$Fe_3O_4$ composite and Pb(II) ions at 298, 308 and 318 K resulted in Pb(II) removal from aqueous solution as shown in Fig 11. The sorption experimental data were modelled onto Langmuir and Freundlich isotherm models. The obtained constants and regression coefficients of Langmuir isotherm are summarised in Table 2. From the data obtained, the Langmuir model gave a better fit than the Freundlich model and the data strongly suggest that the experimental adsorption of Pb(II) onto partially converted mollusk shell-$Fe_3O_4$ composite fit well with the Langmuir isotherm. For this study, the value of $R_L$ calculated for the initial concentrations of Pb(II) is 0.11 which is within the range of 0-1, thus indicating that sorption of Pb(II) onto partially converted mollusk shell-$Fe_3O_4$ composite is favourable.

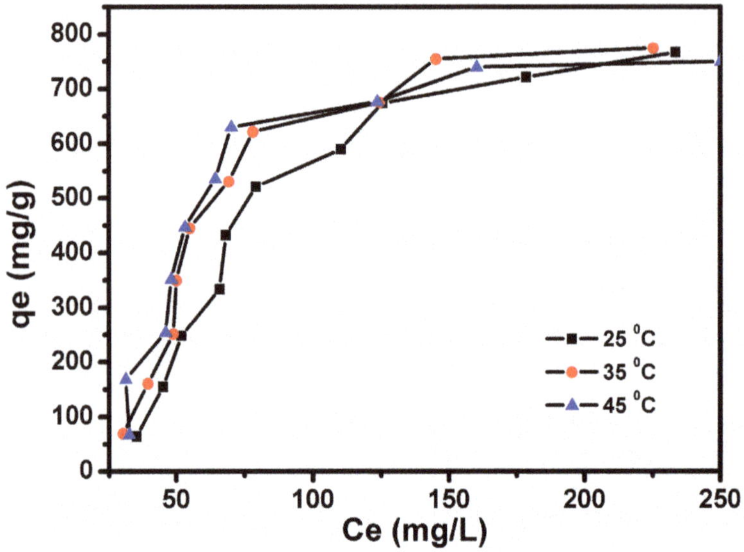


*Fig. 11. Adsorption isotherm of Pb(II) onto partially converted mollusk shell-$Fe_3O_4$ composite at different temperatures.*

*Table 3. Maximum Langmuir adsorption capacities of Zn(II) ions onto different adsorbents.*

| Adsorbent | $q_e$ | Reference |
|---|---|---|
| MMCS | 32.16 mg $g^{-1}$ | [38] |
| MHNPs | 2.151 mmol $g^{-1}$ | [45] |
| MZ | 114 mg $g^{-1}$ | [20] |
| NCT | 133.86 mg $g^{-1}$ | [46] |
| PCMS@$Fe_3O_4$ | 312.5 mg $g^{-1}$ | This study |

Magnetic Modified Chitosan (MMCS); Magnetic zeolite (MZ); Magnetic hydroxyapatite nanoparticles (MHNPs); Nanometer calcium titanate (NCT)

Table 3 & 4 demonstrates the comparison of sorption capacities of Zn(II) and Pb(II) ion to some nano-materials reported in literature. The maximum adsorption capacities obtained in this study are fairly attractive. It however should be stressed that the values presented in Table 3 and 4 have been taken from different studies in which the experimental conditions are not the same as compared to those applied in the current study and, thus, the comparison made here aims at showing the high uptake potential of partially converted mollusk shell@$Fe_3O_4$ composite rather than establishing a quantitative scale of effectiveness among different sorbents.

*Table 4. Comparison of Langmuir adsorption capacities for Pb(II) onto various adsorbents.*

| Adsorbent | pH | T(K) | $q_e$ (mg/g) | Reference |
|---|---|---|---|---|
| CS/MCB | 5 | 298 | 54.8 | [26] |
| | 6 | 298 | 63.33 | |
| GB/MNPs | ? | 298 | 260.13 | [47] |
| CNTs | ? | ? | 17.44 | [43] |
| PCMS@$Fe_3O_4$ | 5.5 | 298 | 769.23 | This study |

Chitosan (CS); magnetite composite beads (MCB); Gellan beads (GB); magnetic nanoparticles (MNPs); Carbon nanotubes(CNTs) ; Partially converted mollusk shell (PCMS)

### 3.2.4 Thermodynamic parameters

When adsorption equilibrium state reached, Zn(II) metal ions adsorbed onto mollusk shell@$Fe_3O_4$ powder remain in equilibrium with the residual Zn(II) concentration in the liquid phase. Thermodynamic factors such as change in free energy $\Delta G°_{ads}$, change in enthalpy of adsorption $\Delta H°_{ads}$ and change in entropy $\Delta S°_{ads}$ provide an insight about the nature and mechanism of adsorption process [48]:

$$\Delta G°_{ads} = -RT \ln b \tag{10}$$

$$\ln b = \frac{\Delta S^{\circ}_{ads}}{R} - \frac{\Delta H^{\circ}_{ads}}{RT} \quad (11)$$

where $R$ is the universal gas constant (8.314 J $mol^{-1}$ $K^{-1}$) and T is the adsorption temperature (K). From the slopes and intercepts of Van't Hoff plot of ln$b$ versus 1/T was linear with y = 294.2x – 12.501 and $R^2$ = 0.9911. Based on Eq.11, $\Delta S^{\circ}_{ads}$ and $\Delta H^{\circ}_{ads}$ were calculated as given in Table 5.

*Table 5. $\Delta G^{o}$ values for adsorption of Pb(II) and Zn(II) on PCMS@$Fe_3O_4$ at different temperatures.*

| | **Temperature (K)** | **$\Delta G^{\circ}$ (kJ $mol^{-1}$)** | **$\Delta H^{o}_{ads}$ (kJ $mol^{-1}$)** | **$\Delta S^{o}_{ads}$ (J $mol^{-1}K^{-1}$)** |
|---|---|---|---|---|
| | 298 | 4.44 | | |
| Pb(II) | 308 | 3.27 | 38.88 | 115.57 |
| | 318 | 2.13 | | |
| | 298 | -6.55 | | |
| Zn(II) | 308 | -7.46 | -24.46 | 103.33 |
| | 318 | -8.63 | | |

The equilibrium constants obtained from the Langmuir model at 298, 308 and 318 K respectively were used to determine the Gibbs free energy changes. A negative change in adsorption standard free energy reveals a spontaneous adsorption process where a positive change reveals a non-spontaneous process. The $\Delta H^{O}_{ads}$ and $\Delta S^{O}_{ads}$ were determined as 38.88 kJ $mol^{-1}$ and 115.57 J $mol^{-1}$ $K^{-1}$, respectively, from Fig. 12 (a & b). A positive standard enthalpy change indicates that adsorption of Pb(II) by the partially mollusk shell-$Fe_3O_4$ composite adsorbent is endothermic. The positive standard entropy change may be caused by an increase in the release of water molecules in the ion exchange reaction between the adsorbate and the functional groups on the surfaces of the PCMS-$Fe_3O_4$ composite. As seen from Table 6, $\Delta G^{\circ}_{ads}$ at all temperatures was negative for Zn(II) ions and increased with an increase in temperature, demonstrating the feasibility and spontaneity of adsorption of Zn(II) onto mollusk shell@$Fe_3O_4$ powder.

The negative value of $\Delta H^{\circ}_{ads}$ confirmed the exothermic nature of the process, while the positive value of $\Delta S^{\circ}_{ads}$ revealed the increase in randomness at the solid/solution interface during the adsorption process. From the $\Delta G^{\circ}_{ads}$ values obtained in this study, it can be concluded that adsorption onto mollusk shell@$Fe_3O_4$ powder is dominated by physisorption.

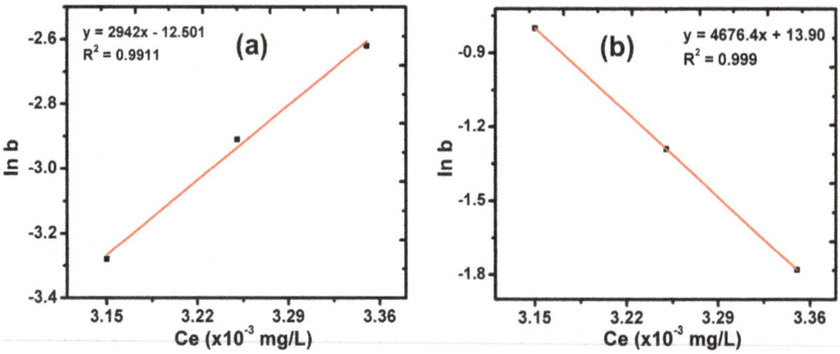


*Fig. 12. Plot of lnb vs 1/T for (a) Zn (II) and (b) Pb(II) adsorption.*

## Conclusion

Partially converted mollusk shell@$Fe_3O_4$ powder was successfully obtained by co-precipitation method. FTIR analysis confirmed the successful binding of $Fe_3O_4$ onto partially converted mollusk shell@$Fe_3O_4$ composite. The magnetic analysis indicated that PCMS@$Fe_3O_4$ composite was super-paramagnetic because of the absence of coercivity and remanence and with a saturation magnetization value of 25.03 emu $g^{-1}$. The point zero charge of PCMS@$Fe_3O_4$ was around 7.1. The removal of Zn(II) and Pb(II) by adsorption onto mollusk shell@$Fe_3O_4$ powder achieved rapid equilibrium and showed enhanced adsorption capacity. Thermodynamically, Zn(II) adsorption onto the mollusk shell@$Fe_3O_4$ powder confirmed the exothermic nature of the process and spontaneous nature of the process. Pb(II) sorption was pH dependent where optimum removal obtained was at pH 5.5. Equilibrium data at different temperatures were well fitted with Langmuir isotherms, indicating a monolayer sorption process. Thermodynamically, $\Delta G^{o}_{ads}$ for Pb(II) is positive for all temperatures, showing the feasibility and non-spontaneously nature of the process, $\Delta H^{o}_{ads}$ was found to be 38.88 kJ $mol^{-1}$ confirming an endothermic process. A positive $\Delta S^{o}_{ads}$ was also obtained showing that there was increased randomness at the solid solution interface. Overall, it is recommended that (i) leaching test be performed before adsorption to determine the toxicity of the material; (ii) regeneration studies be performed in order to recover metals and the adsorbent to improve the economic capability of the system; (iii) seeding time for magnetic separation be shortened for practical reasons such as feasibility of the process.

**Acknowledgement**

Our grateful acknowledgement goes to Council for Scientific and Industrial Research (CSIR) for assistance with characterization facility. M Bopape would also like to thank Tshwane University of Technology and National Research Foundation of South Africa for the financial support.

**References**

[1] D. Bulgariu, L. Bulgariu, Equilibrium and kinetics studies of heavy metal ions biosorption on green algae waste biomass, Bioresour. Technol. 103 (2012) 489-493. https://doi.org/10.1016/j.biortech.2011.10.016

[2] M. Ahmad, A.R.A. Usman, S.S. Lee, S.-C. Kim, J.-H. Joo, J.E. Yang, Y.S. Ok, Eggshell and coral wastes as low cost sorbents for the removal of $Pb^{2+}$, $Cd^{2+}$ and $Cu^{2+}$ from aqueous solutions, J. Ind. Eng. Chem. 18 (2012) 198-204. https://doi.org/10.1016/j.jiec.2011.11.013

[3] R. Paradelo, M.T. Barral, Evaluation of the potential capacity as biosorbents of two MSW composts with different Cu, Pb and Zn concentrations, Bioresour. Technol. 104 (2012) 810-813. https://doi.org/10.1016/j.biortech.2011.11.012

[4] M.H. Abd-Alla, F.M. Morsy, A.-W.E. El-Enany, T. Ohyama, Isolation and characterization of a heavy-metal-resistant isolate of Rhizobium leguminosarum bv. viciae potentially applicable for biosorption of $Cd^{2+}$ and $Co^{2+}$, Int. Biodeterior. Biodegradation 67 (2012) 48-55. https://doi.org/10.1016/j.ibiod.2011.10.008

[5] K. Patil, S.V. Smith, R. Rajkhowa, T. Tsuzuki, X.G. Wang, T. Lin., Milled cashmere guard hair powders: Absorption properties to heavy metal ions., Powder Technol. (2012) 162-168. https://doi.org/10.1016/j.powtec.2011.12.013

[6] M. Fomina, G.M. Gadd, Biosorption: current perspectives on concept, definition and application, Bioresour. Technol. 160 (2014) 3-14. https://doi.org/10.1016/j.biortech.2013.12.102

[7] J. He, J.P. Chen, A comprehensive review on biosorption of heavy metals by algal biomass: Materials, performances, chemistry, and modeling simulation tools,

Bioresour. Technol. 160 (2014) 67-78. https://doi.org/10.1016/j.biortech.2014.01.068

[8] R.W. Coughlin, M.R. Deshaies, E.M. Davis, Chitosan in crab shell wastes purifies electroplating wastewater, Environ. Prog. 9 (1990) 35-39. https://doi.org/10.1002/ep.670090116

[9] S. Pradhan, S.S. Shukla, K.L. Dorris, Removal of nickel from aqueous solutions using crab shells, J. Hazard Mater. 125 (2005) 201-204. https://doi.org/10.1016/j.jhazmat.2005.05.029

[10] L. Dambies, T. Vincent, E. Guibal, Treatment of arsenic-containing solutions using chitosan derivatives: uptake mechanism and sorption performances, Water Res. 36 (2002) 3699-3710. https://doi.org/10.1016/S0043-1354(02)00108-2

[11] R. Laus, T.G. Costa, B. Szpoganicz, V.T. Fávere, Adsorption and desorption of Cu(II), Cd(II) and Pb(II) ions using chitosan crosslinked with epichlorohydrin-triphosphate as the adsorbent, J. Hazard Mater. 183 (2010) 233-241. https://doi.org/10.1016/j.jhazmat.2010.07.016

[12] X.H. Zhou, J.M. Pan, H.X. Ou, X. Wang, W. Guang, C.X. Li, Y.S. Yan, Y.Q. Duan, Adsorptive removal of Cr(III) and Fe(III) from aqueous solution by chitosan/attapulgite composites: Equilibrium, thermodynamics and kinetics., Chem. Eng. J. 167 (2011) 112-121. https://doi.org/10.1016/j.cej.2010.12.009

[13] Y.-C. Chang, S.-W. Chang, D.-H. Chen, Magnetic chitosan nanoparticles: Studies on chitosan binding and adsorption of Co(II) ions, React. Funct. Polym. 66 (2006) 335-341. https://doi.org/10.1016/j.reactfunctpolym.2005.08.006

[14] B. Guan, W. Ni, Z. Wu, Y. Lai, Removal of Mn(II) and Zn(II) ions from flue gas desulfurization wastewater with water-soluble chitosan, Sep. Purif. Technol. 65 (2009) 269-274. https://doi.org/10.1016/j.seppur.2008.10.037

[15] E. Repo, J.K. Warchol, T.A. Kurniawan, M.E.T. Sillanpää, Adsorption of Co(II) and Ni(II) by EDTA- and/or DTPA-modified chitosan: Kinetic and equilibrium modeling, Chem. Eng. J. 161 (2010) 73-82. https://doi.org/10.1016/j.cej.2010.04.030

[16] D.K. Rout, S.K. Pulapura, R.A. Gross, Liquid-crystalline characteristics of site-selectively-modified chitosan, Macromolecules 26 (1993) 5999-6006. https://doi.org/10.1021/ma00074a023

[17] M.S. Masri, F.W. Reuter, M. Friedman, Binding of metal cations by natural substances, J. Appl. Polym. Sci. 18 (1974) 675-681. https://doi.org/10.1002/app.1974.070180305

[18] J. Wang, C. Chen, Chitosan-based biosorbents: Modification and application for biosorption of heavy metals and radionuclides, Bioresour. Technol. 160 (2014) 129-141. https://doi.org/10.1016/j.biortech.2013.12.110

[19] A.M. Muliwa, T.Y. Leswifi, M.S. Onyango, A. Maity, Magnetic adsorption separation (MAS) process: An alternative method of extracting Cr(VI) from aqueous solution using polypyrrole coated $Fe_3O_4$ nanocomposites, Sep. Purif. Technol. 158 (2016) 250-258. https://doi.org/10.1016/j.seppur.2015.12.021

[20] L.C.A. Oliveira, D.I. Petkowicz, A. Smaniotto, S.B.C. Pergher, Magnetic zeolites: a new adsorbent for removal of metallic contaminants from water, Water Res. 38 (2004) 3699-3704. https://doi.org/10.1016/j.watres.2004.06.008

[21] J. Hu, G. Chen, I.M.C. Lo, Removal and recovery of Cr(VI) from wastewater by maghemite nanoparticles, Water Res. 39 (2005) 4528-4536. https://doi.org/10.1016/j.watres.2005.05.051

[22] W. Wan Ngah, L. Teong, M. Hanafiah, Adsorption of dyes and heavy metal ions by chitosan composites: A review, Carbohydr. Polym. 83 (2011) 1446-1456. https://doi.org/10.1016/j.carbpol.2010.11.004

[23] Inamuddin, Y.A. Ismail, Synthesis and characterization of electrically conducting poly-o-methoxyaniline Zr(1V) molybdate Cd(II) selective composite cation-exchanger, Desalination. 250 (2010) 523–529. https://doi.org/10.1016/j.desal.2008.06.033

[24] H. Hu, Z. Wang, L. Pan, Synthesis of monodisperse $Fe_3O_4$@silica core–shell microspheres and their application for removal of heavy metal ions from water, J. Alloys Compd. 492 (2010) 656-661. https://doi.org/10.1016/j.jallcom.2009.11.204

[25] J.-f. Liu, Z.-s. Zhao, G.-b. Jiang, Coating $Fe_3O_4$ Magnetic Nanoparticles with Humic Acid for High Efficient Removal of Heavy Metals in Water, Environ. Sci. Technol. 42 (2008) 6949-6954. https://doi.org/10.1021/es800924c

[26] H.V. Tran, L.D. Tran, T.N. Nguyen, Preparation of chitosan/magnetite composite beads and their application for removal of Pb(II) and Ni(II) from aqueous solution, Mater. Sci. Eng. C 30 (2010) 304-310. https://doi.org/10.1016/j.msec.2009.11.008

[27] M.S. Onyango, Y. Kojima, O. Aoyi, E.C. Bernardo, H. Matsuda, Adsorption equilibrium modeling and solution chemistry dependence of fluoride removal from

water by trivalent-cation-exchanged zeolite F-9, J. Colloid Interface Sci. 279 (2004) 341-350. https://doi.org/10.1016/j.jcis.2004.06.038

[28] P. Chutia, S. Kato, T. Kojima, S. Satokawa, Arsenic adsorption from aqueous solution on synthetic zeolites, J. Hazard Mater. 162 (2009) 440-447. https://doi.org/10.1016/j.jhazmat.2008.05.061

[29] C. Bangyekan, D. Aht-Ong, K. Srikulkit, Preparation and properties evaluation of chitosan-coated cassava starch films, Carbohydr. Polym. 63 (2006) 61-71. https://doi.org/10.1016/j.carbpol.2005.07.032

[30] L. Elena Udrea, D. Hritcu, M.I. Popa, O. Rotariu, Preparation and characterization of polyvinyl alcohol—chitosan biocompatible magnetic microparticles, J. Magn. Magn. Mater. 323 (2011) 7-13. https://doi.org/10.1016/j.jmmm.2010.06.025

[31] G.F. Goya, T.S. Berquó, F.C. Fonseca, M.P. Morales, Static and dynamic magnetic properties of spherical magnetite nanoparticles, J. Appl. Phys. 94 (2003) 3520-3528. https://doi.org/10.1063/1.1599959

[32] M. Mohammad, S. Maitra, N. Ahmad, A. Bustam, T.K. Sen, B.K. Dutta, Metal ion removal from aqueous solution using physic seed hull, J. Hazard Mater. 179 (2010) 363-372. https://doi.org/10.1016/j.jhazmat.2010.03.014

[33] M. Madhava Rao, A. Ramesh, G. Purna Chandra Rao, K. Seshaiah, Removal of copper and cadmium from the aqueous solutions by activated carbon derived from Ceiba pentandra hulls, J. Hazard. Mater. 129 (2006) 123-129. https://doi.org/10.1016/j.jhazmat.2005.08.018

[34] C. Yuwei, W. Jianlong, Preparation and characterization of magnetic chitosan nanoparticles and its application for Cu(II) removal, Chem. Eng. J. 168 (2011) 286-292. https://doi.org/10.1016/j.cej.2011.01.006

[35] Z. Ma, Y. Guan, H. Liu, Synthesis and characterization of micron-sized monodisperse superparamagnetic polymer particles with amino groups, J. Polym. Sci. A Polym. Chem. 43 (2005) 3433-3439. https://doi.org/10.1002/pola.20803

[36] G. Yan, T. Viraraghavan, Heavy-metal removal from aqueous solution by fungus Mucor rouxii, Water Res. 37 (2003) 4486-4496. https://doi.org/10.1016/S0043-1354(03)00409-3

[37] A. Sarı, Ö.D. Uluozlü, M. Tüzen, Equilibrium, thermodynamic and kinetic investigations on biosorption of arsenic from aqueous solution by algae (Maugeotia genuflexa) biomass, Chem. Eng. J. 167 (2011) 155-161. https://doi.org/10.1016/j.cej.2010.12.014

[38] L. Fan, C. Luo, Z. Lv, F. Lu, H. Qiu, Preparation of magnetic modified chitosan and adsorption of $Zn^{2+}$ from aqueous solutions, Colloids Surf., B 88 (2011) 574-581. https://doi.org/10.1016/j.colsurfb.2011.07.038

[39] D. Kołodyńska, Chitosan as an effective low-cost sorbent of heavy metal complexes with the polyaspartic acid, Chem. Eng. J. 173 (2011) 520-529. https://doi.org/10.1016/j.cej.2011.08.025

[40] Y.-S. Ho, Review of second-order models for adsorption systems, J. Hazard Mater. 136 (2006) 681-689. https://doi.org/10.1016/j.jhazmat.2005.12.043

[41] M. Wardell, Z. Wang, J. Ho, J. Robert, F. Ruker, J. Ruble, D. Carter, The atomic structure of human methemalbumin at 1.9 A, 291 (2002) 813 - 819.

[42] J.C. Igwe, A.A. Abia, A. Ibeh, Adsorption kinetics and intraparticulate diffusivities of Hg, As and Pb ions on unmodified and thiolated coconut fiber, Int. J. Environ. Sci. Technol. 5 (2008) 83-92. https://doi.org/10.1007/BF03326000

[43] N.N. Nassar, Rapid removal and recovery of Pb(II) from wastewater by magnetic nanoadsorbents, J. Hazard Mater. 184 (2010) 538-546. https://doi.org/10.1016/j.jhazmat.2010.08.069

[44] K.Y. Foo, B.H. Hameed, Insights into the modeling of adsorption systems, Chem. Eng. J. 156 (2010) 2-10. https://doi.org/10.1016/j.cej.2009.09.013

[45] Y. Feng, J.-L. Gong, G.-M. Zeng, Q.-Y. Niu, H.-Y. Zhang, C.-G. Niu, J.-H. Deng, M. Yan, Adsorption of Cd(II) and Zn(II) from aqueous solutions using magnetic hydroxyapatite nanoparticles as adsorbents, Chem. Eng. J. 162 (2010) 487-494. https://doi.org/10.1016/j.cej.2010.05.049

[46] M. Hua, S. Zhang, B. Pan, W. Zhang, L. Lv, Q. Zhang, Heavy metal removal from water/wastewater by nanosized metal oxides: A review, J. Hazard Mater. 211–212 (2012) 317-331. https://doi.org/10.1016/j.jhazmat.2011.10.016

[47] X. Wang, C. Zhao, P. Zhao, P. Dou, Y. Ding, P. Xu, Gellan gel beads containing magnetic nanoparticles: an effective biosorbent for the removal of heavy metals from aqueous system, Bioresour. Technol. 100 (2009) 2301-2304. https://doi.org/10.1016/j.biortech.2008.10.042

[48] Q. Li, L. Chai, Z. Yang, Q. Wang, Kinetics and thermodynamics of Pb(II) adsorption onto modified spent grain from aqueous solutions, Appl. Surf. Sci. 255 (2009) 4298-4303. https://doi.org/10.1016/j.apsusc.2008.11.024

# Chapter 8

# Waste fruit cortexes for the removal of heavy metals from wastewater

Ritu Payal[1], Arti Jain*

[1]Rajdhani College, Department of Chemistry, University of Delhi, Delhi 110007, India

Daulat Ram College, Department of Chemistry, University of Delhi, Delhi 110007, India

*jainarti21@gmail.com

**Abstract**

There has been an alarming increase in the levels of heavy metals ($Pb^{+2}$, $As^{+2}$, $Hg^{+2}$, $Mn^{+2}$, $Cr^{+2}$, $Zn^{+2}$, $Cu^{+2}$ and $Cd^{+2}$), organic dyes (from textiles, food and beverages industries) and surfactants (linear alkyl benzene sulfonate: LABS, alkyl sulphate: AS, alkyl ethoxylate: AE, sodium dodecyl sulphate: SDS) due to their increased usage in industries.

For the abolition of heavy metal ions, organic dyes and surfactants from polluted areas, cost effective and proficient substitutes to target these impurities is biosorption technique which exploits various fruit cortexes. This chapter briefly deals with a toxicological indication of metal ions, organic dyes, and surfactants in addition to the comprehensive explanation of biosorption techniques using fruit cortexes (which is otherwise waste for us) for the elimination of the impurities. Additionally, the chapter apprises readers about a recent innovation in biosorption protocols such as the use of modified agricultural waste materials in the form of nanoparticles or mesoporous substances.

**Keywords**

Fruit Cortex, Ecosystem, Biosorbent, Wastewater Remediation, Threat

## Contents

## 1. Introduction

Heavy metals, organic compounds (including dyes), pharmaceuticals, *etc*. are responsible pollutants for causing damage to the environment and are also a major culprit for its degradation and induced foremost global concern. All types of water pollution triggered through massive urbanization and industrial activities have been proved to be a major threat to the ecosystem [1]. According to the United Nations organization reports world's many developing countries do not have ample supply of clean and drinking water and face significant health hazards (around 80 % of diseases) due to water pollution [2-9]. Owing to the small size, stability and non-biodegradable nature of organic pollutants, it is a challenging task to remove them from polluted water [10]. Also, the high solubility of heavy metals in the aquatic environments permits them to be absorbed by the living organisms. Once they enter into the food chain, large concentrations of heavy metals may accumulate in the human body. If the metals are ingested beyond the permitted concentration, they can cause serious health disorders [11]. Therefore, it is necessary to treat metal contaminated wastewater prior to its discharge to the environment.

Some of the heavy metals having relevance in the environmental context are listed in Table 1. Heavy metals cause severe health hazards, including reduced growth and development, nervous system damage, cancer, organ damage, and in extreme cases, death. Arsenic is a hazardous heavy metal that causes vascular and nervous system diseases. Individual's immune system outbreaks its own cells due to the contact with toxic metals, such as mercury and lead. Consequently, this leads to joint diseases like rheumatoid arthritis, in addition to numerous kidney, circulatory and nervous system complications. Heavy metals can cause irreversible brain damage at higher doses. Also, children are more prone to higher doses of metals, as they require more food for their body development in comparison to adults. Therefore, in order to minimize human and ecological exposure to hazardous substances, wastewater protocols were established. This consists of limits on the variation and concentration of heavy metals that may be present in the discharged wastewater. The toxic effects and maximum contaminant level (MCL) standards for heavy metals, established by USEPA [12] are summarized in Table 1.

### 2.1 Source of heavy metals in water deposits

In the time of industrialization, urbanization and other activities it is not possible to escape from the exposure of heavy metals (Table 1), which in turn are released into the environment through:

1. Mining activities (metal industries, mine spoils, and tailings, river dredging, smelting, *etc*.).
2. Industries (microelectronics, plastic, refineries, textiles, wood preservatives, *etc*.).
3. Atmospheric degradation (automobile exhausts, fossil fuel combustion, pyrometallurgical industries, urban refuse disposal, *etc*.).
4. Excessive agrochemicals (fertilizers and pesticides) use and waste disposal (sewage sludge, *etc*.).
5. Chemical laboratories.

In ancient times, fabrics have been dyed with natural color extracts obtained from plants, animals, and minerals. Now a day, things began to change with the synthesis of cheaper, easy to put onto the fabrics, brighter, and color fastening synthetic dyes. Human beings are becoming more and more dependent on these dyes for their simple household chores as well as other purposes. Although, these dyes are synthesized using the chemicals which are often carcinogenic, highly toxic, or even explosive. for example, the chemical aniline, used as a reactant to synthesis azo dyes is highly flammable, dangerous to deal and is considered poisonous (gives off carcinogenic amines). Practically every commercial dye process includes a solution of a dye in water, in which the fabrics are dipped or washed. After dying the fabric, it's convenient to dump the used water that is dye effluent, than to recycle that water in the factory. So dye industries across the world are discharging millions of tons of dye effluent into rivers and water sources.

*Table 1. Sources, toxic effect and MCL values of different heavy metals [12].*

| Heavy metal | Source | Toxicities | MCL (mg/L) |
|---|---|---|---|
| Antimony | industries as ammunition, bearing and power transmission equipment, casting, ceramics, enamels, and paints, flame retardants, lead storage batteries and soldering *etc*. | Increase in blood cholesterol and decrease in sugar level [13] | $2\times10^{-4}$ |
| Arsenic | Smelting, mining, rock sediment's, energy creation from fossil fuels. | Skin manifestations, visceral cancers, vascular disease, bone marrow depression, hemolysis, liver tumors, gastrointestinal symptoms, cardiovascular and nervous system functions disturbances [14]. | $5\times10^{-2}$ |
| Cadmium | Electroplating, pigments, smelting, plastic and mining, alloy manufacturing. | Itai–Itai disease, Kidney damage, renal disturbances, lung insufficiency, bone lesions, cancers, hypertension, weight loss [14,15]. | $1\times10^{-2}$ |
| Chromiu m | Canning industry, electroplating, metal processing, paints and pigments, and steel fabrication. | Diarrhea, epigastric pain, headache, mutagenic, nausea, lung tumors, teratogenic, vomiting [16]. | $5\times10^{-2}$ |
| Copper | Electronics plating, paint manufacturing, wire drawing, copper polishing, and printing operations. | Liver damage, Wilson disease, insomnia, Reproductive and developmental toxicity, neurotoxicity, and acute toxicity, dizziness, diarrhea [16, 17]. | $2.5\times10^{-2}$ |
| Lead | Manufacturing of batteries, pigments electroplating. | Kidney diseases, damage to the nervous and circulatory system, anemia, brain damage, anorexia, malaise, loss of appetite [18]. | $6\times10^{-3}$ |
| Mercury | Volcanic eruptions, forest fires, battery. | Rheumatoid arthritis, renal disturbances; kidney, circulatory system, and nervous system diseases, destructive to skin, eyes, muscles [19]. | $3\times10^{-5}$ |
| Selenium | industrial activities such as coal mining | Circulatory problems, hair or fingernail loss; numbness in fingers or toes [20]. | $5\times10^{-5}$ |
| Nickel | Manufacture of copper sulfate, electroplating, and mineral processing. | Dermatitis, nausea, chronic asthma, coughing, reduced lung function, lung cancer, chronic bronchitis [21]. | $2\times10^{-1}$ |
| Zinc | Mining and manufacturing processes, alloys and battery manufacturing, thermoplastics, pigment formation, galvanizing plants. | Gastrointestinal pain, depression, diarrhea, lethargy, neurological disorders, increased thirst, cause short term ''metal fume fever'' [19]. | $8\times10^{-1}$ |

One of the extensively used cationic dyes in various industries is crystal violet (CV), which can easily enter and accumulate into living beings [22]. Therefore, it is today's need to explore eco-friendly, cost effective process for removing heavy metal ions and organic dyes from industrial sewages. Heavy metal ions and organic dyes can easily be removed from waste water through alterable polymeric materials having surface active functional groups [23-25]. Methylene blue (MB) with the molecular formula: $C_{16}H_{18}ClN_3S_3H_2O$ is a toxic heterocyclic aromatic molecule, may produce problems if inhaled, swallowed or made in contact with skin or eye. MB is used in various studies; hence during its use, a substantial amount of dye swamped with wastewater into the oceans and became hazardous for the aquatic lives. Therefore, decolorization and quenching of dyes are important aspects of wastewater remediation before discharge.

We use laundry detergents in our routine life which includes impurities such as surfactants, brighteners, and fragrances. These are suspected toxins, which ran in the sewage system. Hence, there is an urgent requisite to treat the impure water.

## 2.2 Different techniques to treat the waste water

Now a day, different methodologies have been extensively looking forward to exploring cheaper and more efficient technologies, which can reduce the amount of wastewater produced and to improvise the standard of the treated effluent. In the recent years, the search for low-cost adsorbents with metal-binding capacities has intensified, therefore, adsorption arises as an unconventional method of waste water treatment [26]. The adsorbents may possibly be agricultural wastes, biomass, industrial by-products, minerals, organic or biological substances, polymeric materials, zeolites, *etc.* [27]. presently, membrane separation technique has been increasingly used for the treatment of inorganic effluent due to its convenient operation. Different membrane filtration techniques such as nanofiltration (NF), reverse osmosis (RO), ultrafiltration (UF) [28] and electro treatments such as electrodialysis [29] have been contributed widely to environmental safety. The photocatalytic process is similarly an innovative and promising technique for effective destruction of toxins in water [30]. Different water cleansing techniques namely flocculation, activated sludge, chemical coagulation, ion exchange processes, photodegradation, and precipitation have been tested for removing the impurities [31-34]. Although countless approaches are available for the remediation of inorganic pollutants (heavy metal ions), but to meet the mentioned conventional MCL standards, an appropriate treatment should be applicable to the local conditions in a global manner. The advantages and disadvantages of different purification methods have been examined thoroughly and showcased alongside with their removal tendency, operating criteria *i.e* pH and treatment capability (Table 2).

Even though a large number of methods have been reported in the literature, yet, it is a tedious job to search a single ideal method which can remediate all hazardous toxins from contaminated water. Comparative to conservative water remediation strategies, usage of bio-adsorbents could be beneficial due to their high efficiency, low cost, easy accessibility, and their sustainability [35]. Bio-adsorption could be exploited as a reliable biotechnology due to its operational simplicity [36].Bio-sorption can be defined as the removal of contaminants from solution employing biological substances [37]. Various waste parts of fruits such as peels of banana [38], orange [31,38,39], pine bark [40], rice husk [41], sugarcane bagasse, [42], tea waste [43,44], and others [45-52] have been tested as bioadsorbents for the abolition of hazardous heavy metals as well as organic compounds from effluents (Fig. 1).

*Table 2. Demonstrate the merits and demerits of different traditional methods to treat toxic heavy metal.*

| Method | Merits | Demerits |
|---|---|---|
| Adsorption using activated carbon | High efficiency, most of the metals can be removed. | The cost of activated carbon, no regeneration, performance depends upon adsorbent [2]. |
| Chemical coagulation | Dewatering, Sludge settling | depletion of chemicals, high cost [2] |
| Chemical precipitation | The simple and inexpensive technique, the maximum number of the metal ions can be removed. | Production of large amounts of sludge, disposal problems [2]. |
| Electrochemical methods | Pure metals can be achieved, metal selective, non-consumption of chemicals. | High running and capital cost, initial solution pH and current density [53]. |
| Ion-Exchange | Metal selective, high regeneration of materials. | Fewer numbers of metal ions removed, high cost [53]. |
| Membrane process and ultrafiltration | High efficiency, less chemical consumption, less solid waste produced, | Other metals presence decreases removal percentage, expensive methodology, slow flow kinetics [54]. |
| Zeolites | Relatively inexpensive materials, almost all metals can be removed. | Low efficiency [54]. |

*Fig. 1. Various fruit peels/cortex used as bioadsorbents for polluted water.*

## 2.3 Fruit cortex and waste water treatment

Fruit cortex in layman language may be defined as fruit's outer layer (or fruit peel). Generally, all fruit cortexes have polar functionality like hydroxy (−OH), amino (−$NH_2$) and carboxyl (−COOH) on their surface. [55,56]. Literature survey confirms several fruit cortexes like banana peel; kiwi, tangerine [57]; lemon, orange, [58]; dragon fruit, avocado, and hamimelon [59] peels could be employed as a support for the biosorption of heavy metals for the treatment of wastewater (Fig. 2). These materials are widely used as biosorbants due to their easy availability, low cost and to the fact that neither culture nor synthesis is needed for their production.

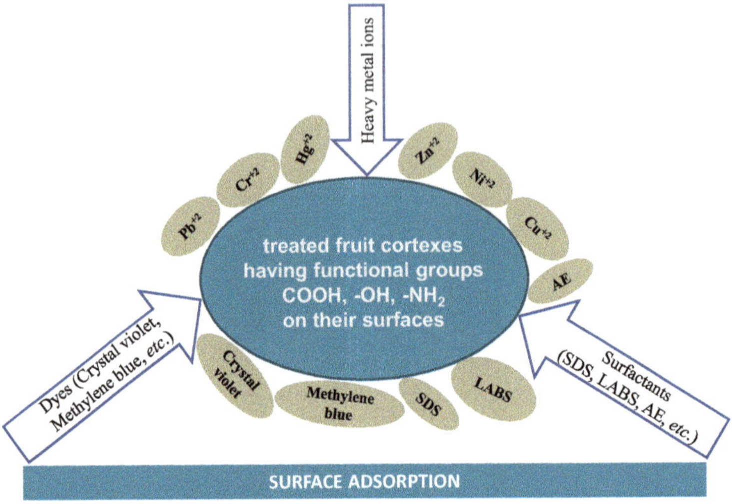


*Fig. 2. Bioadsorption of different metal ions and dyes using fruit cortexes.*

## 2.4 Method for the preparation of biocortex adsorbent

According to literature survey different papers, [58,59] described that various fruit peels were taken, washed thoroughly with water, and cut into small pieces (~ 0.04 $cm^2$ in dimension). Different peels were saponified (saponification term came from soap making; this is a technique through which oils can be converted into soap by the hydrolysis of ester bonds into salts of acids *i.e* soap and alcohol) using base (such as sodium hydroxide) to break ester bonds(-COOR) present on the peel's surface for the generation of more hydroxyl groups (-OH) which are required for better adsorption. Then excess washing was done with deionized water to get rid of the surplus base. At the end, treated peels were ultrasonicated in an organic solvent like *iso*propanol to separate organic solubles which were then filtered using Buckner funnel, and dried.

## 2.5 Characterization of bioadsorbents using FT-IR Spectroscopy

The presence of different functional groups on the treated cortex was confirmed by FT-IR spectrophotometric technique in the range between 4000 - 400 $cm^{-1}$ (Fig. 3). The broad band shown around ~392 $cm^{-1}$ for various bio-peels matched to the O-H stretching frequency. In the spectra the peaks at ~2926, and ~2856 $cm^{-1}$ match with stretching of C-H bonds [60]. The sharp absorption peaks visible at ~1754 $cm^{-1}$ and ~1736 $cm^{-1}$, respectively correspond to the stretching of C-O bonds of carboxylic acids and esters. The peaks observed in the region 1680–1610 $cm^{-1}$ look like C-C bond stretching of aromatic moieties. The peaks witnessed in the range of 1312 to 1000 $cm^{-1}$ account for the angular deformation of the C-H bonds of aromatic rings. To determine the nature of the adsorption and the functionalities responsible for the biosorption of contaminants, the FT-IR spectra of some of the peels were measured after adsorption. As predicted, no considerable alterations in the spectra were noticed before and after biosorption. It confirms that the surface active functional groups of bioadsorbents remained unaltered were recorded upon the adsorption of pollutants.

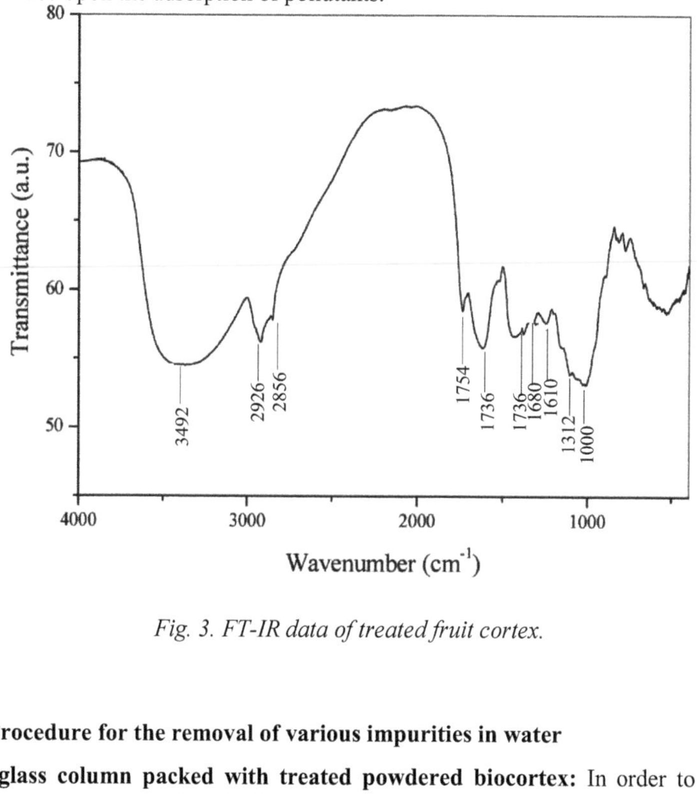


*Fig. 3. FT-IR data of treated fruit cortex.*

## 2.6 Procedure for the removal of various impurities in water

**Use of glass column packed with treated powdered biocortex:** In order to remove scums from waste water, the bioadsorbent was powdered in a mortar and packed to an appropriate height in the glass column. Impure water was passed through column and treated clean water was collected as effluent (Fig. 4).

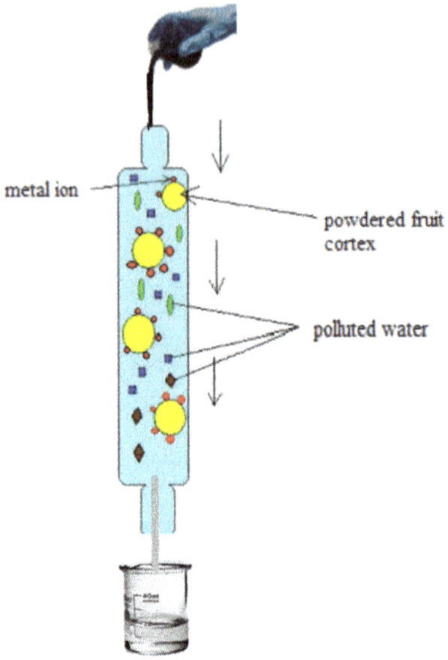


*Fig. 4. Removal of various impurities using glass column packed with treated powdered biocortex.*

**Using batch process**: In order to remove contaminants from waste water, an appropriate amount of powdered biocortex was taken in a beaker then impure water was added into it. Solution was kept at 30 °C using an orbital shaker operating at 200 rpm (Fig. 5).

*Fig. 5. Removal of various impurities with treated powdered biocortex using batch process*

### 2.6.1 Batch adsorption studies

**Variation of Concentration of Pollutants with Contact Time:** Various bio-peels were tested for their adsorption capacity to various pollutants. A known weight of adsorbent was added to 10 mL solution of pollutants by varying their concentrations. All adsorption experimentations were carried out at room temperature using a magnetic stirrer with vigorous stirring.

To estimate the biosorption abilities of the tested samples, different heavy metals were passed through the appropriate height columns and the elution filtrate refilled numerous times to intensify the retention of metal ions, dyes and surfactants. The residual concentrations of toxins were analyzed after scheduled intervals of time prior to equilibrium. The concentration of the adsorbed toxins at equilibrium qe (mg/g) was estimated using eqn. 1:

$$q_{eq} = (C_i - C_{eq})V/M \qquad \text{(Eq. 1)}$$

where $C_i$ and $C_{eq}$ (mg/L) are pollutants concentrations at initial and equilibrium conditions, respectively. V (in L) is volume of solution of the pollutant and M (g) is mass of the peel used. The $C_i$ values of tested polluted solution were in range of 5-200 mg/L, and the investigations were performed at room temperature for ~1 day.

The adsorption capacities of the toxins enhanced with increasing contact time and reached to equilibrium stage within 1 hour. It was found that the adsorption was initially fast, progressively decreases with time progresses and at the end attains saturation at equilibrium stage. High adsorption values were observed for the adsorption of cationic dyes and all cortexes bind cationic dyes more strongly relative to neutral / anionic dyes. This may be attributed to the availability of amino ($-NH_2$), carboxylate ($-COO^-$), and phenoxide ($^-O-Ph$) groups on the surface of the peels, which prompts the extraction of cationic dyes. [61]

**Effect of pH value on the adsorption of various contaminants- removal of dyes and metal ions**

The pH effect on pollutant extraction was studied by changing pH from 2 to 12, where the pH was adjusted by adding either 0.01 N HCl or 0.01 N NaOH solutions. Conversely, other parameters such as adsorbent concentration, agitation speed, and solution temperature were kept constant. The percentage removal of contaminant was calculated as:

$$Percentage\ of\ removal = \left(\frac{C_i - C_f}{C_i}\right)100 \qquad \text{(Eq. 2)}$$

Where $C_i$ and $C_f$ (mg/L) are initial and final pollutant concentrations in water, respectively.

The percentage removal of toxins increased with increasing pH, as a consequence of the effective replacement of cations with proton ($H^+$) of carboxyl and hydroxyl functional groups. With decrease in pH values (acidic medium), the protons and protonate functional groups compete with the adsorption of other cationic contaminants [62]. However, with increase in pH values (basic medium), withdrawal of cations by the negatively charged adsorption sites increased appreciably [63]. Moreover, for high concentrations of $H^+$ at lower pH values, the positively charged adsorbent surface eliminates negatively charged pollutants.

## 3. Conclusions

Recently, various investigations have been performed and several researchers are still carrying rigorous works to find cost-effective, environmental friendly techniques in addition to natural materials to be explored for elimination of heavy metals and organic dyes so that contaminated water can be treated before disposing into the ecosystem. Bio-adsorption is one of the extensively researched methods. Some of the widely used natural adsorbents are activated carbon, clays, agricultural wastes, industrial wastes, *etc*. One of recent approaches is the use of agricultural wastes, specifically fruit waste peels or cortexes (orange, lemon, banana, kiwi, dragon fruit *etc.*) to eradicate heavy metals in polluted effluents.

All these peels are low cost, easily accessible, renewable, nontoxic, competent and effective bio-adsorbents for water cleansing. It was found that as such only treatment with base is required for exposing its active group so that efficient adsorption can occur. The results revealed that recovering of these cortexes is proficient and could be employed in successive contaminant adsorption studies with no evident hampering in their adsorption tendencies.

## References

[1] N. Satya, C.S.P. Ojha, S.K. Mishra, U.C. Chaube, P.K. Sharma, Cadmium and chromium removal by aquatic plant, Int. J. Environ. Sci. 6 (2011) 1297-1304.

[2] K.J. Abaliwano, A.K. Ghebremichael, L.G. Amy, Application of the purified moringa oleifera coagulant for surface water treatment, Water Mill Working Paper Series 5 (2008) 1-22.

[3] A.K. Meena, K. Kadirvelu, G.K. Mishraa, C. Rajagopal, P.N. Nagar, Adsorption of Pb(II) and Cd(II) metal ions from aqueous solutions by mustard husk, J. Hazard. Mater. 150 (2008) 619-625. https://doi.org/10.1016/j.jhazmat.2007.05.011

[4] A. Sarý, Ö.D. Uluozlü, M. Tüzen, Equilibrium, thermodynamic and kinetic investigations on biosorption of arsenic from aqueous solution by algae (Maugeotia genuflexa) biomass, Chem. Eng. J. 167 (2011) 155-161. https://doi.org/10.1016/j.cej.2010.12.014

[5] M. Iqbal, A. Saeed, R.G.J. Edyvean, Bioremoval of antimony(III) from contaminated water using several plant wastes: Optimization of match and dynamic flow conditions for sorption by green bean husk (Vignaradiata). Chem. Eng. J. 225 (2013) 192-201. https://doi.org/10.1016/j.cej.2013.03.079

[6] V. Vinodhini, N. Das, Relevant approach to assess the performance of sawdust as adsorbent of chromium (VI) ions from aqueous solutions, Int. J. Environ. Sci. Technol. 7 (2010) 85-92. https://doi.org/10.1007/BF03326120

[7] W. Guo, R. Chen, Y. Liu, M. Meng, X. Meng, Z. Hu *et. al.*, Preparation of ion-imprinted mesoporous silica SBA-15 functionalized with triglycine for selective adsorption of Co(II). Colloids Surfaces A Physicochem. Eng. Asp. 436 (2013) 693-703. https://doi.org/10.1016/j.colsurfa.2013.08.011

[8] H. Abu Hasan, S.R.S. Abdullah, N.T. Kofli, S.K. Kamarudin, Isotherm equilibria of Mn2+ biosorption in drinking water treatment by locally isolated bacillus species and sewage activated sludge, J. Environ. Manage. 111 (2012) 34-43. https://doi.org/10.1016/j.jenvman.2012.06.027

[9] EI. El-Shafey, Removal of Zn(II) and Hg(II) from aqueous solution on a carbonaceous sorbent chemically prepared from rice husk. J. Hazard. Mater. 175 (2010) 319-327. https://doi.org/10.1016/j.jhazmat.2009.10.006

[10] S. Tunali Akar, S. Arslan, T. Alp, D. Arslan, T. Akar, Biosorption potential of the waste biomaterial obtained from cucumis melo for the removal of Pb2+ ions from aqueous media: Equilibrium, kinetic, thermodynamic and mechanism analysis, Chem. Eng. J. 185-186 (2012) 82-90. https://doi.org/10.1016/j.cej.2012.01.032

[11] S. Babel, T.A. Kurniawan, Cr(VI) removal from synthetic wastewater using coconut shell charcoal and commercial activated carbon modified with oxidizing agents and/or chitosan, Chemosphere 54 (7) (2004) 951–967. https://doi.org/10.1016/j.chemosphere.2003.10.001

[12] S. Babel, T.A. Kurniawan, Low-cost adsorbents for heavy metals uptake from contaminated water: A review. J. Hazard. Mater. B 97 (2003) 219–243. https://doi.org/10.1016/S0304-3894(02)00263-7

[13] M. Iqbal, A. Saeed, R.G.J. Edyvean, Bioremoval of antimony(III) from contaminated water using several plant wastes: Optimization of match and dynamic flow conditions for sorption by green bean husk (Vignaradiata). Chem. Eng. J. 225 (2013) 192-201. https://doi.org/10.1016/j.cej.2013.03.079

[14] M.A. Momodu, C.A. Anyakora, Heavy metal contamination of ground water: The surulere case study, Res. J. Environ. Earth Sci. 2 (2010) 39-43.

[15] A. Sharma, K.G. Bhattacharyya, Azadirachta indica (Neem) leaf powder as a biosorbent for removal of Cd(II) from aqueous medium. J. Hazard. Mater B 125 (2005) 102-112. https://doi.org/10.1016/j.jhazmat.2005.05.012

[16] H. Ali, E. Khan, M.A. Sajad, Phytoremediation of heavy metals-Concepts and applications. Chemosphere 91 (2013) 869-881. https://doi.org/10.1016/j.chemosphere.2013.01.075

[17] M. Bilal, J.A. Shah, T. Ashfaq, S.M.H. Gardazi, A.A. Tahir, A. Pervez, H. Haroon, Q. Mahmood, Waste biomass adsorbents for copper removal from industrial wastewater - A review, J. Hazard. Mater. 263 (2013) 322-333. https://doi.org/10.1016/j.jhazmat.2013.07.071

[18] L.M. Mataka, E.M.T. Henry, W.R.L. Masamba, S.M. Sajidu, Lead remediation of contaminated water using moringa stenopetala and moringa oleifera seed powder. Int. J. Environ. Sci. Technol. 3 (2006) 131-139. https://doi.org/10.1007/BF03325916

[19] U. Farooq, M.A. Khan, M. Athar, J.A. Kozinski, Effect of modification of environmentally friendly biosorbent wheat (triticum aestivum) on the biosorptive removal of cadmium(II) ions from aqueous solution, Chem. Eng. J. 171 (2011) 400-410. https://doi.org/10.1016/j.cej.2011.03.094

[20] L.C. Tan, Y.V. Nancharaiah, E.D.V. Hullebusch, P.N.L. Lens, Biotechnol. Adv. 34(5), (2016) 886-907. https://doi.org/10.1016/j.biotechadv.2016.05.005

[21] J. Febrianto, A.N. Kosasih, J. Sunarso, Y.H. Ju, N. Indraswati, S. Ismadji Equilibrium and kinetic studies in adsorption of heavy metals using biosorbent: A summary of recent studies, J. Hazard. Mater.162 (2009) 616-645. https://doi.org/10.1016/j.jhazmat.2008.06.042

[22] H. He, S. Yang, K. Yu, Y. Ju, C. Sun, L. Wang, Microwave induced catalytic degradation of crystal violet in nano-nickel dioxide suspensions, .J. Hazard. Mater. 173 (2010) 393-400. https://doi.org/10.1016/j.jhazmat.2009.08.084

[23] Z. Chen, M. Deng, Y. Chen, G. He, M. Wu, J. Wang, Preparation and performance of cellulose acetate/polyethyleneimine blend microfiltration membranes and their applications, J. Memb. Sci. 235 (2004) 73-86. https://doi.org/10.1016/j.memsci.2004.01.024

[24] P. Chindaprasirt, S. Rukzon, Strength, porosity and corrosion resistance of ternary blend Portland cement, rice husk ash and fly ash mortar, Constr. Build. Mater. 22 (2008) 1601-1606. https://doi.org/10.1016/j.conbuildmat.2007.06.010

[25] T. Chuah, A. Jumasiah, I. Azni, S. Katayon, S.T. Choong, Rice husk as a potentially low-cost biosorbent for heavy metal and dye removal: an overview, Desalination 175 (2005) 305-316. https://doi.org/10.1016/j.desal.2004.10.014

[26] W.C. Leung, M.F. Wong, H. Chua, W. Lo, C.K. Leung, Removal and recovery of heavy metals by bacteria isolated from activated sludge treating industrial effluents and municipal wastewater, Water. Sci. Technol. 41(12) (2000) 233-240.

[27] T.A. Kurniawan, G.Y.S. Chan, W.H. Lo, S. Babel, Comparisons of low-cost adsorbents for treating wastewaters laden with heavy metals, Sci. Total Environ. 366(2-3) (2005) 409-426. https://doi.org/10.1016/j.scitotenv.2005.10.001

[28] T.A. ChenKurniawan, G.Y.S. Chan, W.H. Lo, S. Babel, Physicochemical treatment techniques for wastewater laden with heavy metals, Chem. Eng. J. 118 (2006) 83-98. https://doi.org/10.1016/j.cej.2006.01.015

[29] A.J. Pedersen, Characterization and electrolytic treatment of wood combustion fly ash for the removal of cadmium, Biomass Bioenergy 25(4) (2003) 447-458. https://doi.org/10.1016/S0961-9534(03)00051-5

[30] L.R. Skubal, N.K. Meshkov, T. Rajh, M. Thurnauer, Cadmium removal from water using thiolactic acid-modified titanium dioxide nanoparticles, J. Photochem. Photobiol. A, Chem. 148 (2002) 393-397. https://doi.org/10.1016/S1010-6030(02)00069-2

[31] K. Mohanty, D. Das, M.N. Biswas, Adsorption of phenol from aqueous solutions using activated carbons prepared from tectona grandis sawdust by ZnCl2 activation, Chem. Eng. J. 115 (2005) 121-131. https://doi.org/10.1016/j.cej.2005.09.016

[32] F. Veglio, F. Beolchini, Removal of metals by biosorption: A review. Hydrometallurgy 44 (1997) 301-316. https://doi.org/10.1016/S0304-386X(96)00059-X

[33] U. Kumar, M. Bandyopadhyay, Sorption of cadmium from aqueous solution using pretreated rice husk, Bioresour. Technol. 97 (2006) 104-109. https://doi.org/10.1016/j.biortech.2005.02.027

[34] S.S. Nawar, H.S. Doma, Removal of dyes from effluents using low-cost agricultural by-products, Sci. Total Environ. 79 (1989) 271-279. https://doi.org/10.1016/0048-9697(89)90342-2

[35] G. Crini, Non-conventional low-cost adsorbents for dye removal: A review, Bioresour. Technol. 97 (2006) 1061-1085. https://doi.org/10.1016/j.biortech.2005.05.001

[36] M. Bilal, J.A. Shah, T. Ashfaq, S.M.H. Gardazi, A.A. Tahir, A. Pervez, H. Haroon, Q. Mahmood, Waste biomass adsorbents for copper removal from industrial wastewater - A review, J. Hazard. Mater. 263 (2013) 322-333. https://doi.org/10.1016/j.jhazmat.2013.07.071

[37] G.M. Gadd, Accumulation and transformation of metals by microorganisms, in: H.J. Rehm, G. Reed, A. Puhler, P. Stadler, (Ed.), Biotechnology, a Multivolume Comprehensive Treatise, Special Processes, Wiley-VCH,Weinheim, 2001, pp. 225-264.

[38] A. Ali, K. Saeed, Decontamination of Cr(VI) and Mn(II) from aqueous media by untreated and chemically treated banana peel: Acomparative study, Desal. Water Treat. 53 (2015) 3586-3591. https://doi.org/10.1080/19443994.2013.876669

[39] G.F. Xuli, Z.X. Shen, R.X. Guo, The kinetic studies for the adsorption of furdan from aqueous solution by orange peel, Adv. Mater. Res. 842 (2014) 187-191.

[40] S. Al-Asheh, F. Banat, R. Al-Omari, Z. Duvnjak, Predictions of binary sorption isotherms for the sorption of heavy metals by pine bark using single isotherm data, Chemosphere 41 (2000) 659-665. https://doi.org/10.1016/S0045-6535(99)00497-X

[41] U. Kumar, M. Bandyopadhyay, Sorption of cadmium from aqueous solution using pretreated rice husk, Bioresour. Technol. 97 (2006) 104-109. https://doi.org/10.1016/j.biortech.2005.02.027

[42] U.K. Garg, M.P. Kaur, V.K. Garg, D. Sud, Removal of hexavalent chromium from aqueous solution by agricultural waste biomass, J. Hazard. Mater. 140 (2007) 60-68. https://doi.org/10.1016/j.jhazmat.2006.06.056

[43] E. Malkoc, Y. Nuhoglu, Investigations of Nickel(II) removal from aqueous solutions using tea factory waste, J. Hazard. Mater., 127 (2005) 120-128. https://doi.org/10.1016/j.jhazmat.2005.06.030

[44] E. Malkoc, Y. Nuhoglu, Fixed bed studies for the sorption of chromium(VI) onto tea factory waste, Chem. Eng. Sci. 61 (2006) 4363-4372. https://doi.org/10.1016/j.ces.2006.02.005

[45] G.S. Agarwal, H.K. Bhuptawat, S. Chaudhari, Biosorption of aqueous chromium(VI) by Tamarindus indica seeds, Bioresour. Technol. 97 (2006) 949-956. https://doi.org/10.1016/j.biortech.2005.04.030

[46] A.G. Devi Prasad, Abdullah, M. Biosorption of Fe(II) from aqueous solution using Tamarind Bark and potato peel waste: Equilibrium and kinetic Studies, J. Appl. Sci. Environ. Sanit. 4 (2009) 273–282.

[47] S.T. Koel Banerjee, R. Ramesh, P.V. Gandhimathi, K.S. Bharathi, A novel agricultural waste adsorbent, watermelon shell for the removal of copper from aqueous solutions, Iran. J. Energy 3 (2012) 143-156.

[48] P. Venkateswarlu, M.V. Ratnam, D.S. Rao, M.V. Rao, Removal of chromium from an aqueous solution using azadirachta indica (neem) leaf powder as an adsorbent, Int. J. Phys. Sci. 2 (2007) 188-195.

[49] T. Altun, E. Pehlivan, Removal of copper(II) ions from aqueous solutions by walnut-, hazelnut- and almond-shells. CLEAN – Soil, Air, Water 35 (2007) 601-606. https://doi.org/10.1002/clen.200700046

[50] A. Saeed, M. Iqbal, M.W. Akhtar, Removal and recovery of lead(II) from single and multimetal (Cd, Cu, Ni, Zn) solutions by crop milling waste (black gram husk), J. Hazard. Mater. 117 (2005) 65-73. https://doi.org/10.1016/j.jhazmat.2004.09.008

[51] Feng, Y.; Yang, F.; Wang, Y.; Ma, L.; Wu, Y.; Kerr, P. G.; Yang, L. Basic dye adsorption onto an agro-based waste material – Sesam hull (Sesamum indicum L.). Bioresour. Technol. 102 (2011)10280-10285. https://doi.org/10.1016/j.biortech.2011.08.090

[52] Mallampati, R.; Valiyaveettil, S. Application of tomato peel as an efficient adsorbent for water purification, Alternative biotechnology? RSC Adv. 2012, 2, 9914–9920.

[53] Rao KS, Mohapatra M, Anand S, Venkateswarlu P. 2010 Review on cadmium removal from aqueous solutions. Int. J. Engineering, Sci. and Technology 2, 81-103. https://doi.org/10.1039/c2ra21108d

[54] F. Fu, Q. Wang, Removal of heavy metal ions from wastewaters: A review. J. Environ. Manage. 92 (2011) 407-418. https://doi.org/10.1016/j.jenvman.2010.11.011

[55] Y. Feng, F. Yang, Y. Wang, L. Ma, Y. Wu, P.G. Kerr, L. Yang, Basic dye adsorption onto an agro-based waste material – Sesame hull (Sesamum indicum L.), Bioresour. Technol. 102 (2011) 10280-10285. https://doi.org/10.1016/j.biortech.2011.08.090

[56] R Negi, G. Satpathy, Y.K. Tyagi, R. K. Gupta, Biosorption of heavy metals by utilising onion and garlic wastes, Int. J. Environ. Pollut. 49 (2012) 179-196. https://doi.org/10.1504/IJEP.2012.050898

[57] K.M. Al-Qahtani, Water purification using different waste fruit cortexes for the removal of heavy metals, J. Taibah Univer. Science 10 (2016) 700-708. https://doi.org/10.1016/j.jtusci.2015.09.001

[58] K. Kelly-Vargas, M. Cerro-Lopez, S. Reyna-Tellez, E.R. Bandala, J.L. Sanchez-Salas, Biosorption of heavy metals in polluted water, using different waste fruit cortex, Physics and Chemistry of the Earth 37-39 (2012) 26–29. https://doi.org/10.1016/j.pce.2011.03.006

[59] R. Mallampati, L. Xuanjun, A. Adin, S. Valiyaveettil, Fruit peels as efficient renewable adsorbents for removal of dissolved heavy metals and dyes from water, ACS Sustainable Chem. Eng. 3 (2015) 1117-1124. https://doi.org/10.1021/acssuschemeng.5b00207

[60] P.Y. Tang, C.J. Wong, K.K. Woo, Optimization of pectin extraction from peel of dragon fruit (Hylocereus polyrhizus), Asian J. Biol. Sci. 4 (2011), 189-195. https://doi.org/10.3923/ajbs.2011.189.195

[61] M.R.H.M. Haris, K. Sathasivam, The removal of methyl red from aqueous solutions using banana pseudostem fibers, Am. J. Appl. Sci. 6 (2009) 1690-1700. https://doi.org/10.3844/ajassp.2009.1690.1700

[62] R. Chand, K. Narimura, H. Kawakita, K. Ohto, T. Watari, K. Inoue, Grape waste as a biosorbent for removing Cr(VI) from aqueous solution, J. Hazard. Mater. 163 (2009) 245-250. https://doi.org/10.1016/j.jhazmat.2008.06.084

[63] P.L. Homagai, K.N. Ghimire, K. Inoue, Adsorption behaviour of heavy metals onto chemically modified sugarcane bagasse, Bioresour. Technol. 101 (2010) 2067-2069. https://doi.org/10.1016/j.biortech.2009.11.073

# Chapter 9

# Advances in desalination for water and wastewater treatment

Aiman Eid Al-Rawajfeh [a*], Ghada A. Al Bazedi [b], Marwa M. Eid [b], Malek G. Hajaya[a]

[a] Tafila Technical University (TTU), Department of Chemical Engineering, P.O. Box 179, 66110 Tafila, Jordan.

[b] National Research Center, Engineering Division, Chemical Engineering and Pilot Plant Department, P.O. Box 12622, Giza, Egypt

[*] aimanr@yahoo.com

**Abstract**

This chapter briefly summarizes the main concepts of conventional and novel/emerging desalination technologies. It also emphasizes the principal process engineering outlays for dual desalination systems, covering simultaneous water/power production and heterogeneous desalination. In addition to schemes that comprises both of the membrane and thermal processes, this chapter distinctively covers the typically overlooked medium and long term environmental impacts on soil and marine environment. Finally, the challenges in this field will be outlined targeting acceptable process economy and triple production of energy, water, and salt.

**Keywords**

Desalination, Water Treatment, Wastewater, Membrane Process, Thermal Desalination

## Contents

## 1. Introduction

The increasing need for desalted water to cope with population growth, development, and industrial applications is highly recognized. Desalination process produces fresh water with competitive cost in comparison to alternative processes. Desalination can be accomplished by various techniques that can be categorized broadly into two main classes: thermal and membrane processes. Thermal processes mainly include multistage flash evaporation (MSF), multiple effect distillation (MED), and vapor compression (VC). Membrane processes commonly used are reverse osmosis (RO), electro-dialysis (ED), and nanofiltration (NF) [1-2].

The number and capacity of large RO plants have increased extensively during the last years due to technological developments in RO desalination. Owing to development in membranes, technological advancement and drastic increase in capacity, systems with product water capacity up to 300,000 $m^3$/day are currently being constructed. Desalted water operating cost has recently decreased from about \$ 2.0/ $m^3$ to \$ 0.5/ $m^3$ [3, 4]. The global installed desalination capacity is increasing. As recent figures have indicated that the production capacity of all desalination plants is projected to be around 44.1 million $m^3/d$ in 2006, representing all facilities listed in the inventory that treat seawater, brackish water, river water, wastewater, and pure water, which were either in construction or online [5].

Desalination plant size distribution within the brackish water and wastewater sectors is about 24% for brackish water and 27% for wastewater of the total water produced in very large-sized plants (capacity > 50,000 $m^3$/day), while in large plants 34% accounts for brackish water and 36% wastewater of the production medium plants for 33% brackish water and 32% wastewater of the production, and almost small plants for 9% brackish water and 5% wastewater, respectively [6].

Shortage of natural fresh water resources and the increasing demand of water forced to consider reusing of wastewater as an alternative water resource. In some cases, conventional water treatment technologies demonstrate their inability to remove all organic pollutants to a level within that permitted by regulations. Development of new water supplies through water reuse and desalination are increasingly considered key components of solutions to urban water supply challenges [7,8]. The sources for desalination and water reuse plants cannot be used as potable water supply without additional purification. Reverse osmosis (RO) treatment is one of the most widely applied processes for these new water supplies. Half of the world's desalination capacity is currently produced through reverse osmosis, with the proportion reaching 70% in the United States [2]. Reverse osmosis technology is even more prevalent in potable water reuse projects [9].

Municipal wastewater often includes different chemical compounds, many of which are toxic to aquatic life and health. Micropollutants removal by traditional treatment methods indicate to be ineffective [10, 11], and accordingly, new technologies are developed. Membrane processes employing nanofiltration (NF) and ultrafiltration (UF) are increasingly used in wastewater reclamation and drinking water to remove micropollutants as well as natural organic matter [12, 13].

Tertiary or advanced treatment is the next treatment step to remove contaminants after using the secondary treatment (nitrogen, phosphorous, heavy metals, bacteria and viruses). Cleaner effluent is produced through this advanced treatment step which includes chemical coagulation, flocculation and sedimentation followed by filtration and adsorption processes. Membrane separation techniques are also used for the treatment and filtration step as well as ion exchange process in case of ion removal technologies [14, 15].

Treatments planned to improve water quality; where reverse osmosis (RO) or nanofiltrations (NF) are always recommended [12]. Recently developed NF membranes provide lower feed pressures and reduced operating costs compaired to RO treatment, for comparable product water quality, resulting in potential savings of \$ 0.03–0.08 $m^3$ of domestic water treated [16].

Recent advances in membrane technology led to significant cost reduction for membrane elements. The cost breakdown of RO plants indicates that a membrane contribution in cost is \$ 1.5-3 per 1000 gallons. Numerous publications on membrane pretreatment systems for RO desalination plants confirmed the technical and economic feasibility of dual membrane systems [3, 17].

Extensive developments on the application of nanofiltration (NF) technology have revealed the technical and economic feasibility of introducing NF in combination with reverse osmosis (RO) desalination [5]. The recently developed nanofiltration (NF) membrane for pretreatment step in combination with conventional filtration systems was effectively employed in a pilot plant and in an operating plant with excellent results. This process prevented reverse osmosis (RO) membrane fouling through turbidity and bacteria removal, as well as about 40% increase in production capacity in the operating plant. The use of NF is justified by the softening ability via reducing calcium and magnesium ions [18].

Different studies on NF confirm the utility of NF as an encouraging treatment alternative for groundwater. NF can reject larger organic contaminants up to the level more than 90% [16]. Ernst and Jekel studied tertiary effluent treatment with DOW NF200, the results showed a reduction in dissolved organic carbon (DOC) and the adsorbable organic halogens concentration to the values less than 2–3 mg /L and 2 mg/L, respectively [19]. Some studies show that DOC values reduced to the level of 0.5 mg/L by integrating NF process with powdered activated carbon (PAC) [20].

One of the options of NF application is to combine with ozonation and slow sand filtration (SSF) to offer a complementary process in dissolved organic matter (DOM) removal, and trihalomethane formation potential (THMFP) can be reduced from $6.5 \pm 1.1$ to $0.7 \pm 0.3$ mg/L and from $267 \pm 24$ to $52 \pm 6$ mg/L, respectively [21].

Membrane materials for (MF, UF, NF and RO) can be prepared organically using organic polymers (hydrophilic as cellulosic esters, poly sulphone, polyamide and polyimide or hydrophobic as polyether, polypropylene and polyvinylidene fluoride) or inorganically using inorganic membranes which are prepared using alumina, zirconia, titania orsilicium carbide.

The main advantages of membrane technology are as follows:

- ♦ Separation can be implemented continuously.
- ♦ Generally, low energy consumption.
- ♦ Easy combined membrane processes with hybrid processing.
- ♦ Separation can be implemented under mild conditions.
- ♦ Membrane properties are variable and can be modified.
- ♦ Not needed additives.

However, the main disadvantages of membrane technology include; short life time, concentration polarization/membrane fouling and scale formation, and up-scaling factor is more or less linear.

## 2. Desalination technologies

Numerous methods have been recommended for seawater desalination, but only a few mentioned below has been commercially used. These include:

- **Thermal processes**.

- "Multistage-Flash distillation" (MSF), "multi effect distillation" (MED), and "vapor compression distillation" (VC)

- **Membrane processes**

- "Reverse osmosis" (RO), "nanofiltration" (NF), and "electro dialysis" (ED)

- **Others**

- Freezing, membrane distillation (MD), capacitive deionization (CDI), and forward osmosis (FO)

### 2.1 Thermal processes

Thermal processes are the oldest among all desalting processes. Thermal processes represent the prime choice for dual-purpose (electric power and potable water production) and for concentrations that cannot be performed by RO or EDR treatment (TDS: greater than 50,000 mg/l) or where the feed water characteristics may damage the membrane structure. Recent developments in material of construction for MED and VC technologies led to lower capital costs and reduced power consumption, leading to economical process affordability [22, 23].

### 2.1.1 Multistage flash distillation

MSF units are widely used in the Middle East (particularly in Saudi Arabia, the United Arab Emirates, and Kuwait) and account for over than 40% of the world's desalination capacity. In the Middle East, MSF plants account for 68% of desalination capacity, while in North Africa, MSF and RO plants acquire about 40% each of the desalination market [23].

MSF process involves evaporation and condensation of water, these steps are coupled so that the latent heat of evaporation is recovered for reuse. The term "performance ratio" is often applied to thermal desalination processes representing the gained output ratio, which is defined as the mass of water product per mass of heating steam. A typical gained output ratio for MSF units is 8. A 20 stage plant has a typical heat requirement of 290 kJ/kg of product [24, 25].

Top brine temperature (TBT) plays a vital role in determining the performance of the MSF systems. It can be increased to 120 °C using additives. However, it is desirable to maintain the TBT at 110° C. One of the most important parameters that affect the performance of MSF is the recirculation brine flow rate. Two important parameters are of concern in MSF operation; the circulating brine flow rate determines the brine flow velocity inside the condenser tubes and the cooling seawater inlet and outlet temperatures affect the efficiency of MSF plant.

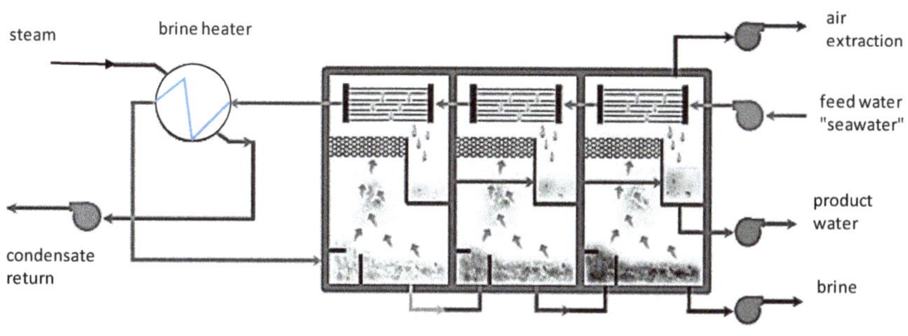


*Fig. 1. Schematic diagram of a basic multi-stage flash desalination process. [26].*

Scaling is the most significant concern in thermal desalination. There are three main compounds forming scales in distillation plants: calcium sulfate, $CaSO_4$; magnesium hydroxide, $Mg(OH)_2$; and calcium carbonate, $CaCO_3$. Scale is particularly undesirable when it forms on the heat transfer surface, thus it can greatly reduce the overall heat transfer. Corrosion and erosion in distillation processes are influenced by: pH, temperature, chloride concentration and dissolved oxygen [22].

### 2.1.2 Multi-effect distillation

Multi-effect distillation (MED) is a distillation process related to MSF. In MED, vapor from each stage is condensed in the next successive stage thereby giving up its heat to drive more evaporation. To increase the performance of MED, each stage is run at a successively lower pressure. MED, is similar to MSF, where it takes place in a series of effects and uses the principle of reducing the ambient pressure in the various effects.

This permits the sea water feed to undergo multiple boiling without supplying additional heat after the first effect. Hybrid system comprising of multi-effect distillation and thermal vapor compression (MED-TVC) may require thermal performance ratios (similar to the gain ratio), energy used to evaporate water in all the stages/ first stage energy input) approaching 17, whereas the combination of MED with a lithium bromide/water absorption heat pump granted a thermal performance ratio of 21 [26].

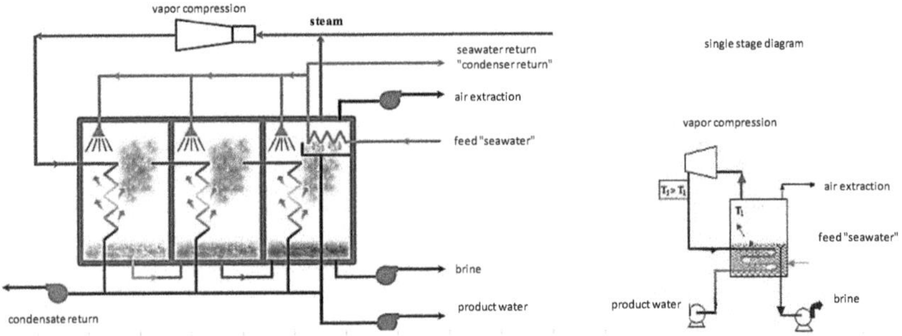


*Fig. 2. Schematic diagram of a basic multi-effect evaporation desalination process. [26].*

Al-Rawajfeh [27] studied hybrid salts precipitation-nanofiltration (SP-NF) process on the scale deposits in thermal and membrane desalination processes. The analysis was carried out to study the scale formation from the Arabian Gulf seawater in MSF and RO reference processes by changing the percentage of pretreatment from 0 to 100%. Four different SP-NF configurations were suggested. A targeted top brine temperature (TBT) of 130°C may be achieved if 30% portion is pretreated by SP and/or NF processes. As a rule of thumb, each 1% pretreatment portion increases the reference TBT of 115°C by 0.6°C. For both MSF and RO, parallel pretreatment of certain percentage of the feed by SP and the rest by NF, showed the lowest scale values. The case showed the best values for sulfate scale prevention and the highest values of increasing the monovalent ions relative to the divalent scale forming ions.

Sulfate scale is significant in MSF process while carbonate scale is significant in RO. Salt precipitation was suggested because it is less costly than nanofiltration, but nanofiltration was used because it is efficient in sulfate ions removal.

MSF units operation could have some adverse environmental implications. Al-Rawajfeh [28] studied the $CO_2$ release and scaling in MSF distillers was investigated for different feed water composition in different intakes from the Arabian Gulf for once-through (OT) and brine recycle (BR) MSF distillers. The results were calculated from a previously developed model. The model was applied to two 20-stage reference MSF once-through and recycles distillers. The $CO_2$ release rates decreased exponentially from the first stage to the last stage. The $CO_2$ release rates increased with increasing top brine temperature (TBT) and so $CaCO_3$ deposition rates did. It also increases with $HCO_3^-$ content and salinity while it decreases with increasing the seawater pH value. The sulfate scale potential increases with increasing top brine temperature. It increases with $Ca^{2+}$, $SO_4^{2-}$ and salinity as well.

### 2.1.3 Vapor compression

Vapor compression (VC) distillation is used for small- and medium-scale seawater desalting units. VC units are usually built for tourist resorts, small industries, and remote sites. Mechanical vapor compression MVC systems generally have only a single stage, while thermal vapor compression TVC systems have several stages. This difference results from the fact that MVC systems have the same specific power consumption (power/unit water produced) despite the number of stages, while the thermal efficiency of TVC systems is improved by adding extra stages.

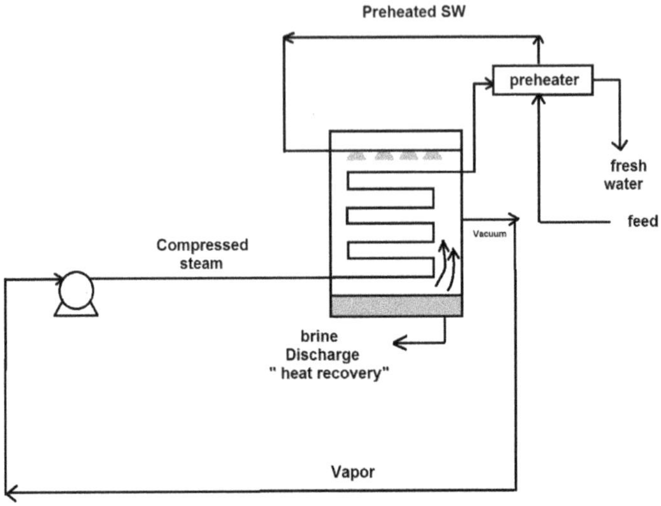


*Fig 3. Schematic diagram of a single stage mechanical vapor compression desalination process [26].*

## 2.2 Membrane processes

There are many types of membrane processes used in desalination; the most popular methods are reverse osmosis, nanofiltration and electrodialysis, and more recently membrane distillation/evaporation. Membrane separation technology offers a promising technique to provide suitable water supply of appropriate quality to meet human needs in addition to environmental and industrial applications. It was considered as a principal separation process in desalination and water treatment since about 35 years ago [29]. Membrane technology plays a key role in advanced wastewater treatment systems because of its potential in reuse and recovery.

### 2.2.1 Reverse osmosis

RO is a pressure-driven non-thermal membrane separation process that recovers water from a pressurized saline solution. The significant feature of the process is that no phase change occurs and it relatively involves low-energy compared to MSF and MED desalination techniques. RO is often the method of choice for brackish water, where only low to intermediate pressures are required.

The operating pressure for brackish water systems ranges from 15 – 25 bar while for seawater systems from 54 to 80 bar (the osmotic pressure of seawater is about 25 bar). Large scale RO systems are equipped with energy recovery systems. Feed pressure of the saline water accounts for the majority of the energy consumed by RO systems, since the pressure needed for separation is directly related to the salt concentration. RO membranes are sensitive to pH, oxidizers, a wide range of organics, algae, bacteria, and any other foulants. Therefore, pretreatment of the feed water is an important consideration that has a significant impact on the cost of RO [25]. Different treatment issues are attained for the product water treatment for example, "post-treatment", includes removing dissolved gasses ($CO_2$), and stabilizing the pH via the addition of Ca or Na salts [30, 31].

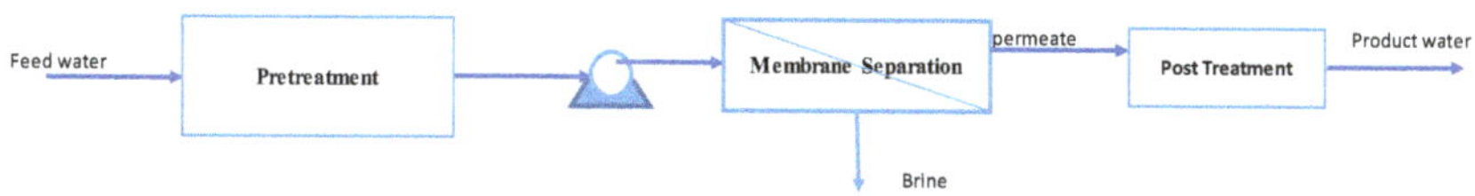


*Fig. 4. Block diagram of reverse osmosis desalination system [26].*

Importantly, on the other hand, membrane characteristics which affect the rejection are; membrane porosity, the surface charge, and membrane polymer composition. While feed water characteristics influence the physicochemical interactions with membrane surfaces include molecular weight, molecular size, acid disassociation constant, and polarity. The rejection of organic contaminants may alter the interaction between solutes and membrane surfaces.

For instance, the presence of fulvic acid in wastewater or organic matter in feed water will decrease the membrane rejection of estrogenic hormones that would adsorb more easily onto membrane surface [30].

### 2.2.2 Nanofiltration

Nanofiltration (NF) membrane is an emerging type of pressure-driven membranes with properties in-between reverse osmosis (RO) and ultrafiltration (UF) membranes. NF membranes are still sometimes denoted as ''softening' 'membranes [32] though they are capable of removing hardness and a wide range of other components [33].

NF has recently gained wide interest in R&D of other applications. NF membranes offer several advantages such as low operating pressure, high flux, and elevated retention of multivalent anion salts and organic molecular size above 300 Da, in addition to relatively low operation and maintenance costs [35]. NF membranes are widely used in industry; this includes treatment of pulp-bleaching effluents from the textile industry [35; 36], separation/demineralization in the dairy industry [37, 38], virus removal and treatment of pharmaceuticals [39], and wastewater treatment [40].

NF membranes being charged partially reject multivalent ions [41, 16]. Different types of integrated membranes were used in the pretreatment of seawater prior to desalination [42, 43] which enhanced the recovery of RO. Using the newly integrated MF/UF+RO+MD, the preliminary experimental results confirmed the possibility of reaching a seawater recovery factor of 87%. Another integrated system (NF+RO+MCr) [32] used a membrane crystallizer (MCr) to achieve the total recovery of desalted water.

The main characteristics of some commercially available NF membranes regarding flux and rejection coefficients of $Ca^{2+}$, $Mg^{2+}$, and $Cl^{-}$ have been also reported [44, 18].

The most important applications of nanofiltration are:

- Removal of monovalent ions from water to certain limit.

- Separation between ions of different valency states.

- Separation of low- and high-molecular weight components

Most NF membranes are thin-film composites of either polymeric or ceramic nature. There is a large variety of polymeric NF membranes, but mainly “cellulose esters”, “aromatic polyamides” (PA) and “polyether sulfones” (PES) are used. “Cellulose esters” as “cellulose acetate” (CA) are appropriate for desalination due to their high permeability [45]. However, the chemical and thermal stabilities of these membranes are quite poor and require frequent cleaning of the membrane modules which is difficult.

Ceramic membranes have good mechanical strength and show thermal and chemical stabilities. Accordingly, they can tolerate high temperature cleaning treatments such as sterilization. The pH stability of alumina membranes is similar to that of polyamides and polyether sulfones [46].

Tertiary treatment of treated wastewater by nanofiltration membrane showed very high organic components removal efficiency (96–99.5%), as indicated by COD (Biochemical Oxygen Demand) removal and the separation of inorganic compounds to be in the range of 25–90% [47].

Many studies conducted on NF system to treat surface water for natural organic matter (NOM) removal [48, 49], humic acids (Has) [50, 51], rejection of disinfection by products (DBPs), and for low turbidity surface water and DOC rejection [52, 53].

Nanofiltration membranes have been investigated in treating brackish water in Australia [54], producing drinking water from brackish groundwater compared to reverse osmosis (RO) and electrodialysis (ED), [38, 55].

Different authors have reported using nanofiltration membranes in the pretreatment of desalination processes; where overall concentration of TDS (Total Dissolved Salts) was reduced by 57.7%. [56].

Integrated membrane systems of MF followed by NF were used to treat color and turbidity of surface water [57, 58, 59]. Also, the studies regarding NF treated agricultural drainage water for reuse purposes [60, 61], and ground water to remove microorganisms, color, pesticides and inorganic matters have been conducted [52, 53, 62].

In advanced treatment for polishing municipal wastewater, the results indicated that NF is appropriate to treat the tertiary effluents to reject dissolved organics. The following case studies for municipal wastewater treatment using NF are reported:

- UF and NF membrane processes were successfully applied in effluent from a municipal wastewater treatment plants to remove micro pollutants as well as natural organic matter (NOM) and COD [63].
- NF and RO membrane systems were found effective for control of pharmaceuticals in secondary treated municipal wastewater, [64].
- Membrane bioreactor (MBR) system in combination with NF membrane for treating municipal wastewater was used to remove nitrogen, phosphorus, pharmaceuticals and personal care products (PPCPs), [65].
- Ultrafiltration-membrane bioreactor (MBR) system in combination with a reverse osmosis (RO) was used to remove COD, nitrogen, phosphorus and other heavy metals [66].
- Adsorption, flocculation and NF system were used to treat biologically treated sewage wastewater in Gwangju (South Korea) [67].
- Wastewater, quality and characteristics of four different nanofiltration (NF) membranes (a molecular weight cut-off (MWCO) in the range of 200- 1000 Da) were investigated for separation of organic constituents in the treated tertiary effluent in order to develop a feasible technology for a safe ground water recharge of the treated effluent [19].

### 2.2.4 Electrodialysis (ED) and electrodialysis reversal (EDR)

Electrodialysis is a voltage driven membrane process which utilizes an electric potential to drive salts selectively through a membrane surface, leaving fresh water behind. ED was originally applied for the production of potable water from brackish water. ED/EDR plants are mainly used for brackish water sources, since energy consumption increases proportionally with salt concentration. ED can handle higher levels of particle pollution than RO plants. Thus, less filtration and pretreatment are needed in ED systems [26]. Recently, ED is used in many industrial applications; e.g. chemical process industry, food industry, pharmaceutical industry and in wastewater treatment.

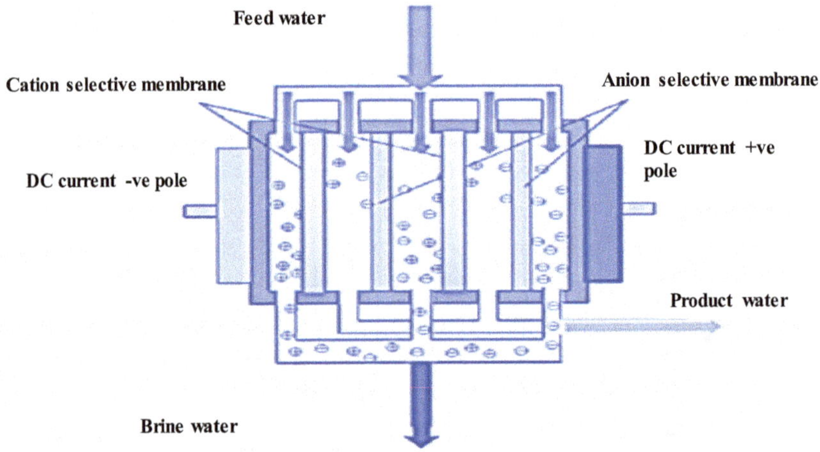


*Fig. 5. Schematic diagram of electrodialysis process.*

Boron removal from water and wastewater using ED has been investigated by several scientists. Kabay et al. studied the removal of boron from water and the effect of water characteristics on separation process [68]. The treatment of boron-containing wastewater by ED was studied by M.Turek et al. [69], as well as by integrated membrane system [70]. Y. Yazicigil studied the removal of boron, by integrated electrodialysis/anion-exchange membranes [71].

Some interesting results were reported for the evaluation of the performance of the ion exchange membranes Electrodialysis of Ni(II) and Co(II) ions showed the removal efficiency of Co(II) and Ni(II) to be 90 and 69%, with initial metal concentrations of 0.84 and 11.72 mg/L, respectively [72]. Electrodialysis treatment of nickel wastewater, electrodialysis and water reuse have been reported and investigated by Tatiane Benvenuti [73]. Kurniawan studied the physico-chemical treatment methods for wastewater laden with heavy metals including electrodialysis process [74].

Removing heavy metals from waste water has been widely investigated. Electrodialysis removal of Cd (II) from wastewater sludge was studied [75]. Modeling-based ED basic process on electrochemistry and copper ion separation experimental investigation have been reported by Mohammadi et al. [76]. Y. Dong studied the combination between microbial desalination cell and electrodialysis system for copper-containing wastewater [77]. ED presents technological advantages for the treatment of heavy metals loaded wastewater, moreover, valuable metals such as Cr and Cu can be recovered [78]. Since ED is a membrane process, it requires clean feed, careful operation, periodic maintenance to prevent any stack damages.

Al-Rawajfeh et al. [79] explored adsorption desalination using low-cost composite synthetic–natural Jordanian materials as a viable option to remove $Cl^-$ ions. It was found that the percentage removal reaches to 25% at 25 min shaking time to 55% at 120 min, in a single batch experiment. The values of $\Delta H^0$, $\Delta S^0$, and $\Delta G^0$ indicate the favorability of physisorption. Zeolite and pozzolana represent potential adsorbents of chloride. Furthermore, two approaches were used to reduce the chlorine corrosion in an electrodeionization unit by Al-Rawajfeh [80]. The first approach was the adsorption of chloride ions on composite materials with layered double hydroxides. A mixed adsorption column containing 15% feldspar, 25% tripoli, 45% pozzolana, 15% Mg–Al layered double hydroxide (LDH) removed about 60% of chloride ion concentration and reduced the chlorine gas release from about 4.8–1.9 ppm. This column composition showed the best percentage removal because it contains the highest percentage of pozzolana. The second approach is using a 60/40 w/w tripoli/LDH filter to protect the positive electrode from the chloride ions. The porosity of the triploi/LDH filter increases with increasing the amount of ammonium carbonate reaching a porosity of about 50% at 5 wt% ammonium carbonate. The filter prevented the chloride from reaching the positive electrode and kept the concentration at the negative electrode without change.

### 2.2.5 Membrane distillation

Membrane distillation (MD) is a thermally driven process, in which only vapor molecules are transported through hydrophobic micro-porous membranes. The driving force is the vapor pressure difference between the hot liquid feed side and the cold permeate side of the membrane [81, 82]. MD systems can be classified into four configurations according to the nature of the cold side of the membrane: Direct contact (DCMD) [83], Air gap (AGMD), Sweeping gas (SGMD), and Vacuum (VMD). Several review papers have provided the fundamentals of MD. Desalination is a major application of MD and the configurations commonly employed are DCMD and AGMD [84, 85].

MD presents some features that make it an attractive option in water treatment and wastewater treatment [86, 87, 88]. It can be used in an integrated system comprising from RO, NF and MD to improve the quality of produce water [89]. MD can be used for wastewater treatment and recovery e.g. Pharmaceutical, textile and metal industries, as well as it can be used in the removal of trace volatile organic compounds from aqueous solutions in chemical and petrochemical applications [87, 90].

### 2.2.6 Capacitive deionization

While studies were not conducted on high-TDS RO concentrate, Capacitive Deionization (CDI) is an innovative technology for removing ionic species from aqueous solutions. This electrochemical process is conducted at ambient conditions and low voltages (1 v) and requires no high-pressure pumps, membranes, distillation columns, or thermal heaters [91].

Avraham used CDI processes with changes in pH near the high-surface-area activated carbon fiber electrodes, in order to remove boron by dissociation of boric acid into borate ions in the CDI cells. Fluoride and nitrate removal from ground water using CDI was studied by Wangwang et al.. Tang et al., use CDI for the removal of copper ions from aqueous solution, and he also examined the disinfection of water using CDI. [92, 93, 94].

### 2.2.7 Forward osmosis

Forward osmosis (FO) is a membrane process used to extract high quality water from low quality feed water source using a high osmotic pressure solution. Forward osmosis process is extensively researched to apply it commercially in different fields, the main reason is the high permeate quality at low energy consumption. FO systems have some complications, when compared to RO systems. In particular, FO process does not produce high quality water for use in a single step as RO systems; but after the forward osmosis step, water is mixed with the osmotic agent [95].

Integrated FO/MBR has been utilized in different applications. Eusebio et al. studied the application of FO-MBR system to utilize RO concentrate as draw solution [96]. Qiu Guanglei, explored hybrid MF-FOMBR configuration to control the salinity and to facilitate the recovery of nutrients [97].

## 3. Recent patents on desalination in water and wastewater treatment

In this section recent patents related to desalination in water and wastewater treatment are presented. We have endeavored to guarantee that all recent technologies are "represented adequately" in this section. Because of the extensive volume of patents issued on this topic, we have restricted the patent time segment to the last three years. We present fundamental data for each patent (such as patent number, inventor names), so it can be easily traced back to the original patent.

### 3.1 Electrodialysis (ED)

Efficient electrodialysis membranes are useful in the desalination of water and waste water. Due to technological advances, membranes are made significantly thinner than prior art commercial electrodialysis membranes. They are used in the desalination of seawater due to their low electrical resistance and high permselectivity. Below are the most recent patents related to technological and engineering advances in electrodialysis systems.

Patent title: Seawater Desalination Process
**Patent number:** US9382135 B2 [98]
Publication date: Jul 5, 2016
**Inventors:** Neil Edwin Moe, John Barber

Patent title: Sea water desalination system
**Patent number:** US9340436 B2 [99]
Publication date: 17 May 2016
**Inventors:** Saroj Kumar Sahu, Francisco E. Torres

Patent title: Ion exchange membranes
**Patent number:** US9403127 B2 [100]
Publication date: 2 Aug 2016
**Inventors:** Juchui Ray Lin, George Y. Gu

Patent title: Electrodialysis
**Patent number:** US20160152492 A1 [101]
Publication date: 2 Jun 2016
**Inventors:** Vinodnarain Bhikhi, Willem Van Baak

Patent title: Anion exchange membrane for electrodialysis applications and process for the preparation thereof
**Patent number:** US20160177043 A1 [102]
Publication date: 23 Jun 2016
**Inventors:** Uma Chatterjee, Suresh Kumar JEWRAJKA, Sreekumaran Thampy

### 3.2 Capacitive Deionization (CDI)

Capacitive deionozation is a recent technology used for desalination of seawater. Below are the most recent patents related to technological advances in CDI systems.

Patent title: Apparatus and method for continuous water desalination and ion separation by flow electrode capacitive deionization
**Patent number:** EP 3 045 431 A1 [103]
Publication date: 20.07.2016
**Inventors:** Wessling, Matthias; Gendel, Youri; Rommerskirchen, Alexandra

Patent title: Single module, flow-electrode apparatus and method for continous water desalination and ion separation by capacitive deionization
**Patent number:** WO2016113139 A1 [104]
Publication date: 20.07.2016
**Inventors:** Matthias Wessling, Youri Gendel, Alexandra Rommerskirchen

Patent title: Desalination and purification system
**Patent number:** US8999132 B2 [105]
Publication date: 7 Apr 2015
**Inventors:** Martin Zdenek Bazant, EthelMae Victoria Dydek, Daosheng Deng, Ali Mani

### 3.3 Forward Osmosis (FO)

A forward osmosis method includes recyclable driving solutes which are easy to be used and recycled, low toxicity and are cost efficient.

Patent title: Forward osmosis using magnetic nanoparticle draw solutions
**Patent number:** US9242213 B1 [106]
Publication date: 26 Jan 2016
**Inventors:** Terrence W. Aylesworth

Patent title: Forward osmosis: recyclable driving solutes
**Patent number:** US9393525 B2 [107]

Publication date: 19 Jul 2016
**Inventors:** Charles D. Moody, Robert L. Riley, John C. Franklin

Patent title: Forward osmosis membrane for seawater desalination and method for preparing the same
**Patent number:** US20150108061 A1 [108]
Publication date: 23 Apr 2015
**Inventors:** Sung Dae Chi, Bong Jun CHA, Jong Hwa Lee, Doo Ri Kim, Su Jeong Lim

Patent title: Osmotic desalination process
**Patent number:** US8753514 B2 [109]
Publication date: 17 Jun 2014
**Inventors:** Robert L. McGinnis

### 3.4 Integrated processes

Different integrated schemes have been propsed to achieve better produced water quality as well as reducing the impact of wastewater disposal.

Patent title: Water processing systems and methods
**Patent number:** US8678080 B2 [110]
Publication date: 25 Mar 2014
**Inventors:** Michael Alvin Curole, Eugene Bruce Greene

Patent title: Multi-use high water recovery process
**Patent number:** US8679347 B2 [111]
Publication date: 25 Mar 2014
**Inventors:** Riad A. Al-Samadi

Patent title: Desalination system and desalination method
**Patent number:** US20140151283 A1 [112]
Publication date: Jun 5, 2014
**Inventors:** Yasunori Sekine, Kazuhiko Noto, Akira Sasaki, Kotaro KITAMURA

## 4. Challenges

Biofouling is a critical issue in membrane water and wastewater treatment as it greatly compromises the efficiency of the treatment processes. It is difficult to control, and significant economic resources have been dedicated to the development of effective biofouling monitoring and control strategies.

Production of fresh water using desalination technologies requires energy availability. However treatment process driven by renewable energy sources (RES) may be a viable solution at remote areas characterized by lack of potable water, lack of an electricity grid and availability of required energy. In recent years there has been intensified R&D effort in this field. Most of them are custom designed for specific locations and utilize solar, wind or geothermal energy to produce fresh water.

New approaches are necessitated to better empathize the current and future value of using water in various ways. Recommended considerations include water reliability, the impacts of water quality, the minimum requirements for human health and the significance of various commercial uses.

## References

[1] F. Salah, R. Oumeddour, Investment and production costs of desalination plants by semi-empirical method, Desalination, 223 (1–3)(2008),457-463

[2] L. F. Greenlee, D. F. Lawler, B. D. Freeman, B. Marrot, and P. Moulin, "Reverse osmosis desalination: Water sources, technology, and today's challenges," Water Res.43(9)(2009) 2317–2348 https://doi.org/10.1016/j.watres.2009.03.010

[3] M. K. Wittholz, B. K. O'Neill, C. B. Colby, and D. Lewis, "Estimating the cost of desalination plants using a cost database," Desalination, 229 (1–3) (2008) 10–20 https://doi.org/10.1016/j.desal.2007.07.023

[4] K. V. Reddy and N. Ghaffour, "Overview of the cost of desalinated water and costing methodologies," Desalination, 205(1–3) (2007) 340–353 https://doi.org/10.1016/j.desal.2006.03.558

[5] A. D. Khawaji, I. K. Kutubkhanah, and J.-M. Wie, "Advances in seawater desalination technologies," Desalination, 221 (1–3) (2008) 47–69 https://doi.org/10.1016/j.desal.2007.01.067

[6] S. Lattemann, M. D. Kennedy, J. C. Schippers, and G. Amy, "Chapter 2 Global Desalination Situation," 2010, 7–39. https://doi.org/10.1201/b10829-4

[7] A. Maurel, Seawater/brackish water desalination and other non-conventional processes for water supply, 2nd edition book, Lavoisier, France (2006).

[8] F. Macedonio, E. Drioli, A. A. Gusev, A. Bardow, R. Semiat, and M. Kurihara, "Efficient technologies for worldwide clean water supply," Chem. Eng. Process. Process Intensif.51 (2012) 2–17 https://doi.org/10.1016/j.cep.2011.09.011

[9] M. K. Chris Binnie, Basic Water Treatment : Edition 5 Title. Royal Society of Chemistry, 2013 https://doi.org/10.1680/bwt.58163

[10] S. Vigneswaran, Waste Water Treatment Technologies - volume I, EOLSS Publications, 2009

[11] M. Kotchen, J. Kallaos, K. Wheeler, C. Wong, and M. Zahller, "Pharmaceuticals in wastewater: Behavior, preferences, and willingness to pay for a disposal program," J. Environ. Manage.90 (3) (2009) 1476–1482 https://doi.org/10.1016/j.jenvman.2008.10.002

[12] T. Wintgens, T. Melin, A. Schäfer, S. Khan, M. Muston, D. Bixio, and C. Thoeye, "The role of membrane processes in municipal wastewater reclamation and reuse," Desalination, 178 (1–3) (2005) 1–11 https://doi.org/10.1016/j.desal.2004.12.014

[13] Y. Yoon, P. Westerhoff, S. A. Snyder, and E. C. Wert, "Nanofiltration and ultrafiltration of endocrine disrupting compounds, pharmaceuticals and personal care products," J. Memb. Sci. 270 (1–2) (2006) 88–100 https://doi.org/10.1016/j.memsci.2005.06.045

[14] Z. A. ALOthman, Inamuddin, and M. Naushad, "Recent Developments in the Synthesis, Characterization and Applications of Zirconium(IV) Based Composite Ion Exchangers," J. Inorg. Organomet. Polym. Mater. 23 (2) (2013) 257–269 https://doi.org/10.1007/s10904-012-9797-2

[15] M. Naushad, Inamuddin, T. A. Rangreez, and Z. A. ALOthman, "A mercury ion selective electrode based on poly-o-toluidine Zr(IV) tungstate composite membrane," J. Electroanal. Chem.713 (2014) 125–130 https://doi.org/10.1016/j.jelechem.2013.12.002

[16] C. Bellona, J. E. Drewes, P. Xu, and G. Amy, "Factors affecting the rejection of organic solutes during NF/RO treatment; a literature review," Water Res. 38 (12) (2004) 2795–2809 https://doi.org/10.1016/j.watres.2004.03.034

[17] Suárez A, Fernández P, Ramón Iglesias J, Iglesias E, Riera F. Cost assessment of membrane processes: A practical example in the dairy wastewater reclamation by reverse osmosis. J. Memb. Sci. (2015) 493, 389-402. https://doi.org/10.1016/j.memsci.2015.04.065

[18] Bazedi, Gh. A., Sh.R. Tewfik, R.S. Ettouney, M.H. Sorour, M.A. El-Rifai, "Prediction of Salt rejection in seawater nanofiltration membrane process," World Appl. Sci. J.17 (1)(2012) 10–19

[19] M. Ernst, "Characterization of the DOC in nanofiltration permeates of a tertiary effluent," Water Res.34 (11), (2000) 2879–2886 https://doi.org/10.1016/S0043-1354(99)00369-3

[20] C. Kazner, T. Wintgens, T. Melin, S. Baghoth, S. Sharma, and G. Amy, "Comparing the effluent organic matter removal of direct NF and powdered activated carbon/NF as high quality pretreatment options for artificial groundwater recharge," Water Sci. Technol.57 (6), (2008) 821 https://doi.org/10.2166/wst.2008.164

[21] L. Wu, Z. Xuan, Z. Meng, Removal of dissolved organic matter in municipal effluent with ozonation, slow sand filtration and nanofiltration as high quality pre-treatment option for artificial groundwater recharge, Chemosphere, 83 (5) (2011) 693–699 https://doi.org/10.1016/j.chemosphere.2011.02.022

[22] L. Watson, I. C.; Morin, O. J.; Henthorne, "Desalting Handbook For Planners (Third Edit.).," 2003

[23] International Energy Agency (IEA), World Energy outlook, 2005.

[24] United States Bureau of Reclamation (USBR), Desalination and Water Purification Technology Roadmap, A report to the executive committee, USA 2003

[25] Ettouney, H.M., El-Dessouky, H.T., Alatiqi, I. Understand thermal desalination. Chem Eng Prog. 95: (1999) 43–54

[26] J. E. MILLER, "Review of Water Resources and Desalination Technologies," Albuquerque, NM, and Livermore, CA, Mar. 2003.

[27] A. E. Al-Rawajfeh, "Hybrid salts precipitation-nanofiltration pretreatment of MSF and RO seawater desalination feed," Membr. Water Treat.3 (4)(2012) 253–266 https://doi.org/10.12989/mwt.2012.3.4.253

[28] A. E. Al-Rawajfeh, "Influence of Seawater Composition on CO2 Release and Scaling in Multi-Stage Flash (MSF) Distillers from Different Arabian Gulf Intakes," Int. J. Chem. Eng. Appl.(2010) 43–48

[29] Membrane Technologies for Water Applications, Highlights from a selection of European Research Projects, EUR 24552 , 2010

[30] A. I. Schafer, L. D. Nghiem, and T. D. Waite, "Removal of the Natural Hormone Estrone from Aqueous Solutions Using Nanofiltration and Reverse Osmosis," Environ. Sci. Technol.37 (1) (2003) 182–188 https://doi.org/10.1021/es0102336

[31] H. Ozaki and H. Li, "Rejection of organic compounds by ultra-low pressure reverse osmosis membrane," Water Res.36 (1) (2002) 123–130 https://doi.org/10.1016/S0043-1354(01)00197-X

[32] N. Hilal, H. Al-Zoubi, N. A. Darwish, A. W. Mohammad, and M. Abu Arabi, "A comprehensive review of nanofiltration membranes:Treatment, pretreatment, modelling, and atomic force microscopy," Desalination, 170 (3) (2004) 281–308 https://doi.org/10.1016/j.desal.2004.01.007

[33] J. Schaep, B. Van der Bruggen, S. Uytterhoeven, R. Croux, C. Vandecasteele, D. Wilms, E. Van Houtte, and F. Vanlerberghe, "Removal of hardness from groundwater by nanofiltration," Desalination, 119 (1–3) (1998) 295–301 https://doi.org/10.1016/S0011-9164(98)00172-6

[34] W. S. Z. Kov'acs, "Characterization of nanofiltration membranes with uncharged solutes," Membrántechnika, 12 (2) (2008) 22–36

[35] R. Maria João, Maria Norberta de Pinho, Nanofiltration of textile industry wastewater using a physicochemical process as a pre-treatment, Desalination, 178 (1-3)(2005) 343-349 https://doi.org/10.1016/j.desal.2004.11.044

[36] J.M Gozalvez-Zafrilla., A. Santafé-Moros, "Nanofiltration modeling based on the extended Nernst Plank equation under different physical modes," in Comsol Confrence, 2008.

[37] B. Balannec, M. Vourch, M. Rabiller-Baudry, and B. Chaufer, "Comparative study of different nanofiltration and reverse osmosis membranes for dairy effluent treatment by dead-end filtration," Sep. Purif. Technol.42 (2)(2005) 195–200 https://doi.org/10.1016/j.seppur.2004.07.013

[38] M. Frappart, O. Akoum, L. Ding, and M. Jaffrin, "Treatment of dairy process waters modelled by diluted milk using dynamic nanofiltration with a rotating disk module," J. Memb. Sci.282 (1–2)(2006) 465–472 https://doi.org/10.1016/j.memsci.2006.06.005

[39] Y. Yoon, P. Westerhoff, S. A. Snyder, E. C. Wert, and J. Yoon, "Removal of endocrine disrupting compounds and pharmaceuticals by nanofiltration and ultrafiltration membranes," Desalination, 202 (1–3)(2007) 16–23 https://doi.org/10.1016/j.desal.2005.12.033

[40] H. Ivnitsky, I. Katz, D. Minz, G. Volvovic, E. Shimoni, E. Kesselman, R. Semiat, and C. Dosoretz, "Bacterial community composition and structure of biofilms developing on nanofiltration membranes applied to wastewater treatment," Water Res.41 (17)(2007) 3924–3935 https://doi.org/10.1016/j.watres.2007.05.021

[41] M. Teixeira, M. Rosa, and M. Nystrom, "The role of membrane charge on nanofiltration performance," J. Memb. Sci.265 (1–2) (2005) 160–166 https://doi.org/10.1016/j.memsci.2005.04.046

[42] E. Curcio and E. Drioli, "Membranes for Desalination," 2009, 41–75

[43] E. Curcio, X. Ji, A. M. Quazi, S. Barghi, G. Di Profio, E. Fontananova, T. Macleod, and E. Drioli, "Hybrid nanofiltration?membrane crystallization system for the treatment of sulfate wastes," J. Memb. Sci.360 (1–2) (2010) 493–498 https://doi.org/10.1016/j.memsci.2010.05.053

[44] M. Turek, P. Dydo, and R. Klimek, "Salt production from coal-mine brine in NF ? evaporation ? crystallization system," Desalination, 221 (1–3) (2008) 238–243 https://doi.org/10.1016/j.desal.2007.01.080

[45] M. Mulder: Basic principles in membrane technology, Kluwer Academic Publishers, 1997.

[46] R. Weber, H. Chmiel, and V. Mavrov, "Characteristics and application of new ceramic nanofiltration membranes," Desalination, 157 (1–3) (2003) 113–125 https://doi.org/10.1016/S0011-9164(03)00390-4

[47] G. E. Üstün, S. K. A. Solmaz, F. Çiner, and H. S. Başkaya, "Tertiary treatment of a secondary effluent by the coupling of coagulation–flocculation–disinfection for irrigation reuse," Desalination, 277(1–3)(2011) 207–212 https://doi.org/10.1016/j.desal.2011.04.032

[48] J. A. Nilson, F. A. DiGiano, "Influence of NOM composition on NF", J. AWWA, 53-66, 1996.

[49] P. Laurent, P. Servais, D. Gatel, G. Randon, P. Bonne and Cavard J., " Microbiological quality before and after NF", J. AWWA 9 (1999) 62-72

[50] S.-H. Yoon, C.-H. Lee, K.-J. Kim, and A. G. Fane, "Effect of calcium ion on the fouling of nanofilter by humic acid in drinking water production," Water Res.32 (7) (1998) 2180–2186 https://doi.org/10.1016/S0043-1354(97)00416-8

[51] S.-H. Yoon and J.-S. Kim, "Effect of the origin of humic acids on nanofilter fouling in drinking water production," Chem. Eng. Sci.55 (21)(2000) 5171–5175 https://doi.org/10.1016/S0009-2509(00)00136-6

[52] K. M. Agbekodo, B. Legube, and S. Dard, "Atrazine and simazine removal mechanisms by nanofiltration: Influence of natural organic matter concentration," Water Res.30 (11)(1996) 2535–2542 https://doi.org/10.1016/S0043-1354(96)00128-5

[53] K. Agbekodo, B. legude, P. Cote, " Organics in NF permeate", J. AWWA,88 (5) (1996) 67-74

[54] A. I. Richards, L. ; Richards, B.S. ; Schäfer, "Renewable energy powered membrane technology: Effect of contaminant speciation on retention and membrane deposition," J. Memb. Sci.(2010) 188–195

[56] H. Al-Zoubi and W. Omar, "Rejection of salt mixtures from high saline by nanofiltration membranes," Korean J. Chem. Eng.26 (3) (2009) 799–805 https://doi.org/10.1007/s11814-009-0133-7

[57] J. Lozier, G. Jones, and W. Bellamy, "Integrated membrane treatment in Alaska", J. AWWA, (1997) 50-64

[58] S. Cheliam, J. Jacangelo, T. Bonacquisti, and B. Schauer, " Effect of pretreatment on surface water NF" J. AWWA, 89 (1997)77-89

[59] C. Robert Reiss, J. S. Taylor, and C. Robert, "Surface water treatment using nanofiltration?pilot testing results and design considerations," Desalination, 125 (1–3) (1999) 97–112 https://doi.org/10.1016/S0011-9164(99)00127-7

[60] M. H. Sorour, A. G. Abulnour, and H. A. Talaat, "Desalination of agricultural drainage water," Desalination, 86 (1) (1992) 63–75 https://doi.org/10.1016/0011-9164(92)80024-4

[61] H. Talaat, M. . Sorour, N. . Rahman, and H. . Shaalan, "Pretreatment of agricultural drainage water (ADW) for large-scale desalination," Desalination, 152 (1–3) (2003) 299–305 https://doi.org/10.1016/S0011-9164(02)01077-9

[62] W. E. Betrand S, Lemaitre I, "Performance of a nanofiltration plant on hard and highly sulphated water during two years of operation," Desalination, 113 (2–3) (1997) 277–281

[63] J. L. Acero, F. J. Benitez, F. Teva, and A. I. Leal, "Retention of emerging micropollutants from UP water and a municipal secondary effluent by ultrafiltration and nanofiltration," Chem. Eng. J.163 (3) (2010) 264–272 https://doi.org/10.1016/j.cej.2010.07.060

[64] K. Kimura, T. Iwase, S. Kita, and Y. Watanabe, "Influence of residual organic macromolecules produced in biological wastewater treatment processes on

removal of pharmaceuticals by NF/RO membranes," Water Res.43 (15) (2009) 3751–3758 https://doi.org/10.1016/j.watres.2009.05.042

[65] K. Chon, H. KyongShon, and J. Cho, "Membrane bioreactor and nanofiltration hybrid system for reclamation of municipal wastewater: Removal of nutrients, organic matter and micropollutants," Bioresour. Technol. 122 (2012) 181–188 https://doi.org/10.1016/j.biortech.2012.04.048

[66] E. Dialynas and E. Diamadopoulos, "Integration of a membrane bioreactor coupled with reverse osmosis for advanced treatment of municipal wastewater," Desalination, 238 (1–3) (2009) 302–311 https://doi.org/10.1016/j.desal.2008.01.046

[67] H. K. Shon, S. Vigneswaran, R. Ben Aim, H. H. Ngo, I. S. Kim, and J. Cho, "Influence of Flocculation and Adsorption as Pretreatment on the Fouling of Ultrafiltration and Nanofiltration Membranes:? Application with Biologically Treated Sewage Effluent," Environ. Sci. Technol.39 (10) (2005) 3864–3871 https://doi.org/10.1021/es040105s

[68] N. Kabay, O. Arar, F. Acar, A. Ghazal, U. Yuksel, and M. Yuksel, "Removal of boron from water by electrodialysis: effect of feed characteristics and interfering ions," Desalination, 223 (1–3) (2008) 63–72 https://doi.org/10.1016/j.desal.2007.01.207

[69] M. Turek, P. Dydo, J. Ciba, J. Trojanowska, J. Kluczka, and B. Palka-Kupczak, "Electrodialytic treatment of boron-containing wastewater with univalent permselective membranes," Desalination, 185 (1–3) (2005) 139–145 https://doi.org/10.1016/j.desal.2005.03.077

[70] M. Turek, J. Was, and P. Dydo, "Brackish water desalination in RO?single pass EDR system," Desalin. Water Treat.7 (1–3) (2009) 263–266 https://doi.org/10.5004/dwt.2009.710

[71] Z. Yazicigil and Y. Oztekin, "Boron removal by electrodialysis with anion-exchange membranes," Desalination, 190 (1–3) (2006) 71–78 https://doi.org/10.1016/j.desal.2005.07.016

[72] N. Tzanetakis, W. . Taama, K. Scott, R. J. . Jachuck, R. . Slade, and J. Varcoe, "Comparative performance of ion exchange membranes for electrodialysis of nickel and cobalt," Sep. Purif. Technol.30 (2) (2003) 113–127 https://doi.org/10.1016/S1383-5866(02)00139-9

[73] T. Benvenuti, M. Antônio Siqueira Rodrigues, A. Moura Bernardes and J. Zoppas Ferreira, Chapter 12: Electrodialysis Treatment of Nickel Wastewater, Electrodialysis and Water Reuse, Topics in Mining, Metallurgy and Materials Engineering, 2014, Part of the series Topics in Mining, Metallurgy and Materials Engineering, Springer Internatiol,pp 133-144

[74] T. A. Kurniawan, G. Y. S. Chan, W.-H. Lo, and S. Babel, "Physico-chemical treatment techniques for wastewater laden with heavy metals," Chem. Eng. J.118 (1–2) (2006) 83–98 https://doi.org/10.1016/j.cej.2006.01.015

[75] M. Jakobsen, "Electrodialytic removal of cadmium from wastewater sludge," J. Hazard. Mater.106 (2–3) (2004) 127–132 https://doi.org/10.1016/j.jhazmat.2003.10.005

[76] T. Mohammadi, A. Moheb, M. Sadrzadeh, and A. Razmi, "Modeling of metal ion removal from wastewater by electrodialysis," Sep. Purif. Technol.41 (1) (2005) 73–82 https://doi.org/10.1016/j.seppur.2004.04.007

[77] Y. Dong, J. Liu, M. Sui, Y. Qu, J. J. Ambuchi, H. Wang, and Y. Feng, "A combined microbial desalination cell and electrodialysis system for copper-containing wastewater treatment and high-salinity-water desalination," J. Hazard. Mater.321 (2017) 307–315 https://doi.org/10.1016/j.jhazmat.2016.08.034

[78] H. A. Aziz, M. N. Adlan, and K. S. Ariffin, "Heavy metals (Cd, Pb, Zn, Ni, Cu and Cr(III)) removal from water in Malaysia: Post treatment by high quality limestone," Bioresour. Technol.99 (6) (2008) 1578–1583 https://doi.org/10.1016/j.biortech.2007.04.007

[79] A. E. Al-Rawajfeh, E. M. Al-Shamaileh, K. Al-Whoosh, A. Al-Ma'abrah, R. Al-Zorqan, R. Zanoon, K. Rawajfeh, and S. Al-Jufout, "Adsorption desalination of chloride ions on composite natural–synthetic materials: An approach for the reduction of chlorine corrosion in electrodeionization units," J. Ind. Eng. Chem.19 (6) (2013) 1895–1902 https://doi.org/10.1016/j.jiec.2013.02.035

[80] A. E. Al-Rawajfeh, "Enhancement of hardness and chloride removal and reduction of Cl2 release and corrosion in electrodeionization units," J. Water Process Eng.5 (2015) 160–165 https://doi.org/10.1016/j.jwpe.2014.06.010

[81] M. . García-Payo, C. . Rivier, I. . Marison, and U. von Stockar, "Separation of binary mixtures by thermostatic sweeping gas membrane distillation," J. Memb. Sci.198 (2) (2002) 197–210 https://doi.org/10.1016/S0376-7388(01)00649-4

[82] M. . Garcia-Payo, M. . Izquierdo-Gil, and C. Fernandez-Pineda, "Air gap membrane distillation of aqueous alcohol solutions," J. Memb. Sci.169 (1) (2000) 61–80 https://doi.org/10.1016/S0376-7388(99)00326-9

[83] G. W. Meindersma, C. M. Guijt, and A. B. de Haan, "Desalination and water recycling by air gap membrane distillation," Desalination, 187 (1–3) (2006) 291–301 https://doi.org/10.1016/j.desal.2005.04.088

[84] H. Chang, G.-B. Wang, Y.-H. Chen, C.-C. Li, and C.-L. Chang, "Modeling and optimization of a solar driven membrane distillation desalination system," Renew. Energy, 35 (12) (2010) 2714–2722 https://doi.org/10.1016/j.renene.2010.04.020

[85] S. T. Hsu, K. T. Cheng, and J. S. Chiou, "Seawater desalination by direct contact membrane distillation," Desalination, 143 (3) (2002) 279–287 https://doi.org/10.1016/S0011-9164(02)00266-7

[86] M. Gryta, M. Tomaszewska, and K. Karakulski, "Wastewater treatment by membrane distillation," Desalination, 198 (1–3) (2006) 67–73 https://doi.org/10.1016/j.desal.2006.09.010

[87] Mohamed Khayet Souhaimi Takeshi Matsuura, Membrane Distillation 1st Edition, View on ScienceDirect Membrane Distillation 1st Edition Principles and Applications. 2011

[88] A. Alkhudhiri, N. Darwish, and N. Hilal, "Membrane distillation: A comprehensive review," Desalination, 287 (2012) 2–18 https://doi.org/10.1016/j.desal.2011.08.027

[89] E. Curcio, A. Criscuoli, and E. Drioli, "Membrane Crystallizers," Ind. Eng. Chem. Res.40 (12) (2001) 2679–2684 https://doi.org/10.1021/ie000906d

[90] M. S. EL-Bourawi, M. Khayet, R. Ma, Z. Ding, Z. Li, and X. Zhang, "Application of vacuum membrane distillation for ammonia removal," J. Memb. Sci.301 (1–2) (2007) 200–209

[91] T. J. Welgemoed and C. F. Schutte, "Capacitive Deionization Technology?: An alternative desalination solution," Desalination, 183 (1–3) (2005) 327–340 https://doi.org/10.1016/j.desal.2005.02.054

[92] E. Avraham, M. Noked, A. Soffer, and D. Aurbach, "The feasibility of boron removal from water by capacitive deionization," Electrochim. Acta, 56 (18) (2011) 6312–6317 https://doi.org/10.1016/j.electacta.2011.05.037

[93] W. Tang, P. Kovalsky, B. Cao, D. He, and T. D. Waite, "Fluoride Removal from Brackish Groundwaters by Constant Current Capacitive Deionization (CDI),"

Environ. Sci. Technol.50 (19) (2016) 10570–10579 https://doi.org/10.1021/acs.est.6b03307

[94] W. Tang, P. Kovalsky, D. He, and T. D. Waite, "Fluoride and nitrate removal from brackish groundwaters by batch-mode capacitive deionization," Water Res.84 (2015) 342–349 https://doi.org/10.1016/j.watres.2015.08.012

[95] L.-C. Shen and N. P. Hankins, "Forward Osmosis for Sustainable Water Treatment," in Emerging Membrane Technology for Sustainable Water Treatment, Elsevier, 2016, pp 55–76. https://doi.org/10.1016/B978-0-444-63312-5.00003-6

[96] R.C. Eusebio, M.A. Promentilla and H.S. Kim, Application of FO-MBR System to Utilize RO Concentrate as Draw Solution, International Conference "Industrial waste and wastewater treatment and valorization, Greece IWWATV 2015

[97] Q. Guanglei, Forward osmosis membrane bioreactor (FOMBR): More than a novel membrane bioreactor process, March 2017, Seminar SBS CR3 (Level 1), Singapore Centre for Environmental Life Sciences Engineering (SCELSE)

[98] N. Edwin Moe, J. Barber, US9382135 B2, (2016)

[99] S. Kumar Sahu, F. E. Torres. US9340436 B2, (2016)

[100] J. Ray Lin, G. Y. Gu. US9403127 B2, (2016)

[101] V. Bhikhi, W. Van Baak, US20160152492 A1, (2016)

[102] U. Chatterjee, S. Kumar JEWRAJKA, S. Thampy. US20160177043 A1, (2016)

[103] M. Wessling, Y. Gendel, A. Rommerskirchen, EP 3 045 431 A1, (2016)

[104] M. Wessling, Y. Gendel, A. Rommerskirchen, WO2016113139 A1, (2016)

[105] M. Zdenek Bazant, E.M Victoria Dydek, D. Deng, A. Mani, US8999132 B2, (2015)

[106] T. W. Aylesworth, US9242213 B1, (2016)

[107] C. D. Moody, R. L. Riley, J. C. Franklin, US9393525 B2, (2016)

[108] S. Dae Chi, B. Jun CHA, J. Hwa Lee, D. Ri Kim, S. Jeong Lim, US20150108061 A1, (2015)

[109] R. L. McGinnis, US8753514 B2, (2014)

[110] M. Alvin Curole, E. Bruce Greene, US8678080 B2, (2014)

[111] Riad A. Al-Samadi, US8679347 B2, (2014)

[112] Y. Sekine, K. Noto, . Sasaki, K. KITAMURA. US20140151283 A1, (2014)

# Chapter 10

# Advanced membrane materials for desalination: carbon nanotube and graphene

Rasel Das*

Leibniz Institute of Surface Modification, Permoserstraße 15, 04318 Leipzig, Germany

rasel.das@iom.leipzig.de

**Abstract**

In this chapter, we discuss the fundamental knowledge underlying each carbon nanotube (CNT) and graphene based desalting membrane systems; their methods of fabrication, and separation performances. Challenges of these novel membrane technologies are highlighted, based on which future research can be undertaken. The CNT membrane shows higher water permeability, but their salt retention capacity is compromised. On the contrary, the nanoporous graphene and graphene oxide (GO) laminate membranes are effective for salts retention, but they have low water permeability. Doping of functionalized CNT and graphene/GO into other supportive materials is important to increase the stability and reusability of membranes.

**Keywords**

Carbon Nanotube, Graphene, Graphene Oxide, Membrane, Desalination

## Contents

## 1. Introduction

Clean and safe water facilities are inevitable for our daily activities such as domestic, agriculture, and industry. World population is increasing at a rate of 80 million per year, putting an additional demand of 64 billion cubic meters of potable water per annum [1, 2]. Around the world, 780 million people cannot access to safe and clean drinking water to use [3]. By 2025 about 3 billion people will likely be forced to survive under water scarcity. These can affect economic growth and human health, since these jeopardize the industrial production and shrink the availability of hygienic foods, and drinks and cause various epidemic diseases such as diarrhea. Deficiency of clean and safe water presently counts almost 3.1% death worldwide. Further, water crisis causes poverty especially in developing countries, which weaken the Millennium Development Goals (MDG) depends on the major improvements in access to water.

Since most of the freshwater resources are polluted at an extreme level, alternative sources of clean water are required to address this growing need. Abundant sea water (~ 98%) of the globe can be a useful stock for water desalination. Over fourteen thousand desalination plants were implemented in 2009 worldwide and 244 plants are under consideration for implementation [4]. The plants could purify 60 million cubic meter of water per day from which 52, 16, 12, 13, 4, 3 and 0.3% are used in the Middle East, North America, Asia, Europe, Africa, Central America, and Australia, respectively. The implanted plants have been prepared by membrane-based reverse osmosis (RO) or thermal-based multi-stage flash (MSF) and multi-effect distillation (MED) processes. Desalination capacities of RO and MSF plants are 20000 and 76000 $m^3\ d^{-1}$, respectively [5]. Electrodialysis has also contributed a small fraction (5.6%) of the desalination capacity.

A major roadblock to expand the uses of current desalination technologies is high-cost and energy requirement. RO, MSF and MED require energies in the scale of 3.6–5.7, 23.9-96, and 26.4-36.7 kWh, respectively. Moreover, additional infrastructure and maintenance costs for these technologies are a must. It was shown that in order to desalinate 1 $m^3$ of seawater and brackish water, it costs US $0.53–$1.50 and the US $0.10–$1.00, respectively [6]. Furthermore, pollutant precipitation reduces the lifetime and modules of the membranes and causes fouling and pore blocking [7]. Additionally, membrane technologies are less robust and incapable of self-cleaning, necessitating the chemical treatments for cleaning and recycling. Since water desalination requires operating at a big scale, its total cost should be minimized by following any of the following strategies: a) decreasing energy requirement, b) minimizing membrane fouling, ultimately increasing membrane lifespan, c) increasing water permeability and d) maximizing salt-rejection efficiency of the membranes.

Although the advent of nanotechnology has given many impressive nanomaterials (NM), special thanks to both carbon nanotube (CNT)- and graphene-based membrane technologies for their remarkable capacities in water desalination [8, 9]. This chapter presents a review of two emerging CNT and graphene-based membrane material concepts intended for use in sea and brackish water purification. We have selected these membrane nanotechnologies because each is touted as a possible low-energy replacement for conventional RO membranes. Our objective is to review the basic concepts underlying each technology, the respective materials and methods of fabrication, separation performance, and current commercialization efforts.

## 2. Carbon nanotube: as a novel membrane

CNT is one of the quintessential and fascinating NMs of the 21$^{st}$ century. Although it was a fortuitous discovery at the beginning [10], Iijima (1991) [11] first defined the material atomically. Three CNT synthesis methods such as chemical vapor deposition (CVD), arc discharge (AD) and laser ablation (LA) are popular [12]. CNT is composed of hexagonal arrayed of carbon atoms (graphite sheet) which rolled up in a cylindrical tube-like shape cum the two ends typically capped by fullerene-like tips [11]. Based on graphitic layers presence, one can categorize the CNT as single-walled carbon nanotube (SWCNT) which consists of a single graphene shell [13], whereas multi-walled carbon nanotube (MWCNT) which is composed of multiple graphite sheets [11]. The length of CNT is several hundred nanometers with the diameters of SWCNT (0.2-2.0 nm) and MWCNT (2-100 nm).

## 3. Carbon nanotube-based membranes for water desalination

Being hydrophobic conduit with smooth molecular surface and controlled pore geometry, CNTs have emerged feasible to novel membrane filtration technology. In general, CNT membrane shows 10 times higher water permeability than the RO membrane [14], but it depends on CNT diameter. Lee et al. have reported [15] that a CNT membrane with inner pore diameter (6 nm) had water permeability 30,000 L m$^{-2}$ h$^{-1}$ bar$^{-1}$, which is extremely high than classical membranes [16]; such filtration technique is not suitable for desalination technologies. CNT needs to have an inner pore diameter of 0.6-1.1 nm in order to pass water through it as well as to reject most of the ions from solution. High water flux of CNT is because of its hydrophobic inner core and smooth surface, which accelerate frictionless transport of water molecules (flow velocity) with little absorption (weak interfacial force) [17]. Recent molecular dynamics (MD) simulation has suggested that the oscillation forces generated by the frictions of CNT longitudinal phonons with confined water molecules enhance the diffusion of water molecules >300% [18].

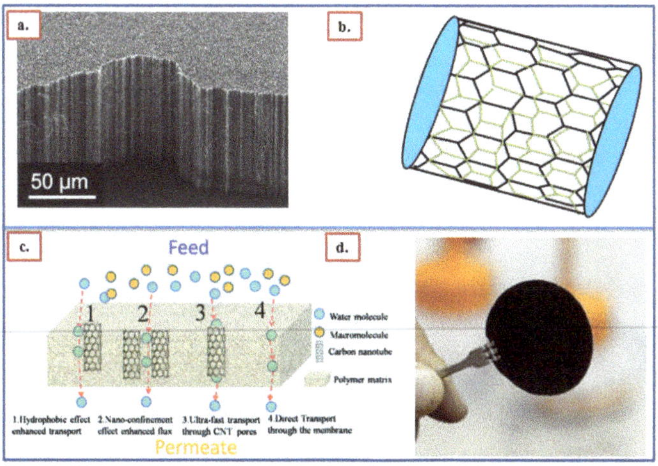


*Fig. 1. Some prototypes of CNT-based membranes. The figures are adapted with permissions from references [19] (Fig. a) and [20](Fig. c).*

The cross-sectional scanning electron microscope (SEM) images of a pristine CNT membrane (a), CNT-based water filter with cylindrical geometry (b), possible pathways for water transportation in a CNT/polymer blend membrane (c) and scattered NaCl nanocrystals on CNT membrane surface (d) are illustrated in Fig. 1.CNT membrane performances often rely on its processing and fabrication methods. Currently, two types of nanotube membranes are available: (i) vertically aligned (VA) and (ii) mixed matrix (MM) or thin-film composite (TFC) membranes. Differences of VACNT and CNT-based MM/TF composite membranes are shown in Fig. 2. The VA-CNT membranes are synthesized by arranging perpendicular CNTs with supportive filler contents between the tubes. The CVD is probably the best method to synthesize VA-CNT membranes. The use of catalysts in CVD method makes uniform CNT membranes of 20–50 nm in diameter and 5–10 μm in length. The fabrication process of VA-CNT membrane is demonstrated in Fig. 3 [21]. These membranes are high molecular sieves with intercalated filler matrix such as polymer between them (Fig. 2 (A). The fillers may be epoxy, silicon nitride and others with no water permeability. The VA-CNT membrane was first introduced by Hinds et al. [22] with polystyrene as filler material between CNTs. The fabrication procedure was simple (Fig. 3) but the pore sizes were irregular.

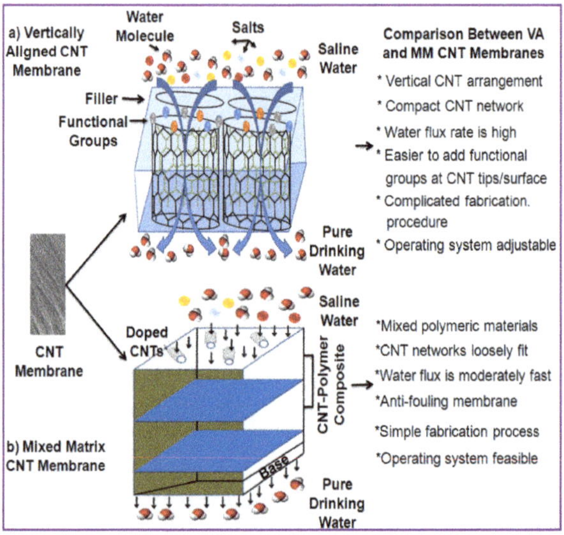


*Fig. 2. Schematic diagram of a VA-CNT (a) and MM-CNT (b) membranes. The figure is adapted with permission from references [16].*

Holt et al. [23] introduced a micro-electromechanical method for synthesizing VA-CNT membranes with stimulated nanofluidic functions. They incorporated silicon nitride ($Si_3N_4$) fillers between the nanotube spaces to inhibit water flow between the nanotube gaps and create stress to stimulate water to flow through the tube. The water flux was increased by 3 folds and enhanced ion selectivity. The membrane allowed to transport $Ru^{2+}(bipyr)_3$ species with sizes 1.3 nm, but blocked 2 nm Au particles, suggesting that their pore sizes were between 1.3 and 2 nm. Improved ions selectivity, high water flux and low energy consumption were advantageous features of these CNT membranes.

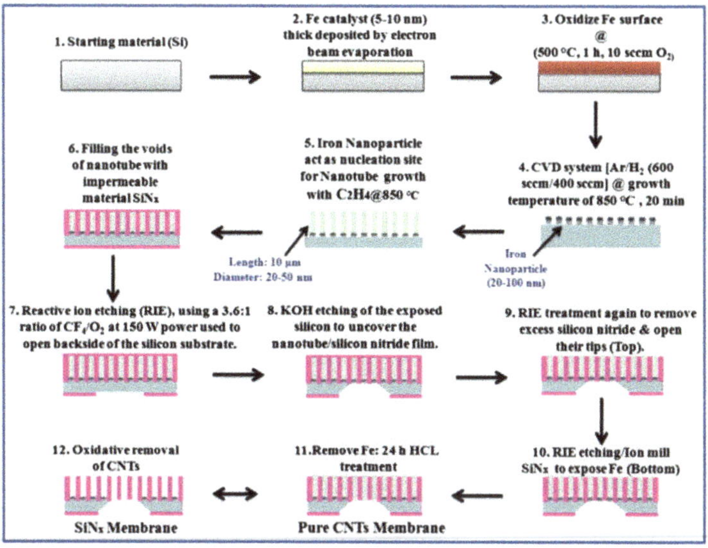


*Fig. 3. Process flow chart for the fabrication of VA-CNT membrane. Redrawn with permission from ref. [21].*

Most of the examined VA-CNT membranes share almost similar pore diameters with ultrafiltration (UF) and microfiltration (MF) techniques. Therefore, VA-CNT membranes can be used as a replacement of UF and MF because of their high water permeability and low fouling effects. Lee and Park [24] fabricated a VA-CNT membrane with diameter 4.1 nm which showed 938 times high water permeability than the UF membrane. However, most of the fabricated VA-CNT membranes have pore diameters >2.0 nm and hence are not useable for desalination (hydrated radius of $Na^+$ ion: 0.716 nm, $Cl^-$ ion: 0.664 nm) of water. Therefore, shrinking CNT pore size in sub-nanometer scale is important for water deionization.

On the other hand, an MM-CNT membrane consists of several layers of polymers or other composite materials (Fig. 2 (b)). Such membrane can be easily fabricated at reduced cost while overcoming the drawbacks including membrane fouling and pollutant precipitation. Compared to MM-CNT, TFC membrane is also popular for high flux and sea water desalination. TFC membrane consists of a thin active layer followed by a porous support layer on a substrate e.g. polymers. The purpose of these CNT membranes is because of keeping "trade-off" between permeability and selectivity and also to decrease membrane fouling, which is caused by the microbes and colloids accumulation. These two drawbacks are considered as major hurdles for commercial membrane technologies. However, this membrane has significantly strengthened the water purification ability of the existing membranes. Both the RO and nanofiltration (NF) based TFC membranes are used for the removal of dissolved organics and water softening.

Functionalization can add positive ($-NH_3^+$), negative ($-COO^-$, sulfonic acids) and hydrophobic (aromatic rings) groups on CNT surfaces as shown in Fig. 4. These make CNT membranes selective for particular pollutant retention and increase water influx through the nanotube hole [2]. Functionalized CNT membranes show good water permeability, mechanical strength, thermal stability, fouling resistance, pollutant degradation and self-cleaning capacity. Tip functionalized CNT membranes have selective functional groups on the nanotube mouth [25] and the core functionalized CNT have functionalities at the sidewall or interior core (Fig. 4).

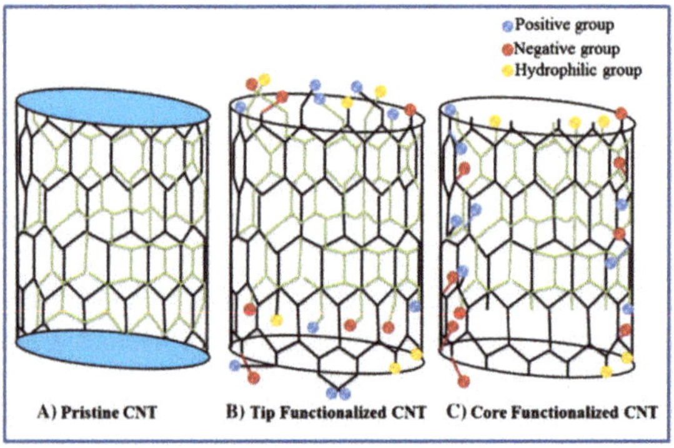


*Fig. 4. Functionalization of CNT membranes. The figure is adapted with permission from references [16].*

Both types demonstrate increase water fluxing and selective rejection of pollutants. Functionalization decreases energy consumption through increased permeability and physical adjustability. Additionally, CNTs are capped into hemisphere like fullerene type curvature during synthesis and purification. These capped CNTs are unzipped into open tips which could be oxidized into specific functional groups to trap desired pollutants. Walther et al. [26] assumed that extreme pressure (>120 bar) is required to pump water at hydrophobic virgin CNT mouths. Therefore, functionalization of CNT vestibules is a must to make the entrance hydrophilic. Fornasiero et al. [27] anchored some negatively charged functional groups (OH, C=O and COOH) at the CNT mouth. These create steric hindrance as well as electrostatic repulsion, and hence the membrane retains $Cl^-$ (45%) and $K^+$ (37%) at0.69 bar, which is comparable with commercial NF membrane (Filmtec NF90). The salts rejection behavior is in the order: $K_3F_3(CN)_6$> $K_2SO_4$> $CaSO_4$>KCl> $CaCl_2$. Similar data are comparable with other studies [28, 29] where almost 100% ion rejection was possible. Functionalized MWCNT TFC membrane shows 20 and 90% enhanced water flux in the sea and brackish water RO, respectively without compromising NaCl rejection (96.10%) [30]. Functionalized MWCNT increases negative surface charges of TFC, so that the organic foulant alginate loosely bound on the TFC surface, slowing down the fouling tendencies. Incorporating NPs into these CNT composite membranes would further help to make the membrane versatile in water purification. For example, doping of $TiO_2$ and Ag NPs into the CNT membrane facilitates the removal of organic [31] and biological pollutants [32], respectively.

## 4. Research gaps and future recommendations

Several research gaps are identified from CNT membrane preparation to application processes. First, current CVD technique is not suitable to synthesize the VA-CNT with >1.1 nm in diameter which is required for salt rejection. Use of SWCNT having sub-nanometer pore diameter is not practical for membrane fabrication because of its high-cost. Although MWCNT is less expensive, it is difficult to synthesize MWCNTs with inner diameters close to 1.1 nm. Implementation of carbon nanorings [33] as a template may be useful for synthesizing the desired sub-nanometer CNT pore architectures for water desalination. Alternatively, CNT tip-ends could be functionalized to reject the salt ions; but they might create a steric-blockage or temporarily pause of water molecules. An unexpected pore blocking and opening phenomena have been observed for SWCNT (>1.0 nm in diameter) in salt solutions of $Na^+$, $Li^+$, $K^+$ and $Cs^+$ [34]. Functionalizing CNT tip-ends with a saturated group (e.g. $-CH_2$-$CH_2$-), applying electric [35] and thermal fields [36] could prevent ion blockage to maintain a fast flow rate with 100% ion rejection. Second, most of the MM-CNT membranes have been prepared by using phase inversion process where CNT is physically attached to polymers, and such system is not stable to withstand in the high-flux separation process. Such loosely bound CNT will leach from the WWTP during operation. Third, since sea and brackish water are mixed with both monovalent and multivalent ions (cations and anions), one needs to check the capability of prepared membrane selectivity to a particular water pollutant in a mixture even for the neutral molecule. Fourth, Lee et al. [15] claimed that CNT has antiadhesion property towards CNT surface which is against the findings of earlier reports suggesting the edges of CNTs impede into the bacterial cell membrane leading to physical death [37-39]. At last, it is worth to mention that most of the CNT membrane data for water desalination rely on MD simulations rather than experimental observations of the membrane performances. Therefore, they are not conclusive and theoretical ion rejection might be insufficient for commercial uses. As a corollary, more study is necessary to clarify these all challenges before commercializing the CNT membranes for water purification.

## 5. Graphene family members

Pristine graphene (PG) is single atom thick $sp^2$ hybridized carbon in a two-dimensional honeycomb lattice and is the fundamental building block of graphite. Boehm et al. [40] first experimentally identified PG in 1962 and a complete characterization was eventually completed by Novoselov et al.[41] in 2004. PG is synthesized by exfoliation of graphite in presence of specific organic solvents under ultrasonications [42]. Based on the presence of a number of layers, PG classifies as single, a few (2-5), multi- (2-10) graphene monolayer. Graphene layers >10, but <100 nm in thickness or lateral dimensions constitute a graphite nanoplate/sheet. Two alternative graphene family members (GFM) are GO and reduced GO (rGO) synthesized by the oxidative exfoliation of graphite [43-46] and reduction of the GO, respectively [46-48]. After the CNT membrane, GFM has shown extraordinary productivity in terms of high water flux and salts retention.

## 6. Graphene-based membrane nanotechnology

While CNT membrane has shown high water permeability, its poor salt selection can be compensated by developing graphene-based membrane system. But the virgin graphene is not permeable to water so that introducing nanopores in single layer graphene or a neat stack of GO sheets (laminate) is a must to create water pathways. Fig. 5 shows the nanoporous graphene (a) and GO (b) membranes. The water molecules permeate through the deliberately created nanopores in the graphene membrane, whereas the interlayer space (oxidized area) and non-oxidized parts of GO laminate are the major routes for water flux. In both cases, hydrated salt molecules are rejected/blocked.

Creating nanochannel in graphene membrane is important for which certain methods such as electron-beam bombardment, helium ion beam drilling and chemical etching introducing nanopores with appropriate sizes for ion sieving and water permeability have been reported [50]. Removing one carbon atom from a 2D compact graphene sheet has been assumed to create a mono-vacancy of pore diameter 0.26 nm [50]. After achieving this, the generic hydrophobicity of graphene generates capillary force for water transport.

Recent evidence suggests that the nanoporous graphene is comparable to RO in terms of salt rejection and water flux. Cohen-Tanugi and Grossman [9] simulated and introduced the nanochannel pores into a graphene monolayer for water desalination. Water flux was about 129 $L.m^{-2}.day^{-1}.MPa^{-1}$, which is several order of magnitudes higher than the RO and NF membranes while maintaining 100% of salt rejection. Later O'Hern et al. [50] experimentally proved the transport system of nanoporous graphene with pore diameter 0.40 nm fabricated by the ion bombardment followed by chemical etching. At short etching time, the created pores were cation selective, but for longer etching time retaining salts and large organic molecules were possible. It opens a window of opportunity where gate-keeper controlled separation is possible at the molecular level. Plasma etching was also found effective to generate nanopores for water desalination (water flux: $10^6$ $g/m^2S$) as proposed by Surwade et al. [51].

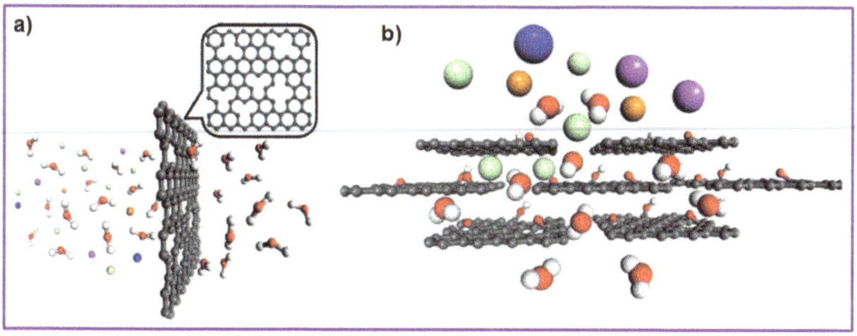


*Fig. 5. Schematic representation of nanoporous graphene (a) and stacked GO layer (laminate) (b) membranes. The symbols in the figure mean: C: black, H: white, O: red, $K^+$: blue, $Na^+$: purple, $Mg^{2+}$: orange, $Cl^-$: cyan). Adapted with permission from RSC 2016 [49].*

Another nanoporous graphene-like single-atom-thick material called "Graphyne", has been used for water desalination [53]. The γ-graphyne is made by reacting between acetylene and phenyl groups which exhibit triangular pores (Fig. 6) [52]. Compared to nanoporous graphene, it is possible to control pore diameter in graphyne membrane by changing the numbers of acetylene bonds (n) among phenyl rings (n=0, graphene). Different types of graphynes (graphynes 1-4) are available, among which graphyne-4 is good for water desalination (Fig. 6). Water molecules are capable of permeating through graphyne when $n \geq 3$. For example, the graphyne 4 shows high water flux 13 $L/cm^2/day/MPa$, which is three orders higher than the state-of-the-art RO membrane

(0.026 13 L/cm$^2$/day/MPa). On the other hand, graphyne (n=3, 4) shows highest salt rejection efficiency (100%) (Fig. 6(c)). This suggests that carbon conjugated network is an excellent tuning factor through which a selected desalination membrane technology could be developed. Such membrane will be helpful where controlled separation is required.

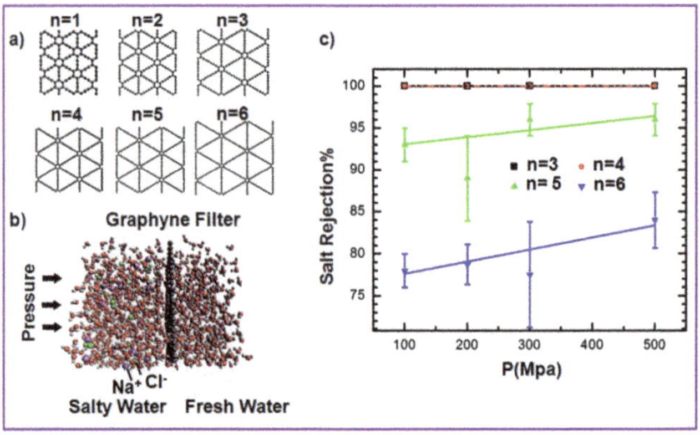


*Fig. 6. Different graphyne types based on "n sheets" with pore sizes of n = 1-6 (a), MD simulation of graphyne-4 (side view) (b), here color indicates: C (grey), O (red), H (white), $Na^+$ (purple), and $Cl^-$ (green); and salt rejection percentage of graphyne-4 as a function of applied pressure (c). Copyright Nature 2013 [52].*

Nair et al. [54] were the first to demonstrate the pathway in GO film through which water transport was possible ($10^{10}$ times faster than helium). After that, many studies have recently been published on using graphene as a membrane for water desalination [9, 51, 55]. GO acts as a freestanding membrane, MM and TFC using methods such as drop-casting, vacuum pressure was driven, spin-coatings, self-assembly and so on [56]. Freestanding GO rejects 100% of salt molecules and simultaneously capable of maintaining high water flux, even double than the RO membrane. The strong H-bonds among GO sheets might endow the sufficient mechanical strength of GO films for a freestanding membrane, but may not be suitable under high operation pressure. Utilization of the interaction forces between the GO membrane functionalities and pollutants (organic and inorganic) is possible to develop a membrane with great selectivity to a particular water pollutant. Sun et al. [55] developed GO membrane, which was impermeable to copper sulfate and RhB because of strong interactions with GO.

However, the slow and high permeability of heavy metals and Na salts were possible, respectively. It is due to the coordination chemistry of metals/salt with GO along with the nanocapillary formed within the lameller structures. But, the functionalities of GO may cause the membrane damage in aqueous solution and also hinder the water passage due to hydrogen bonding. In order to overcome this drawback, some cross-linkers/intercalators such as dopamine [57], dicarboxylic acids [58], polyetheramine [59] have been anchored which help to tune water channels by changing the size, structure, hydrophilicity and mechanical stability of the membrane. Jia and Shi [56] functionalized the GO edges and basal plane by anchoring two crosslinkers such as dicarboxylic and diamine for selective isolation of KCl, NaCl, $MgCl_2$ and $NiCl_2$. It was observed that the fluxes of the single salt solution are governed by the swelling degrees of GO membrane, whereas radii of hydrated cations are the major determinant for mixed salt permeations. The fluxes of the salts were decreased by increasing the hydrophilicity of permeation channels. On the other hand, MWCNT (diameter: 50 nm) has been used as an intercalator of GO membrane [60]. It shows high water permeability (11.3 L/m$^2$h bar) that is two times than the pristine NF membrane, and it simultaneously rejects salts (83.5% $Na_2SO_4$ and 51.4% NaCl) and dyes (>99% direct yellow and >96% methyl orange).

Freestanding GO membranes are not stable and cannot withstand high pressure. They often broke down by gentle finger touch and released from the support upon water rinsing. Such membrane is not suitable for real-world operation. Therefore, reinforced GO with existing polymeric membranes is essential for viable applications. Apart from CNTs, newly emerging GO membranes have also been used to dope into other polymers as MM or TFC membranes for improving water permeability, antifouling, and surface charge properties. Zhu et al. [61] functionalized GO with zwitterion (phosphatidylcholine) and doped it into poly(sulfobetaine methacrylate) (PSBMA) using phase inversion method. GO-PSBMA enhances water permeability from 6.44 to 11.98 L/m$^2$h bar because of high hydrophilicity, lower compactness and more porous macro-voids formations due to GO dispersion. More than 95% of dyes (Reactive black and red) rejection was possible for GO-PSBMA. Doping of negatively charged GO into neutral PSBMA makes the composite membrane surface negative. It increases the hydrophilicity, free from H-bond donors and containing H-bond receptors of composite materials. As a corollary, it has been shown that there is no adsorption of bovine serum albumin (BSA) foulant and maximum flux recovery (>94%) can be obtained from the membrane. Similar observations have also been realized during other studies published by Bano et al. (water flux: 22 L/m$^2$h, salt rejection 80%) [62], Wang et al. (water flux: 24 L/m$^2$h, salt rejection 93%) [63] and Zhang et al. (water flux: 5 L/m$^2$h, metal ions rejection >90%) [64]. In order to achieve the best performances of GO composite membrane, doping NPs such as

$TiO_2$, ZnO, $Fe_2O_3$ and Ag have been well examined for simultaneous filtration and photocatalytic removal of water pollutants [65]. Fig. 7 shows the deposition of $TiO_2$ on GO and polysulfone substrates using LBL approach [66]. The $TiO_2$-GO better photocatalytic degradation of the methylene blue (MB) under UV radiation (about 60–80% faster) compared with polysulfone membranes.

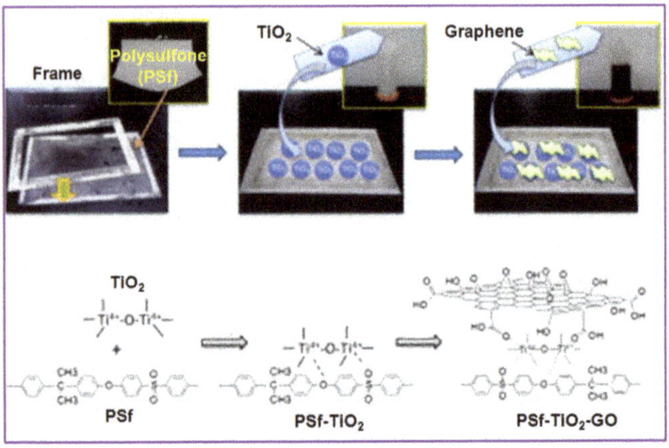


*Fig. 7. Illustration of fabrication procedures of $TiO_2$-GO-PSf composite membrane (top) with their reaction scheme (down). The figure is adapted from Copyright 2014 Elsevier [66].*

Most of the synthesized GO membranes have higher negative charged potential (from -30 to -58 mV) at pH 5-8 than the CNTs (-8.3 mV) at pH 7. Therefore, GO modified TFC membranes would have a trade-off between water permeability and salt rejection. The TFC membranes require a small amount of GO and hence are cost effective for commercial uses. Lai et al. [67] fabricated TFC using GO membrane, and only 0.3 wt% of GO was sufficient to achieve high water permeability of about 400 $L/m^2h$ bar and salt rejections: 95.2 ($Na_2SO_4$), 91.1 ($MgSO_4$), 62.1 ($MgCl_2$) and 59.5 % (NaCl). Another research group has deposited GO as a dual action barrier coating layer to protect the TFC membrane from chlorine and foulant attacks [68]. Although the TFC-GO membrane shows quite similar water flux 14 $L/m^2h$ and salt rejection 96% to polyamide 12 $L/m^2h$ and 97%, GO decreases preferential attraction between hydrophobic BSA by polyamide leading to improve antifouling performance. Meanwhile, polyamide membrane lost around 80% of salt rejection in 4 h of chlorination treatment which is 60% for the GO coated polyamide membrane. Here GO acts as a barrier layer to protect underlying polyamide from chlorine attack, suggesting an important role in preventing the selectivity of deterioration.

## 7. Research gaps and future recommendations

It is believed that a lot of opportunities exist to mature the graphene-based membrane technology. Current MD simulation results are good to experimentally proof the performance of the membrane. The highly porous membrane is necessary to develop high-flux separation technology. But increasing pore density in nanoporous graphene and GO membranes results in a decrease of mechanical stabilities that make the membrane susceptible for destruction. Ironically, designing such a flexible membrane for large scale water purification is a dream. Our suggestion is to use GO membrane as a support for reinforced composite (strong interaction with other polymers) or as an electrode material in CD techniques for water desalination. This will prevent unexpected leaching from the membrane system and require a small amount of material to be used (cost effective). Conventional polymeric membranes are thick in sizes and their operation based on solution-diffusion mechanism. Although scientists assumed that graphene-based membrane follows the similar trend, but this assumption seems not valid for graphene membrane which is typically 250 times thinner than the commercial RO membrane [69].

## Conclusions

In this chapter, we have discussed two most prominent nanotechnology-based membrane technologies, such as CNT and GFM as low-energy replacements for classical RO membranes. The membranes are economically-viable and have dramatic improvement of water desalting abilities with improved stabilities. Functionalized CNT and graphene-based membranes reveal an opportunity for performance enhancements in terms of high water flux and salt retention cum reusability. However, practical fabrications of CNT and graphene-based desalting membrane technologies are undocumented and need to be explored. Currently, there is no evidence of CNT or graphene based desalting membrane commercialization. Therefore, the transformation of technology from lab-bench to field or pilot level needs to be done on an urgent basis which might insulate the people who are suffering from severe water shortage problems, especially in developing countries.

## Abbreviations

CNT, carbon nanotube; GO, graphene oxide; MDG, millennium development goals; RO, reverse osmosis; MSF, multi-stage flash; MED, multi-effect distillation; NM, nanomaterials; CVD, chemical vapor deposition; AD, arc discharge; LA, laser ablation; SWCNT, single-walled carbon nanotube; MWCNT, multi-walled carbon nanotube; MD, molecular dynamics; SEM, scanning electron microscope; VA, vertically aligned; MM, mixed matrix; TFC, thin-film composite; UF, ultrafiltration; MF, microfiltration; PG, pristine graphene; GFM, graphene family members; rGO, reduced GO; PSBMA, poly(sulfobetaine methacrylate); BSA, bovine serum albumin and MB, methylene blue.

## References

[1] WWAP, The United Nations World Water Development Report 4: Managing Water Under Uncertainty and Risk, UNESCO, Paris, 2012.

[2] R. Das, S.B.A. Hamid, M.E. Ali, A.F. Ismail, M. Annuar, S. Ramakrishna, Multifunctional carbon nanotubes in water treatment: The present, past and future, Desalination 354 (2014) 160-179. https://doi.org/10.1016/j.desal.2014.09.032

[3] C. Sealy, Cleaning up water on the nanoscale, Nano Today 8 (2013) 337-338. https://doi.org/10.1016/j.nantod.2013.06.004

[4] T. Humplik, J. Lee, S. O'hern, B. Fellman, M. Baig, S. Hassan, M. Atieh, F. Rahman, T. Laoui, R. Karnik, Nanostructured materials for water desalination, Nanotechnology 22 (2011) 292001. https://doi.org/10.1088/0957-4484/22/29/292001

[5] J.E. Miller, Review of water resources and desalination technologies, Sandia national labs unlimited release report SAND-2003-0800 (2003).

[6] L.F. Greenlee, D.F. Lawler, B.D. Freeman, B. Marrot, P. Moulin, Reverse osmosis desalination: water sources, technology, and today's challenges, Water Res. 43 (2009) 2317-2348. https://doi.org/10.1016/j.watres.2009.03.010

[7] X. Qu, P.J. Alvarez, Q. Li, Applications of nanotechnology in water and wastewater treatment, Water Res. 47 (2013) 3931-3946. https://doi.org/10.1016/j.watres.2012.09.058

[8] W.-F. Chan, H.-y. Chen, A. Surapathi, M.G. Taylor, X. Shao, E. Marand, J.K. Johnson, Zwitterion functionalized carbon nanotube/polyamide nanocomposite membranes for water desalination, ACS nano 7 (2013) 5308-5319. https://doi.org/10.1021/nn4011494

[9] D. Cohen-Tanugi, J.C. Grossman, Water desalination across nanoporous graphene, Nano Lett. 12 (2012) 3602-3608. https://doi.org/10.1021/nl3012853

[10] M. Monthioux, V.L. Kuznetsov, Who should be given the credit for the discovery of carbon nanotubes?, Carbon 44 (2006) 1621-1623. https://doi.org/10.1016/j.carbon.2006.03.019

[11] S. Iijima, Helical microtubules of graphitic carbon, nature 354 (1991) 56-58. https://doi.org/10.1038/354056a0

[12] R. Das, Z. Shahnavaz, M.E. Ali, M.M. Islam, S.B.A. Hamid, Can We Optimize Arc Discharge and Laser Ablation for Well-Controlled Carbon Nanotube Synthesis?, Nanoscale Research Letters 11 (2016) 510. https://doi.org/10.1186/s11671-016-1730-0

[13] S. Iijima, T. Ichihashi, SINGLE-SHELL CARBON NANOTUBES OF 1-NM DIAMETER, Nature 363 (1993) 603-605. https://doi.org/10.1038/363603a0

[14] A. Subramani, J.G. Jacangelo, Emerging desalination technologies for water treatment: A critical review, Water Res. 75 (2015) 164-187. https://doi.org/10.1016/j.watres.2015.02.032

[15] B. Lee, Y. Baek, M. Lee, D.H. Jeong, H.H. Lee, J. Yoon, Y.H. Kim, A carbon nanotube wall membrane for water treatment, Nature communications 6 (2015). https://doi.org/10.1038/ncomms8109

[16] R. Das, M.E. Ali, S.B.A. Hamid, S. Ramakrishna, Z.Z. Chowdhury, Carbon nanotube membranes for water purification: A bright future in water desalination, Desalination 336 (2014) 97-109. https://doi.org/10.1016/j.desal.2013.12.026

[17] W. Sparreboom, A. Van Den Berg, J. Eijkel, Principles and applications of nanofluidic transport, Nature nanotechnology 4 (2009) 713-720. https://doi.org/10.1038/nnano.2009.332

[18] M. Ma, F. Grey, L. Shen, M. Urbakh, S. Wu, J.Z. Liu, Y. Liu, Q. Zheng, Water transport inside carbon nanotubes mediated by phonon-induced oscillating friction, Nature nanotechnology 10 (2015) 692-695. https://doi.org/10.1038/nnano.2015.134

[19] K. Sears, L. Dumée, J. Schütz, M. She, C. Huynh, S. Hawkins, M. Duke, S. Gray, Recent developments in carbon nanotube membranes for water purification and gas separation, Materials 3 (2010) 127-149. https://doi.org/10.3390/ma3010127

[20] K. Gethard, O. Sae-Khow, S. Mitra, Water desalination using carbon-nanotube-enhanced membrane distillation, ACS applied materials & interfaces 3 (2010) 110-114. https://doi.org/10.1021/am100981s

[21] J.K. Holt, A. Noy, T. Huser, D. Eaglesham, O. Bakajin, Fabrication of a Carbon Nanotube-Embedded Silicon Nitride Membrane for Studies of Nanometer-Scale Mass Transport, Nano Lett. 4 (2004) 2245-2250. https://doi.org/10.1021/nl048876h

[22] B.J. Hinds, N. Chopra, T. Rantell, R. Andrews, V. Gavalas, L.G. Bachas, Aligned multiwalled carbon nanotube membranes, Science 303 (2004) 62-65. https://doi.org/10.1126/science.1092048

[23] J.K. Holt, H.G. Park, Y. Wang, M. Stadermann, A.B. Artyukhin, C.P. Grigoropoulos, A. Noy, O. Bakajin, Fast mass transport through sub-2-nanometer carbon nanotubes, Science 312 (2006) 1034-1037. https://doi.org/10.1126/science.1126298

[24] K.-J. Lee, H.-D. Park, The most densified vertically-aligned carbon nanotube membranes and their normalized water permeability and high pressure durability, Journal of Membrane Science 501 (2016) 144-151. https://doi.org/10.1016/j.memsci.2015.12.009

[25] R. Das, S.B.A. Hamid, M. Ali, M. Annuar, E.M.B. Samsudin, S. Bagheri, Covalent Functionalization Schemes for Tailoring Solubility of Multi-Walled Carbon Nanotubes in Water and Acetone Solvents, Science of Advanced Materials 7 (2015) 2726-2737. https://doi.org/10.1166/sam.2015.2694

[26] J.H. Walther, K. Ritos, E.R. Cruz-Chu, C.M. Megaridis, P. Koumoutsakos, Barriers to Superfast Water Transport in Carbon Nanotube Membranes, Nano Lett. 13 (2013) 1910-1914. https://doi.org/10.1021/nl304000k

[27] F. Fornasiero, H.G. Park, J.K. Holt, M. Stadermann, C.P. Grigoropoulos, A. Noy, O. Bakajin, Ion exclusion by sub-2-nm carbon nanotube pores, Proceedings of the National Academy of Sciences 105 (2008) 17250-17255. https://doi.org/10.1073/pnas.0710437105

[28] B. Corry, Water and ion transport through functionalised carbon nanotubes: implications for desalination technology, Energy Environ. Sci. 4 (2011) 751-759. https://doi.org/10.1039/c0ee00481b

[29] W.F. Chan, H.Y. Chen, A. Surapathi, M.G. Taylor, X.H. Hao, E. Marand, J.K. Johnson, Zwitterion Functionalized Carbon Nanotube/Polyamide Nanocomposite Membranes for Water Desalination, Acs Nano 7 (2013) 5308-5319. https://doi.org/10.1021/nn4011494

[30] M. Son, H.-g. Choi, L. Liu, E. Celik, H. Park, H. Choi, Efficacy of carbon nanotube positioning in the polyethersulfone support layer on the performance of thin-film composite membrane for desalination, Chem. Eng. J. 266 (2015) 376-384. https://doi.org/10.1016/j.cej.2014.12.108

[31] A. Moslehyani, A.F. Ismail, M.H.D. Othman, T. Matsuura, Design and performance study of hybrid photocatalytic reactor-PVDF/MWCNT nanocomposite membrane system for treatment of petroleum refinery wastewater, Desalination 363 (2015) 99-111. https://doi.org/10.1016/j.desal.2015.01.044

[32] Ihsanullah, T. Laoui, A.M. Al-Amer, A.B. Khalil, A. Abbas, M. Khraisheh, M.A. Atieh, Novel anti-microbial membrane for desalination pretreatment: A silver nanoparticle-doped carbon nanotube membrane, Desalination 376 (2015) 82-93. https://doi.org/10.1016/j.desal.2015.08.017

[33] H. Omachi, T. Nakayama, E. Takahashi, Y. Segawa, K. Itami, Initiation of carbon nanotube growth by well-defined carbon nanorings, Nature chemistry 5 (2013) 572-576. https://doi.org/10.1038/nchem.1655

[34] C.Y. Lee, W. Choi, J.-H. Han, M.S. Strano, Coherence resonance in a single-walled carbon nanotube ion channel, Science 329 (2010) 1320-1324. https://doi.org/10.1126/science.1193383

[35] J. Liu, G. Shi, P. Guo, J. Yang, H. Fang, Blockage of Water Flow in Carbon Nanotubes by Ions Due to Interactions between Cations and Aromatic Rings, Phys. Rev. Lett. 115 (2015) 164502. https://doi.org/10.1103/PhysRevLett.115.164502

[36] K. Zhao, H. Wu, Fast water thermo-pumping flow across nanotube membranes for desalination, Nano Lett. 15 (2015) 3664-3668. https://doi.org/10.1021/nl504236g

[37] S. Kang, M.S. Mauter, M. Elimelech, Physicochemical determinants of multiwalled carbon nanotube bacterial cytotoxicity, Environ. Sci. Technol. 42 (2008) 7528-7534. https://doi.org/10.1021/es8010173

[38] E. Celik, H. Park, H. Choi, H. Choi, Carbon nanotube blended polyethersulfone membranes for fouling control in water treatment, Water Res. 45 (2011) 274-282. https://doi.org/10.1016/j.watres.2010.07.060

[39] G.S. Ajmani, D. Goodwin, K. Marsh, D.H. Fairbrother, K.J. Schwab, J.G. Jacangelo, H. Huang, Modification of low pressure membranes with carbon nanotube layers for fouling control, Water Res. 46 (2012) 5645-5654. https://doi.org/10.1016/j.watres.2012.07.059

[41] K.S. Novoselov, A.K. Geim, S. Morozov, D. Jiang, Y. Zhang, S.a. Dubonos, I. Grigorieva, A. Firsov, Electric field effect in atomically thin carbon films, science 306 (2004) 666-669. https://doi.org/10.1126/science.1102896

[42] Y. Hernandez, V. Nicolosi, M. Lotya, F.M. Blighe, Z. Sun, S. De, I. McGovern, B. Holland, M. Byrne, Y.K. Gun'Ko, High-yield production of graphene by liquid-phase exfoliation of graphite, Nature nanotechnology 3 (2008) 563-568. https://doi.org/10.1038/nnano.2008.215

[43] W.S. Hummers Jr, R.E. Offeman, Preparation of graphitic oxide, J. Am. Chem. Soc. 80 (1958) 1339-1339. https://doi.org/10.1021/ja01539a017

[44] L. Staudenmaier, Method for the preparation of graphitic acid, Ber Dtsch Chem Ges 31 (1898) 1481-1487. https://doi.org/10.1002/cber.18980310237

[45] B. Brodie, Note sur un nouveau procédé pour la purification et la désagrégation du graphite, Ann. Chim. Phys 45 (1855) 351-353.

[46] D.R. Dreyer, S. Park, C.W. Bielawski, R.S. Ruoff, The chemistry of graphene oxide, Chem. Soc. Rev. 39 (2010) 228-240. https://doi.org/10.1039/B917103G

[47] S. Abdolhosseinzadeh, H. Asgharzadeh, H.S. Kim, Fast and fully-scalable synthesis of reduced graphene oxide, Scientific reports 5 (2015). https://doi.org/10.1038/srep10160

[48] Y. Zhu, S. Murali, W. Cai, X. Li, J.W. Suk, J.R. Potts, R.S. Ruoff, Graphene and graphene oxide: synthesis, properties, and applications, Adv. Mater. 22 (2010) 3906-3924. https://doi.org/10.1002/adma.201001068

[49] Y. You, V. Sahajwalla, M. Yoshimura, R.K. Joshi, Graphene and graphene oxide for desalination, Nanoscale 8 (2016) 117-119. https://doi.org/10.1039/C5NR06154G

[50] S.C. O'Hern, M.S. Boutilier, J.-C. Idrobo, Y. Song, J. Kong, T. Laoui, M. Atieh, R. Karnik, Selective ionic transport through tunable subnanometer pores in single-layer graphene membranes, Nano Lett. 14 (2014) 1234-1241. https://doi.org/10.1021/nl404118f

[51] S.P. Surwade, S.N. Smirnov, I.V. Vlassiouk, R.R. Unocic, G.M. Veith, S. Dai, S.M. Mahurin, Water desalination using nanoporous single-layer graphene, Nature nanotechnology 10 (2015) 459-464. https://doi.org/10.1038/nnano.2015.37

[52] C. Zhu, H. Li, X.C. Zeng, E. Wang, S. Meng, Quantized water transport: Ideal desalination through graphyne-4 membrane, Scientific reports 3 (2013). https://doi.org/10.1038/srep03163

[53] M. Xue, H. Qiu, W. Guo, Exceptionally fast water desalination at complete salt rejection by pristine graphyne monolayers, Nanotechnology 24 (2013) 505720. https://doi.org/10.1088/0957-4484/24/50/505720

[54] R. Nair, H. Wu, P. Jayaram, I. Grigorieva, A. Geim, Unimpeded permeation of water through helium-leak–tight graphene-based membranes, Science 335 (2012) 442-444. https://doi.org/10.1126/science.1211694

[55] P. Sun, M. Zhu, K. Wang, M. Zhong, J. Wei, D. Wu, Z. Xu, H. Zhu, Selective ion penetration of graphene oxide membranes, Acs Nano 7 (2012) 428-437. https://doi.org/10.1021/nn304471w

[56] Z. Jia, W. Shi, Tailoring permeation channels of graphene oxide membranes for precise ion separation, Carbon 101 (2016) 290-295. https://doi.org/10.1016/j.carbon.2016.02.016

[57] Y. Tian, Y. Cao, Y. Wang, W. Yang, J. Feng, Realizing ultrahigh modulus and high strength of macroscopic graphene oxide papers through crosslinking of mussel-inspired polymers, Adv. Mater. 25 (2013) 2980-2983. https://doi.org/10.1002/adma.201300118

[58] Z. Jia, Y. Wang, Covalently crosslinked graphene oxide membranes by esterification reactions for ions separation, Journal of Materials Chemistry A 3 (2015) 4405-4412. https://doi.org/10.1039/C4TA06193D

[59] L. Chen, L. Huang, J. Zhu, Stitching graphene oxide sheets into a membrane at a liquid/liquid interface, Chem. Commun. 50 (2014) 15944-15947. https://doi.org/10.1039/C4CC07558G

[60] Y. Han, Y. Jiang, C. Gao, High-flux graphene oxide nanofiltration membrane intercalated by carbon nanotubes, ACS applied materials & interfaces 7 (2015) 8147-8155. https://doi.org/10.1021/acsami.5b00986

[61] J. Zhu, M. Tian, J. Hou, J. Wang, J. Lin, Y. Zhang, J. Liu, B. Van der Bruggen, Surface zwitterionic functionalized graphene oxide for a novel loose nanofiltration membrane, Journal of Materials Chemistry A 4 (2016) 1980-1990. https://doi.org/10.1039/C5TA08024J

[62] S. Bano, A. Mahmood, S.-J. Kim, K.-H. Lee, Graphene oxide modified polyamide nanofiltration membrane with improved flux and antifouling properties, Journal of Materials Chemistry A 3 (2015) 2065-2071. https://doi.org/10.1039/C4TA03607G

[63] J. Wang, X. Gao, J. Wang, Y. Wei, Z. Li, C. Gao, O-(Carboxymethyl)-chitosan nanofiltration membrane surface functionalized with graphene oxide nanosheets

for enhanced desalting properties, ACS applied materials & interfaces 7 (2015) 4381-4389. https://doi.org/10.1021/am508903g

[64] Y. Zhang, S. Zhang, T.-S. Chung, Nanometric graphene oxide framework membranes with enhanced heavy metal removal via nanofiltration, Environ. Sci. Technol. 49 (2015) 10235-10242. https://doi.org/10.1021/acs.est.5b02086

[65] J.H. Jhaveri, Z. Murthy, A comprehensive review on anti-fouling nanocomposite membranes for pressure driven membrane separation processes, Desalination 379 (2016) 137-154. https://doi.org/10.1016/j.desal.2015.11.009

[66] Y. Gao, M. Hu, B. Mi, Membrane surface modification with TiO2–graphene oxide for enhanced photocatalytic performance, Journal of Membrane Science 455 (2014) 349-356. https://doi.org/10.1016/j.memsci.2014.01.011

[67] G. Lai, W. Lau, P. Goh, A. Ismail, N. Yusof, Y. Tan, Graphene oxide incorporated thin film nanocomposite nanofiltration membrane for enhanced salt removal performance, Desalination 387 (2016) 14-24. https://doi.org/10.1016/j.desal.2016.03.007

[68] W. Choi, J. Choi, J. Bang, J.-H. Lee, Layer-by-layer assembly of graphene oxide nanosheets on polyamide membranes for durable reverse-osmosis applications, ACS applied materials & interfaces 5 (2013) 12510-12519. https://doi.org/10.1021/am403790s

[69] D.-Y. Koh, R.P. Lively, Nanoporous graphene: Membranes at the limit, Nat Nano 10 (2015) 385-386. https://doi.org/10.1038/nnano.2015.77

# Chapter 11

# A review of the progress of desalination technologies: application to wastewater treatment

Lavanya Madhura and Shalini Singh*

*Department of Operations and Quality Management, Durban University of Technology, Durban 4000, South Africa.

shalinis@dut.ac.za

**Abstract**

Due to rapid industrialization, water is one of the key components for the world's economic and social growth. The supply of clean water to the community is a big challenge. This chapter is aimed to discuss the prospects of nanotechnology in solving the water challenges.

**Keywords**

Desalination, Wastewater Treatment, Clean- water

## Contents

## 1. Introduction

Clean water is indispensable to living systems and is an important element in a variety of industries such as food, electronics, and pharmaceuticals, to name a few. The rising demands for clean water in the world is a difficult challenge due to (a) population growth (b) droughts for extended periods (c) strict health-based guidelines, and (d) challenging demands from various consumers. On the planet, only 3 % of available water is fresh, making seawater the alternative source for drinking and industrial purposes in many regions. This spearheaded the research in the development of novel technologies to desalinate the seawater for domestic, industrial and agricultural uses. In the last decade, seawater reverse osmosis (SWRO) has become the most reliable technology in South African coastal areas for the policy makers in water security system. In South Africa, researchers are exploiting the feasibility of SWRO to meet the accumulative demand and relieve the increasing pressure on the present water supply infrastructure. Keeping this in mind, it is necessary for the strategic plan to highlight the advantages and limitations of the desalination process within the national water and energy policy framework. Alternatively, local authorities or organizing bodies will be encouraged to establish SWRO plants in the recognized regions [1].

Throughout the world, desalination technologies have been employed in the last few decades to generate fresh drinking water from the sea and brackish groundwater systems to enhance the quality of the water. On the other hand, these techniques have improved the existing water quality for drinking and industrial use and treated municipal and industrial wastewaters prior to discharge or reuse [2].

A survey of the literature showed that there were nearly 225 desalination plants worldwide in 1950 which produced about 27 million gallons of drinking water per day. Recent studies revealed that about 3,500 desalination plants were established worldwide to produce 3,000 million gallons of drinking water per day. As the demand for freshwater increases and the quality of existing supplies deteriorates, the use of desalination technologies [2,3]. Seawater distillation plants dominated the early desalination market, which was primarily overseas. However, due to the cost factor, recently desalination process has been modified and reinvented as a reverse osmosis (RO), which appears to be more economical than the plants designed for desalination of seawater. For effective desalination of brackish water, a new process known as electrodialysis (ED) has been developed. The ED and RO processes are identical in terms of their working principle, with ED cheaper than the RO system [3].

The selection and application of suitable technology are always based on the composition, site-specific factor and desired quality of the water system which is to be desalinated. On large scale (for example 1million gallons per day) brackish water can be desalinated using well-designed RO or ED plants at an average cost of about $1.50 to $2.50 per 1,000 gallons. In the case of seawater, desalination process carried out at large scale using both RO and ED cost about $4 to $6 per 1,000 gallons. Fig. 1 schematically outlines the key drivers in the development and investment for the water production in the forthcoming years [3].

Literature suggests that there are no recent developments in desalination technologies [4]. However, water industry experts believe that the cost of RO or ED may be decreased by designing the effective and selective membranes, suitable treatment equipment and proper operational protocols [4].

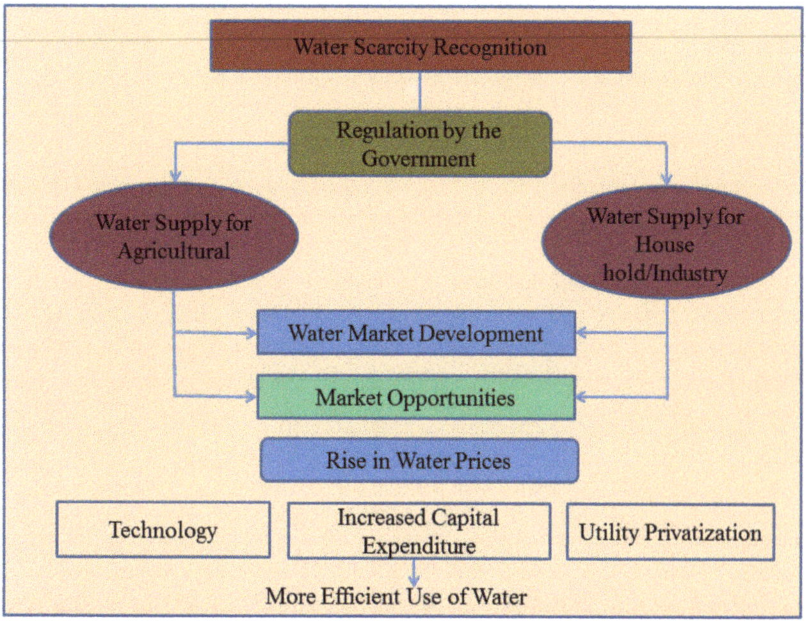


*Fig 1. Key drivers in the development of water production (adapted from [4]).*

**Key issues in desalination [5]:**

1. To enhance the performance of a desalination plant, the pre-treatment of feed water is necessary before entering the RO system. It is used to determine the lifetime of the plant and also reduces membrane replacement and chemical cleaning steps.
2. In order to maintain international water quality standards, the RO process can be performed in two passes. The RO process depends on the salinity of seawater and its temperature. To achieve the European Union standards, one pass is sufficient. However, to achieve the World Health Organization Standards, the second pass is also required, especially to remove the boron (0.5 mg $L^{-1}$, as per WHO boron guidelines) from the water (1 $mgL^{-1}$).
3. Energy consumption is the key operational parameters in desalination plants. Therefore, it is necessary to fix the energy recovery device to monitor the electrical cost as per the local energy and environmental policies.
4. After the pre-treatment step, it is essential to perform the post-treatment step to condition the water after RO to generate suitable water for different applications.
5. At the end, the brine obtained from the desalination plant must be disposed off in a proper manner to avoid an adverse effect on local bio/ecosystem.

## 2. Membrane technology in water quality maintenance

From the global water reserve, only 2.5% is fresh water and the rest is saline. From the 2.5%, the largest part is frozen in polar region and 30% are also in remote aquifers of difficult access. As a result, only 0.007% of the total global water is directly accessible for use. Unfortunately, part of this water is polluted by industrial plants, mining, oil or gas exploration, fertilizer and pesticide residue used in agriculture. In addition, the uneven distribution of water over the globe causes even more severe water scarcity in some regions. Desalination and water reclamation are of paramount importance in water security, where desalination happens to be one of the main life supports in many arid regions.

### 2.1. The desalination process

Desalination plays a vital role in maintaining the water quality in many countries around the world. For example, in Kuwait and Qatar, 100 % domestic and industrial water supply is generated by desalination [6]. The total global capacity (63.7 %) of desalted water is obtained by a membrane process. As per literature reports, the world comprises of 58.9 % seawater, 21.2 % brackish ground water and remaining 19.9 % surface and wastewater [7] which clearly indicates the importance of the membrane technology in desalination market [7].

### 2.1.1 Produced water from oil and gas industries

Oil and gas industries generate large quantities of produced water. Globally, the amount of wastewater co-produced from the oil and gas industries yield about 210 million barrels per day, which is three times greater than the production of oil [8]. The production of wastewater increases in an attempt to recover the oil completely from the matured fields with substantial environmental concerns. One of the most stimulating issues of oil and gas industries is to design and implement the Produced Water Management for reuse and recycling rather than disposing it into the environment. This treatment is required due to toxic nature of the produced water to humans and the environment. The Produced Water contains dissolved organic matter and suspended solid substances. This physiochemical structure makes treating "Produced Water" very difficult. In this case, membrane technology shows a promising platform for the production of quality water for desirable applications [8].

The application of membrane techniques depends on the purpose and reuse of "Produced Water". Ultrafiltration (UF) and microfiltration (MF) or their hybrid (UF-MF) have been effectively adopted for the separation of suspended solid substances and macromolecules. With the combination of nanofiltration (NF) and reverse osmosis (RO), high standards of water quality have been achieved. The water produced by hybrid (NF+RO) technique can be used for drinking and irrigation purposes [9-21]. However, this chapter focuses on the modern membrane techniques for the treatment of high salinity water produced from gas wells which contain total dissolved solids (TSD) in the range from 8000 to 360,000 mg $L^{-1}$[22-24]. Depending on the TDS value recorded, desalination technologies can be used. Wastewater samples with TDS more than 35,000 mg $L^{-1}$ can be treated with commercially available RO membrane technique. However, this membrane is not appropriate due to the high hydraulic pressure which is required to overcome the osmotic pressure of high-salinity Produced Water. The generated hydraulic pressure in this system may exceed the allowable pressure of the membrane and finally make the application unfeasible [22].

Forward osmosis (FO) and membrane distillation (MD) are the emerging techniques which can be effectively used to treat high salinity water with very low energy consumption. On the other hand, MD and FO have a lower tendency for fouling due to the absence of an applied hydraulic pressure. In desalination technology, MD is the only method which uses the separation process at the low-grade heat. Hydrophobic and microporous membranes are employed in the MD to separate the analyte and the filtrate.

The transportation of analytes across the membrane is delayed by the hydrophobic nature of the membrane. The difference in vapor pressure established by the difference in temperature on the both sides of the membrane initiates the movement of water vapor through the pores to the filtrate side [24].

Various physical parameters such as sweeping gas, direct contact, air gap, vacuum and direct contact will stimulate the vapor pressure gradient in MD [25-29]. This is an energy efficient method in which minimum auxiliary energy is used to circulate the solutions. As mentioned earlier, Produced Water is a complex matrix, where dissolved gases and small organic molecules may penetrate through the membrane. Then after this water pollute the filtrate stream or certain molecules such as alcohols or surfactants decrease the sample surface tension and promote the wetting of the membrane. As a result, the sample solution is allowed to pass through the pores and enhances the permeation quality. Pre- and post- treatment steps may be employed to remove the dissolved gases and organic molecules [27, 30] from Produced Water. The separation in FO process takes place due to the difference in osmotic pressure between the sample solution and the concentrated Draw Solution, which has higher osmotic pressure than the sample solution. The Draw Solution is diluted by the water flow across the membrane and lowers the osmotic pressure gradient. Thus, an additional regeneration of the Draw Solution step is required to recover the Draw Solute and accumulate the Produced Water [31,32]. The selection of Draw Solution is very important for FO, especially when dealing with high salinity water and its higher osmatic pressure than the sample solution [32].

Magnetic nanoparticles [33], polyelectrolytes [34] and stimuli-responsive hydrogels [35] are not suitable for FO desalination of high salinity Produced Water because these materials generate very low osmotic pressure. Though the dissolved salts can offer high osmotic pressures, the regeneration step in RO makes the system unsuitable for the treatment of Produced Water [36-38]. The most favorable Draw Solutes are thermolytic salts (most popular ammonia-carbon dioxide), which undergo vaporization due to the change in solution temperature. In this process, distillation due to thermal energy is used for its regeneration [39-41]. In 1964, Neff first identified a Draw Solution for desalination of water [42]. This solution could generate high osmotic pressure (200 atm) and easily regenerated at 60 $^{o}$C [41]. Trimethylamine–carbon dioxide [43] and switchable polarity solvents [44, 45] were reported as efficient candidates besides ammonia–carbon dioxide for the desalination of Produced Water in the FO process.

### 2.1.2 Treatment of desalinated water for agricultural purposes

In the past, desalinated water was not used in the agricultural sector, because water from the direct sources was cheap [46]. However, desalinated water is becoming more competitive because its cost is decreasing, while the cost to treat surface water and groundwater is increasing gradually. It is now clear that the use of groundwater is not sustainable and new regulations are expected to mitigate their use compared to desalinated water. In spite of this evolution, the cost of desalinated water is still too high to be utilized in irrigation. However, it might be affordable for intensive horticulture sector with high-value crops such as vegetables and flowers are grown in greenhouse or coastal areas (where safe disposal of brine is easier than in inland areas) [46, 47]. Literature suggests that the desalinated water of high quality offers less negative effect on soil than the direct use of brackish water [48]. The main benefits of desalinated water for irrigation purpose are (a) supplementary and ecological water supply does not depend on the weather (b) improvement in the yield and (c) worthy agricultural harvest with minimum consumption of water [48, 49]. Moreover quality drinking water is not really necessary for the irrigation. Fractional or half-done desalination can also be used in several cases, particularly for salt-resistance crops [50].

Literature reveals that there is a growing demand for the use of desalinated water in agriculture globally [51]. For example, Spain is focussing on the desalination of water by producing 1.4 million cubic meters per day, out of which 22 % is used for agriculture. In this country, most of the brackish water is fed into the desalination plants to get cleaned and used for agricultural purpose. Interestingly, all the desalination plants in Spain are located in the coastal areas (60 km from the sea) to provide a safe disposal of brine solution. Kuwait is also concentrating on the production of desalinated water. It is found that 13 % of the total desalinated water is used in agriculture sector [51]. Saudi Arabia is the leader in the production of desalinated water, however, only 0.5 % is used in the agricultural sector. Countries like Australia, Chile, and China are also assessing the viability of desalination tools to support water supply for agricultural purposes. Reverse osmosis (RO) is the perfect tool due to its recent advantageous features in terms of quality and cost for the agricultural use of other thermal desalination treatments methods [49]. The salinity levels of the sea (35,000 mg/L) and brackish water (1600mg/L) are equivalent to the osmotic pressure of 2800 kPa and 140 kPa respectively [49]. These data indicate that the higher hydraulic pressure is required when RO system is applied to the seawater. Therefore, there is high energy consumption in this case thus the cost of the process increases drastically. Literature highlighted that the RO process decreases the concentration of calcium and other important salts in the treated water with slightly acidic pH. Water with these features can damage the structure of the soil [48]. Therefore, to retain the desired features, re-mineralization is necessary to adjust the ratios of the minerals in the treated water. The cost of the re-mineralization step is insignificant when compared to the cost of desalination [48]. To produce water with a high mineral content with lower hydraulic pressure can be achieved with nanofiltration for desalination water rather than using RO system. Another major concern with desalinated water is the presence of Boron, which is highly toxic (0.3-4 mg/L) to the various crops [52]. In the desalination process, Boron ions pass across NF [48]. To reduce the concentration, Boron selective exchange resin is used in the desalination plant. This increases the production cost of clean water for agriculture [48].

### 2.1.3 Treatment of desalinated water for mining purposes

In the mining, industry water plays an important role due to its wide applications such as flotation, gold or copper ores leaching, as a coolant and in dust suppression [53]. Mining water have two major concerns: (a) involves a stable supply of water to the mining sites[53], (b) water from mining process contains many explosive residues, acids and thus requires suitable treatment before recycling or disposal [54]. Ground and seawaters (high salinity water) are the main sources for mining industries situated in remote and unfertile areas [53]. The salinity of ground water in mine sites may exceed 100,000 ppm TDS or even higher in Western Australia [53]. These water resources alone or blended with mine water runoff from stockpile, waste or mine dewatering can be directly used for some processing steps in mining such as copper leaching, flotation or cooling [55-58]. However, to alleviate negative impacts on local aquifers, protect the health of miners and to minimize corrosion of underground equipment in underground mines, desalinated water produced from sea water and ground water are preferred[59, 60]. Recent desalination technology suggests that RO is the perfect desalination process to treat ground water in several mining sites to produces clean and fresh water [53].

Generally, the water spent from the mining industry requires a specific proper desalination treatment prior to their disposal into the environment or recycling or reuse. The spent mine water is categorized into two types depending on their capacity to form calcium sulfate scale. Electrodialysis (EDR) and Tubular reverse osmosis (TRO) are two conventional membrane-based desalination methods which were used to produce non-scaling water in early 1990's [54, 61]. In the TRO process; the tubular membrane module was used, due to high suspended solid mining waters were treated effectively [61]. Electric current is used in the EDR to migrate the dissolved salts across the electrodialysis disc which is a multi-layered structure with cationic and anionic membranes. Electrodialysis has several advantages over RO owing that include: (i) low capital cost (ii) sensitive to lower effluent pH and (iii) temperature [62]. In South Africa, about 80% salts and 84% recycled water were achieved by the Beatrix gold mine industry using an EDR pilot plant [63]. In New South Wales vacuum membrane distillation process was employed using a hallow-fiber membrane to separate nearly 99.5% of the total dissolved solids from mining water [64]. Recently, novel desalination techniques such as slurry precipitation recycle reverse osmosis (SPARRO) and seeded reverse osmosis (SRO) have been reported for the treatment of mine scaling water [54, 61, 60, 65, 66]. Calcium sulfate was removed by the SRO process before membrane treatment to reduce the scaling. This removal was based on the calcium sulfate seed crystals suspension fed into the desalination system through recycling of waste slurry. The SRO has a very high rejection rate towards the salt and fair recovery of water. But this system consumes large amounts of high electricity with poor control over calcium sulfate seed formation. Based on research performed during1988-1993, the SPARRO process was developed based on the principle of SRO and Chamber of Mines Research Organization (COMRO) has patented this technique [66]. The schematic diagram of a pilot plant was depicted in Fig. 2. This system has a capability to generate high-quality water with a recovery more than 94%. But the quartzitic materials which are suspended will be prone to fouling and thereby decline the flux high quality.

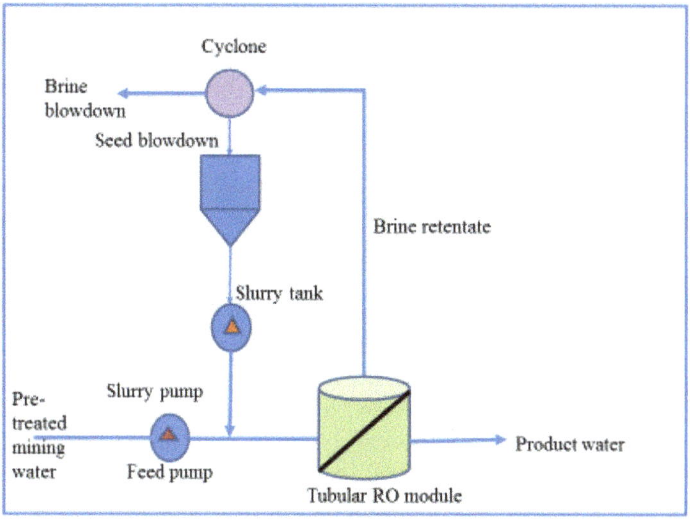


*Fig. 2. The pilot plant of SPARRO (adapted with permission from [66]).*

## 3. Reclamation and reuse of wastewater

### 3.1. Application to municipal wastewater

Sewage is a municipal waste derived from daily activities such as shower, toilet sewage, kitchen sink and laundry. These wastewaters contain a variety of limited chemicals and several biodegradable compounds, which are treated with activated carbon beds integrated with effluent treatment plants. On the other hand membrane bioreactor integrated with either ultrafiltration or microfiltration membranes are also used in routine activated sludge reactor to accomplish the separation of sludge from wastewaters.

In the 1990s, membrane bioreactor process was introduced for the removal of large amounts of biochemical oxygen demand, total organic compounds and to remove total suspended solid [67]. When compared with routine activated sludge, the membrane bioreactor process has secondary benefits of low turbidity and less slit density index, which produces feed water for the RO system. Lozier and co-worker [68] have designed a pilot-scale plant to feed membrane bioreactor effluent water for reverse osmosis [68]. This method is gaining more acceptance from researchers and marketing applications all over the world.

For the first time, the Public Utilities Board of Singapore commissioned a membrane bioreactor in December 2006. This bioreactor delivered useful information on the design and operational conditions of membrane bioreactor system under tropical environment. This is advantageous because of its low energy consumption (0.55 $kwh/m^3$) and did not require any special chemical cleaning since its start-up. Globally, more than 1500 membrane bioreactors were situated in Europe, Japan and North America [70-76].

The waterless hygienic toilet (Nano Membrane Toilet-NMT) has been designed by the Cranfield University with the support of Bill and Melinda Gates Foundation for the Water, sanitation and hygiene program [77, 78]. In this toilet, the human waste is converted into pathogen free water and condensed lumps, which can be used for agriculture and fertilizer respectively. The NMT works without the use of energy and water. In NMT, a bundle of hollow fiber membranes is adapted to separate loosely bounded water (especially from urine) using a membrane distillation mechanism as shown in Fig. 3. In this process, the sweep gas is pumped from the permeate side of the membrane to generate the vapor pressure difference between the feed and permeate which allows water to penetrate through the membrane in the form of vapor [78]. These hallow-fibers have a good capacity to reject pathogens and nano-coated beads are employed as a condenser to collect 90 % of permeates. The volatile organic compounds present in the condensed water cannot be stored. Alternatively, this water can be used for domestic purposes in the homes [77].

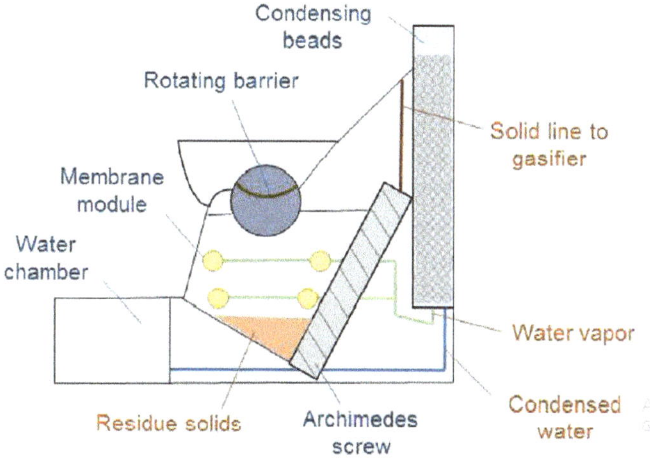


*Fig. 3. The schematic diagram of nanomembrane toilet (reproduced with permission from [78]).*

### 3.2 Application to industrial wastewater

Industrial wastewater is generated from various industries including chemical, mining food, pulp and paper, iron & steel and pharmaceuticals. The membrane bioreactor is a perfect technique to treat most of the industrial wastewaters [76, 79-83], similar to the municipal wastewaters. Due to the spontaneous expansion of industrialization and urbanization or from specific industries, wastewater contains hazardous chemicals like trace organic contaminants and heavy metals. Pharmaceutically active compounds, endocrine disrupting chemicals, pesticides and other persistent organic compounds fall under trace organic contaminants [84]. Trace organic contaminants are biologically active compounds. Therefore the presence of toxic molecules at low concentrations is highly toxic and harmful to humans. Generally, in raw wastewater, the trace organic contaminants range from100 μg/L to 100 ng/L. Different membrane technologies such as membrane bioreactor, single membrane processes (nanofiltration, reverse osmosis and forward osmosis), dual membrane processes (ultrafiltration-reverse osmosis, forward osmosis reverse osmosis) and integrated process (membrane bioreactor/bioreactor coupled to nanofiltration or reverse osmosis or forwarded osmosis) have been used effectively for the treatment of trace organic contaminants [85].The processes mentioned above have characteristic merits and demerits in respect of the removal of trace organic contaminants, therefore researchers are focussing on developing selective and effective methodologies in this field. Literature suggests that the membrane bioreactor completely removed some of the trace organic contaminants [85]. The removal process of these contaminants is administrated by the adsorption on the sludge engaged by the membrane and thereby successive degradation in the reactor by the biomass. Their removal mechanism is governed by the adsorption of the trace organic contaminants on the sludge retained by a membrane and subsequent degradation by the biomass in the reactor. Due to this, removal of trace organic contaminants mainly depends on the inherent biodegradability and physiochemical properties such as molecular weight, hydrophobicity, electron donating/withdrawing groups and functional groups [86-90]. The parameters such as temperature [91-93], pH [94-97], hydraulic retention time [98-105], the retention time of sludge and concentration of dissolved oxygen have various effects depending on the physiochemical properties of trace organic contaminants. Sorption diffusion mechanism [84] or charge repulsion and size exclusion are useful in the separation of trace organic contaminants in nanofiltration, reverse osmosis and forward osmosis processes [106-123]. It is observed that the trace organic contaminants cannot undergo degradation process instead. Thus increase the concentration and therefore additional treatment is necessary to recover or nullify them. Reports [125-17] show that the dual membrane process is used to treat trace organic contaminants, in which one membrane process acted as a supplement to the other membrane. For example, UF was used to decrease the concentration of foulant in the RO process [124] or RO is used to collect the water from the forwarded process [125-127]. Using the single membrane process, it is very difficult to remove a wide range of trace organic contaminants. Therefore, hybrid processes have been considered highly efficient in the removal of trace organic contaminants more than 90% [128-135]. For example, membrane bioreactor integrated with NF or RO enhanced the removal capability of trace organic contaminants of hydrophobic and hydrophilic nature. The adsorbed trace organic contaminants are successively pass through NF or RO membranes and retain into the membrane bioreactor. Nanofiltration or reverse osmosis can be used for the removal of hydrophobic trace organic contaminants efficiently due to their low interaction behavior with the sludge. Additionally, the membrane bioreactor may reduce the fouling challenging of NF and RO because of their holding capacity towards large organic or colloidal compounds. Nguyen and co-workers [133, 170] have designed a model that can be used to remove 22 different trace organic contaminants with a recovery range of 90-100% from synthetic wastewater. Likewise, membrane bioreactor integrated with RO was adopted to remove 11 trace organic contaminants from wastewater [134,135]. The hybrid membrane bioreactor-forwarded osmosis has generated promising results [136-

141] with a less fouling tendency. In the forwarded osmosis process, high retention time enhances the biodegradation due to higher holding ability by the membrane for small and persistent trace organic contaminants. Alturki and co-workers [141] have achieved higher removal percentage of 25 out of 50 trace organic contaminants using hybrid membrane bioreactor-forwarded osmosis. Whereas, Lay and co-workers have also achieved high quality water after removing pharmaceutical pollutants from the wastewater [141]. Recently, Phattaranawik and co-workers [142] have designed a hybrid of bioreactor and membrane distillation system. In this system, initially, organic matter (carbohydrates or proteins) undergo digestion by the bioreactor before they reach to the distillation membrane pores and depreciate its performance [143]. The thermophilic microorganism is used in the bioreactor to enhance the biodegradation of organic matter, as the membrane distillation system was operated in the temperature range from 30 to 80 $^{o}$C [144]. Wijekoon and his team [145] for the first time have designed a salinity component in the reactor and found that the higher removal capacity (95 %) was achieved for 25 trace organic contaminants.

## 4. Conclusion

In desalination of seawater, membranes have been proved to be a powerful tool for several decades to generate sustainable water quality using reverse osmosis. The designing of novel nanomaterials and membranes with good specificity and selectivity are required to achieve high flux and salt rejection capacity. On the other hand, attention should be focussed more on membranes which can provide functionalized transport channels for small molecules. The membranes with these features can achieve a water recovery of more than 95 % from industrial and municipal wastewaters. The partially desalinated water obtained from the membranes could be used for agricultural lands and mining industries when they are situated in isolated regions. In future industrial effluent treatment can offer ample opportunities for membrane technology. Membranes with good thermal and chemical properties, especially those which can resist strong chemicals during the treatment enhance the lifetime of the membrane while separating the toxic organic and inorganic molecules and Produced Water. In summary, there are a large number of open challenges and opportunities for new materials in membrane technology. With the advent of better materials in this area, new sustainable technologies and separation processes will be implemented, a task for the current and new generation of material scientists.

**Acknowledgements**

Authors are highly grateful to Durban University of Technology and National Research Foundation of South Africa for their financial support in the form of NRF-DST Innovation Doctoral Scholarship (UID: 107643) to carry out this work.

**References**

[1] [G. O. Schreiner, R. C. van Ballegooyen and W. Osman, Seawater desalination as an option to alleviate water scarcity in South Africa: the need for a strategic approach to planning and environmental decision-making, J. Water Reuse Desal. 4 (2014) 287-293. https://doi.org/10.2166/wrd.2014.035

[2] [G. M. Rios, Nanotechnologies and membrane engineering at the core of water and related questions in the Mediterranean and Middle-East countries, Options Méditerranéennes, A n° 88, 2009 - Technological Perspectives for Rational Use of Water Resources in the Mediterranean Region.

[3] [Using Desalination Technologies for Water Treatment, Background paper, March 1988,NTIS order #PB88-193354.

[4] [L. Wagner, "Water desalination – Tap into the liquid Gold", Research report, MORA ASSOCIATES, December 2007.

[5] [http://www.solaqua.com/solstilbas.html

[6] [N. Ghaffour, The challenge of capacity-building strategies and perspectives for desalination for sustainable water use in MENA, Desalin. Water Treat. 5 (2009) 48–53. https://doi.org/10.5004/dwt.2009.564

[7] [G.W.I.G.I. DesalData, Market profile and desalination markets, 2009–2012 yearbooks and GWI website, http://www.desaldata.com/.

[8] [Z. Khatib, P. Verbeek,Water to value-produced water management for sustainable field development of mature and green fields, SPE International Conference on Health Safety and Environment in Oil and Gas Exploration and Production, Society of Petroleum Engineers, 2002. https://doi.org/10.2118/73853-MS

[9] [M. Çakmakce, N. Kayaalp, I. Koyuncu, Desalination of produced water from oil production fields by membrane processes, Desalination 222 (2008) 176–186. https://doi.org/10.1016/j.desal.2007.01.147

[10] [C.H. Webb, L. Nagghappan, G. Smart, J. Hoblitzell, R. Franks, Desalination of oilfieldproducedwater at the San Ardo water reclamation facility, CA, SPE Western Regional Meeting, Society of Petroleum Engineers, 2009.

[11] [Y. He, Z.W. Jiang, Technology review: treating oilfield wastewater, Filtr. Sep. 45 (2008) 14–16. https://doi.org/10.1016/S0015-1882 (08)70174-5

[12] [N. Liu, J. Lu, L. Li, R. Lee, Factors determining the reverse osmosis performance of zeolite membranes on produced water purification, SPE International Symposium on Oilfield Chemistry, Houston, Texas, USA, 2007. https://doi.org/10.2118/106168-MS

[13] [A. Zaidi, K. Simms, S. Kok, The use of micro/ultrafiltration for the removal of oil and suspended solids from oilfield brines, Water Sci. Technol. 25 (1992) 163–176.

[14] [P. Xu, J.E. Drewes, D. Heil, Beneficial use of co-produced water through membrane treatment: technical-economic assessment, Desalination 225 (2008) 139–155. https://doi.org/10.1016/j.desal.2007.04.093

[15] [D.B. Burnett, Potential for beneficial use of oil and gas produced water, Global Petroleum Institute, Texas Water Resources Institute, 2004 1–11.

[16] [A. Chen, J. Flynn, R. Cook, A. Casaday, Removal of oil, grease, and suspended solids from produced water with ceramic crossflow microfiltration, SPE Prod. Eng. 6 (1991) 131–136. https://doi.org/10.2118/20291-PA

[17] [K.S. Ashaghi, M. Ebrahimi, P. Czermak, Ceramic ultra-and nanofiltration membranes for oilfield produced water treatment: a mini review, Open Environ. J. 1 (2007) 1–8. https://doi.org/10.2174/1874233500701010001

[18] [M. Bader, Seawater versus produced water in oil-fields water injection operations, Desalination 208 (2007) 159–168. https://doi.org/10.1016/j.desal.2006.05.024

[19] [J.M. Lee, T.C. Frankiewicz, Treatment of produced water with an ultrafiltration (UF) membrane-a field trial, SPE Annual Technical Conference and Exhibition, Society of Petroleum Engineers, 2005. https://doi.org/10.2118/95735-MS

[20] [T. Bilstad, E. Espedal, Membrane separation of produced water, Water Sci. Technol. 34 (1996) 239–246. https://doi.org/10.1016/S0273-1223(96)00810-4

[21] [Y.S. Li, L. Yan, C.B. Xiang, L.J. Hong, Treatment of oily wastewater by organic–inorganic composite tubular ultrafiltration (UF) membranes, Desalination 196 (2006) 76–83. https://doi.org/10.1016/j.desal.2005.11.021

[22] [K.L. Hickenbottom, N.T. Hancock, N.R. Hutchings, E.W. Appleton, E.G. Beaudry, P. Xu, T.Y. Cath, Forward osmosis treatment of drilling mud and fracturing wastewater from oil and gas operations, Desalination 312 (2013) 60–66. https://doi.org/10.1016/j.desal.2012.05.037

[23] [D.L. Shaffer, L.H. Arias Chavez, M. Ben-Sasson, S. Romero-Vargas Castrillón, N.Y. Yip,M. Elimelech, Desalination and reuse of high-salinity shale gas produced water: drivers, technologies, and future directions, Environ. Sci. Technol. 47 (2013) 9569–9583. https://doi.org/10.1021/es401966e

[24] [G. Breit, J. Otton, Produced waters database, United States Geological Survey, Available:http://energy. cr. usgs. gov/prov/prodwat/data2. html (access on 2002, June 12).

[25] [A. Alklaibi, N. Lior, Membrane-distillation desalination: status and potential, Desalination 171 (2005) 111–131. https://doi.org/10.1016/j.desal.2004.03.024

[26] [E. Curcio, E. Drioli, Membrane distillation and related operations—a review, Sep. Purif. Rev. 34 (2005) 35–86. https://doi.org/10.1081/SPM-200054951

[27] [M.K. Souhaimi, T. Matsuura, Membrane Distillation: Principles and Applications, Elsevier, 2011.

[28] [A. Alkhudhiri, N. Darwish, N. Hilal, Membrane distillation: a comprehensive review, Desalination 287 (2012) 2–18. https://doi.org/10.1016/j.desal.2011.08.027

[29] [P. Wang, T.S. Chung, Recent advances in membrane distillation processes: membrane development, configuration design and application exploring, J. Membr. Sci. 474 (2015) 39–56. https://doi.org/10.1016/j.memsci.2014.09.016

[30] [S. Al-Obaidani, E. Curcio, F. Macedonio, G. Di Profio, H. Al-Hinai, E. Drioli, Potential of membrane distillation in seawater desalination: thermal efficiency, sensitivity study and cost estimation, J. Membr. Sci. 323 (2008) 85–98. https://doi.org/10.1016/j.memsci.2008.06.006

[31] [T.Y. Cath, A.E. Childress, M. Elimelech, Forward osmosis: principles, applications, and recent developments, J. Membr. Sci. 281 (2006) 70–87. https://doi.org/10.1016/j.memsci.2006.05.048

[32] [Q. Ge, M. Ling, T.S. Chung, Draw solutions for forward osmosis processes: developments, challenges, and prospects for the future, J.Membr. Sci. 442 (2013) 225–237. https://doi.org/10.1016/j.memsci.2013.03.046

[33] [M. MingáLing, Facile synthesis of thermo-sensitive magnetic nanoparticles as "smart" draw solutes in forward osmosis, Chem. Commun. 47 (2011) 10788–10790. https://doi.org/10.1039/c1cc13944d

[34] [Q. Ge, J. Su, G.L. Amy, T.S. Chung, Exploration of polyelectrolytes as draw solutes in forward osmosis processes, Water Res. 46 (2012) 1318–1326. https://doi.org/10.1016/j.watres.2011.12.043

[35] [D. Li, X. Zhang, G.P. Simon, H.Wang, Forward osmosis desalination using polymer hydrogels as a draw agent: influence of draw agent, feed solution and membrane on process performance, Water Res. 47 (2013) 209–215. https://doi.org/10.1016/j.watres.2012.09.049

[36] [D.L. Shaffer, N.Y. Yip, J. Gilron, M. Elimelech, Seawater desalination for agriculture by integrated forward and reverse osmosis: improved product water quality for potentially less energy, J. Membr. Sci. 415 (2012) 1–8. https://doi.org/10.1016/j.memsci.2012.05.016

[37] [R.W. Holloway, A.E. Childress, K.E. Dennett, T.Y. Cath, Forward osmosis for concentration of anaerobic digester centrate, Water Res. 41 (2007) 4005–4014. https://doi.org/10.1016/j.watres.2007.05.054

[38] [C.R. Martinetti, A.E. Childress, T.Y. Cath, High recovery of concentrated RO brines using forward osmosis and membrane distillation, J. Membr. Sci. 331 (2009) 31–39. https://doi.org/10.1016/j.memsci.2009.01.003

[39] [M. Elimelech, W.A. Phillip, The future of seawater desalination: energy, technology, and the environment, Science 333 (2011) 712–717. https://doi.org/10.1126/science.1200488

[40] [Y.C. Kim, S.-J. Park, Experimental study of a 4040 spiral-wound forward-osmosis membrane module, Environ. Sci. Technol. 45 (2011) 7737–7745. https://doi.org/10.1021/es202175m

[41] [J.R. McCutcheon, R.L. McGinnis, M. Elimelech, A novel ammonia–carbon dioxide forward (direct) osmosis desalination process, Desalination 174 (2005) 1–11. https://doi.org/10.1016/j.desal.2004.11.002

[42] [R.A. Neff, Solvent extractor, Google Patents, 1964.

[43] [C. Boo, Y.F. Khalil, M. Elimelech, Performance evaluation of trimethylamine–carbon dioxidethermolytic draw solution for engineered osmosis, J. Membr. Sci. 473 (2015) 302–309. https://doi.org/10.1016/j.memsci.2014.09.026

[44] [A.D. Wilson, F.F. Stewart, M.L. Stone, Methods and systems for treating liquids using switchable solvents, Google Patents, 2012.

[45] [M.L. Stone, C. Rae, F.F. Stewart, A.D. Wilson, Switchable polarity solvents as draw solutes for forward osmosis, Desalination 312 (2013) 124–129. https://doi.org/10.1016/j.desal.2012.07.034

[46] [J.M. Beltrán, S. Koo-Oshima, Water desalination for agricultural applications, FAO Land and water discussion paper, 5 2006, p. 48.

[47] [O. Barron, R. Ali, G. Hodgson, D. Smith, E. Qureshi, D. McFarlane, E. Campos, D. Zarzo, Feasibility assessment of desalination application in Australian traditional agriculture, Desalination 364 (2014) 33–45. https://doi.org/10.1016/j.desal.2014.07.024

[48] [D. Zarzo, E. Campos, P. Terrero, Spanish experience in desalination for agriculture, Desalin. Water Treat. 51 (2013) 53–66. https://doi.org/10.1080/19443994.2012.708155

[49] [S. Burn, M. Hoang, D. Zarzo, F. Olewniak, E. Campos, B. Bolto, O. Barron, Desalination techniques—a review of the opportunities for desalination in agriculture, Desalination 364 (2015) 2–16. https://doi.org/10.1016/j.desal.2015.01.041

[50] [M. Tester, P. Langridge, Breeding technologies to increase crop production in a changing world, Science 327 (2010) 818–822. https://doi.org/10.1126/science.1183700

[51] [Desaldata, http://www.idadesal.org/publications/desaldata-com/2012.

[52] [M.J.R. Guerreiro, M.d.M. de la Fuente, E.M. Camacho, Toxicidad del boroen las plantas, Encuentros Biol. 82 (2002) 1.

[53] [R. Dunne, Water everywhere and not a drop to drink, nor do i know its whereabouts, Water in Mineral Processing2012.

[54] [G. Juby,Membrane desalination of service water from goldmines, J. South Afr. Inst. Min. Metall. 92 (1992) 69.

[55] [L. Haig-Smillie, Sea water flotation, Proceedings of the 6th Annual Meeting of Canadian Mineral Processors 1974, pp. 263–281.

[56] [G. Poling, D. Ellis, Importance of geochemistry: the black angel lead-zinc mine, Greenland, Mar. Georesour. Geotechnol. 13 (1995) 101–118. https://doi.org/10.1080/10641199509388280

[57] [J. Weirtz, When best water use efficiency is not enough, what can the mining industry do? Water in Mining 2009 Conference Proceedings, Australasian Institute of Mining and Metallurgy Perth, Carlton (Australia) 2009, pp. 15–17.

[58] [J. Chadwick, Copper recovery, Min. Mag. (2009) 23–30.

[59] [V. Lévy, R. Fabre, B. Boebel, C. Hertle, Water use in the mining industry—threats and opportunities, Proceedings of Water in Mining, 2006.

[60] [R. Harries, A field trial of seeded reverse osmosis for the desalination of a scaling type mine water, Desalination 56 (1985) 227–236. https://doi.org/10.1016/0011-9164(85)85027-X

[61] [G.Du. Plessis, J. Swartz, Tubular reverse osmosis treatment of Secunda mine water: a pilot plant investigation, Water Sci. Technol. 25 (1992) 193–201.

[62] [R. Bowell, A review of sulfate removal options for mine waters, Proceedings of Mine Water 2004, pp. 75–88.

[63] [G. Juby,W. Pulles, Evaluation of electrodialysis reversal for desalination of brackish mine water, WRC Report, 1791990 1190.

[64] [M. Sivakumar, M. Ramezanianpour, G. O'Halloran, Mine water treatment using a vacuum membrane distillation system, APCBEE Proc. 5 (2013) 157–162. https://doi.org/10.1016/j.apcbee.2013.05.028

[65] [G. Juby, C. Schutte, J. Van Leeuwen, Desalination of calcium sulphate scaling mine water: design and operation of the SPARRO process, Water S. A. 22 (1996) 161–172.

[66] [W. Pulles, G. Juby, R. Busby, Development of the slurry precipitation and recycle reverse osmosis (SPARRO) technology for desalinating scaling mine waters, Water Sci. Technol. 25 (1992) 177–192.

[67] [N. Cicek, J.P. Franco, M.T. Suidan, V. Urbain, Using a membrane bioreactor to reclaim wastewater, J. Am. Water Works Assoc. 90 (1998) 105–113.

[68] [J. Lozier, A. Fernandez, B. Hines, K. Carns, Using a membrane bioreactor/reverse osmosis system for indirect water reuse, Proc. 2000 AWWA Water Reuse Conference, San Antonio, Texas, 2000.

[69] [T. Guihe, K. Kiran, V. Bala, H.O. Maung, S. Harry, Membrane bioreactor for water reclamation in Singapore, Water Pract. Technol. 3 (2008) 11-17.

[70] [N. Cicek, A review of membrane bioreactors and their potential application in the treatment of agricultural wastewater, Can. Biosyst. Eng. 45 (2003) 6.37–36.37.

[71] [S. Rosenberger, U. Krüger, R. Witzig, W. Manz, U. Szewzyk, M. Kraume, Performance of a bioreactor with submerged membranes for aerobic treatment of municipal waste water, Water Res. 36 (2002) 413–420. https://doi.org/10.1016/S0043-1354(01)00223-8

[72] [K.H. Ahn, H.Y. Cha, K.G. Song, Retrofitting municipal sewage treatment plants using an innovative membrane–bioreactor system, Desalination 124 (1999) 279–286. https://doi.org/10.1016/S0011-9164(99)00113-7

[73] [X.J. Fan, V. Urbain, Y. Qian, J. Manem, Nitrification and mass balance with a membrane bioreactor for municipal wastewater treatment, Water Sci. Technol. 34 (1996) 129–136. https://doi.org/10.1016/0273-1223(96)00502-1

[74] [E. Muller, A. Stouthamer, H.W.V. van Verseveld, D. Eikelboom, Aerobic domestic waste water treatment in a pilot plant with complete sludge retention by cross-flow filtration, Water Res. 29 (1995) 1179–1189. https://doi.org/10.1016/0043-1354(94)00267-B

[75] [W. Lorenz, T. Cunningham, J.P. Penny, Phosphorus removal in a membrane reactor system: a full-scale wastewater demonstration study, Proc. Water Environ. Fed. 2002 (2002) 406–414. https://doi.org/10.2175/193864702784247864

[76] [J. Manam, E. Sanderson, Membrane bioreactors, in: Megaw Hill (Ed.), Water Treatment: Membrane Process, 1996.

[77] [A. Parker, Membrane technology plays key role in waterless hygienic toilet, Membr. Technol. 2014 (2014) 8. https://doi.org/10.1016/S0958-2118(14)70255-1

[78] [E. Perez Lopez, Design and testing of a novel human-powered generator device as a backup solution to power Cranfield' s Nano-Membrane Toilet, 2014.

[79] [A. Hogetsu, T. Ishikawa, M. Yoshikawa, T. Tanabe, S. Yudate, J. Sawada, High rate anaerobic digestion of wool scouring wastewater in a digester combined with membrane filter, Water Sci. Technol. 25 (1992) 341–350.

[80] [K. Minami, A trial of high performance anaerobic treatment on wastewater from a kraft pulp mill, Desalination 98 (1994) 273–283. https://doi.org/10.1016/0011-9164(94)00152-9

[81] [M. Knoblock, P. Sutton, P. Mishra, K. Gupta, A. Janson, Membrane biological reactor system for treatment of oily wastewaters, Water Environ. Res. (1994) 133–139. https://doi.org/10.2175/WER.66.2.6

[82] [R. Zaloum, S. Lessard, D. Mourato, J. Carriere, Membrane bioreactor treatment of oily wastes from a metal transformation mill, Water Sci. Technol. 30 (1994) 21–27.

[83] [W. Ross, J. Barnard, N. Strohwald, C. Grobler, J. Sanetra, Practical application of the ADUF process to the full-scale treatment of a maize-processing effluent, Water Sci. Technol. 25 (1992) 27–39.

[84] [L.D. Nghiem, A. Schäfer, Trace contaminant removal with nanofiltration, in: A.I. Schäfer, T.D. Waite, A.G. Fane (Eds.), Nanofiltration—Principles and Applications, Elsevier 2004, pp. 479–520 (Chapter 8).

[85] [F.I. Hai, L.N. Nguyen, L.D. Nghiem, B.-Q. Liao, I. Koyuncu, W.E. Price, Trace organic contaminants removal by combined processes for wastewater reuse, The Handbook of Environmental Chemistry, 45 2016, pp. 39–78.

[86] [M. Cirja, P. Ivashechkin, A. Schäffer, P.F. Corvini, Factors affecting the removal of organic micropollutants from wastewater in conventional treatment plants (CTP) and membrane bioreactors (MBR), Rev. Environ. Sci. Biotechnol. 7 (2008) 61–78. https://doi.org/10.1007/s11157-007-9121-8

[87] [O.H. Jones, N. Voulvoulis, J. Lester, Human pharmaceuticals in wastewater treatment processes, Crit. Rev. Environ. Sci. Technol. 35 (2005) 401–427. https://doi.org/10.1080/10643380590956966

[88] [F.I. Hai, N. Tadkaew, J.A. McDonald, S.J. Khan, L.D. Nghiem, Is halogen content the most important factor in the removal of halogenated trace organics by MBR treatment? Bioresour. Technol. 102 (2011) 6299–6303. https://doi.org/10.1016/j.biortech.2011.02.019

[89] [K. Kimura, H. Hara, Y. Watanabe, Removal of pharmaceutical compounds by submerged membrane bioreactors (MBRs), Desalination 178 (2005) 135–140. https://doi.org/10.1016/j.desal.2004.11.033

[90] [N. Tadkaew, F.I. Hai, J.A. McDonald, S.J. Khan, L.D. Nghiem, Removal of trace organics by MBR treatment: the role of molecular properties, Water Res. 45 (2011) 2439–2451. https://doi.org/10.1016/j.watres.2011.01.023

[91] [S. Castiglioni, R. Bagnati, R. Fanelli, F. Pomati, D. Calamari, E. Zuccato, Removal of pharmaceuticals in sewage treatment plants in Italy, Environ. Sci. Technol. 40(2006) 357–363. https://doi.org/10.1021/es050991m

[92] [F.I. Hai, K. Tessmer, L.N. Nguyen, J. Kang, W.E. Price, L.D. Nghiem, Removal of micropollutants by membrane bioreactor under temperature variation, J. Membr. Sci. 383 (2011) 144–151. https://doi.org/10.1016/j.memsci.2011.08.047

[93] [N.M. Vieno, T. Tuhkanen, L. Kronberg, Seasonal variation in the occurrence of pharmaceuticals in effluents from a sewage treatment plant and in the recipient water, Environ. Sci. Technol. 39 (2005) 8220–8226. https://doi.org/10.1021/es051124k

[94] [A.S. Stasinakis, S. Kotsifa, G. Gatidou, D. Mamais, Diuron biodegradation in activated sludge batch reactors under aerobic and anoxic conditions, Water Res. 43 (2009) 1471–1479. https://doi.org/10.1016/j.watres.2008.12.040

[95] [N. Tadkaew, M. Sivakumar, S.J. Khan, J.A. McDonald, L.D. Nghiem, Effect of mixed liquor pH on the removal of trace organic contaminants in a membrane bioreactor, Bioresour. Technol. 101 (2010) 1494–1500. https://doi.org/10.1016/j.biortech.2009.09.082

[96] [T. Urase, C. Kagawa, T. Kikuta, Factors affecting removal of pharmaceutical Substances and estrogens in membrane separation bioreactors, Desalination 178 (2005) 107–113. https://doi.org/10.1016/j.desal.2004.11.031

[97] [C. Zwiener, F. Frimmel, Short-term tests with a pilot sewage plant and biofilm reactors for the biological degradation of the pharmaceutical compounds clofibric acid, ibuprofen, and diclofenac, Sci. Total Environ. 309 (2003) 201–211. https://doi.org/10.1016/S0048-9697(03)00002-0

[98] [S. Kim, P. Eichhorn, J.N. Jensen, A.S. Weber, D.S. Aga, Removal of antibiotics in wastewater: effect of hydraulic and solid retention times on the fate of tetracycline in the activated sludge process, Environ. Sci. Technol. 39 (2005) 5816–5823. https://doi.org/10.1021/es050006u

[99] [M. Bernhard, J. Müller, T.P. Knepper, Biodegradation of persistent polar pollutants in wastewater: comparison of an optimised lab-scale membrane bioreactor and activated sludge treatment, Water Res. 40 (2006) 3419–3428. https://doi.org/10.1016/j.watres.2006.07.011

[100] [M. Clara, N. Kreuzinger, B. Strenn, O. Gans, H. Kroiss, The solids retention time—a suitable design parameter to evaluate the capacity of wastewater treatment plants to remove micropollutants, Water Res. 39 (2005) 97–106. https://doi.org/10.1016/j.watres.2004.08.036

[101] [N. Kreuzinger,M. Clara, B. Strenn, H. Kroiss, Relevance of the sludge retention time(SRT) as design criteria for wastewater treatment plants for the removal of endocrinedisruptors and pharmaceuticals from wastewater, Water Sci. Technol. 50 (2004) 149–156.

[102] [C. Abegglen, A. Joss, C.S. McArdell, G. Fink, M.P. Schlüsener, T.A. Ternes, H. Siegrist, The fate of selected micropollutants in a single-house MBR, Water Res. 43 (2009)2036–2046. https://doi.org/10.1016/j.watres.2009.02.005

[103] [M. Clara, B. Strenn, O. Gans, E. Martinez, N. Kreuzinger, H. Kroiss, Removal of selectedpharmaceuticals, fragrances and endocrine disrupting compounds in a membranebioreactor and conventional wastewater treatment plants, Water Res. 39 (2005) 4797–4807. https://doi.org/10.1016/j.watres.2005.09.015

[104] [F.I. Hai, X. Li,W.E. Price, L.D. Nghiem, Removal of carbamazepine and sulfamethoxazolebyMBR under anoxic and aerobic conditions, Bioresour. Technol. 102 (2011)10386–10390. https://doi.org/10.1016/j.biortech.2011.09.019

[105] [A. Joss, H. Andersen, T. Ternes, P.R. Richle, H. Siegrist, Removal of estrogens in municipal wastewater treatment under aerobic and anaerobic conditions: consequences for plant optimization, Environ. Sci. Technol. 38 (2004) 3047–3055. https://doi.org/10.1021/es0351488

[106] [Y. Kiso, A. Mizuno, R.A.A. Binti Othman, Y.-J. Jung, A. Kumano, A. Ariji, Rejection properties of pesticides with a hollow fiber NF membrane (HNF-1), Desalination143 (2002) 147–157. https://doi.org/10.1016/S0011-9164(02)00236-9

[107] [Y. Kiso, Y. Sugiura, T. Kitao, K. Nishimura, Effects of hydrophobicity and molecularsize on rejection of aromatic pesticides with nanofiltration membranes, J. Membr. Sci. 192 (2001) 1–10. https://doi.org/10.1016/S0376-7388(01)00411-2

[108] [M. Reinhard, N.L. Goodman, P.L. McCarty, D.G. Argo, Removing trace organics by reverse osmosis using cellulose acetate and polyamide membranes, J. Am. Water Works Assoc. 78 (1986) 163–174.

[109] [P. Berg, G. Hagmeyer, R. Gimbel, Removal of pesticides and other micropollutants by nanofiltration, Desalination 113 (1997) 205–208. https://doi.org/10.1016/S0011-9164(97)00130-6

[110] [R. Boussahel, A. Montiel, M. Baudu, Effects of organic and inorganic matter on pesticide rejection by nanofiltration, Desalination 145 (2002) 109–114. https://doi.org/10.1016/S0011-9164(02)00394-6

[111] [B. Van der Bruggen, J. Schaep,W. Maes, D. Wilms, C. Vandecasteele, Nanofiltration as a treatment method for the removal of pesticides from ground waters, Desalination117 (1998) 139–147. https://doi.org/10.1016/S0011-9164(98)00081-2

[112] [K.M. Agbekodo, B. Legube, S. Dard, Atrazine and simazine removal mechanisms by nanofiltration: influence of natural organic matter concentration, Water Res. 30 (1996) 2535–2542. https://doi.org/10.1016/S0043-1354(96)00128-5

[113] [L. Nghiem, A. Manis, K. Soldenhoff, A. Schäfer, Estrogenic hormone removal from wastewater using NF/RO membranes, J. Membr. Sci. 242 (2004) 37–45. https://doi.org/10.1016/j.memsci.2003.12.034

[114] [L.D. Nghiem, A.I. Schäfer, M. Elimelech, Removal of natural hormones by nanofiltration membranes: measurement, modeling, and mechanisms, Environ. Sci. Technol. 38 (2004) 1888–1896. https://doi.org/10.1021/es034952r

[115] [K. Agenson, J. Oh, T. Kikuta, T. Urase, Rejection mechanisms of plastic additives and natural hormones in drinking water treated by nanofiltration, Water Supply 3 (2003) 311–319.

[116] [S. Weber, M. Gallenkemper, T. Melin, W. Dott, J. Hollender, Efficiency of nanofiltration for the elimination of steroids from water, Water Sci. Technol. 50 (2004) 9–14.

[117] [T.A. Ternes, M. Meisenheimer, D. McDowell, F. Sacher, H.J. Brauch, B. Haist-Gulde,G. Preuss, U.Wilme, N. Zulei-Seibert, Removal of pharmaceuticals during drinking water treatment, Environ. Sci. Technol. 36 (2002) 3855–3863. https://doi.org/10.1021/es015757k

[118] [E.M. Golet, A.C. Alder, W. Giger, Environmental exposure and risk assessment of fluoroquinolone antibacterial agents in wastewater and river water of the Glatt Valley Watershed, Switzerland, Environ. Sci. Technol. 36 (2002) 3645–3651. https://doi.org/10.1021/es0256212

[119] [T.A. Ternes, R. Hirsch, Occurrence and behavior of X-ray contrast media in sewage facilities and the aquatic environment, Environ. Sci. Technol. 34 (2000) 2741–2748. https://doi.org/10.1021/es991118m

[120] [S.J. Khan, J.E. Ongerth, Modelling of pharmaceutical residues in Australian sewage by quantities of use and fugacity calculations, Chemosphere 54 (2004) 355–367. https://doi.org/10.1016/j.chemosphere.2003.07.001

[121] [K. Kimura, G. Amy, J.E. Drewes, T. Heberer, T.-U. Kim, Y.Watanabe, Rejection of organic micropollutants (disinfection by-products, endocrine disrupting compounds, and pharmaceutically active compounds) by NF/RO membranes, J.Membr. Sci. 227 (2003) 113–121. https://doi.org/10.1016/j.memsci.2003.09.005

[122] [S.P. Sun, T.A. Hatton, S.Y. Chan, T.S. Chung, Novel thin-film composite nanofiltration hollow fiber membranes with double repulsion for effective removal of emerging organic matters from water, J. Membr. Sci. 401 (2012) 152–162. https://doi.org/10.1016/j.memsci.2012.01.046

[123] [R.W. Holloway, J. Regnery, L.D. Nghiem, T.Y. Cath, Removal of trace organic chemicals and performance of a novel hybrid ultrafiltration-osmotic membrane bioreactor, Environ. Sci. Technol. 48 (2014) 10859–10868. https://doi.org/10.1021/es501051b

[124] [T.Y. Cath, N.T. Hancock, C.D. Lundin, C. Hoppe-Jones, J.E. Drewes, A multi-barrierosmotic dilution process for simultaneous desalination and purification of impaired water, J. Membr. Sci. 362 (2010) 417–426. https://doi.org/10.1016/j.memsci.2010.06.056

[125] [N.T. Hancock, P. Xu, D.M. Heil, C. Bellona, T.Y. Cath, Comprehensive bench- and pilot-scale investigation of trace organic compounds rejection by forward osmosis, Environ. Sci. Technol. 45 (2011) 8483–8490. https://doi.org/10.1021/es201654k

[126] [R.V. Linares, V. Yangali-Quintanilla, Z. Li, G. Amy, Rejection of micropollutants by clean and fouled forward osmosis membrane, Water Res. 45 (2011) 6737–6744. https://doi.org/10.1016/j.watres.2011.10.037

[127] [J.H. Al-Rifai, H. Khabbaz, A.I. Schäfer, Removal of pharmaceuticals and endocrine disrupting compounds in a water recycling process using reverse osmosis systems, Sep. Purif. Technol. 77 (2011) 60–67. https://doi.org/10.1016/j.seppur.2010.11.020

[128] [A.A. Alturki, N. Tadkaew, J.A. McDonald, S.J. Khan, W.E. Price, L.D. Nghiem, CombiningMBR and NF/RO membrane filtration for the removal of trace organics in indirect potable water reuse applications, J. Membr. Sci. 365 (2010) 206–215. https://doi.org/10.1016/j.memsci.2010.09.008

[129] [C. Bellona, J.E. Drewes, P. Xu, G. Amy, Factors affecting the rejection of organic solutes during NF/RO treatment—a literature review, Water Res. 38 (2004) 2795–2809. https://doi.org/10.1016/j.watres.2004.03.034

[130] [P. Cartagena, M. El Kaddouri, V. Cases, A. Trapote, D. Prats, Reduction of emerging micropollutants, organic matter, nutrients and salinity from real wastewater by combined MBR–NF/RO treatment, Sep. Purif. Technol. 110 (2013) 132–143. https://doi.org/10.1016/j.seppur.2013.03.024

[131] [D. Dolar, M. Gros, S. Rodriguez-Mozaz, J. Moreno, J. Comas, I. Rodriguez-Roda, D. Barceló, Removal of emerging contaminants from municipal wastewater with an integrated membrane system, MBR–RO, J. Hazard. Mater. 239 (2012) 64–69. https://doi.org/10.1016/j.jhazmat.2012.03.029

[132] [L.N. Nguyen, F.I. Hai, J. Kang,W.E. Price, L.D. Nghiem, Removal of emerging trace organic contaminants by MBR-based hybrid treatment processes, Int. Biodeterior. Biodegrad. 85 (2013) 474–482. https://doi.org/10.1016/j.ibiod.2013.03.014

[133] [E. Sahar, I. David, Y. Gelman, H. Chikurel, A. Aharoni, R. Messalem, A. Brenner, The use of RO to remove emerging micropollutants following CAS/UF or MBR treatment of municipal wastewater, Desalination 273 (2011) 142–147. https://doi.org/10.1016/j.desal.2010.11.004

[134] [F. Zaviska, P. Drogui, A. Grasmick, A. Azais,M. Héran, Nanofiltrationmembrane bioreactorfor removing pharmaceutical compounds, J. Membr. Sci. 429 (2013) 121–129. https://doi.org/10.1016/j.memsci.2012.11.022

[135] [A. Achilli, T.Y. Cath, E.A. Marchand, A.E. Childress, The forward osmosis membranebioreactor: a low fouling alternative to MBR processes, Desalination 239 (2009)10–21. https://doi.org/10.1016/j.desal.2008.02.022

[136] [A. Alturki, J. McDonald, S.J. Khan, F.I. Hai, W.E. Price, L.D. Nghiem, Performance of a novel osmotic membrane bioreactor (OMBR) system: flux stability and removal of trace organics, Bioresour. Technol. 113 (2012) 201–206. https://doi.org/10.1016/j.biortech.2012.01.082

[137] [A.A. Alturki, J.A. McDonald, S.J. Khan,W.E. Price, L.D. Nghiem,M. Elimelech, Removal of trace organic contaminants by the forward osmosis process, Sep. Purif. Technol. 103 (2013) 258–266. https://doi.org/10.1016/j.seppur.2012.10.036

[138] [J.L. Cartinella, T.Y. Cath, M.T. Flynn, G.C. Miller, K.W. Hunter, A.E. Childress, Removal of natural steroid hormones from wastewater using membrane contactor processes, Environ. Sci. Technol. 40 (2006) 7381–7386. https://doi.org/10.1021/es060550i

[139] [F.I. Hai, K. Yamamoto, Membrane biological reactors, in: P. Wilderer (Ed.), Treatiseon Water Science 2011, pp. 571–613. https://doi.org/10.1016/B978-0-444-53199-5.00096-8

[140] [W.C. Lay, Q. Zhang, J. Zhang, D. McDougald, C. Tang, R. Wang, Y. Liu, A.G. Fane, Effect of pharmaceuticals on the performance of a novel osmotic membrane Bioreactor(OMBR), Sep. Sci. Technol. 47 (2012) 543–554. https://doi.org/10.1080/01496395.2011.630249

[141] [J. Phattaranawik, A.G. Fane, A.C. Pasquier, W. Bing, A novel membrane bioreactor based on membrane distillation, Desalination 223 (2008) 386–395. https://doi.org/10.1016/j.desal.2007.02.075

[142] [S. Goh, J. Zhang, Y. Liu, A.G. Fane, Fouling and wetting in membrane distillation (MD) and MD-bioreactor (MDBR) for wastewater reclamation, Desalination 323 (2013) 39–47. https://doi.org/10.1016/j.desal.2012.12.001

[143] [T.M. LaPara, J.E. Alleman, Thermophilic aerobic biological wastewater treatment, Water Res. 33 (1999) 895–908. https://doi.org/10.1016/S0043-1354(98)00282-6

[144] [K.C.Wijekoon, F.I. Hai, J. Kang, W.E. Price, T.Y. Cath, L.D. Nghiem, Rejection and fate of trace organic compounds (TrOCs) during membrane distillation, J. Membr. Sci.453 (2014) 636–642. https://doi.org/10.1016/j.memsci.2013.12.002

[145] [K.C. Wijekoon, F.I. Hai, J. Kang, W.E. Price,W. Guo, H.H. Ngo, T.Y. Cath, L.D. Nghiem, A novel membrane distillation–thermophilic bioreactor system: biological Stability and trace organic compound removal, Bioresour. Technol. 159 (2014) 334–341. https://doi.org/10.1016/j.biortech.2014.02.088

# Chapter 12

# Photo-Fenton oxidation technology for the treatment of wastewater

V. Muelas-Ramos[1], R. Muñoz-Mansilla[2], V. Calvino-Casilda[3], E. Muñoz-Camacho[3], R.M. Martín-Aranda[1]* and I.A. Santos-López[4]*

[1] Departamento de Química Inorgánica y Química Técnica. Facultad de Ciencias. Universidad Nacional de Educación a Distancia (UNED). Paseo Senda del Rey, 9. 28040 Madrid (Spain)

[2] Departamento de Informática y Automática. ETS Ingeniería Informática. Universidad Nacional de Educación a Distancia (UNED). C/ Juan del Rosal, 16. 28040 Madrid (Spain)

[3] Departamento de Ingeniería Eléctrica, Electrónica, Control, Telemática y Química Aplicada a la Ingeniería. ETS Ingenieros Indutriales (UNED). C/ Juan del Rosal, 12. 28040 Madrid (Spain)

[4] Ingeniería Química, Facultad de Ciencias Químicas. Universidad Autónoma de Nuevo León (UANL). Av. Universidad s/n, Cd. Universitaria, 66455 San Nicolás de los Garza, Nuevo León, (México)

ivan.santoslp@uanl.edu.mx; rmartin@ccia.uned.es

**Abstract**

The main recent advances in photo-Fenton oxidation treatment for recalcitrant wastewaters have been reviewed. Biodegradable intermediates and mineralized pollutants are obtained from recalcitrant compounds when this oxidation technology is employed in wastewater. However, the cost of the technology is one of the main limitations. Several strategies, such as new nanotechnologies, heterogeneous catalysts or chelating agents have been used to reduce the cost and enhance the efficiency of the photo-Fenton process. Moreover, applying solar energy cost can be reduced integrating biological treatments in the oxidation process. Direct UV photolysis is able to degrade pollutants (chemicals, pharmaceuticals, agrochemicals, endocrine-disrupting compounds, and herbicides) in water. In this scenery, photo-Fenton is presented as an attractive and very effective alternative for the removal of recalcitrant compounds and emerging contaminants present in wastewater.

**Keywords**

Photo-Fenton Oxidation, Emerging Pollutants, Recalcitrant Compounds, Wastewater Treatment; Pollutant Degradation

## Contents



## 1. Introduction

In the last years, the environment legislative regulations [1] have promoted a fast investigation on water protection due to the special attention paid to the elimination of biologically recalcitrant substances and pollutants. New technologies and the combination of technologies have been rapidly developed. These technologies are mainly phased separation processes like adsorption processes, stripping methodologies, and other methods (chemical oxidation and reduction). Chemical oxidation involves the transformation of contaminants into less toxic products, or in the total decomposition of the pollutants to inorganics, water and $CO_2$. These chemical methods afford a total solution to the problem of pollutants in water. However, in other cases, the use of reactive systems which are much more effective than conventional processes are necessary. Many research groups have been working during the last years to develop new oxidation techniques defined as advanced oxidation processes. Normally, these processes take place at ambient temperature and pressure [2]. The advanced oxidation processes (AOPs) are efficient methods to remove pollutants from wastewaters [3].

Generally, the *in situ* generation of a powerful oxidizing agent, such as hydroxyl radicals (•OH), is the main characteristic of the AOPs in order to remove the pollutants. Different types of AOPs include chemical, photochemical, sonochemical and electrochemical reactions. The pristine and most employed one is the Fenton reaction. During a Fenton process, a matrix containing $H_2O_2$ and a soluble iron (II) salt (Fenton's reagent) are used to degrade pollutants [4]. The oxidation efficiency of pristine Fenton can be enhanced in the photo-Fenton method (irradiation with ultraviolet light) or solar photo-Fenton method (sunlight irradiation) (Fig. 1) [5, 6].

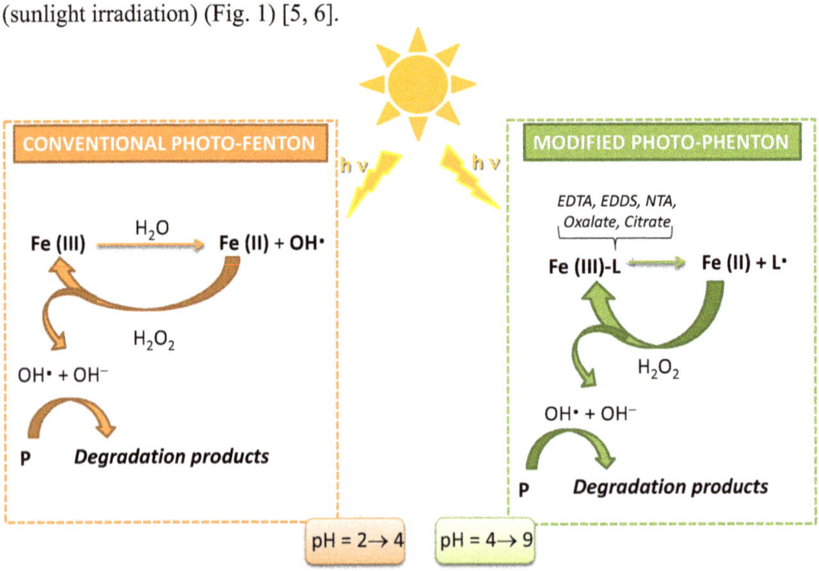


*Fig. 1. Solar photo-Fenton process modified by adding chelating agents at neutral pH (adapted from reference [6]).*

Another important improvement combines the Fenton method with electrochemical reactions. Lately, an important number of electrochemical advanced oxidation processes have been described [7, 8].

The publications on Fenton oxidation in wastewater treatment confirm that it is one of the most effective methods for the degradation and mineralization of recalcitrant substances. However, some critical points with regards to pH control, the cost associated to $H_2O_2$ and catalyst consumption have to be considered in the Fenton process. In addition, the sludge disposal limits its application on full-scale [9]. Some alternatives have been proposed to improve the efficiency of the Fenton oxidation under reduced cost [10-13].

The recent review on the removal of pharmaceuticals from water published by Mirzaei et al. [14], studies on the advanced oxidation processes (AOPs) in the oxidation of organic compounds by Fenton-based processes has been documented. The impacts of some controlling parameters including the $H_2O_2$ and catalyst dosage, solution pH, initial contaminants concentrations, temperature, type of catalyst, the intensity of irradiation, reaction time, and feeding mode on the removal efficiencies of hetero/homogeneous Fenton processes have been highlighted. Moreover, the combination of Fenton-type processes with biological systems as the pre/post treatment stages has been analyzed. Other recent studies include the elimination of recalcitrant compounds in water even at micro quantities [15, 16]. The high chemical oxygen demand (COD) and low biological oxygen demand (BOD) are clear characteristics of wastewater when these types of compounds are present. AOPs are excellent methodologies for degradation of pollutants since these are capable of removing most of the organic compounds, antibiotics, dyes [17-21], micropollutants [22] and persistent pollutants [23].

The photo-Fenton process has been established as an energetically improved way to treat wastewaters. Lately, numerous studies have been performed of photo-Fenton as individual technology or linked with conventional biological methods for wastewaters application. The objective of this chapter is to review the different photo-Fenton applications in order to treat a number of toxic compounds in wastewaters (Table 1). The evaluation of its effectiveness in the mineralization of pollutants and biodegradability enhancement have been considered. The influence of operational conditions on the efficiency of photo-Fenton oxidation has been assessed. Recent advances for degradation of organic contaminants and cost minimization factors are also pointed out.

In Fenton and combined processes, the main disadvantage is the requirement of 50–80 ppm ferrous ion. Homogeneous AOPs may produce a large amount of sludge in the final neutralization step due to the large quantity of treated water. These disadvantages can be avoided by using nano zero valent iron as an alternative route to induce Fenton oxidation.

Zero-valent state metals such as $Fe^0$, $Al^0$, $Zn^0$ and $Sn^0$ are very active agents in the processes [24, 25].

*Table 1. Homogeneous and heterogeneous advanced oxidation processes (AOPs) employed in wastewater.*

| Type of system | Example |
|---|---|
| **Homogeneous** | Fenton-based reaction<br>Fenton: $H_2O_2 + Fe^{2+}$<br>Fenton-like: $H_2O_2 + Fe^{3+}/M_n^+$<br>Sono-Fenton: $US/H_2O_2 + Fe^{2+}$<br>Photo-Fenton: $UV/H_2O_2 + Fe^{2+}$<br>Electro-Fenton<br>Sono-electro-Fenton<br>Photo-electro-Fenton<br>Sono-photo-Fenton<br>$O_3$ based processes<br>$O_3$ + UV<br>$O_3$ +UV+$H_2O_2$<br>$H_2O_2 + Fe^{2+}/Fe^{3+}/ M_n^+$solid |
| **Heterogeneous** | $TiO_2$/ZnO/CdS + UV<br>$H_2O_2 + Fe^0$/Fe (nano-zero valent iron)<br>$H_2O_2$ + immobilized nano-zero valent iron |

## 2. Fenton-based technologies

The Fenton reaction was developed by H.J.H. Fenton in 1894. He reported that $H_2O_2$ could be activated by ferrous ($Fe^{2+}$) salts to oxidize tartaric acid. In the recent past, since the beginning, Fenton reaction was employed to remove toxic organic substances from wastewater [26]. Fenton reactions use iron ions and hydrogen peroxide to generate active oxygen species that oxidize organic or inorganic compounds.
In the photo-Fenton processes, the simplified reaction sequence to generate the hydroxyl radical from $H_2O_2$ is as follows:

$$Fe^{3+} + h\upsilon + H_2O_2 \rightarrow Fe^{2+} + HO\bullet + H^+$$

$$Fe^{2+} + H_2O_2 \rightarrow k\ Fe^{3+} + HO\bullet + OH^-$$

During the photo-Fenton processes, pollutants are broken down by hydroxyl radicals and several types of contaminants, such as pharmaceuticals, pesticides, dyes, insecticides, and bacterias are removed. Furthermore, the use of solar energy in photo-Fenton processes improves the economic and environmental sustainability of the technology [27, 28]. However, the application of such processes implies the need to work in a narrow range of pH between 2.8–3.5 to ensure that Fe(II) and Fe(III) species exert their catalytic role, maximizing the concentration of photoactive species and avoiding the precipitation of inactive iron oxyhydroxides. The possibility of working at neutral or near neutral pH has aroused great interest, but the low efficiency of photo-Fenton processes at pH 7 is due to iron precipitation and can be prevented by adding iron complexing agents.

Fenton reaction is run by an arsenal of heterogeneous catalysts, such as zeolites, alumina, ferrites, clays, and nanomaterials [29]. During the electro-Fenton process, the catalytic effect of $Fe^{2+}$ is enhanced by irradiating with UV light. The use of electrochemical and photochemical combined process joined to the Fenton process is called photo-electro-Fenton (PEF) process and generates a greater quantity of free radicals due to the combination effect as illustrated in Fig. 2 [30]. Thus, the degradation of target toxic substances can be accelerated when the water matrix is irradiated with UV light in addition to the electro-Fenton process. The photo-electro-Fenton systems are more efficient due to the photochemical regeneration of $Fe^{2+}$ by the photo-reduction of $Fe^{3+}$ ions and photo-activation of complexes [31].

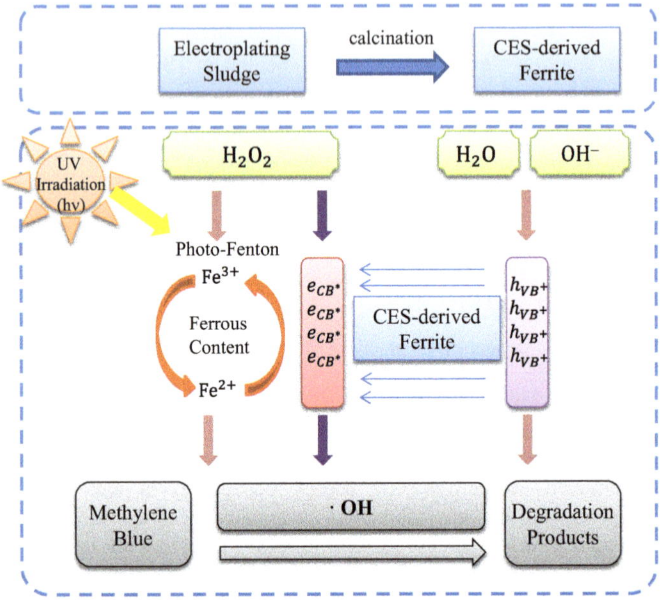


*Fig. 2. The process proposed for UV/CES/$H_2O_2$ system (adapted from reference [30]).*

## 3. Behavior in wastewater systems

Application of photo-Fenton treatment of several industrial wastewaters containing recalcitrant compounds at laboratory or pilot plant scales has been extensively investigated. Degradation of dyes [32, 33], pesticides [34], bacterias [35] and also micropollutants [36] has recently been reported. The treatment of industrial wastewaters, such as pulp [37] or pharmaceuticals [38] has been investigated by photo-Fenton application exclusively or in combination with other methodologies. We will summarize these oxidation processes used for pharmaceuticals, pesticides, pathogens, bacteria, chemicals, agrochemicals, petroleum refinery plants and landfill leachates.

### 3.1 Pharmaceuticals and pesticides

The so-called emerging contaminants (pharmaceuticals, pesticides, personal care products, and industrial contaminants) are increasing in frequency and concentrations

[39]. To develop strategies to reduce their presence in water systems has been an active research area at present

Combining of photo-Fenton method with conventional treatment technique for pharmaceutical wastewaters significantly enhances biodegradability of wastewaterTokumura et al. [40] performed a comprehensive study on effects of water matrices on removal of pharmaceuticals by three different kinds of advanced oxidation processes, the photo-Fenton process on the removal of pharmaceuticals (carbamazepine and diclofenac), the $TiO_2$ photocatalytic oxidation, and the combined ozone and $H_2O_2$ oxidation process. A comparative study of the models, using pure water and wastewater, was evaluated.

Gimeno et al. [41] studied the removal of emerging contaminants from primary effluents of municipal wastewater by sequential biological and solar-degradation-photocatalytic oxidation processes. A mixture of nine pharmaceuticals model substances (acetaminophen, antipyrine, caffeine, ketorolac, metoprolol, sulfamethoxazole, carbamazepine, hydrochlorothiazide, and diclofenac) was analyzed. Only acetaminophen and caffeine were totally removed from municipal wastewater in aerobic biological experiments. Under the optimum experimental conditions, 80-100 % degradation of the emerging compounds was obtained in the photocatalytic ozonation. Solar light (SL) Fe(III) photocatalytic ozonation, SL/$O_3$/Fe (III) during 180 min and ozonation in 45 min systems reduced the concentrations of the studied emerging contaminants below their detection limits. In this study, 41.3 % mineralization is obtained by photocatalytic ozonation and 34 % is achieved by single ozonation.

In another investigation, Exposito et al. [42] developed a semi-industrial autonomous plant for the solar photo-Fenton degradation of pharmaceutical wastewater effluent with a parabolic collector. Ferrioxalate was employed in a photo-Fenton process and over 79 % of TOC was removed in 120 min. The extent of the photo-Fenton process was directly dependent on the initial conditions.

In the work developed by Perini et al., the photo-Fenton degradation of two frequent antibiotic and antidepressant; ciprofloxacin and fluoxetine were analyzed in an anaerobic pre-treated hospital effluent [43]. Parameters such as the concentration of $H_2O_2$, iron, pH and the effect of the iron citrate complex were followed during the degradation of the pharmaceuticals. The authors concluded that it was possible to achieve over 50 % degradation of both substances after 90 min.

Anjana Anand´s group employed a fluidized bed solar photo-Fenton process for the treatment of hospital wastewater in order to reduce the chemical oxygen demand (COD) of wastewater and enhance the biodegradability [44]. The ideal operational conditions were determined at pH 3, $Fe^{2+}$ dosage of 5 mM, the $H_2O_2$ dosage of 50 mM and silica carrier amount of 40 g/L. The maximum COD removal was 98 % in 90 min and the enhancement of biodegradability (evaluated in terms of $BOD_5$/COD ratio) increased from 0.16 to 0.7. During the fluidized bed solar photo-Fenton process, the COD efficiency was 92 % whereas in the solar photo-Fenton process the COD removal efficiency was 67 % during one hour.

Serna-Galvis et al. [45], showed the removal of antibiotic cloxacillin in by electrochemical oxidation, $TiO_2$ photocatalyst, and the photo-Fenton system. Cloxacillin was totally removed under these treatments, and photo-Fenton enabled 15% of mineralization within 240 min.

Degradation of antibiotic norfloxacin using advanced oxidation processes has been investigated by Shankaraiah et al. [46]. They studied the effect of UV, initial pH of the recalcitrant solution, UV/$H_2O_2$, Fenton, photo-Fenton process, UV/$TiO_2$ and UV/$TiO_2$ immobilization with glass beads. Photo-Fenton process was effective for removing around 96 % of norfloxacin. For an initial drug concentration of 150 ppm, under photo-Fenton process, optimum conditions were pH 3, $H_2O_2$ concentration of 200 mg/L and an iron concentration of 30 mg/L.

Endocrine disruptors like progesterone, among others, are present in water and wastewater. The mechanisms for UV photo-assisted Fenton-like degradation of progesterone were investigated by Ifelebuegu et al. [47]. The UV photo-assisted process completely degrades progesterone by 60 %.

New nanomaterials based on Cu-doped α-FeOOH (α-(Fe, Cu)OOH) nanoflowers have been used in the catalytic degradation of diclofenac sodium by the photo-Fenton process [48]. The results demonstrated that α-(Fe, Cu)OOH nanomaterials exhibited better catalytic behavior than pure α-FeOOH. The sample doped with 3 % Cu (molar ratio of Cu to Fe) showed the optimal visible-light photocatalytic activity. Moreover, the degradation efficiency of diclofenac increased a number of catalysts and $H_2O_2$ dosages but decreased with the increasing initial pH between 5 and 8. The α-(Fe, Cu)OOH nanoflowers possess high degradation efficiency and excellent stability even after five cycles. The high catalytic activity of α-(Fe, Cu)OOH could be attributed to the synergistic activation effects of Fe and Cu in α-(Fe, Cu)OOH towards $H_2O_2$.

Considering that paracetamol is a representative compound of emerging pollutants of pharmaceutical origin, a recent work from Villota el al. [49], established photo-Fenton as an interesting technology to oxidize wastewater with this contaminant.

In other investigation [50], the degradation of ciprofloxacin and paracetamol was compared. The formation of hydroquinone was detected only during paracetamol degradation in absence of irradiation. The degradation rate of paracetamol increased drastically since hydroquinone formed was able to reduce $Fe^{3+}$. When hydroquinone was initially added, the degradation rates ciprofloxacin and paracetamol, in the dark, were much higher in comparison to the rates in the absence of hydroquinone. It was due to the faster formation of $Fe^{2+}$ at the beginning of the reaction. In the absence of hydroquinone, degradation of ciprofloxacin was not observed. However, after 30 min, the degradation rate increased drastically when hydroquinone was added. When the paracetamol degradation was observed in absence of hydroquinone, ten paracetamol hydroxylated intermediates were recognized; which have a contribution to the reduction of $Fe^{3+}$ to $Fe^{2+}$ and in a suitable way the paracetamol degradation.

The ibuprofen also appears in the wastewater from pharmaceutical industries. Chakma et al. [51], studied the degradation of ibuprofen using combined techniques of sono-enzymatic treatments. A positive synergy between sonolysis and enzyme treatment was noticed due to the formation of hydrophilic intermediates during degradation. The degradation of ibuprofen under sono-enzymatic treatment was comparable with other hybrid methodologies like photo-Fenton, sono-Fenton and sono-photocatalysis.

Removal of mefenamic acid, which is a non-steroidal anti-inflammatory drug, was investigated by different oxidative processes ($H_2O_2$, $H_2O_2$/UV, Fenton, and photo-Fenton). The influence of the main variables in the process (concentration of hydrogen peroxide, type of catalyst, pH) was also investigated [52]. In a comparative study, it was found that photo-Fenton process was the best technology as it was able to remove 95 % of mefenamic acid. The authors concluded that this methodology could be applied in water quality control of municipal or industrial wastewater.

Iovino et al. [53] studied the presence of traces of pharmaceutical substances in drinking, superficial and ground waters. The removal of several compounds of the type of ibuprofen was investigated in a lab-scale batch reactor equipped with a monochromatic UV lamp (Fig. 3). Two series of experiments were carried out; the first one studied the concentration of ibuprofen and the second one explored the pH influence on ibuprofen degradation. From the results, it was possible to propose a reaction mechanism.

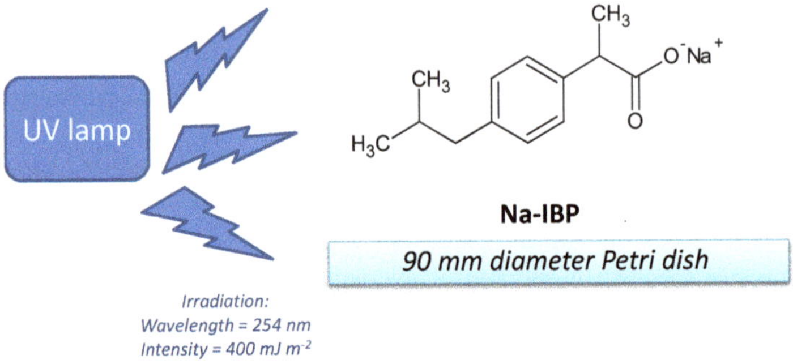


*Fig. 3. The system used for the degradation of Ibuprofen with UV light (adapted from reference [53]).*

Degradation of β-lactam antibiotic by $TiO_2$ photocatalysis, sonochemistry, photo-Fenton process, and electrochemistry was conducted by Serna-Galvis et al. The four oxidation processes were able to remove the pollutant. Nevertheless, important differences with respect to the degree of mineralization were observed. The removal of total organic carbon in $TiO_2$ photocatalysis and in the photo-Fenton system was around 90 and 35 % respectively, but with the electrochemical and sonochemical treatments, the antibiotic was not degraded [54].

Mineralization of metronidazole (the most important drug of the group of 5-nitroimidazole) can be accelerated by sunlight via photo-Fenton mechanism under gradual incorporation of $H_2O_2$. The solar photo-Fenton system leads to 96 % of COD removal for 960 mg $L^{-1}$ at 60 °C and pH 3 in 12 min. An increase of 16 % of mineralization was obtained when the sunlight was applied to the photo-Fenton reactor [55].

Moreover, a recent study on the biological hazard evaluation of pharmaceutical effluents, before and after a photo-Fenton treatment [56], concluded that the photo-Fenton process catalyzed with an iron-pillared clay decreased the oxidative stress, genotoxic damage and LC50 in Hyalella azteca, and the concentration of paracetamol.

During several years, the food and cosmetic industries have employed parabens, which are apparently innocuous preservatives. The group of Cuerda-Correa [57], studied the oxidation of different parabens, such as methylparaben (MP), ethylparaben (EP), propyl paraben (PP), and butylparaben (BP) in water using $O_3$. The tested processes were $O_3$/Fenton, $O_3/H_2O_2$, $O_3$/UV, $O_3/UV/H_2O_2$, $O_3$/photo-Fenton, $O_3/UV/TiO_2$, and

$O_3/UV/TiO_2/H_2O_2$. The results demonstrated that single ozonation was more effective than $O_3/H_2O_2$ and $O_3$/Fenton processes. It was observed that ultraviolet irradiation afforded faster removal of parabens, due to the effect of ozone photolysis in the radical-mediated mechanism. From the results, it was concluded that $O_3/UV/TiO_2/H_2O_2$ was the most efficient method for the degradation of this type of emerging pollutants.

Degradation of ethylparaben (EPB), a potential endocrine disrupting chemical was studied by Zuñiga-Benítez et al. The combination of the conventional Fenton process with visible light radiation (photo-Fenton) was used in this treatment. In this way, the effectiveness of the photo-Fenton process for EPB degradation was assessed using an experimental design. Results indicated that after 180 min of reaction almost all EPB was eliminated [58].

In a similar study, the photodegradation aspects of benzophenone and methylparaben have been reported by the same authors [59]. Removal of these endocrine-disrupting chemicals from aqueous samples using photo-Fenton technology was examined. Under optimal conditions, benzophenone and methylparaben were completely degraded in 2 h. Dissolved organic carbon studies confirmed that more than 60 % of contaminants were mineralized. Moreover, the biodegradability analysis suggested that this factor increased gradually during 300 min of photo-Fenton treatment.

Zhang et al. investigated the photodegradation of bisphenol A (BPA) during photo-Fenton-like reaction using α-FeOOH as a catalyst [60]. The effects of pH and the addition of organic acids were examined. The incorporation of several organic acids influenced the generation of free radicals and ferric-carboxylate complexes, causing modifications in the photodegradation efficacy. The results indicated that oxalic acid (OA) was the most effective in enhancing the photodegradation. The level of degradation strongly depends on the OA concentration. Interestingly, the addition of OA could extend the operational range of pH from acidic to neutral during the process. An interesting degradation of BPA was observed at pH 6.0, due to the formation of ferric oxalate complexes and OH radicals as a result of synergistic interactions of OA and α-FeOOH. This results pointed out that photo-Fenton-like process using the α-FeOOH catalyst and promoted by oxalate could be applied in wastewater treatment without pH adjustment.

Another study evaluated the degradation of metoprolol (MET) by a photo-Fenton process using different photoreactors (Fig. 4) [61]. MET was completely eliminated and 83.7 % of TOC conversion was achieved. In addition, the enhancement of Fenton and photo-Fenton processes for the degradation of MET was investigated by Romero group [62]. The authors added resorcinol to avoid acidification during the photo-Fenton processes, and a total MET degradation was achieved within 3 min with the highest hydrogen peroxide and iron concentrations.

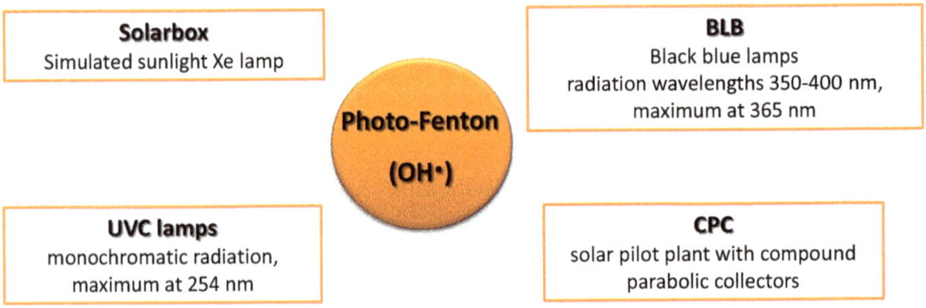


*Fig. 4. The process efficiency is dependent on the used light, different irradiation sources can be used at lab-scale (adapted from reference [61]).*

The elimination of radiocontrast agent has been also investigated. The radiocontrast, diatrizoic acid (DIA) was treated by a combination of photo-Fenton and Electro-Fenton processes. In an initial step of the process, it was determined the influence of the $H_2O_2$ dosage on degradation efficiency under ultraviolet (UV) irradiation. A maximum mineralization degree of 60 % total organic carbon (TOC) was obtained after 4 h of reaction. To complete the study, the influence of different concentrations of Fenton's reagents was evaluated for homogeneous Fenton process, giving rise an 80 % TOC removal efficiency at 4 h treatment. The heterogeneous treatment was also carried out using iron-activated carbon as catalyst. It was demonstrated that the selected catalyst was an interesting option, affording 67 % of TOC removal in 4 h without formation of iron leachate in the medium. To improve the efficiency, the coupling of photo-Fenton with the electro-Fenton process in two different ways: photoelectric-Fenton or photo-Fenton followed by electro-Fenton processes was proposed. In both cases the complete mineralization of DIA within 2 h was possible. At last, total mineralization by both proposed ways was obtained [63].

Bianco Prevot et al. [64] studied the applicability of urban biological waste materials for caffeine photodegradation after aerobic aging. Caffeine photodegradation was performed under several test reactions. Degradation results were good when $H_2O_2$ was present without an additional Fe(II). A Photo-Fenton-like process was selected for this study using synthetized hybrid magnetite-bio-based substance nanoparticles to improve the reusability of the catalysts. In the presence of such nanoparticles, $H_2O_2$ and Fe(II), caffeine degradation was completed in very short time. Both experiments, under homogeneous and heterogeneous conditions, were run at pH = 5, milder conditions compared to the classic photo-Fenton process.

Another important improvement combines photo-Fenton with biological reactions. By applying only biological process the removal of Imazalil (IMZ, a fungicide for the post-harvest treatment of citrus) was no satisfactory compared to coupling photo-Fenton and biological degradation process. Thus it requires a microbial consortium which plays an important role in the biological process by enhancing the capability to resist and degrade high concentrations of IMZ; $H_2O_2$ concentration leads to degrading IMZ by photo-Fenton while have been found a reduction on TOC and COD by biological treatment [65].

Magnetically recyclable Bi/Fe-based hierarchical nanostructures (BFO-M) have been employed for environmental decontamination of pesticides and pharmaceutical substances [66]. Even at a low catalyst loading, the removal rate of organic pollutants (5-fluorouracil and isoproturon) was around 99 % in 100 min under visible light irradiation.

Antipyrine is an emerging contaminant present in a municipal wastewater effluents, and it has been removal by a homogeneous sono-photo-Fenton process (UV/$H_2O_2$/Fe(II)/ultrasound). An optimum value of $Fe^{2+}$ plays a key role in the Fenton oxidation mechanism by enhancing the mineralization, whereas an excess of $Fe^{2+}$ involves a scavenging of hydroxyl radicals which affects the degree of mineralization.

The overall mineralization reaction was optimized and the TOC removal was 79 % in 50 min. The radical reaction in the bulk solution was found to be the primary mineralization pathway (94.8 %), followed by photolysis (3.65 %), direct reaction with $H_2O_2$ (0.86 %), and reaction by ultrasonically generated oxidative species (0.64 %).

In a laboratory scale study, degradation of 2,4-dichlorophenoxyacetic acid (2,4-D) in groundwater samples was examined under simulated solar irradiation [68]. A 75.2 % mineralization of 2,4-D was measured and the effects of iron concentration, hydrogen peroxide concentration and pH on the 2,4-D degradation in groundwater were evaluated. Results showed that the pH was the main parameter affecting the photo-Fenton process. This study demonstrated that the photo-Fenton process at circumneutral pH and in the presence of iron seems to be a competitive method to remove pesticides from groundwater.

Conte´s group also studied the photo-Fenton degradation of 2,4-D in the aqueous medium near neutral pH [69]. A kinetic model was proposed employing the ferrioxalate complex as an iron source at pH = 5, T = 50 °C, and low level of radiation. Under experimental conditions, 95.6 % conversion of 2,4-D was obtained in 3 h. The 2,4-D conversion was almost 100 % within 2 h when the reactor was operated under the same operating conditions but at higher radiation level.

In a recent study, the photo-Fenton oxidation of tributyltin, dibutyltin, and monobutyltin was reported the developed was used for the treatment of sediment contaminated by organotin compounds following homogeneous and heterogeneous photocatalysis. Photo-Fenton process and photocatalysis of organotins in water were carried out. Almost total degradation of tributyltin in water (99.8 %) was reached in 30 and 7 min for photocatalytic and photo-Fenton treatment, respectively. The investigation demonstrated that the use of UV and the generation of hydroxyl radicals were efficient routes to treat adsorbed organotins onto marine sediment [70].

Among neonicotinoid class, acetamiprid is a pesticide, which has been removed from wastewater by the Fenton and photo-Fenton processes using the oxidizing power of hydroxyl radicals. Acetamiprid removal by the Fenton process required initial concentrations of 190 mg/L $H_2O_2$ and 19 mg/L $Fe^{2+}$, and UV enhanced the Fenton process reducing the requirement of time and amount of reagents. The results and analysis of kinetic parameters calculated from the kinetic model demonstrated that degradation of acetamiprid follows pseudo-first order kinetics [71].

Other insecticides like fipronil have been mineralized during the photo-Fenton process. Under the optimum experimental conditions, 57 % of DOC and 74 % of COD were removed only by one hour of irradiation. With these values, the integration of coagulation/flocculation/decantation and photo-Fenton processes was presented as an alternative to the pre- or complete treatment of wastewater containing fipronil and related compounds [72].

In a recent communication by Gozzi et al., on the kinetic study of the reaction of pesticides with hydroxyl radicals generated by laser flash photolysis, the rate constants of different pesticides (carbaryl, propoxur, fenoxycarb, ethoxysulfuron and chlorimuron-ethyl) with $HO^{\bullet}$ in degassed acetonitrile were reported [73]. The values were about an order of magnitude smaller than the diffusion controlled rate and correlated with the relative rates of degradation of the pesticides in the photo-Fenton reaction. This study allowed to establish the correlation between relative rate constants with the relative photo-Fenton degradation rates of pesticides (Table 2).

*Table 2. The reaction between hydroxyl radical and the pesticides: Kinetic constants (adapted from reference [72]).*

| Pesticide | Slope, LFP | $k_{OH\text{-}LFP}$ ($M^{-1}s^{-1}$) | $K_{photo\text{-}Fenton}$ ($s^{-1}$) |
|---|---|---|---|
| Naphtalene[a] | 0,0142 | $1{,}8 \times 10^{9}$ | - |
| Carbaryl (CB) | 0,0126 | $1{,}6 \times 10^{9}$ | $12{,}3 \times 10^{-4}$ |
| Propoxur (PP) | 0,0050 | $0{,}6 \times 10^{9}$ | $4{,}8 \times 10^{-4}$ |
| Fenoxycarb (FX) | 0,0095 | $1{,}2 \times 10^{9}$ | b |
| Ethoxyfulfuron (ES) | 0,0192 | $2{,}4 \times 10^{9}$ | $15{,}8 \times 10^{-4}$ |
| Chlorimuron-ethyl (CE) | 0,0173 | $2{,}2 \times 10^{9}$ | $13{,}8 \times 10^{-4}$ |

[a] Standard compound. [b] Too insoluble in water.LFP stands for laser flash photolysis.

Recently, García-Ballesteros et al. [74] analyzed the photo-Fenton degradation of phenolic compounds, which are commonly appeared in food processing industries. A matrix with eight compounds such as 2,4-dinitrophenol, tannic, ellagic, gallic, protocatechuic, vanillic, syringic and sinapic acids, was treated via photo-Fenton process under real and simulated sunlight. Under solar irradiation, entire removal of the compounds was achieved in less than 3 min at pH 3.9.

A comparative study on the degradation of diatrizoate (DTZ) using different sources such as solar radiation, $H_2O_2$, solar radiation/$H_2O_2$, $K_2S_2O_8$, solar radiation/$K_2S_2O_8$, Fenton, solar radiation/Fenton, Fenton-Like, and solar radiation/Fenton-Like systems has been conducted [75]. Direct DTZ photolysis resulted in low degradation. Interestingly, the degradation extension depended on irradiance level but it was independent of the pH. However, when radical-promoting species result in total DTZ degradation, it was shown that the generation of peroxo and iron hydroperoxy complexes may foster a generation of additional OH radicals, necessary for DTZ remediation.

Photo-Fenton may also serve as pre-treatment method to enhance pesticide (atrazine) biodegradability in wastewater [76]. Optimized levels of $H_2O_2$, $Fe^{3+}$ and irradiation allowed the removal of 85 and 100% of atrazine within 2 h by Fenton and photo-Fenton processes, respectively. This work indicated that combined photo-Fenton-biological treatment may deliver total degradation of biorecalcitrant pollutants, such as atrazine.

Removal of chlorophenoxypropionic and mecoprop (MCPP) herbicides from surface water using photo-Fenton oxidation were investigated using Photo-Fenton process [77]. The optimum conditioning neutral pH, 20 mg/L of $H_2O_2$ and 20 mg/L $Fe^{2+}$ for effective degradation.

In another study [78], calcined layered double hydroxides (cLDHs) Ti-doped and undoped MgFe were tested for photodegradation (photocatalytic and photo-Fenton-like) of herbicide 2,4,5-trichlorophenoxyacetic acid (2,4,5-T) using visible light irradiation. Under the experimental conditions employed, an excellent photo-Fenton-like activity of the samples was observed which allowed 80-95 % degradation of 2,4,5-T at an initial concentration of 100 mg $L^{-1}$ with 4 g $L^{-1}$ calcined LDHs. The degradation was accomplished in 360 min.

Natural iron oxide has been reported to remediate linuron 3-[3,4-(dichlorophenyl)-1-methoxy-1-methylurea] herbicide in aqueous solutions via photo-Fenton–like oxidation [79]. Natural iron oxide contains hematite as an iron source and its dissolution provides $Fe^{3+}$ cations for the Fenton-like reaction. An initial linuron concentration of 4.0 × 10-5 mol $L^{-1}$ was completely mineralized within 45 min.

In an interesting study, the mineralization of sulfamethizole in photo-Fenton and photo-Fenton-Like systems was carried out with good results [80]. Under the best experimental conditions and after 60 min of reaction, the SFZ mineralization percentages in UV/$H_2O_2$, UV/$H_2O_2$/$Fe^{2+}$ and UV/$H_2O_2$/$Fe^{3+}$ systems were 16, 90 and 88 %, respectively.

Very recently, Roongkarn's group investigated the ability of iron nanoparticles in the degradation of paraquat and other pesticides under heterogeneous Fenton and photo-Fenton processes [81]. Iron nanoparticles at pH 3 showed that initial concentrations of paraquat (in the range of 60-100 ppm) were degraded with the removal percentages within the range of 43.7-75.8 %. In the same study, the photo-Fenton process at pH 3 was able to remove paraquat perce up to 70.9-99.1 % at initial paraquat concentration of 100-300 ppm. Thus, the photo-Fenton method using iron nanoparticles provided higher efficiency in paraquat removal than the Fenton process.

There are many other studies concerning the degradation of pharmaceutical and pesticides. Continuously, improved advances are appearing for the mineralization of pollutants to find applications in a large-scale wastewater treatment, at a lower cost.

### 3.2 Pathogens and bacterias

A review by Giannakis et al. [82] discussed the photo-Fenton reaction at near-neutral pH, in respect of bacterial inactivation. The fundamental concepts controlling the near-neutral photo-Fenton reaction and the mechanism, besides the UVB and UVA influence on the bacterial cell has been analyzed. It has been found that in presence of light there is an indirect oxidative inactivation mechanism into the microorganisms. Thus, it is proved that the solar photo-Fenton reactions are an enhanced disinfection process. They focus on the effect of different parameters that affect the photo-catalytic process efficiency, such as the fundamental physical, chemical and biological.

In a second part of the study, the authors presented the bacteria inactivation under solar-photo-Fenton and the first application of Fe-citrate-based photo-Fenton for the inactivation of *Escherichia coli.* At neutral and alkaline pH, interesting results were obtained. The efficiency of the homogeneous photo-Fenton process using Fe-citrate complex strongly improved bacterial inactivation as compared with the heterogeneous photo-Fenton treatment ($FeSO_4$ and goethite as sources of iron). The bacterial inactivation rate order observed: goethite$<FeSO_4<$Fe-citrate, follows the trend for the HO radical formation. This treatment was applied in natural water samples from Lake Geneva (Switzerland) at pH 8.5. No bacterial reactivation was detected when photo-Fenton treatment was used [83, 84].

A comparative study of different AOPs ($H_2O_2$/sunlight, $TiO_2$/sunlight, $H_2O_2/TiO_2$/sunlight, natural photo-Fenton) for the inactivation of a series of antibiotics such as ampicillin, ciprofloxacin, and tetracycline was carried out by Ferro et al. [85] using the urban wastewater. Different concentrations of $H_2O_2$, $TiO_2$ $H_2O_2/TiO_2$ and $Fe^{2+}/H_2O_2$ were evaluated in a parabolic collector reactor with real biologically treated wastewater at pilot-scale. All processes afforded a total inactivation of bacteria. In terms of time, best disinfection method was photo-Fenton at pH 4.

A paper by Valero et al. describes how different parameters as culture media (PCA, LBA, T-7, T-7 + TTC, VRBA, and MAC), compositions, photo-treatments (SODIS, SODIS + $H_2O_2$ and photo-Fenton) affect the immediate recovery and regrowth of bacteria *E. Coli.* The addition of catalase or sodium pyruvate in plate count agar (PCA) enhanced the recovery of photo-treated *E. Coli*, whereas photo-Fenton cause cellular damages it has been suggested to use PCA to support the growth of injured *E. coli* bacteria [85]. Several kinetic approaches and mechanistic studies on the photocatalytic inactivation of resistant microorganisms (Bacillus subtilis spores) in water have been carried out in a photoreactor by Fenton reaction. The inactivation process was affected by the generation and a specific amount of hydroxyl radicals required to cause cell death. In addition, it was observed that a number of hydroxyl radicals generated is correlated with radiative intensity and reagent concentration. The relationship between a number of hydroxyl radicals and the inactivation was supported by adding chloride ions as radical scavengers. Under experimental conditions, even the lowest chloride ion concentration was able to produce a significant inactivation delay [87, 88].

For the wastewater disinfection, the role of solar UVA irradiance and accumulated solar UVA energy dose per unit of volume of treated water were evaluated as two solar energy parameters. A systematic study on the influence of three UVA irradiance levels (10, 20 and 30 $Wm^{-2}$) and different accumulated energy doses (0.6, 1.4, 1.8, 2.5, 3.2 and 3.7 kJ $L^{-1}$) was carried out during the inactivation of *E. faecalis* by solar photo-Fenton at 25 ºC. It was observed that an increase in radiation intensity enhanced the bacterial inactivation rate. Under experimental conditions, *E. faecalis* inactivation rate by solar photo-Fenton at neutral pH was limited by solar UVA irradiance when using a simulated secondary effluent from a municipal wastewater [89]. In a parallel study, Pulgarin group studied Saccharomyces cerevisiae photo-Fenton inactivation at near-neutral pH. A significant loss of cultivability was monitored during the application of all the different photo-Fenton systems, attributed to the oxidative stresses applied [90].

Nanotechnology and nanomaterials are introducing interesting innovations in the eco-friendly approaches on the development of new materials for the water treatment. In a recent study, the synthesis of hierarchical CuO nanostructures was investigated. It is an eco-friendly hydrothermal route for the controllable synthesis of various morphologies CuO nanostructures. Toxic additives traditionally used were replaced by biocompatible 4-(2-hydroxyethyl)-1-piperazineethane-sulfonic acid (HEPES). The growth rate and orientation of CuO nanoparticles were tuned adjusting the pH and concentration of HEPES. The analogous photo-Fenton assay demonstrated that the CuO nanostructures enhanced dye removal capability when compared with CuO synthesized in pure water. This nanomaterial was also investigated on *S. aureus* inactivation. CuO nanoparticles with the smallest size displayed the best antibacterial performance [91].

### 3.3 Chemicals and agrochemicals

Chemicals and agrochemicals are important pollutants, which have to be removed from industries and municipal wastewaters. Pesticides, herbicides, insecticides, and fungicides are toxic and carcinogenic. Pesticides are typically removed from industrial wastewaters by physical–chemical treatments. Many reviews confirm that photo-Fenton is effective to degrade these agrochemicals. Recently, several approaches have appeared to optimize the pollutants removal by the photo-Fenton process. For example, decontamination of wastewater containing cyanobacterial blooms, oxidation of geosmin and 2-

methylisoborneol by the photo-Fenton process have been studied by Park et at[92]. The taste and odor caused by these compounds have become a matter of great concern for drinking water treatment plants.

Malato et al. [93] compared the efficiencies of titania photo-Fenton with photocatalysis and Fenton process for degradation of a number of pesticides (alachlor, atrazine, chlorfenvinphos, diuron, isoproturon, and pentachlorophenol) and combinations thereof. They concluded that photo-Fenton was faster to mineralize these compounds and therein mixtures. The photo-reactivity of the pesticides followed the order: diuron > alachlor > isoproturon > chlorfenvinphos > atrazine.

In another study, degradation of phenol using graphene oxide–iron oxide ($Fe_3O_4$-GO) nanocomposites as catalysts was investigated. It was considered as one of the most promising routes to enhance the performance of nanomaterials employed in the Photo-Fenton technology. Zero-dimensional nanoparticles with two-dimensional graphene oxide nanosheets were hybridized and the resulting graphene oxide–iron oxide ($Fe_3O_4$-GO) nanocomposites were used to degrade phenol in a heterogeneous photo-Fenton process. Different parameters, such as the amount of catalyst, GO loading, the concentration of $H_2O_2$ and pH of the phenol solution were investigated. The authors showed that under experimental conditions with UV-light irradiation, about 98.8 % phenol and 81.3 % TOC of a phenol solution could be removed after 120 min photo-Fenton degradation. In all cases, UV-light improved the phenol removal in the photo-Fenton process [94].

Magnetic materials based on mesoporous materials ($Fe_3O_4$/C/Cu) have been recently investigated by several groups for their high performance as Fenton-like catalysts. Li et al. [95] prepared Fe-Cu composites with different compositions and carried out the Fenton catalytic performance of $Fe_3O_4$/C/Cu composite. They observed that only proper amounts of Fe and Cu exhibited a synergistic enhancement in Fenton catalytic activity. Cu reduced $Fe^{3+}$ to $Fe^{2+}$, which accelerated the $Fe^{3+}/Fe^{2+}$ cycles and favored $H_2O_2$ decomposition to produce more hydroxyl radicals for methylene blue oxidation. The Fenton catalytic performance was greatly improved and methylene blue was removed in 60 min. The catalysts showed good recyclability and could be easily separated by a magnetic field. This methodology is cheap and sustainable for the synthesis of mesoporous composites as excellent Fenton-like catalysts. Compared with conventional methods for mesoporous composites, it has the advantages of low cost and easy preparation.

Soltani and Lee [96] improved the heterogeneous photo-Fenton catalytic degradation of toluene by doping Ba in $BiFeO_3$ nanoparticles via the sol-gel procedure. They found that

Ba-incorporation in $BiFeO_3$ magnetic nanoparticles (BFO MNPs) played an important role in the degradation of toluene. Increasing the Ba doping level up to 12 %, greatly affect the iron redox cycling and oxygen vacancies as compared to pure BFO MNPs. The iron redox cycling and existence of oxygen on the surface of Ba doped BFO had been affected by the photo-Fenton process. The degradation of toluene was partial in the dark but almost complete under visible light irradiation by the photo-Fenton catalytic degradation reaction. The highest photo-Fenton efficiency (toluene removal of 98 %, total organic carbon (TOC) reduction of 85 % and chemical oxygen demand (COD) reduction of 94 %) was detected after 40 min of irradiation with visible light.

In a parallel study, the same authors [97] described the enhanced effect of Ba-doped $BiFeO_3$ nanoparticles for the total degradation of benzene in wastewater. Doping of $Ba^{2+}$ into pure BFO decreased the band-gap energy of pure BFO and increased the magnetic saturation. In consequence, a 97 % photocatalytic benzene removal was observed within 60 min under visible light irradiation, with 93 % reduction in chemical oxygen demand (COD) and 82 % total organic carbon (TOC). In this study, major intermediate products of photodegradation were predicted and enhanced photocatalytic activity of benzene removal was confirmed as a consequence of both photocatalytic and photo-Fenton catalytic degradation mechanisms.

The catalytic behavior of a new nanoporous copper doped aluminum pillared montmorillonite (Cu-Alpill-MMT) was investigated in the treatment of dye-containing wastewater [98]. The catalysts were synthesized applying high power ultrasonic treatment to the suspension of $Cu^{2+}$ and aluminum pillared montmorillonite and calcining at 500 °C to obtaining CuO. The Cu-Alpill-MMT exhibited the superior adsorption capacity for basic yellow 1 but it hardly adsorbed reactive orange 16. The degradation of reactive orange 16 was successfully achieved when the $H_2O_2$ or $H_2O_2$/UV was incorporated in combination with the Cu-Alpill-MMT. This nanomaterial is an interesting heterogeneous catalyst for both Fenton and photo-Fenton processes. It was also observed that the catalyst exhibits antibacterial activity during *Escherichia coli* treatment.

Other dyes like methylene blue were also decolorized using iron tetrasulphophthalocyanine immobilized layered double hydroxides [99] as a heterogeneous catalyst. The photocatalytic activity of catalysts remain unchanged for three cycles, and under optimum conditions, the methylene blue decolorization efficiency was 99.80 % and total organic carbon removal of 76.50 % was calculated.

Soares et al. [100] reported an integrated treatment for wastewater containing polyester-cotton dye combining biological and photochemical oxidation processes as remediation strategy for synthetic textile wastewater. It was possible to achieve a 76 % dissolved organic carbon (DOC) removal.

Jaramillo et al. [101] prepared mixed oxides ($\alpha$-$Fe_2O_3/Bi_2WO_6$) to degrade a selected dye (methyl orange) as well as the transparent substrate Phenol at pH around 5.5. Sun-like illumination and UV in the absence and presence of $H_2O_2$ were selected to study the photocatalytic activity, of single and mixed catalysts. It was demonstrated that the use of a $\alpha$-$Fe_2O_3$-$Bi_2WO_6/H_2O_2$ system possesses higher photocatalytic efficiency to degrade both contaminants than pristine $Bi_2WO_6$ or $\alpha$-$Fe_2O_3$ single or mixed. Using the system $\alpha$-$Fe_2O_3$-$Bi_2WO_6/H_2O_2$, ca. 85 % of methyl orange was degraded in one hour under sun-like illumination whereas total degradation, at the same time was obtained under UV-illumination. In the case of Phenol, just around 30 % was degraded in two hours under sun-like illumination whereas around 95 % was degraded within 90 min. It was concluded that under UV-illumination, the generation of hydroxyl radicals was favorable but under sun-like illumination, only a small portion of UV can generate hydroxyl radicals.

More recently $Bi_{25}FeO_{40}$ microcrystals as reusable heterogeneous photo-Fenton-like catalyst have been proposed by Ren et al. [102]. Rhodamine B (RhB) and 2,4-dichlorophenol (2,4-DCP) were degraded in an aqueous medium by using the $Bi_{25}FeO_{40}/H_2O_2$/sunlight system. The higher degradation of RhB was observed at pH 3.0 with the degradation rate of 95 % under 4 h simulated sunlight irradiation. Total photocatalytic degradation of 2,4-DCP at pH 6.0 was achieved within 90 min. Degradation of rhodamine B was also investigated using a novel millimetric-scale sponge iron (s-$Fe^0$) particles (IV). The photo-Fenton-like system with millimetric sponge iron (s-Fe0), $H_2O_2$, visible light (vis, $\lambda \geq 420$ nm) and rhodamine B (RhB) was investigated. A new insight into the visible light photo-Fenton-like process with an optimum dosage of $H_2O_2$ and RhB photosensitizers was described [103].

Zhang et al. [104] developed a series of biopolymer complexes for the degradation of 4-chlorophenol in the presence of $H_2O_2$. The synthesis of the complex was achieved by coordinating iron ions to sulfur- and amino- groups in a modified wool. The study demonstrated that the strong coordination between iron and hydroxamic acid in the modified complex, favored the least iron leaching during the runs. Hydroxyl radical species were confirmed to be the dominant reactive oxidant in the decontamination process. The use of these novel biopolymer-based photocatalysts provides a new alternative for efficient degradation of toxic organic pollutants.

Other potent chemical pollutants (aniline and benzotriazole) were selected to carry out the catalytic degradation on mesoporous composite PAC@$FeIIFe_2IIIO_4$ under photo-

Fenton oxidation process. The optimum conditions indicated that the simultaneous presence of aniline and benzotriazole had a synergistic effect on the removal efficiency, and the degradation performance. After 5 cycles, the removal efficiencies of 70.4 and 99.5 % were obtained for aniline and benzotriazole. The mineralization experiments revealed that photo-Fenton system improved the biodegradability, and provide adequate conditions for the biological process. Therefore, PAC@FeIIFe$_2$IIIO$_4$ could be considered as a recyclable and low-cost catalyst for the purification of multi-pollutants contaminated waters [105].

Zhang et al. [106] described the high performance of the three-dimensionally ordered macroporous (3DOM) $BiVO_4$ and its supported iron oxide (x$Fe_2O_3$/3DOM $BiVO_4$, x = 0.18, 0.97, and 3.40 wt%). Photocatalytic activity of these materials was evaluated for the degradation of 4-nitrophenol under visible light illumination. The x$Fe_2O_3$/3DOM $BiVO_4$ samples enhanced the performance of 3DOM $BiVO_4$ sample, and 0.97$Fe_2O_3$/3DOM $BiVO_4$ showed the best photocatalytic behavior affording 98 % of 4-nitrophenol degradation in the presence of 0.6 mL of $H_2O_2$, during 30 min of visible light illumination. The incorporation of increasing amounts of $H_2O_2$ to the reaction media promoted the photodegradation of 4-nitrophenol by providing the active hydroxyl radicals generated via the reaction of photoinduced electrons and $H_2O_2$. The enhanced photocatalytic performance of 0.97$Fe_2O_3$/3DOM $BiVO_4$ was correlated with its porous architecture, high surface area, good light-harvesting ability, and high adsorbed oxygen species concentration. In addition, excellent photocatalytic stability was observed.

Polyphenolic compounds in wastewaters from the industrial processing of natural products, such as cork are capable of forming complexes with ferric and ferrous iron favoring the reduction of ferric iron. Malato et al. [107] studied the regeneration of iron during the Fenton and photo-Fenton processes, in order to reuse the wastewater from this type of industries. In the study, the degradation of imidacloprid insecticide, as well as a mixture of contaminants (phenol/methomyl/imidacloprid) at near-neutral pH, were performed and 70 % mineralization of such compounds was also achieved at pH 5. The observed effects of the reused cork boiling wastewater were comparable to the ones achieved by the addition of ethylenediamine-N, N′-disuccinic acid (EDDS), well-known commercial iron-chelating agents of the photo-Fenton process.

Photo-Fenton is proving to be one of the most appropriate AOPs for chemical and agrochemicals remediation (> 60 %). New nanomaterials and mesoporous catalysts have been implemented. Alternatively, solar photo-Fenton process combined with nanofiltration was applied in the removal of microcontaminants in real municipal effluents. In this context, the results indicated is suitable to use a combination of photo-Fenton with nanofiltration. Particular attention must be paid to use iron complexing agents because are not cheap chemicals, and the operating costs will be higher than with classical photo-Fenton at pH 3 [108].

### 3.4 Petroleum refinery plants

In the petroleum refinery plants, crude oil is transformed into a variety of refined products and many effluents are formed causing water and air pollution [109]. During the crude oil treatment, a large amount of water is used for refining processes and this wastewater contains several contaminants. Petroleum Refinery Wastewaters (PRW) contain high concentrations of aliphatic and aromatic compounds especially polycyclic hydrocarbons and grease, where the aromatic part is more toxic and recalcitrant than an aliphatic fraction. Degradation of these compounds is very necessary due to their toxicity. Traditional approaches embrace the application of adsorption, coagulation, and flocculation, or reactive treatments, such as bioremediation, electrochemical processes, membrane technology or different advanced oxidation processes. In recent years, important advances to treat PRW have been developed [110]. Some of the most efficient technologies include biological, photocatalytic, electro- and photo-Fenton, etc. The use of microbial fuel cell is one of the newest routes to treat PRW. Bagheri-Esfe et al. [111] investigated the treatment of PRW using oxygen and permanganate as cathodic electron acceptors in a microbial fuel cell. The effects of external resistance and temperature on the cell performance and PRW treatment were studied and the electrochemical model was proposed to evaluate the activity. The authors detected COD removal efficiency of 49.27 % during 44 h. In order to enhance the power of the cell, they used potassium permanganate as a cathodic electron acceptor and observed COD removal efficiency of 78 % during the same time, at the 33 °C.

Aljuboury et al. studied the solar photo-Fenton to treat petroleum wastewater from Sohar oil refinery. They evaluated the relationship among experimental conditions such as pH, $H_2O_2$ dosage, $Fe^{2+}$ dosage, and reaction time to determine the optimal conditions. The experimental results of the maximum TOC and COD removal rates were 59.3 and 74.7 %, respectively. This method achieved well degradation of the existing pollutants [112, 113].

Only a few investigations related to PRW treatment, have been reported in literature during recent years. This chapter is therefore limited to specific contaminants found in PRW. Phenol and xylene are the main compounds present in the effluents of petrochemicals and petroleum refinery plants, due to their high solubility in water. Estrada-Arriaga et al. [114] have reported the post-treatment of real oil refinery effluents (200 mg phenols/L) with photo-ferrioxalate reaction and Fenton's reaction. This methodology was an interesting alternative to simultaneously reduce the chemical oxygen demand (COD), phenols, and others pollutants such as chlorides, sulfates, and sulfides from petroleum refinery wastewater (Table 3). The higher removals of COD (84 %) and phenol (100 %) were obtained by photo-ferrioxalate reaction at pH 5 in two hours. Photo-ferrioxalate reaction was followed by the filtration of wastewater in an ultrafiltration hollow fiber module, to obtain removal efficiencies of 66.3 % for COD. The total removal of phenol, sulfides, color, and turbidity after a photo-ferrioxalate-ultrafiltration membrane treatment was 94 %.

*Table 3. Chlorides, sulfates and sulfides removal (%) after photo-ferrioxalate y Fenton's reactions (adapted from [114]).*

| | **Photo-ferrioxalate reaction** | | | | **Fenton´s reaction** | | |
|---|---|---|---|---|---|---|---|
| Test | 1 | 2 | 3 | 4 | 1 | 2 | 3 |
| Substances | Oxalate/$Fe^{2+}$ ratio-$H_2O_2$[(a)] | Oxalate/$Fe^{2+}$ ratio-$H_2O_2$[(a)] | Oxalate/$Fe^{2+}$ ratio-$H_2O_2$[(a)] | Oxalate/$Fe^{2+}$ ratio-$H_2O_2$[(a)] | $H_2O_2$/$Fe^{2+}$ ratio | $H_2O_2$/$Fe^{2+}$ ratio | $H_2O_2$/$Fe^{2+}$ ratio |
| | 10-without $H_2O_2$ (pH 3) | 10-without $H_2O_2$ (pH 3) | 10-500 (pH 3) | 10-500 (pH 3) | 15 (pH 3) | 15 (pH 3) | 15 (pH 4) |
| Chlorides | 3 | 2 | 14 | 5 | 11 | 12 | 12 |
| Sulfates | 0 | 0 | 0 | 0 | 0 | 0 | 0 |
| Sulphides | 91 | 89 | 82 | 93 | 78 | 67 | 88 |
| Reaction time (h) | 2 | 2 | 2 | 2 | 2 | 2 | 2 |

[(a)]$H_2O_2$ concentration (mg/L)

Fenton treatment of recalcitrant wastewaters associated with the problem of sludge generation. Diya'uddeen et al. [115] reported a 53 % mineralization of petroleum refinery effluent, producing 0.16 L of wet sludge per liter of wastewater. They also addressed the potential reuse of sludge.

The efficiency of Fenton and photo-Fenton processes in treating real groundwater contaminated with gasoline and diesel was investigated [116]. In the study, the mineralization results of waters contaminated with hydrocarbons showed that photo-Fenton was 34% more efficient than Fenton. Besides main contaminant removals as COD, diesel range organics, gasoline range organics, and total petroleum hydrocarbons were up to 95% by Fenton and photo-Fenton processes.

Hussian et al. [117] also studied the pollution of groundwater and soil appeared due to the fuel leakage from underground storage tanks and spills during transferring. Huge quantities of benzene, toluene, xylenes and ethylbenzene (BTEX) have been detected in water resources and the leaching of hydrocarbons into water generates many environmental problems. Paraffin, olefins, aliphatic hydrocarbons, aromatic compounds and also molecules containing sulfur, nitrogen, and metal oxides are toxic and carcinogenic. Therefore, the development of efficient methods for the removal of pollutants from contaminated water is very important. Photo-Fenton reaction is an efficient environment-friendly method in which hydroxyl radicals are used to oxidize recalcitrant organic pollutants and convert them to harmless end-products such as $H_2O$ and $CO_2$.

More studies on simulated and real PRW with complex pollutant mixtures are required to develop a general methodology to clean the complex PRW. The photo-Fenton treatment with a series of pre and post treatment processes would be a helpful tool to achieve acceptable results with a low investment [118, 119].

### 3.5 Leachates from landfills

Landfill waste disposal requires a wise management strategy to optimize costs. The landfill leachate formed by organic wastes decomposition is one of the most relevant issues on sanitary landfills. The leachate contains large amounts of organic substances, toxic materials, ammonia and heavy metals and its specific properties depend on many variables [120].

Hydroxyl radicals are key performers for AOPs degradation of pollutants in (solar photo) Fenton oxidations. Hydroxyl radicals degrade refractory organic matter into more biodegradable compounds that can be further treated with more economic biological-based methods. This process is considered as one of the most promising AOPs for remediation of refractory organic compounds in wastewater. The high investment and operation costs can be reduced by the use of sunlight. Silva et al. [121] compared the costs of a photo-Fenton process, using solar and/or artificial radiation, for the treatment of 100 $m^3$ per day of a sanitary landfill leachate that had previously been oxidized by a biological system. The scale-up of the photo-oxidation system was performed and the optimal conditions, to achieve the target COD values were identified.

Zha et al. [122] employed sonication in the treatment of landfill leachate in presence of $Fe^{2+}$ (US/$Fe^{2+}$) and achieved the highest total organic carbon (TOC) removal efficiency among the screened processes by photo-Fenton. Under the optimal conditions for the US-photo-Fenton process at pH at 3, 68.3 % the removal efficiency of TOC and 79.6 % chemical oxygen demand (COD) were achieved. Using US-photo-Fenton treatment, 36 of a total of 56 pollutants were totally degraded.

In a recent study, the role of the composition of sanitary landfill leachate on the removal of dissolved organic carbon (DOC) by the photo-Fenton process was investigated [123]. It was observed that the composition of the matrix did not reduce the efficiency of DOC removal, but significantly influenced the reaction time and $H_2O_2$ consumption. The pre-treatment of the effluent by coagulation-flocculation before the photo-Fenton process was found useful.

An innovative methodology for the treatment of landfill leachates has been recently proposed by Silva et al. [124]. This methodology involves: a) aerobic activated sludge biological pre-oxidation, b) coagulation/sedimentation step and c) photo-oxidation through photo-Fenton combining solar and artificial light. The biological process downstream from the photocatalytic system would promote the 60% mineralization. The cost of photo-Fenton step to treat 100 $m^3$/day of leachate was 6.41 €/$m^3$.

Similarly, removal efficiencies of Fenton and photo-Fenton processes for the post-treatment of composting leachate have been studied [125]. It is noteworthy that they used response surface methodology (RSM) with central composite design (CCD) to determine the maximum response on COD, color and turbidity removal. Furthermore, it was afforded efficiencies of 55.91 % and 75.53 % in Fenton process and 69.64 and 84.09 % in a photo-Fenton process. The results showed that photo-Fenton process had a better performance in COD and color removal for composting leachate than Fenton process.

## 4. Influence of operational conditions

The optimization of the reaction conditions in the photo-Fenton oxidation system is a key step to achieving good results in degradation processes [126]. Among the different factors that influence the degradation, the pH, temperature, concentration of the contaminant and Fenton reagents are the most relevant.

In recent years, many contributions have been described to enhance and optimize the experimental conditions. In addition, many theoretical studies have been proposed in order to improve the yields of degradation and to decrease the economic costs related to the Fenton technology. In this sense, contaminated water with anthracene was treated by Fenton processes [127]. The removal of the organic pollutants from the water and soil was afforded. In this study, the influences of different parameters on the degradation have been reported such as the concentration of anthracene, the amount of $FeSO_4$, pH, temperature, and amount of hydrogen peroxide. The authors concluded that the best conditions for degradation of anthracene by Fenton process were: anthracene concentration 16 mg $L^{-1}$, the mass of $FeSO_4$ = 120 mg, pH 4, temperature 50 °C. In another study, the impacts of pH, temperature, sulfur dioxide and caffeic acid, were investigated in the photoproduction of glyoxylic acid in model wine [128].

In the following sections, the influences of the main factors such as the concentration of contaminant, Fenton reagents, temperature, pH media, Ferrous ion concentration, $H_2O_2$ concentration, the composition of reaction media and chelating agents are discussed separately.

### 4.1 Contaminant concentration

Contaminant concentration is a key parameter for the photo-Fenton process. As expected, an increase in the concentration of pollutants has negative effects on its degradation, mainly due to the inner filtration effect related to high concentrations of substances to be absorbed. The research group of Rodriguez and Casas [129] published an interesting review of the trends in the intensification of Fenton process (Scheme 1). Contaminant concentration influences are discussed in deep and several experimental advances on it were discussed and compared by the authors. Sharma et al. [130] recently discussed the catalytic performance of $CoFe_2O_4$ with yttrium ($Y^{3+}$) inclusion in the spinel structure, analyzing the influence of the concentration of contaminants. In another investigation, the removal of β-lactam antibiotics from pharmaceutical wastewaters using the photo-Fenton process at near-neutral pH was investigated focusing the study on the effect of the antibiotics concentration [131].

In the photo-Fenton system, the $H_2O_2$, UV-light/$H_2O_2$, and Fe(II)/$H_2O_2$ systems presented a significant activity on antibiotic removal depending on the pollutant concentration. Interestingly, the existence of pharmaceutical additives (glucose, isopropanol, and oxalic acid) had a moderate influence on the efficiency of the pollutant removal. The authors also carried out the degradation of the β-lactam antibiotic, at pilot scale, in a complex mixture of pharmaceutical substances under solar radiation. The total degradation of the pollutant was attained in only 5 min. These results confirmed the applicability of photo-Fenton process to treat wastewaters from pharmaceutical industries loaded with different amounts of antibiotics. Koltsakidou et al. [132] studied the photo-Fenton and Fenton-like processes for the mineralization of the antineoplastic drug, 5-fluorouracil (5-FU). They optimized photo-Fenton treatment under simulated solar light radiation to investigate the effects of several parameters. Concentration effect of 5-FU on the treatment efficiency was assessed and the speed of degradation and mineralization percentages was correlated with the amount of drug present in the treated water. Generally, degradation efficiency at pH 7 is higher when the concentration of pollutant was less.

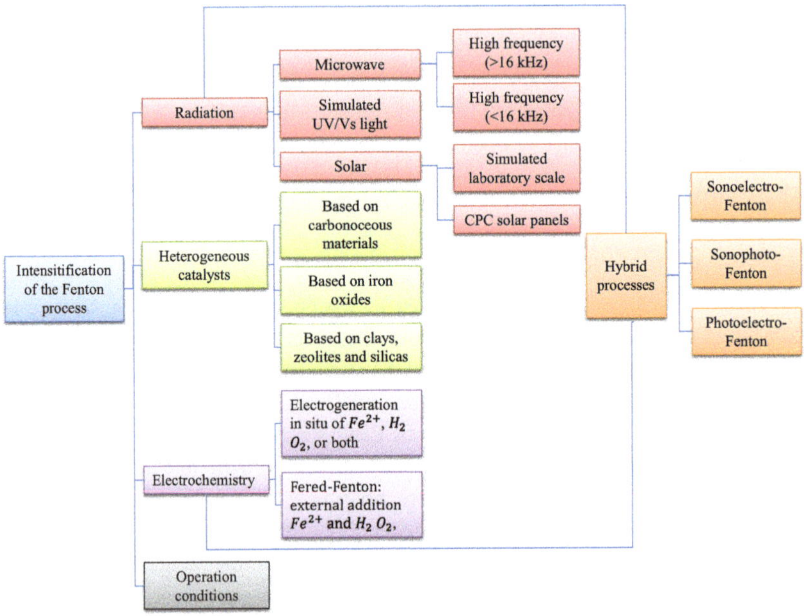


*Scheme 1. Strategies for the intensification of the Fenton process (adapted from reference [129]).*

## 4.2 Fenton reagents

Hydrogen peroxide and iron concentrations directly affect the reaction efficiency and cost. The presence of both reagents in the adequate concentrations is fundamental because $H_2O_2$ concentration determines the quantitative degradation of pollutants and the doses of iron control the reaction kinetics. For example, Wei et al. [133] studied the performance of iron-montmorillonite (Fe-Mt) in Fenton process. The Fe-Mt was prepared in delaminated structures via the introduction of iron oxides into Na-montmorillonite. Fe-Mt showed significant increases in the available iron content. The solid was efficient for phenol removal under visible light irradiation. From the study, it was concluded that the irradiation promoted certain intermediates (1,4-benzoquinone, hydroquinone, and catechol) when hydroquinone or catechol was added to the Fe-Mt/$H_2O_2$ system. Phenol degradation was accelerated because catechol and hydroquinone were capable of reducing Fe(III) into Fe(II). It was also observed that the concentrations of dissolved total Fe were increased with the increased in concentration oxalic acid, which has a strong chelating tendency to Fe(III).

In another recent report, photo-Fenton treatment of valproate was carried out using three different light sources UVA (black light blue lamps, BLB reactor), UVC (UVC reactor) and simulated sunlight in a solar box (SB). The effect of iron concentration was studied. Using the highest concentration of $Fe^{2+}$ (10 mg $L^{-1}$) and $H_2O_2$ (150 mg $L^{-1}$), total degradation of VA was observed in BLB and UVC systems, and 89.7 % in the solar box. Regarding mineralization, 67.4 % and 76.4 % of TOC conversion were achieved in BLB and UVC, respectively. Mineralization was negligible in solar box [134].

Iron sources determine the relative populations of $Fe^{2+}$, $Fe^{3+}$, and Fe-Ox in the photo-Fenton treatment of wastewater. The optimum ratio of Fenton reagents, rather than their absolute amounts, was assessed in a number of studies.These studies were aimed to assess the role paid by the generated hydroxyl radicals.

### 4.3 Temperature and pH media

$H_2O_2$ stability (up to 50 °C) imposes a temperature limit on photo-Fenton tests. This phenomenon accelerated in alkaline media. Taking into account the Arrhenius theory of rate constants in relation to temperature, it could be expected that an increase of the temperature provokes a higher generation of hydroxyl radicals, but their stability would be increasingly compromised, thus requiring an accurate determination of the operation temperature. Fenton process is also very dependent on the pH of media due to iron and hydrogen peroxide speciation factors. It is considered that the optimum pH for the Fenton reaction is around 3. Karam et al. [135] applied Fenton, Photo-Fenton and electro-Fenton processes to remove several organic pollutants from the water. In this study, the influence of some parameters on the mineralization has been reported. The ideal temperature for anthracene, mineralization was 50 °C, working at pH 4.

In another study, the pH in raceway pond reactors (RPR) was investigated in the removal of microcontaminants [136]. RPR is an alternative photoreactor for solar photo-Fenton applications, ideal for the treatment of large volumes of water with low-cost. A mixture of pharmaceuticals (in carbamazepine, flumequine, ibuprofen, ofloxacin and sulfamethoxazole) in a municipal wastewater effluent at circumneutral pH was analyzed; the micropollutant removal efficiency was higher (four times) in tubular photoreactor with compound parabolic collectors (CPC) than RPRs reactors [136]. It was evident the improvement into the removal of pharmaceuticals with CPC reactors increasing from 0.26 mg/kJ to 1.07 mg/kJ in the 15 cm-deep RPR.

A new catalyst based on ferrous ion onto graphene oxide composites was employed to study the influence of pH in Fenton process between pH 4 and 8. The composite was supported on Al-MCM-41, MCM-41 and $\gamma$-$Al_2O_3$ mesoporous materials [137]. The stability and activity for phenol degradation were studied for a wide pH range and the catalytic mechanism in acid and alkaline conditions was also determined. The visible light irradiation dramatically accelerated the rate of Fenton reactions and also extended the operating pH from 4.0 to 8.0. The Al-MCM-41 supported material was the highly energy-efficient sample with catalytic activity under wide operating pHs.

Franzoso et al. [138] studied Fenton-like or photo-Fenton-like experiments at circumneutral pH in order to evaluate the photo-activity of magnet-sensitive nanoparticles (NPs) functionalized with bio-based substances (BBS) obtained from composted urban biowaste. BBS-functionalized NPs were employed as catalysts in the removal of caffeine as a model emerging pollutant. The effect of pH was investigated and the use of new magnet-sensitive nanoparticles (NPs) functionalized with bio-based substances (BBS) was presented as a green alternative for the preparation of smart materials.

Furthermore, the photo-Fenton based processes are affected by several process parameters, the pH plays an important role to achieve the best degradation efficiency. A removal of reactive black 5 (RB5) textile dye with 100% efficiency was reached after 60 min and at optimum pH 3 value by heterogeneous photo-Fenton. It is known the pH affects the long-term stability of the catalyst and the photocatalytic performance [138].

## 4.4 Concentrations of ferrous ion and hydrogen peroxide

Ferrous ion concentration usually increases the rate of degradation. However, the extent of increase is still not well correlated. Thus, more laboratory studies are needed to be able to establish the optimum loading of ferrous ions to mineralize the organic pollutants. The concentration of hydrogen peroxide also presents important effects on the degradation reactions. Generally, the proportion of mineralized pollutant increases with an increase of hydrogen peroxide. The presence of hydrogen peroxide affects the overall degradation. Hydrogen peroxide addition must be controlled in order to its complete utilization. In a recent study, heterogeneous $Fe_3O_4$ catalyst and homogeneous $FeSO_4$ catalyst were compared for the magenta MB degradation in textile industries wastewater [140]. The effects of different system variables such as $H_2O_2$ dosage, the initial concentration of the dye, pH, and catalyst concentration were studied. The influence of ligands in the photo-Fenton system was examined and the stability of $Fe_3O_4$ catalyst was determined.

## 5. Economic evaluation

Irradiation and Fenton reagents are the two key cost parameters for photo-Fenton treatments. One of the most important factors affecting the application of advanced oxidation processes (AOPs) at full scale is the high operating cost, especially those associated with the cost of the reagents [141]. The optimization of operating parameters, such as a reduction in reagent consumption to achieve the partial or complete decontamination of waters, is crucial in AOP application. The present study is focused on the comparison of solar and UV photo-Fenton technologies for the removal of micro-contaminants from water at very low concentrations. In addition, economic assessment of the implementation of several AOPs was also carried out. It was concluded that consumption of reagents and electricity costs coming from lamp operation in the UV process had a significant impact on the total operating cost of both technologies. New strategies for reducing cost by using solar photo-Fenton treatment combined with nanofiltration to remove microcontaminants from real municipal effluents were chalked out. Furthermore, it must be stressed that solar and UV processes promoted by $H_2O_2$ are cost-effective.

In the photo-Fenton process, increase in the intensity of incident light contributes to increasing the degradation of contaminants. Because there is a direct influence on the removal rate of the contaminant with the configuration of the photoreactor, and the distribution of light inside the reactor which rises the probabilities of excitation of electrons as well as the re-excitation of recombined electrons. Recently, studies on light emitting diode (LED) has been considered as an alternative UVA of the radiation source in photoreactors due to low power consumption, feasibility, and sustainability. UVA-LED was used in the photo-Fenton reaction for a water containing pesticide, acetamiprid. Experiments were performed in a LED reactor at pH 2.8. The relationships among energy supply, LED consumption, UVA irradiance and reaction rate were evaluated and optimum experimental conditions were found on LED radiation [142].

Degradation of carbamazepine (CBZ) by photo-Fenton process showed low efficiencies (below 10%), but a combination of photo-Fenton with ultrasound radiation has a synergistic effect improving the removal of carbamazepine. A mineralization value of 60% was obtained at a pilot plant level and the cost for conventional is 2.1 and 3.8 €/$m^3$ for thin film plant. Although, the degradation of CBZ is faster when use ultrasound coupled to photo-Fenton due to more hydroxyl radical are available, thus makes the process more expensive. In the solar process, electric consumption accounts for a maximum of 33 % of total costs. Thus, for a TOC removal of 80 %, the cost of this treatment is about 1.36 €/m$^3$ [143].

## 6. Conclusions and future perspectives

During the past decades, many investigations have been carried out on the wastewater purification by advanced oxidation processes (AOPs). Latest studies regarding photo-Fenton technology revealed the influence of many factors in the degradation of pollutants. The treatment of wastewaters from pharmaceutical, agrochemical, petroleum refinery plants and landfill leachate is challenging to conventional approaches due to their recalcitrant nature. Different AOPs have been developed to select the most appropriate technique for treatment of wastewaters of different composition. Photo-Fenton oxidation is a useful technique for treating wastewaters, albeit its cost may be a hurdle. The reduction of the cost may come from the use of solar energy, heterogeneous catalysts, chelating agents, and by coupling with biological treatments.

Future studies will be focussed on wastewaters from industries and petroleum refinery plants, where the only little application of photo-Fenton technology has been reported. Moreover, the studies on the degradation of emerging compounds and endocrine disruptors are hoped to be undertaken in, pH and temperature dependence of the reactions has to be minimized.

## Acknowledgements

This work has been supported by MICINN (CTM 2014-56668-R project). RMMA thanks the Spanish Ministerio de Educación, Cultura y Deporte by the "Salvador de Madariaga" sabatical financial support (Ref. PRX16/00390) in the Operando Molecular Spectroscopy and Catalysis Laboratory of Prof. Israel E. Wachs at Lehigh University, Bethlehem, PA (EEUU).

## References

[1] A.R. Ribeiro, O.C. Nunes, M.F.R Pereira, A.M.T. Silva, A.M.T, An overview on the advanced oxidation processes applied for the treatment of water pollutants defined in the recently launched Directive 2013/39/EU, Environment International, 75 (2015) 33-51. https://doi.org/10.1016/j.envint.2014.10.027

[2] L. Clarizia, D. Russo, I, Di Somma, R. Marotta, R. Andreozzi, Homogeneous photo-Fenton processes at near neutral pH: A review, Applied Catalysis B: Environmental, 209 (2017) 358-371. https://doi.org/10.1016/j.apcatb.2017.03.011

[3] N. Wang, Y. Du, W. Ma, P. Xu, X. Han, Rational design and synthesis of SnO2-encapsulated α-Fe2O3 nanocubes as a robust and stable photo-Fenton catalyst, Applied Catalysis B: Environmental, 210 (2017) 23-33. https://doi.org/10.1016/j.apcatb.2017.03.037

[4] J. Alexander, P. Karaolia, D. Fatta-Kassinos, T. Schwartz, Impacts of Advanced Oxidation Processes on Microbiomes during Wastewater Treatment, Handbook of Environmental Chemistry,45 (2016) 129-144.

[5] D. Panda, S. Manickam, Recent advancements in the sonophotocatalysis (SPC) and doped-sonophotocatalysis (DSPC) for the treatment of recalcitrant hazardous organic water pollutants, Ultrasonics Sonochemistry, 36 (2017) 481-496. https://doi.org/10.1016/j.ultsonch.2016.12.022

[6] L. Clarizia, D. Russo, I. Di Somma, R. Marotta, R. Andreozzi, Homogeneous photo-Fenton processes at near neutral pH: A review, Applied Catalysis B: Environmental, 209 (2017) 358-371. https://doi.org/10.1016/j.apcatb.2017.03.011

[7] H. Wang, Y. Xu, L. Jing, S. Huang, Y. Zhao, M. He, H. Xu, H. Li, Novel magnetic BaFe12O19/g-C3N4 composites with enhanced thermocatalytic and photo-Fenton activity under visible-light, Journal of Alloys and Compounds, 710 (2017) 510-518. https://doi.org/10.1016/j.jallcom.2017.03.144

[8] S. Garcia-Segura, J. Anotai, S. Singhadech, M. Lu, Enhancement of biodegradability of o-toluidine effluents by electro-assisted photo-Fenton treatment, Process Safety and Environmental Protection, 106 (2017) 60-67. https://doi.org/10.1016/j.psep.2016.12.011

[9] N. López, S. Plaza, A. Afkhami, P. Marco, J. Giménez, S. Esplugas, Treatment of Diphenhydramine with different AOPs including photo-Fenton at circumneutral pH, Chemical Engineering Journal, 318 (2017) 112-120. https://doi.org/10.1016/j.cej.2016.05.127

[10] N. Wang, T. Zheng, G. Zhang, P. Wang, A review on Fenton-like processes for organic wastewater treatment, Journal of Environmental Chemical Engineering, 4 (1) (2016) 762-787. https://doi.org/10.1016/j.jece.2015.12.016

[11] S. Rahim Pouran, A.R. Abdul Aziz, W.M.A Wan Daud, W.M.A, Review on the main advances in photo-Fenton oxidation system for recalcitrant wastewaters, Journal of Industrial and Engineering Chemistry, 21 (2015) 53-69. https://doi.org/10.1016/j.jiec.2014.05.005

[12] A. Cruz, A., L. Couto, S. Esplugas, C. Sans, Study of the contribution of homogeneous catalysis on heterogeneous Fe(III)/alginate mediated photo-Fenton process, Chemical Engineering Journal, 318 (2017) 272-280. https://doi.org/10.1016/j.cej.2016.09.014

[13] Z. Jia, J. Kang, W.C. Zhang, W.M. Wang, C. Yang, H. Sun, D. Habibi, L.C. Zhang, Surface aging behaviour of Fe-based amorphous alloys as catalysts during heterogeneous photo Fenton-like process for water treatment, Applied Catalysis B: Environmental, 204 (2017) 537-547. https://doi.org/10.1016/j.apcatb.2016.12.001

[14] A. Mirzaei, Z. Chen, F. Haghighat, L. Yerushalmi, Removal of pharmaceuticals from water by homo/heterogonous Fenton-type processes-A review, Chemosphere 174 (2017) 665-688. https://doi.org/10.1016/j.chemosphere.2017.02.019

[15] M. Yoon, Y. Oh, S. Hong, J.S. Lee, R. Boppella, S.H. Kim, F. Marques Mota, S.O. Kim, D.H. Kim, Synergistically enhanced photocatalytic activity of graphitic carbon nitride and WO3 nanohybrids mediated by photo-Fenton reaction and H2O2, Applied Catalysis B: Environmental, 206 (2017) 263-270. https://doi.org/10.1016/j.apcatb.2017.01.038

[16] Y. Liu, W. Jin, Y. Zhao, G. Zhang, W. Zhang, Enhanced catalytic degradation of methylene blue by α-Fe2O3/graphene oxide via heterogeneous photo-Fenton reactions, Applied Catalysis B: Environmental, 206 (2017) 642-652. https://doi.org/10.1016/j.apcatb.2017.01.075

[17] A. Dutta, S. Datta, M. Ghosh, D. Sarkar, S. Chakrabarti, Sunlight-assisted photo-fenton process for removal of insecticide from agricultural wastewater, Trends in Asian Water Environmental Science and Technology, (2016) 23-33.

[18] C. Trellu, E. Mousset, Y. Pechaud, D. Huguenot, E.D. van Hullebusch, G. Esposito, M.A. Oturan, Removal of hydrophobic organic pollutants from soil washing/flushing solutions: A critical review, Journal of Hazardous Materials, 306 (2016) 149-174. https://doi.org/10.1016/j.jhazmat.2015.12.008

[19] S. Giannakis, S. Liu, A. Carratalà, S. Rtimi, M. Bensimon, C. Pulgarin, Effect of Fe(II)/Fe(III) species, pH, irradiance and bacterial presence on viral inactivation in wastewater by the photo-Fenton process: Kinetic modeling and mechanistic interpretation, Applied Catalysis B: Environmental, 204 (2017) 156-166. https://doi.org/10.1016/j.apcatb.2016.11.034

[20] B. Khennaoui, M.A. Malouki, M.C. López, F. Zehani, N. Boutaoui, Z.R. Salah, A. Zertal, Heterogeneous photo-Fenton process for degradation of azo dye: Methyl

orange using a local cheap material as a photocatalyst under solar light irradiation, Optik, 137 (2017) 6-16. https://doi.org/10.1016/j.ijleo.2017.02.081

[21] M.J. Lima, C.G. Silva, A.M.T. Silva, J.C.B. Lopes, M.M. Dias, J.L. Faria, Homogeneous and heterogeneous photo-Fenton degradation of antibiotics using an innovative static mixer photoreactor, Chemical Engineering Journal, 310(2017) 342-351. https://doi.org/10.1016/j.cej.2016.04.032

[22] M. Castro-Alférez, M.I. Polo-López, J. Marugán, P. Fernández-Ibá-ez, Mechanistic modeling of UV and mild-heat synergistic effect on solar water disinfection, Chemical Engineering Journal, 316 (2017) 111-120. https://doi.org/10.1016/j.cej.2017.01.026

[23] L. Santos-Juanes, F.S. García Einschlag, A.M. Amat, A. Arques, Combining ZVI reduction with photo-Fenton process for the removal of persistent pollutants, Chemical Engineering Journal, 310(2017) 484-490 https://doi.org/10.1016/j.cej.2016.04.114

[24] M. Minella, E. Sappa, K. Hanna, F. Barsotti, V. Maurino, C. Minero, D. Vione, Considerable Fenton and photo-Fenton reactivity of passivated zero-valent iron, RSC Advances, 6(89) (2016) 86752-86761. https://doi.org/10.1039/C6RA17515E

[25] A. D. Bokare, W. Choi, Review of iron-free Fenton-like systems for activating H2O2 in advanced oxidation processes, Journal of Hazardous Materials 275 (2014) 121-135. https://doi.org/10.1016/j.jhazmat.2014.04.054

[26] A. Babuponnusami, K. Muthukumar, A review on Fenton and improvements to the Fenton process for wastewater treatment, Journal of Environmental Chemical Engineering, 2 (2014) 557-572. https://doi.org/10.1016/j.jece.2013.10.011

[27] O. Tsydenova, V. Batoev, A. Batoeva, Solar-enhanced advanced oxidation processes for water treatment: Simultaneous removal of pathogens and chemical pollutants, International Journal of Environmental Research and Public Health, 12(8) (2015) 9542-9561. https://doi.org/10.3390/ijerph120809542

[28] S.O. Ganiyu, E.D. Van Hullebusch, M. Cretin, G. Esposito, M.A. Oturan, Coupling of membrane filtration and advanced oxidation processes for removal of pharmaceutical residues: A critical review, Separation and Purification Technology, 156 (2015) 891-914. https://doi.org/10.1016/j.seppur.2015.09.059

[29] Q. Liu, T. Chen, Y. Guo, Z. Zhang, X. Fang, Grafting Fe(III) species on carbon nanodots/Fe-doped g-C3N4 via interfacial charge transfer effect for highly

improved photocatalytic performance, Applied Catalysis B: Environmental, 205 (2017) 173-181 https://doi.org/10.1016/j.apcatb.2016.12.028

[30] Z. Cao, J. Zhang, J. Zhou, X. Ruan, D. Chen, J. Liu, Q. Liu, G. Qian, Electroplating sludge derived zinc-ferrite catalyst for the efficient photo-Fenton degradation of dye. Journal of Environmental Management, 193 (2017) 146-153. https://doi.org/10.1016/j.jenvman.2016.11.039

[31] Y. Jin, Y. Wang, Q. Huang, L. Zhu, Y. Cui, L. Cui, C. Lin, The experimental study of a hybrid solar photo-Fenton and photovoltaic system for water purification, Energy Conversion and Management, 135 (2017) 178-187. https://doi.org/10.1016/j.enconman.2016.12.073

[32] P. Abdi, A. Farzi, A. Karimi, Application of a hybrid enzymatic and photo-fenton process for investigation of azo dye decolorization on TiO2/metal-foam catalyst, Journal of the Taiwan Institute of Chemical Engineers, 71, (2017) 137-144. https://doi.org/10.1016/j.jtice.2016.11.022

[33] S. Chen, Y. Wu, G. Li, J. Wu, G. Meng, X. Guo, Z. Liu, A novel strategy for preparation of an effective and stable heterogeneous photo-Fenton catalyst for the degradation of dye, Applied Clay Science, 136 (2017) 103-111. https://doi.org/10.1016/j.clay.2016.11.016

[34] P.V.L. Reddy,K-H. Kim, A review of photochemical approaches for the treatment of a wide range of pesticide, Journal of Hazardous Materials, 285 (2015) 325-335. https://doi.org/10.1016/j.jhazmat.2014.11.036

[35] P. Villegas-Guzman, S. Giannakis, R.A.Torres-Palma, C. Pulgarin, Remarkable enhancement of bacterial inactivation in wastewater through promotion of solar photo-Fenton at near-neutral pH by natural organic acids, Applied Catalysis B: Environmental, 205 (2017) 219-227. https://doi.org/10.1016/j.apcatb.2016.12.021

[36] S. Arzate, J.L. García Sánchez, P. Soriano-Molina, J.L. Casas López, M.C. Campos-Ma-as, A. Agüera, J.A. Sánchez Pérez, Effect of residence time on micropollutant removal in WWTP secondary effluents by continuous solar photo-Fenton process in raceway pond reactors, Chemical Engineering Journal, 316 (2017) 1114-1121. https://doi.org/10.1016/j.cej.2017.01.089

[37] D. Hermosilla, N. Merayo, A. Gascó, Á. Blanco, The application of advanced oxidation technologies to the treatment of effluents from the pulp and paper industry: a review, Environmental science and pollution research international, 22(1) (2015) 168-191. https://doi.org/10.1007/s11356-014-3516-1

[38] P. Verlicchi, M. Al Aukidy, E. Zambello, What have we learned from worldwide experiences on the management and treatment of hospital effluent? - An overview and a discussion on perspectives, Science of the Total Environment, 514 (2015) 467-491. https://doi.org/10.1016/j.scitotenv.2015.02.020

[39] A. Talib, T.O. Randhir, Managing emerging contaminants in watersheds: Need for comprehensive, systems-based strategies, Sustainability of Water Quality and Ecology, (2016). DOI:10.1016/j.swaqe.2016.05.002.

[40] M. Tokumura, A. Sugawara, M. Raknuzzaman, M. Habibullah-Al-Mamun, S. Masunaga, Comprehensive study on effects of water matrices on removal of pharmaceuticals by three different kinds of advanced oxidation processes, Chemosphere, 159 (2016) 317-325. https://doi.org/10.1016/j.chemosphere.2016.06.019

[41] O. Gimeno, J.F. García-Araya, F.J. Beltrán, F.J. Rivas, A. Espejo, Removal of emerging contaminants from a primary effluent of municipal wastewater by means of sequential biological degradation-solar photocatalytic oxidation processe, Chemical Engineering Journal, 290 (2016) 12-20. https://doi.org/10.1016/j.cej.2016.01.022

[42] A.J. Expósito, A. Durán, J.M. Monteagudo, A. Acevedo, Solar photo-degradation of a pharmaceutical wastewater effluent in a semi-industrial autonomous plant, Chemosphere, 150 (2016) 254-257. https://doi.org/10.1016/j.chemosphere.2016.02.044

[43] J.A.D.L Perini, B.C.E. Silva, A.L. Tonetti, R.F.P. Nogueira, Photo-Fenton degradation of the pharmaceuticals ciprofloxacin and fluoxetine after anaerobic pre-treatment of hospital effluent, Environmental Science and Pollution Research, 24(7) (2016) 6233-6240. https://doi.org/10.1007/s11356-016-7416-4

[44] A.S. Anjana Anand, S. Adish Kumar, J. Rajesh Banu, G. Ginni, The performance of fluidized bed solar photo Fenton oxidation in the removal of COD from hospital wastewaters, Desalination and Water Treatment, 57 (18) (2016) 8236-8242. https://doi.org/10.1080/19443994.2015.1021843

[45] E.A. Serna-Galvis, A.L. Giraldo-Aguirre, J. Silva-Agredo, O.A. Flórez-Acosta, R.A. Torres-Palma, Removal of antibiotic cloxacillin by means of electrochemical oxidation, TiO2 photocatalysis, and photo-Fenton processes: analysis of degradation pathways and effect of the water matrix on the elimination of antimicrobial activity, Environmental Science and Pollution Research, 24(7) (2016) 6339-6352. https://doi.org/10.1007/s11356-016-6257-5

[46] G. Shankaraiah, S. Poodari, D. Bhagawan, V. Himabindu, S. Vidyavathi, Degradation of antibiotic norfloxacin in aqueous solution using advanced oxidation processes (AOPs)—A comparative study, Desalination and Water Treatment, 57(57) (2016) 27804-27815. https://doi.org/10.1080/19443994.2016.1176960

[47] A.O. Ifelebuegu, J. Ukpebor, B. Nzeribe-Nwedo, Mechanistic evaluation and reaction pathway of UV photo-assisted Fenton-like degradation of progesterone in water and wastewater, International Journal of Environmental Science and Technology, 13(12) (2016) 2757-2766. https://doi.org/10.1007/s13762-016-1103-3

[48] J. Xu, L. Li, B. Yuan, C. Shen, M. Fu, H. Cui, W. Sun, Large scale preparation of Cu-doped α-FeOOH nanoflowers and their photo-Fenton-like catalytic degradation of diclofenac sodium, Chemical Engineering Journal, 291 (2016) 174-183. https://doi.org/10.1016/j.cej.2016.01.059

[49] N. Villota, J.M. Lomas, L.M. Camarero, Study of the paracetamol degradation pathway that generates color and turbidity in oxidized wastewaters by photo-Fenton technology, Journal of Photochemistry and Photobiology A: Chemistry, 329 (2016) 113-119. https://doi.org/10.1016/j.jphotochem.2016.06.024

[50] E. Costa, B. Silva, J.A. de Lima Perini, R.F.P. Nogueira, Influence of dihydroxybenzenes on paracetamol and ciprofloxacin degradation and iron(III) reduction in Fenton processes, Environmental Science and Pollution Research, 24(7) (2016) 6157-6164. https://doi.org/10.1007/s11356-016-6402-1

[51] S. Chakma, V.S. Moholkar, Investigations in sono-enzymatic degradation of ibuprofen, Ultrasonics Sonochemistry, 29 (2016) 485-494. https://doi.org/10.1016/j.ultsonch.2015.11.002

[52] R. Colombo, T.C.R. Ferreira, R.A. Ferreira, M.R.V. Lanza, Removal of Mefenamic acid from aqueous solutions by oxidative process: Optimization through experimental design and HPLC/UV analysis, Journal of Environmental Management, 167 (2016) 206-213. https://doi.org/10.1016/j.jenvman.2015.11.029

[53] P. Iovino, S. Chianee, S. Canzano, M. Prisciandaro, D. Musmarra, Degradation of Ibuprofen in Aqueous Solution with UV Light: the Effect of Reactor Volume and pH, Water Air Soil Pollut, 227 (2016) 194-201. https://doi.org/10.1007/s11270-016-2890-3

[54] E.A. Serna-Galvis, J. Silva-Agredo, A.L. Giraldo, O.A. Flórez, R.A. Torres-Palma, Comparison of route, mechanism and extent of treatment for the

degradation of a β-lactam antibiotic by TiO2 photocatalysis, sonochemistry, electrochemistry and the photo-Fenton system, Chemical Engineering Journal, 284 (2016) 953-962. https://doi.org/10.1016/j.cej.2015.08.154

[55] H.B. Ammar, M.B. Brahim, R. Abdelhédi, Y. Samet, Enhanced degradation of metronidazole by sunlight via photo-Fenton process under gradual addition of hydrogen peroxide, Journal of Molecular Catalysis A: Chemical, 420 (2016) 222-227. https://doi.org/10.1016/j.molcata.2016.04.029

[56] K.A. Novoa-Luna, A. Mendoza-Zepeda, R. Natividad, R. Romero, M. Galar-Martínez, L.M. Gómez-Oliván, Biological hazard evaluation of a pharmaceutical effluent before and after a photo-Fenton treatment, Science of the Total Environment, 569 (2016) 830-840. https://doi.org/10.1016/j.scitotenv.2016.06.086

[57] E.M. Cuerda-Correa, J.R. Domínguez, M.J. Mu-oz-Pe-a, T. González, Degradation of Parabens in Different Aqueous Matrices by Several O3-Derived Advanced Oxidation Processes, Industrial and Engineering Chemistry Research, 55(18) (2016) 5161-5172. https://doi.org/10.1021/acs.iecr.6b00740

[58] H. Zú-iga-Benítez, G.A. Pe-uela, Degradation of ethylparaben under simulated sunlight using photo-Fenton, Water Science and Technology,73(4) (2016) 818-826.

[59] H. Zú-iga-Benítez, C. Aristizábal-Ciro, G.A. Pe-uela, Photodegradation of the endocrine-disrupting chemicals benzophenone-and methylparaben using Fenton reagent: Optimization of factors and mineralization/biodegradability studies, Journal of the Taiwan Institute of Chemical Engineers, 59 (2016) 380-388. https://doi.org/10.1016/j.jtice.2015.09.004

[60] G. Zhang, Q. Wang, W. Zhang, T. Li, Y. Yuan, P. Wang, Effects of organic acids and initial solution pH on photocatalytic degradation of bisphenol A (BPA) in a photo-Fenton-like process using goethite (α-FeOOH), Photochemical and Photobiological Sciences,15(8), (2016) 1046-1053. https://doi.org/10.1039/C6PP00051G

[61] V. Romero, O. González, B. Bayarri, P. Marco, J. Giménez, S. Esplugas, Degradation of Metoprolol by photo-Fenton: Comparison of different photoreactors performance, Chemical Engineering Journal, 283 (2016) 639-648. https://doi.org/10.1016/j.cej.2015.07.091

[62] V. Romero, S. Acevedo, P. Marco, J. Giménez, S. Esplugas Enhancement of Fenton and photo-Fenton processes at initial circumneutral pH for the degradation

of the β-blocker metoprolol, Water Research, 88 (2016) 449-457. https://doi.org/10.1016/j.watres.2015.10.035

[63] E. Bocos, N. Oturan, M. Pazos, M.Ã Sanroman, M.A. Oturan, Elimination of radiocontrast agent diatrizoic acid by photo-Fenton process and enhanced treatment by coupling with electro-Fenton process, Environmental Science and Pollution Research, 23(19) (2016) 19134-19144. https://doi.org/10.1007/s11356-016-7054-x

[64] A. Bianco Prevot, F. Baino, D. Fabbri, F. Franzoso, G. Magnacca, R. Nisticò, A. Arques, A. Urban biowaste-derived sensitizing materials for caffeine photodegradation, Environmental Science and Pollution Research, 24(14) (2017) 12599-12607. https://doi.org/10.1007/s11356-016-7763-1

[65] E. López-Loveira, F. Ariganello, M.S. Medina, D. Centrón, R. Candal, G. Curutchet, Degradation alternatives for a commercial fungicide in water: biological, photo-Fenton, and coupled biological photo-Fenton processes Environmental Science and Pollution Research, (2016) 1-11. https://doi.org/10.1007/s11356-016-7602-4

[66] Z-T. Hu, Z. Chen, R. Goei, W. Wu, T-T. Lim, Magnetically recyclable Bi/Fe-based hierarchical nanostructures: Via self-assembly for environmental decontamination, Nanoscale, 8(25) (2016) 12736-12746. https://doi.org/10.1039/C6NR03677E

[67] A.J. Expósito, J.M. Monteagudo, A. Durán, A. Fernández, Dynamic behavior of hydroxyl radical in sono-photo-Fenton mineralization of synthetic municipal wastewater effluent containing antipyrine, Ultrasonics Sonochemistry, 35 (2017) 185-195. https://doi.org/10.1016/j.ultsonch.2016.09.017

[68] H.M. Gutiérrez-Zapata, K.L. Rojas, J. Sanabria, J.A. Rengifo-Herrera, 2,4-D abatement from groundwater samples by photo-Fenton processes at circumneutral pH using naturally iron present. Effect of inorganicions, Environmental Science and Pollution Research, (2017) 1-10.

[69] L.O. Conte, A.V. Schenone, O.M. Alfano, Photo-Fenton degradation of the herbicide 2,4-in aqueous medium at pH conditions close to neutrality, Journal of Environmental Management, 170 (2016) 60-69. https://doi.org/10.1016/j.jenvman.2016.01.002

[70] S. Brosillon, C. Bancon-Montigny, J. Mendret, Photo-oxidation of tributyltin, dibutyltin and monobutyltin in water and marine sediments, International Journal

of Chemical Reactor Engineering, 14(3) (2016) 719-726. https://doi.org/10.1515/ijcre-2014-0127

[71] P.A. Fasnabi, G. Madhu, P.A. Soloman, Removal of Acetamiprid from Wastewater by Fenton and Photo-Fenton Processes – Optimization by Response Surface Methodology and Kinetics, Clean - Soil, Air, Water, 44(6) (2016) 728-737. https://doi.org/10.1002/clen.201400131

[72] B.M. da Costa Filho, V.M. da Silva, J.O. Silva, A.E. da Hora Machado, A.G. Trovó, Coupling coagulation, flocculation and decantation with photo-Fenton process for treatment of industrial wastewater containing fipronil: Biodegradability and toxicity assessment, Journal of Environmental Management, 174 (2016) 71-78. https://doi.org/10.1016/j.jenvman.2016.03.019

[73] F. Gozzi, S.C. Oliveira, R.F. Dantas, V.O. Silva, F.H. Quina, A.Jr. Machulek, Kinetic studies of the reaction between pesticides and hydroxyl radical generated by laser flash photolysis, Journal of the Science of Food and Agriculture, 96(5) (2016) 1580-1584. https://doi.org/10.1002/jsfa.7258

[74] S. García-Ballesteros, M. Mora, R. Vicente, C. Sabater, M.A. Castillo, A. Arques, A.M. Amat, Gaining further insight into photo-Fenton treatment of phenolic compounds commonly found in food processing industry, Chemical Engineering Journal, 288 (2016) 126-136. https://doi.org/10.1016/j.cej.2015.11.031

[75] A.M.S. Polo, J.J. López-Pe-alver, M. Sánchez-Polo, J. Rivera-Utrilla, I. Velo-Gala, J.J. Salazar-Rábago, Oxidation of diatrizoate in aqueous phase by advanced oxidation processes based on solar radiation, Journal of Photochemistry and Photobiology A: Chemistry, 319-320 (2016) 87-95. https://doi.org/10.1016/j.jphotochem.2015.12.009

[76] T.B. Benzaquén, M.A. Isla, O.M. Alfano, Combined chemical oxidation and biological processes for herbicide degradation, Journal of Chemical Technology and Biotechnology, 91(3) (2016) 718-725. https://doi.org/10.1002/jctb.4635

[77] S. Martinez, M. Delgado, P. Jarvis, Removal of herbicide mecoprop from surface water using advanced oxidation processes (AOPS) (2016).

[78] N.T. Kim Phuong, M.W. Beak, B.T. Huy, Y.-I. Lee, Adsorption and photodegradation kinetics of herbicide 2,4,5-trichlorophenoxyacetic acid with MgFeTi layered double hydroxides, Chemosphere, 146 (2016) 51-59. https://doi.org/10.1016/j.chemosphere.2015.12.008

[79] H. Mechakra, T. Sehili, M.A. Kribeche, A.A. Ayachi, S. Rossignol, C. George, Use of natural iron oxide as heterogeneous catalyst in photo-Fenton-like oxidation of chlorophenylurea herbicide in aqueous solution: Reaction monitoring and degradation pathways, Journal of Photochemistry and Photobiology A: Chemistry, 317 (2016) 140-150. https://doi.org/10.1016/j.jphotochem.2015.11.019

[80] C.H. Wu, J.T. Wu, Y.H. Lin, Mineralization of sulfamethizole in photo-Fenton and photo-Fenton-like systems, Water Science and Technology, 73(4) (2016) 746-750.

[81] A.-O. Roongkarn, P. Aphaiphak, J. Ananpattarachai, P. Kajitvichyanukul, Y.-T. Hung, Heterogeneous Fenton and photo-Fenton reactions in paraquat removal using iron nanoparticles, International Journal of Environment and Waste Management, 18(1) (2016) 1-12. https://doi.org/10.1504/IJEWM.2016.080255

[82] S. Giannakis, M. I. Polo López, D. Spuhler, J. A. Sánchez Pérez, P. Fernández Ibá-ez, C. Pulgarin, Solar disinfection is an augmentable, in situ-generated photo-Fenton reaction—Part 1: A review of the mechanisms and the fundamental aspects of the process, Applied Catalysis B: Environmental 199 (2016) 199-223. https://doi.org/10.1016/j.apcatb.2016.06.009

[83] S. Giannakisa, M. I. Polo López, D. Spuhler, J. A. Sánchez Pérez, P. Fernández Ibá-ez, C. Pulgarin, Solar disinfection is an augmentable, in situ-generated photo-Fenton reaction—Part 2: A review of the applications for drinking water and wastewater disinfection, Applied Catalysis B: Environmental 198 (2016) 431-446. https://doi.org/10.1016/j.apcatb.2016.06.007

[84] C. Ruales-Lonfat, .J.F. Barona, A. Sienkiewicz, J. Vélez, L.N. Benítez, C. Pulgarín, Bacterial inactivation with iron citrate complex: A new source of dissolved iron in solar photo-Fenton process at near-neutral and alkaline pH, Applied Catalysis B: Environmental, 180 (2016) 379-390. https://doi.org/10.1016/j.apcatb.2015.06.030

[85] G. Ferro, A. Fiorentino, M.C. Alferez, M.I. Polo-López, L. Rizzo, P. Fernández-Ibá-ez, Urban wastewater disinfection for agricultural reuse: Effect of solar driven AOPs in the inactivation of a multidrug resistant E. coli strain, Applied Catalysis B: Environmental, 178 (2014), DOI: 10.1016/j.apcatb.2014.10.043. https://doi.org/10.1016/j.apcatb.2014.10.043

[86] P. Valero, S. Giannakis, R. Mosteo, M.P. Ormad, C. Pulgarin, Comparative effect of growth media on the monitoring of E. coli inactivation and regrowth after solar

and photo-Fenton treatment, Chemical Engineering Journal, 313 (2017) 109-120. https://doi.org/10.1016/j.cej.2016.11.126

[87] T.F.C.V. Silva, E. Vieira, A.R. Lopes, O.C. Nunes, A. Fonseca, I. Saraiva, R.A.R. Boaventura, V.J.P. Vilar, How the performance of a biological pre-oxidation step can affect a downstream photo-Fenton process on the remediation of mature landfill leachates: Assessment of kinetic parameters and characterization of the bacterial communities Separation and Purification Technology, 175 (2017) 274-286. https://doi.org/10.1016/j.seppur.2016.11.011

[88] L.L.D.C. Huesca-Espitia, V. Aurioles-López, I. Ramírez, J.L. Sánchez-Salas, E.R. Bandala, Photocatalytic inactivation of highly resistant microorganisms in water: A kinetic approach, Journal of Photochemistry and Photobiology A: Chemistry, 337 (2017) 132-139. https://doi.org/10.1016/j.jphotochem.2017.01.025

[89] E. Ortega-Gómez, M.M.B. Martín, B.E. García, J.A.S Pérez, P.F. Ibá-ez, Wastewater disinfection by neutral pH photo-Fenton: The role of solar radiation intensity, Applied Catalysis B: Environmental, 181 (2016) 1-6. https://doi.org/10.1016/j.apcatb.2015.06.059

[90] S. Giannakis, C. Ruales-Lonfat, S. Rtimi, S. Thabet, P. Cotton, C. Pulgarin, Castles fall from inside: Evidence for dominant internal photo-catalytic mechanisms during treatment of Saccharomyces cerevisiae by photo-Fenton at near-neutral pH, Applied Catalysis B: Environmental, 185 (2016) 150-162. https://doi.org/10.1016/j.apcatb.2015.12.016

[91] S. Wang, Q. Li, F. Chen, J. Ke, R. Chen, HEPES-mediated controllable synthesis of hierarchical CuO nanostructures and their analogous photo-Fenton and antibacterial performance, Advanced Powder Technology, (2017), DOI:10.1016/j.apt.2017.03.001. https://doi.org/10.1016/j.apt.2017.03.001

[92] J.-A. Park, H.-L. Nam, J.-W. Choi, J. Ha, S.-H. Lee, Oxidation of geosmin and 2-methylisoborneol by the photo-Fenton process: Kinetics, degradation intermediates, and the removal of microcystin-LR and trihalomethane from Nak-Dong River water, South Korea, Chemical Engineering Journal, 313 (2017) 345-354. DOI: 10.1016/j.cej.2016.12.086. https://doi.org/10.1016/j.cej.2016.12.086

[93] V.J. Vilar, C.C.Amorim, E.Brillas, G. Li Puma, S. Malato, D.D. Dionysiou, AOPs: recent advances to overcome barriers in the treatment of water, wastewater and air, Environ Sci Pollut Res 24 (2017) 5987-5990. https://doi.org/10.1007/s11356-017-8425-7

[94] L. Yu, J. Chen, Z. Liang, W. Xu, L. Chen, D. Ye, Degradation of phenol using Fe3O4-GO nanocomposite as a heterogeneous photo-Fenton catalyst, Separation and Purification Technology, 171 (2016) 80-87. https://doi.org/10.1016/j.seppur.2016.07.020

[95] K. Li, Y. Zhao, M.J. Janik, C. Song, X. Guo, Facile preparation of magnetic mesoporous Fe3O4/C/Cu composites as high performance Fenton-like catalysts, Applied Surface Science, 396 (2017) 1383-1392. https://doi.org/10.1016/j.apsusc.2016.11.170

[96] T. Soltani, B.-K. Lee, Improving heterogeneous photo-Fenton catalytic degradation of toluene under visible light irradiation through Ba-doping in BiFeO3 nanoparticles, Journal of Molecular Catalysis A: Chemical, 425 (2016) 199-207. https://doi.org/10.1016/j.molcata.2016.10.009

[97] T. Soltani, B.-K. Lee, Novel and facile synthesis of Ba-doped BiFeO3 nanoparticles and enhancement of their magnetic and photocatalytic activities for complete degradation of benzene in aqueous solution, Journal of Hazardous Materials, 316 (2016) 122-133. https://doi.org/10.1016/j.jhazmat.2016.03.052

[98] P. Tepmatee, P. Siriphannon, Nanoporous copper doped aluminium pillared montmorillonite for dye-containing wastewater treatment, Water Environment Research, 88(11) (2016) 2015-2022. https://doi.org/10.2175/106143015X14362865227076

[99] X. Tang, Y. Liu, Heterogeneous photo-Fenton degradation of methylene blue under visible irradiation by iron tetrasulphophthalocyanine immobilized layered double hydroxide at circumneutral pH, Dyes and Pigments, 134 (2016) 397-408. https://doi.org/10.1016/j.dyepig.2016.07.026

[100] P.A. Soares, R. Souza, J. Soler, T.F.C.V. Silva, S.M.A.G.U. Souza, R.A.R. Boaventura, V.J.P.Vilar, Remediation of a synthetic textile wastewater from polyester-cotton dyeing combining biological and photochemical oxidation processes, Separation and Purification Technology, 172 (2017) 450-462. https://doi.org/10.1016/j.seppur.2016.08.036

[101] C. Jaramillo-Páez, J.A. Navío, M.C. Hidalgo, A. Bouziani, M.E. Azzouzi, Mixed α-Fe2O3/Bi2WO6 oxides for photoassisted hetero-Fenton degradation of Methyl Orange and Phenol, Journal of Photochemistry and Photobiology A: Chemistry, 332 (2017) 21-533. https://doi.org/10.1016/j.jphotochem.2016.09.031

[102] L. Ren, S.Y. Lu, J.Z. Fang, Y. Wu, D.Z. Chen, L.Y. Huang, Y.F. Chen, C. Cheng, Y. Liang, Z.Q. Fang, Enhanced degradation of organic pollutants using

Bi25FeO40 microcrystals as an efficient reusable heterogeneous photo-Fenton like catalyst, Catalysis Today, 281 (2017) 656-661. https://doi.org/10.1016/j.cattod.2016.06.028

[103] Y. Ju, Y. Yu, X. Wang, M. Xiang, L. Li, D. Deng, D.D. Dionysiou, Environmental application of millimetre-scale sponge iron (s-Fe0) particles (IV): New insights into visible light photo-Fenton-like process with optimum dosage of H2O2 and RhB photosensitizers, Journal of Hazardous Materials, 323 (2017) 611-620. https://doi.org/10.1016/j.jhazmat.2016.09.064

[104] L. Zhang, Q. Zhai, X. Zhao, X. Min, Q. Zhu, J. Li, Modified wool-iron biopolymer-based complex as an active heterogeneous decontamination photocatalyst, Journal of Energy Chemistry, 25(6) (2016) 1064-1069. https://doi.org/10.1016/j.jechem.2016.10.010

[105] M. Ahmadi, B. Kakavandi, S. Jorfi, M. Azizi, Oxidative degradation of aniline and benzotriazole over PAC@FeIIFe2 IIIO4: A recyclable catalyst in a heterogeneous photo-Fenton-like system, Journal of Photochemistry and Photobiology A: Chemistry, 336 (2017) 42-53. https://doi.org/10.1016/j.jphotochem.2016.12.014

[106] K. Zhang, Y. Liu, Deng, S. Xie, H. Lin, X. Zhao, J. Yang, Z. Han, H. Dai, Fe2O3/3DOM BiVO4: High-performance photocatalysts for the visible light-driven degradation of 4-nitrophenol, Applied Catalysis B: Environmental, 202 (2017)569-579. https://doi.org/10.1016/j.apcatb.2016.09.069

[107] S. Papoutsakis, C. Pulgarin, I. Oller, R. Sánchez-Moreno, S. Malato, Enhancement of the Fenton and photo-Fenton processes by components found in wastewater from the industrial processing of natural products: The possibilities of cork boiling wastewater reuse, Chemical Engineering Journal, 304 (2016) 890-896. https://doi.org/10.1016/j.cej.2016.07.021

[108] S. Miralles-Cuevas, I. Oller, A.Agüera, J.A.Sanchez Pérez and S.Malato, Strategies for reducing cost by using solar photo-Fenton treatment combined with nanofiltration to remove microcontaminants in real municipal effluents: Toxicity and economic assessment, Chemical Engineering Journal 318 (2017) 161-170. https://doi.org/10.1016/j.cej.2016.06.031

[109] M. Ebrahimi, H. Kazemi, S.A. Mirbagheri, T.D. Rockaway, An optimized biological approach for treatment of petroleum refinery wastewater, Journal of Environmental Chemical Engineering, 4(3) (2016) 3401-3408. https://doi.org/10.1016/j.jece.2016.06.030

[110] Z. Ghasemi, H. Younesi, A.A. Zinatizadeh, Preparation, characterization and photocatalytic application of TiO2/Fe-ZSM-5 nanocomposite for the treatment of petroleum refinery wastewater: Optimization of process parameters by response surface methodology, Chemosphere, 159 (2016) 552-564. https://doi.org/10.1016/j.chemosphere.2016.06.058

[111] M. Bagheri-Esfe, H. Bagheri-Esfe, Petroleum refinery wastewater treatment in two-chamber microbial fuel cells (MFCs), Scientia Iranica, 23(6) (2016) 2791-2799.

[112] D.A. Aljuboury, P. Palaniandy, H.B.A. Aziz, S. Feroz, Evaluation of the solar photo-Fenton process to treat the petroleum wastewater by response surface methodology (RSM), Environmental Earth Sciences, 75(4) (2016) 333:1-12.

[113] D.A.D.A. Aljuboury, P. Palaniandy, H.B.A. Aziz, S. Feroz, S.S.A Amr, Evaluating photo-degradation of COD and TOC in petroleum refinery wastewater by using TiO2/ZnO photo-catalyst, Water Science and Technology, 74(6) (2016) 1312-1325. https://doi.org/10.2166/wst.2016.293

[114] E.B. Estrada-Arriaga, J.A. Zepeda-Aviles, L. García-Sánchez, Post-treatment of real oil refinery effluent with high concentrations of phenols using photo-ferrioxalate and Fenton's reactions with membrane process step, Chemical Engineering Journal, 285 (2016) 508-516. https://doi.org/10.1016/j.cej.2015.10.030

[115] B.H. Diya'uddeen, S. Rahim Pouran, A.R. Abdul Aziz, W.M.A.W. Daud, Fenton oxidative treatment of petroleum refinery wastewater: process optimization and sludge characterization. RSC Advances, 5(83) (2015) 68159-68168. https://doi.org/10.1039/C5RA08079G

[116] G.F. Torres, J.A. Ortega Mãndez, D.L. Tinoco, E.D. Marin, J. Aranda, J.A. Herrera-Melián, J.M. Do-a RodrÃigez, J. Pérez Pe-a, Detoxification of synthetic and real groundwater contaminated with gasoline and diesel using Fenton, photo-Fenton, and biofilters, Desalination and Water Treatment, 57(50) (2016) 23760-23769. https://doi.org/10.1080/19443994.2016.1141710

[117] Z. Hussain, S.A. Mohammed, H. Ibraheem, N. Khan, B.M. Abdullah, E. Yousif, Photo-Fenton reaction: How to remediation waste-water using (Fe+2/UV), Research Journal of Pharmaceutical, Biological and Chemical Sciences, 7(4) (2016) 1072-1078.

[118] A. Moslehyani, A.F. Ismail, M.H.D. Othman, T. Matsuura, Design and performance study of hybrid photocatalytic reactor-PVDF/MWCNT

nanocomposite membrane system for treatment of petroleum refinery wastewater, Desalination, 363 (2015) 99-111. https://doi.org/10.1016/j.desal.2015.01.044

[119] B.H. Diya'uddeen, S. Rahim Pouran, A.R. Abdul Aziz, S.M. Nashwan, W.M.A. Wan Daud, M.G. Shaaban, Hybrid of Fenton and sequencing batch reactor for petroleum refinery wastewater treatment, Journal of Industrial and Engineering Chemistry, 25 (2015) 186-191. https://doi.org/10.1016/j.jiec.2014.10.033

[120] C. Amor, De Torres-Socías, J. A. Peres. M. Maldonado, I. Oller, S. Malato, M. S.Lucas, Mature landfill leachate treatment by coagulation/flocculation combined with Fenton and solar photo-Fenton processes, Journal of Hazardous Materials, 286 (2015) 261-268. https://doi.org/10.1016/j.jhazmat.2014.12.036

[121] T.F.C.V. Silva, A. Fonseca, I. Saraiva, R.A.R. Boaventura, V.J.P. Vilar, Scale-up and cost analysis of a photo-Fenton system for sanitary landfill leachate treatment, Chemical Engineering Journal, 283 (2016) 76-88. https://doi.org/10.1016/j.cej.2015.07.063

[122] F.G. Zha, D.X. Yao, Y.B. Hu, L.M. Gao, X.M. Wang, Integration of US/Fe2+ and photo-Fenton in sequencing for degradation of landfill leachate, Water Science and Technology, 73(2) (2016) 260-266. https://doi.org/10.2166/wst.2015.487

[123] J.O. Silva, V.M. Silva, V.L. Cardoso, A.E.H. Machado, A.G. Trovó, Treatment of sanitary landfill leachate by photo-Fenton process: Effect of the matrix composition, Journal of the Brazilian Chemical Society, 27(12) (2016) 2264-2272. https://doi.org/10.5935/0103-5053.20160120

[124] T.F.C.V. Silva, P.A. Soares, D.R. Manenti, A. Fonseca, I. Saraiva,R.A.R. Boaventura, V.J.P. Vilar, V.J.P, An innovative multistage treatment system for sanitary landfill leachate depuration: Studies at pilot-scale, Science of the Total Environment, 576 (2017) 99-117. https://doi.org/10.1016/j.scitotenv.2016.10.058

[125] F. Mahdad, H. Younesi, N. Bahramifar, M. Hadavifar, Optimization of Fenton and photo-Fenton-based advanced oxidation processes for post-treatment of composting leachate of municipal solid waste by an activated sludge process, KSCE Journal of Civil Engineering,20(6) (2016) 2177-2188. https://doi.org/10.1007/s12205-015-1045-1

[126] P. Villegas-Guzman, Jj. Silva-Agredo, O. Florez, A.L. Giraldo-Aguirre, C. Pulgarin, R.A. Torres-Palma, Selecting the best AOP for isoxazolyl penicillins degradation as a function of water characteristics: Effects of pH, chemical nature of additives and pollutant concentration, Journal of Environmental Management, 190 (2017) 72-79. https://doi.org/10.1016/j.jenvman.2016.12.056

[127] F.F. Karam, F.H. Hussein, S.J. Baqir, A.F. Alkaim, Optimal conditions for treatment of contaminated waters with anthracene by Fenton processes in close system reactor, Journal of Chemical and Pharmaceutical Sciences, 9(3) (2016) 1111-1115.

[128] P. Grant-Preece,L.M. Schmidtke,C. Barril, A.C. Clark, Photoproduction of glyoxylic acid in model wine: Impact of sulfur dioxide, caffeic acid, pH and temperature, Food Chemistry, 215 (2017) 292-300. https://doi.org/10.1016/j.foodchem.2016.07.131

[129] G. Pliego, J.A. Zazo, P. Garcia-Mu-oz, M. Munoz, J.A. Casas, J.J. Rodriguez, Trends in the Intensification of the Fenton Process for Wastewater Treatment: An Overview, Critical Reviews in Environmental Science and Technology, 45(24) (2015) 2611-2692. https://doi.org/10.1080/10643389.2015.1025646

[130] R. Sharma, Komal, V. Kumar, S. Bansal, S. Singhal, Boosting the catalytic performance of pristine CoFe2O4 with yttrium (Y3+) inclusion in the spinel structure Materials Research Bulletin, 90 (2017) 94-103. https://doi.org/10.1016/j.materresbull.2017.01.049

[131] A.L. Giraldo-Aguirre, E.A. Serna-Galvis, E.D. Erazo-Erazo, J. Silva-Agredo, H. Giraldo-Ospina, O.A. Flórez-Acosta, R.A. Torres-Palma, Removal of β-lactam antibiotics from pharmaceutical wastewaters using photo-Fenton process at near-neutral pH, Environmental Science and Pollution Research, (2017) 1-11. DOI:10.1007/s11356-017-8420-z https://doi.org/10.1007/s11356-017-8420-z

[133] X. Wei, H. Wu, G. He, Y. Guan, Y, Efficient degradation of phenol using iron-montmorillonite as a Fenton catalyst: Importance of visible light irradiation and intermediates, Journal of Hazardous Materials, 321(2017) 408-416. https://doi.org/10.1016/j.jhazmat.2016.09.031

[134] D.H. Funai, F. Didier, J. Giménez, S. Esplugas, P. Marco, A. Machulek, Junior, Photo-Fenton treatment of valproate under UVC, UVA and simulated solar radiation, Journal of Hazardous Materials, 323 (2017) 537-549. https://doi.org/10.1016/j.jhazmat.2016.06.034

[135] F.F. Karam, F.H. Hussein, S.J. Baqir, A.F. Alkaim, Optimal conditions for treatment of contaminated waters with anthracene by Fenton processes in close system reactor, Journal of Chemical and Pharmaceutical Sciences, 9(3) (2016) 1111-1115.

[136] I. de la Obra, L. Ponce-Robles, S. Miralles-Cuevas, I. Oller, S. Malato, J.A. Sánchez Pérez, Microcontaminant removal in secondary effluents by solar photo-

Fenton at circumneutral pH in raceway pond reactors, Catalysis Today, 287 (2017) 10-14. https://doi.org/10.1016/j.cattod.2016.12.028

[137] Y. Wang, J. Fang, J.C. Crittenden, C. Shen, Novel RGO/α-FeOOH supported catalyst for Fenton oxidation of phenol at a wide pH range using solar-light-driven irradiation, Journal of Hazardous Materials, 329 (2017) 321-329. https://doi.org/10.1016/j.jhazmat.2017.01.041

[138] F. Franzoso, R. Nisticò, F. Cesano, I. Corazzari, F. Turci, D. Scarano, A. Bianco Prevot, G. Magnacca, L. Carlos, D.O. Mártire, Biowaste-derived substances as a tool for obtaining magnet-sensitive materials for environmental applications in wastewater treatments, Chemical Engineering Journal, 310 (2017) 307-316. https://doi.org/10.1016/j.cej.2016.10.120

[139] F.F. Dias, A.A.S. Oliveira, A.P. Arcanjo, F.C.C. Moura, J.G.A. Pacheco, Residue-based iron catalyst for the degradation of textile dye via heterogeneous photo-Fenton, Applied Catalysis B: Environmental, 186 (2016) 136-142. https://doi.org/10.1016/j.apcatb.2015.12.049

[140] S. Xavier, R. Gandhimathi, P.V. Nidheesh, S.T. Ramesh, Comparative removal of Magenta MB from aqueous solution by homogeneous and heterogeneous photo-Fenton processes. Desalination and Water Treatment, 57(27), (2016) 12832-12841. https://doi.org/10.1080/19443994.2015.1054887

[141] S. Miralles-Cuevas, I. Oller, A. Agüera, J.A. Sánchez Pérez, S. Malato, Strategies for reducing cost by using solar photo-Fenton treatment combined with nanofiltration to remove microcontaminants in real municipal effluents: Toxicity and economic assessment, Chemical Engineering Journal, 318 (2017) 161-170. https://doi.org/10.1016/j.cej.2016.06.031

[142] I. de La Obra, B. Esteban García, J.L. García Sánchez, J.L. Casas López, J.A, Sánchez Pérez, Low cost UVA-LED as a radiation source for the photo-Fenton process: a new approach for micropollutant removal from urban wastewater, Photochemical and Photobiological Sciences, 16(1) (2017) 72-78. https://doi.org/10.1039/C6PP00245E

[143] A.J. Expósito, D.A. Patterson, J.M. Monteagudo, A. Durán, Sono-photo-degradation of carbamazepine in a thin falling film reactor: Operation costs in pilot plant, Ultrasonics Sonochemistry,34 (2017) 496-503. https://doi.org/10.1016/j.ultsonch.2016.06.029

# Chapter 13

# Electrochemical technologies for produced water treatment

Dina T. Moussa and Muftah H. El-Naas*

Gas Processing Center, College of Engineering, Qatar University,

P.O. Box 2713, Doha, Qatar

muftah@qu.edu.qa

**Abstract**

Produced water is the byproduct generated by the oil and gas industries, and it is becoming a global concern, due to its complex composition and extreme salinity. On the other hand, such large volumes of water can be treated and utilized to replenish the freshwater supply that is becoming increasingly scarce, making the potential reuse of treated produced water an attractive opportunity. This chapter provides a detailed characterization of produced water from different sources as well as a comprehensive overview of the well-established and emerging electrochemical technologies of produced water treatment. The electrochemical technologies covered in this chapter include anodic oxidation, electrodialysis, and electrodialysis reversal, electro-deionization, capacitive-deionization, electro-coagulation, and electrodeposition, Finally, opportunities of reusing treated produced water are discussed.

**Keywords**

Produced Water, Electrochemical Technologies, Wastewater Treatment, Electrocoagulation

## Contents

## 1. Introduction

Many countries depend on oil and gas industries as a major source of revenue and their economy is greatly affected by the ups and downs in this industry. To maximize the production of oil and gas wells, large volumes of water is injected at high pressure to help the oil/ gas production process. When such injected water is extracted along with the oil and gas, it is usually highly contaminated with hydrocarbons, solids, gases and other impurities. Besides injected water; the water, which occurs naturally in the reservoir formation within the pores of the rocks (formation water), is also extracted. Formation water together with injection water is known as produced water, a term that gained considerable attention recently as will be discussed next.

Almeida et al. reported that by 2030, oil will account for 32% of the global energy supply with the oil demand rising to 107 mbpd compared to 85 mbpd back in 2006 [2]. This means that oil related effluents, specifically produced water which accounts for more than 80% of liquid waste, will be generated at larger volumes. According to Duraisamy et al., the volume of produced water extracted along with oil and gas exceeds 77 billion bbl globally [3]. Oilfields alone account for more than 60% of the total volume of produced water globally [4].

### 1.1 Overview of produced water composition

Produced water is characterized by a very complex composition that varies depending on the reservoir type, the geographical location, reservoir depth, the geological formation and even the lifetime of a specific well. Therefore, samples of produced waters don't have identical composition. However, the typical constituents of produced water can be categorized into seven groups as described in the following sections [3, 5-10].

#### 1.1.1 Total dissolved solids (TDS)

Dissolved solids or dissolved inorganic minerals are classified as anions and cations; the most abundant inorganic ions in produced water are sodium, chloride, calcium, magnesium, potassium, sulfate, bromide, and bicarbonate. Depending on the location and nature of the reservoir, other minerals including iodide, boron, lithium and strontium salts may exist as well [6]. Sodium and chloride ions are responsible for the extreme salinity of produced water, which exceeds seawater salinity. Typically, produced water salinity ranges from few parts per thousand to as much as 300 part per thousand, which is similar to concentrated brine. Sulfate ions usually exist in small concentrations in produced water compared to seawater. The presence of sulfate ions affects the solubility of other inorganic ions such as barium and calcium; an elevated concentration of sulfates may lead to scale formation due to the precipitation of calcium and barium sulfates [10]. Table 1 represents the typical range of the most common inorganic ions in oil and gas produced waters compared to seawater.

*Table 1. Inorganic constituents in mg/l of oil & gas produced waters (PW) compared to seawater.*

| Constituent | Oil field PW [11] | Gas field PW [11] | Seawater [10] |
|---|---|---|---|
| TDS/ salinity | - | 139,000- 360,000 | 35,000 |
| Sodium | 132-97,000 | 37,500- 120,000 | 10,760 |
| Chloride | 80- 200,000 | 81,500- 167,448 | 19,353 |
| Sulfate | 1,650 | ND- 19 | 2,712 |
| Potassium | 24- 4,300 | 149- 3,8750 | 387 |
| Calcium | 13- 25,800 | 9,400- 51,300 | 416 |
| Barium | 1.3- 650 | 9.65- 1,740 | - |
| Magnesium | 8- 6000 | 1,300- 3,900 | 1,294 |

ND (not detected)

### 1.1.2 Total petroleum hydrocarbons

Total petroleum hydrocarbons (TPHs) are defined as all organic compounds consisting of hydrogen and carbon atoms. TPH could be classified as shown in Fig. 1 according to the type of bond (single, double or triple) or the shape of the compound (cyclic or straight). The solubility of petroleum hydrocarbons in water depends on their polarity and molecular weight. Polar compounds tend to dissolve in water and lower molecular weight compounds dissolve better than heavier compounds. In general, aromatic hydrocarbons tend to dissolve more in water compared to aliphatic compounds of the same molecular weight. Polyaromatic hydrocarbons PAHs (hydrocarbons with more than one benzene ring), phenols, organic acids and BTEX (benzene, toluene, ethylbenzene, xylene) are examples of soluble hydrocarbons. Other hydrocarbons which are not soluble in water are found in the form of dispersed oil such as high molecular weight PAHs and alkyl phenols [6, 10, 11]. Existing produced water treatment technologies can separate dispersed oils efficiently but not the dissolved hydrocarbons. Hence, dissolved hydrocarbons are discharged into the environment. TPHs can cause serious environmental concerns, as they contribute to the biological oxygen demand (BOD) which affects aquatic life besides being a toxic and persistent contaminant. In Qatar for instance, TPHs are the only regulated parameter in case of produced water discharge [6, 10, 11].

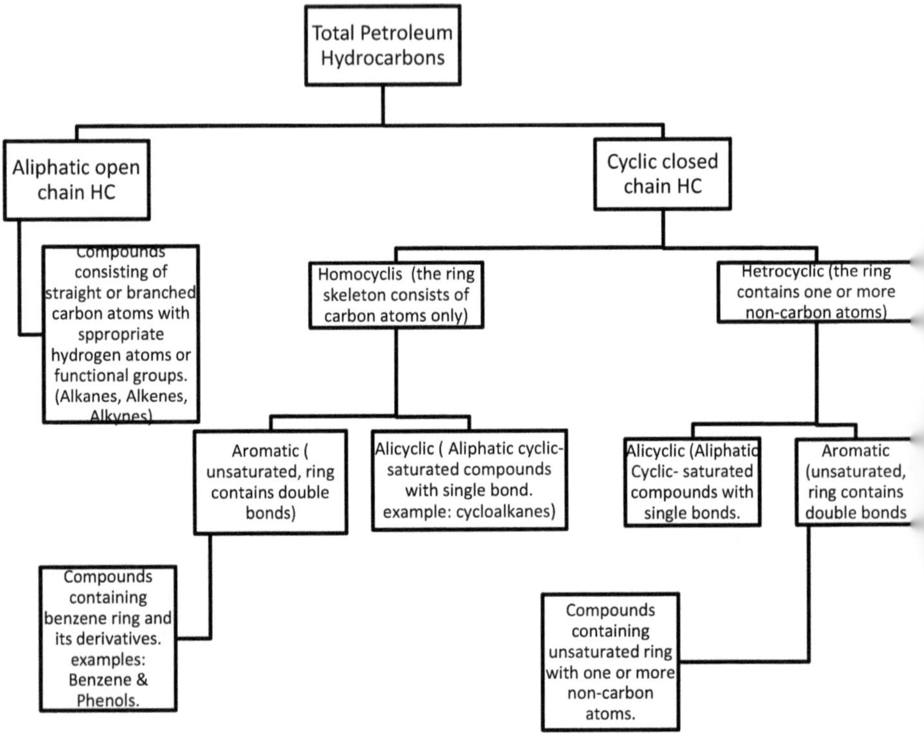


*Fig. 1 Classification of hydrocarbons.*

### 1.1.3 Dissolved & dispersed gasses

Dissolved gases occur in produced water due to bacterial activities and chemical reactions. The most abundant dissolved gases in produced water are carbon dioxide, oxygen, and hydrogen sulfide [6, 12]. The solubility of gases in produced water is governed by Henry's law, which relates solubility to the temperature and partial pressure of the gas over the liquid (produced water). The pressure increases along the depth of the reservoir and so does the amount of dissolved gasses [6].

### 1.1.4 Production chemicals

Production chemicals are those compounds that are added to produced water for several purposes, such as scale & corrosion inhibition, wax deposition, gas dehydration, breaking emulsions, prevention of bacterial growth and for other means [6, 12]. Depending on nature, chemical additives can, be more soluble in the oil phase than produced water and therefore will remain dissolved in oil and those more soluble in produced water are discharged with it. Due to the toxicity of some of the production chemicals and the environmental threats associated with their discharge, they should only be used if required and should not exceed the required optimum amount[10]. Table 2 illustrates the typical concentrations of various production chemicals in oil and gas produced waters.

*Table 2. Concentrations of typical production chemicals in oil and gas produced waters, adapted from [4].*

| Chemical | Concentration in oil field PW [mg/L] | Concentration in gas field PW [mg/L] |
|---|---|---|
| Corrosion inhibitor | 2-10 | 2-10 |
| Demulsifier | 1-2 | N.A |
| Scale inhibitor | 4-30 | N.A |
| Polyelectrolyte | 0-10 | N.A |
| Methanol | N.A | 1000-15000 |
| Di-ethylene glycol (DEG) | N.A | 500-2000 |

N.A (Not Applicable)

### 1.1.5 Produced solids

Produced water is extracted along with various types of production solids represented as clay, sand, silt, scale and corrosion products, wax and suspended solids. The removal of these solids is crucial to prevent clogging of pipes, the formation of oily sludge and other operational problems [3, 10, 12].

*Table 3. Concertation of heavy metals in oil and gas produced waters, adapted from [11].*

| Heavy metal | Concertation in oil field PW [mg/L] | Concertation in gas field PW [mg/L] | Concentration in seawater [mg/L] |
|---|---|---|---|
| Arsenic | 0.005- 0.3 | 0.005-151 | 1-3 |
| Lead | 0.002- 8.8 | 0.2- 10.2 | 0.001-0.1 |
| Mercury | 0.001-0.002 | - | 0.00007-0.006 |
| Cadmium | 0.005-0.2 | 0.02-1.21 | 3-34 |
| Chromium | 0.02-1.1 | 0.03 | 0.1-0.55 |
| Iron | 0.1-100 | 39-860 | 0.008-2 |
| Zinc | 0.01-35 | 0.02-5 | 0.006-0.12 |
| Silver | 0.001-0.15 | 0.047-7 | - |
| Copper | 0.002-1.5 | 0.02-5 | 0.03-0.35 |

### 1.1.6 Heavy metals

Produced water may contain a wide range of heavy metals depending on the source (oil/gas fields), geological formation, the age of the well, geographic location and other parameters. Typically, the following heavy metals are usually found in produced waters: cadmium, chromium, mercury, lead, arsenic, iron, zinc, manganese, silver, and copper [3, 6, 11]. The presence of heavy metals in produced water and discharging of such metals into the environment can have detrimental effects on aquatic life. Table 3 summarizes the concentration of heavy metals in oil and gas produced waters compared to seawater.

### 1.1.7 Naturally occurring radioactive materials (NORM)

Naturally occurring radioactive materials (NORM) are usually present in produced water as a result of the decay of certain rocks and clays in the hydrocarbon reservoir which contains uranium- 238 and thorium-232. The decay of these two radioactive elements results in the presence of radium-236 & radium-238 in produced water which are the most abundant types of NORM nucleoids. The rate of radioactive decay of these elements is measured in picocuries/L (pCi/L); one pCi is equivalent to 2.22 disintegrations per minute (dpm). Table 4 lists the typical activity of Ra-226 and Ra-228 in produced water compared to seawater [10].

*Table 4. Typical activities of radium isotopes in produced water and sea water [10].*

| Radionuclide | Produced water | Sea water |
|---|---|---|
| Ra-226 | 0.054 – 32,400 | 0.027-0.04 |
| Ra-228 | 8.1 - 4860 | 0.005-0.03 |

## 1.2 Environmental regulations for produced water discharge

The overview of produced water composition discussed in the previous section illustrates that produced water discharge must be strictly regulated, as it is generated in huge quantities in addition to being heavily polluted. In courtiers/ regions with water scarcity issues, the problem is even more critical due to the limited available water resources[2]. For these countries, discharging of PW without proper treatment will not only pollute the available water resources, but will also waste the large amounts of PW that could be reused for irrigation, landscaping, and other beneficial uses after proper treatment. The most common approach used for PW treatment prior to discharge is oil/water separation, which is an efficient technique for removing suspended oils as mentioned previously. However, the dissolved contaminants including oils, heavy metals and the production chemicals when discharged along with PW into surface water, pose toxic effect to aquatic life. These contaminants can also contaminate soil and underground water [12].

Although it is agreed worldwide that PW discharge can cause serious environmental concerns, the standards/ regulations of PW discharge vary from one country to another. On the international level, the United States Environmental Protection Agency (USEPA) is the most recognized regulatory institution that regulates and enforces laws for produced water extraction and discharge. On the national level, each country should have an institution that regulates and enforces laws related to PW discharge for the country and must be followed by all operating companies [6]. According to EPA, the pollutants are divided into three major groups: "conventional pollutants (five-day biochemical oxygen demand, total suspended solids, pH, fecal coliform, and oil and grease), toxic or priority pollutants (including metals and man-made organic compounds), and nonconventional (including ammonia, nitrogen, phosphorus, chemical oxygen demand, and whole effluent toxicity)" [13]. EPA has developed effluent limitation guidelines (ELGs) tailored to each industry and represent the maximum pollutant reductions based on the best available technology. For oil and gas industry, ELGs are available for onshore, offshore, and coastal operations as well. For PW generated onshore, the maximum allowable oil & grease content is *35 mg/L* prior to discharge; moreover, the quality of PW must be reasonably good to be used for irrigation or for livestock. In cases where the PW generated from onshore facilities will not be used for livestock or agriculture, thenPW discharge from such facilities is prohibited by EPA and a "zero-discharge" is implemented. For PW generated from offshore, the EPA has set a maximum daily oil & grease limit of *42 mg/L* and a monthly average limit of*29 mg/L* [13]. For PW reinjection in reservoirs or aquifer, EPA limits the oil & grease content to 30 mg/L. It could be noticed that the only regulated parameter when it comes to PW reinjection or discharge is oil & grease, where it is considered as a representative of other pollutants. When oil and grease are controlled, other pollutants will also be controlled [14]. On the regional level, several conventions/ treaties were held with the aim of limiting the discharge of pollutants into the environment. The major conventions such as the Oslo and Paris Commission for the North-East Atlantic (OSPAR), the Barcelona Convention for Mediterranean countries, the Kuwait Convention and the Helsinki Convention (HELCOM Convention) for the Baltic are listed in Table 5 [15].

*Table 5. Maximum allowable oil and grease limit for PW discharge according to regional conventions.*

| Legal Basis | Oil and grease limit for PW discharge |
|---|---|
| OSPAR Convention(North Sea countries) [16] | 30 [mg/L] |
| Baltic Sea Convention and HELCOM standards [17] | 15 [mg/L] max; 40 mg/l if BAT cannot achieve 15 [mg/L] |
| KUWAIT Convention and Protocols (Red Sea region) [15] | 40 [mg/l], 100 [mg/L] max |
| Barcelona Convention and Protocols (Mediterranean countries) [15] | 40 [mg/l], 100 [mg/L] max |

### 1.3 Conventional produced water treatment options

Produced water technologies could be categorized into three major groups: physical, chemical, and biological processes. A typical produced water treatment process consists of a combination of physical, chemical, and biological unit operations to target the removal of different constituents/ pollutants. Physical unit operations depend purely on the physical separation of pollutants from wastewater without causing a significant change in the chemical or biological characteristics of the treated water [18].Chemical processes are referred to as additive processes since they combine the addition of chemicals to react with the desired contaminants and remove their removal. The additive nature of chemical processes makes them less favorable compared to other processes. Biological unit processes utilize microorganisms for the biodegradation of contaminants in wastewater, and the main aim of these processes is to reduce the organic content and nutrients in wastewater. Biological units are generally classified into aerobic, anaerobic or facultative depending on the availability of dissolved oxygen in wastewater [19].

As mentioned earlier, the existing technologies have limited efficiencies to deal with small droplets of suspended oils and/ or dissolved contaminants. Moreover, each of the above- mentioned processes has their own limitations; physical process normally requires large surface area, huge capital cost and in the case of an adsorptive physical process, the regeneration of adsorbent is quite expensive. Chemical additives, on the other hand, require the addition of chemicals which generate hazardous sludge and could have high operational cost. Biological treatment is considered efficient for the degradation of organic matter but it is very sensitive to the inlet feed concentration and salinity, a major problem in produced water [12].

There is a wide variation in the produced water composition, quality, and throughout the lifetime of the oil/gas well; therefore, a robust technology that could adapt to the varying operating conditions in produced water is required. Other than the above-mentioned processes, promising, relatively new technologies that utilize the concepts of electrochemistry are also available. A review of the current state of electrochemical technologies for the treatment of produced water is presented in the following sections of this chapter.

## 2. Electrochemical technologies for produced water treatment

Recently there has been growing interest towards electrochemical techniques for the treatment of produced water. The extreme salinity of produced water favors the electrochemical reaction as the conductivity is high. A detailed review of the existing and emerging electrochemical technologies is presented below.

## 2.1 Anodic oxidation/ Electro-oxidation

Electro-oxidation or anodic oxidation technology employs the in-situ generation of oxidants, which are responsible for oxidizing organic contaminants. Among the various electrochemical techniques available, the anodic oxidation is widely used to achieve complete mineralization of organic pollutants. Pollutants are completely mineralized into $CO_2$ and $H_2O$ as a result of the reaction with hydroxyl radical ($OH^{\cdot}$) formed from water oxidation (Eq 1-3).

$$H_2O \rightarrow OH_{ads} + H^+ + e^- \quad (1)$$

$$mOH_{ads}\text{+organic pollutants}\rightarrow\text{intermediates} \quad (2)$$

$$nOH_{ads}\text{+intermediates}\rightarrow xH_2O\text{+}yCO_2n \quad (3)$$

$$2OHads\rightarrow 1/2\ O_2\text{+}H_2O \quad (4)$$

Anodic oxidation has several advantages over conventional wastewater treatment technologies as the electrogenerated hydroxyl radical is the second strongest oxidant after fluorine and capable to oxidize almost all organic contaminants including persistent species. Moreover, the lifetime of hydroxyl radicals in water is in the range of few nanoseconds, therefore they are self-eliminated from the water environment unlike other additive chemical oxidants [20]. Along with the main and desirable reaction of complete mineralization of organic contaminants into $CO_2$ and $H_2O$, side competing reactions such as oxygen evolution reactions could occur at higher potentials (Eq.4) which can decrease the efficiency of the process by consuming the hydroxyl radicals.

Anodic oxidation of organic pollutants can be classified as direct & indirect oxidation processes (Fig. 2). In the case of direct oxidation, organics are oxidized at the anodic surface through the physically adsorbed hydroxyl radicals or the chemisorbed oxygen on the metal oxide lattice. Indirect oxidation (mediated oxidation), on the other hand, occurs in the bulk solution away from the anodic surface due to the presence of other strong oxidants such as active chlorine species, hydrogen peroxide, ozone, and peroxidisulfate [2, 9, 21-33].

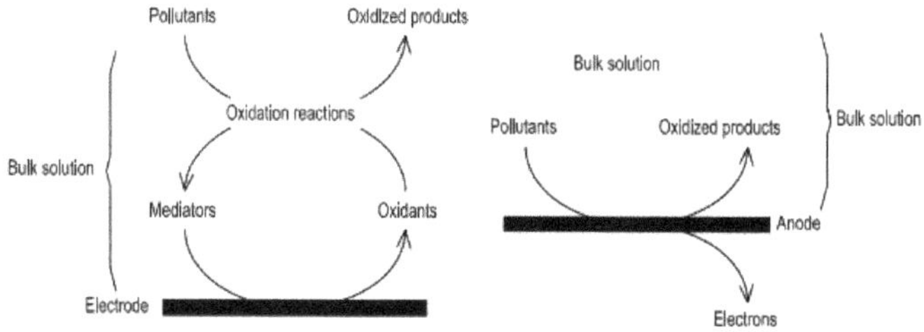


*Fig. 2. Direct and indirect anodic oxidation schematic diagram [1].*

Due to the possibility of oxygen evolution reactions, anodic oxidation normally requires special electrode materials with high oxygen overpotential. Anodic oxidation has been tested for several applications using a wide range of electrodes, such as dimensionally stable anodes, noble metals, and carbon-based anodes. The electrode materials could be generally grouped into two categories: active (a corrosion resistant metal base coated with metal oxide) & non-active anodes; active anodes such as $IrO_2$, which are characterized by a strong interaction of hydroxyl radicals with the electrode surface and low oxygen overpotential. Therefore, they are mainly used in the presence of chloride ions to facilitate indirect oxidation with mediated species. Non-active anodes such as boron doped diamond (BDD) are mainly used for direct oxidation due to the weak interaction between the electrode surface and hydroxyl radical and the high oxygen over potential[30, 34]. BDD is an excellent anode material for achieving promising results and high removal efficiencies. However, the cost of manufacturing of BDD anode is more than 10 times higher than metal oxide anodes. BDD is also not practical for scale up applications as it becomes brittle [23, 25, 35, 36].

## 2.2 Electrocoagulation

Electrocoagulation has gained an increasing attention over the past years as it proved to be efficient for the removal of a wide range of contaminants. It is a unique technology that combines the principles of coagulation, flotation, and electrochemistry, each of which has been extensively studied in a separate manner [18]. Electrocoagulation is favored over chemical coagulation, as coagulants are generated in-situ eliminating the addition of chemicals that generates high sludge volumes and add to the cost of the treatment [37-41]. A basic electrocoagulation cell normally consists of two metal electrodes connected to an external power supply and immersed in the wastewater to be treated (Fig. 3). Once power is supplied, the anode starts dissociating into the wastewater in the form of metal ion which reacts with water to form a metal hydroxide. Metal hydroxides are good coagulants as they trap colloidal particles by sweep coagulation/ precipitation; the formed sludge can then be removed leaving the clean water. Hydrogen generated at the cathode also acts through floatation mechanism to bring lighter contaminants to the water surface for easier separation. Several authors have examined the theory of electrocoagulation in detail[18, 42-44]. Although electrocoagulation has been applied to treat a wide range of wastewaters, it was seldom examined for produced water treatment[45-48].

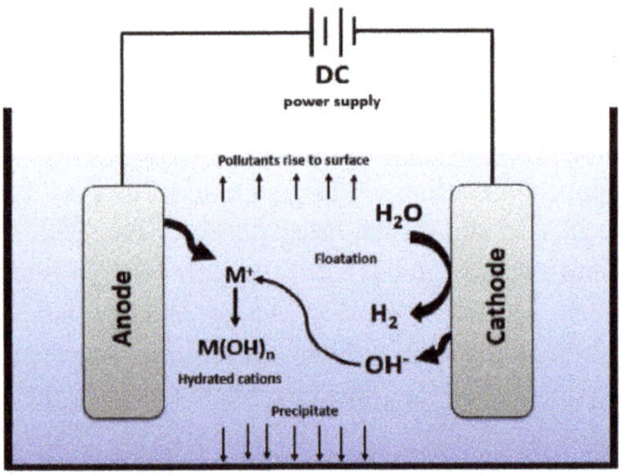


*Fig. 3. A schematic diagram of an Electrocoagulation cell [18].*

## 2.3 Electrodeposition

The concept of electrodeposition aims at recovering metals from produced water rather than removing it through a cathodic reaction that converts metal ions into metal after gaining electrons. Electrodeposition is a well-established technology but has only been reported for produced water metal recovery in a couple of studies. Igunnu and Chen reported that they were able to recover copper, lead, nickel and iron from synthetic produced water either separately or as a mixture of metals deposited on an inert electrode [12, 49]. Two sets of experiments were conducted; first synthetic wastewater with one metal contaminant at a time was studied then synthetic produced water with a mixture of metals was examined. Experiments revealed that several factors affect metal recovery/ deposition such as metal reactivity, surface area of the cathode, applied potential, pH, and the metal ions concentration. Metal reactivity and cathode surface area were found to be directly proportional to the rate of metal deposition, meaning that as the surface area increases, better deposition is achieved. Platinum, titanium and glassy carbon electrodes were tested as the inert electrodes for metal deposition, where platinum exhibited the best performance for individual metal deposition excluding Ni, which was better recovered on titanium electrode. Glassy carbon and platinum electrodes were also found to be effective for simultaneous recovery of multiple metals. Moreover, it was found that metals were recovered from synthetic produced water as a mixture of metals rather than an alloy. [12, 49]

## 2.4 Electrodialysis and electrodialysis reversal

Electrodialysis is basically a desalination technology that combines electrochemistry with membrane technology and utilizes the electrophoresis of dissolved inorganic ions in the solution and the selective permeability of ion exchange membranes to desalinate the solution. An electrodialysis cell consists of alternative anion and cation exchange membranes stacked between an anode and a cathode; electric current facilitates the movement of ions in the solution towards the electrodes of opposite charge. Positive ions will move towards the anode and will only pass through the selective cation exchange membrane. Negative ions, on the other hand, will try to reach the cathode and will selectively pass through the anion exchange membranes, which will result in two product streams: the concentrated brine and the desalinated water (Fig. 4) [50]. Electrodialysis reversal is the same as electrodialysis; however, the polarity is periodically reversed to prevent membrane fouling [8, 50-52]. It should be noted that electrodialysis is the most well-established electrochemical technology for produced water treatment and also available in commercial scale units.

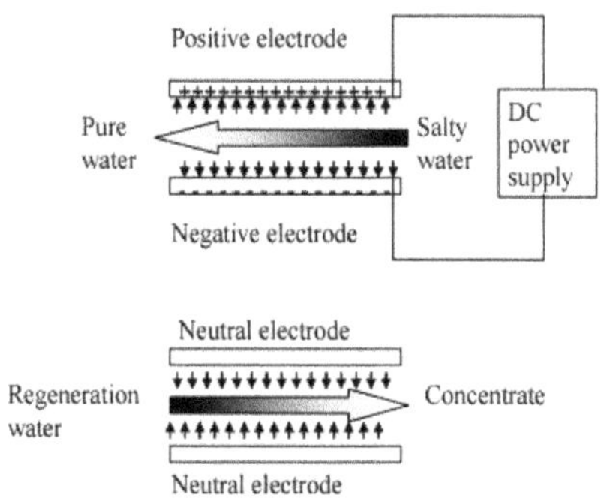


*Fig. 4. Electro-dialysis cell [2].*

## 2.5 Electro-deionization

One of the problems of electrodialysis is that the efficiency of treatment decreases as the water becomes more diluted. Therefore, electro-deionization is more powerful as it combines the electrochemical cell, ion exchange membranes, and ion exchange resins all in one cell. Anion and cation exchange resins are packed between alternative ion exchange membranes where they act as a bridge to facilitate the movement of ions through the selective membranes. Electro-deionization technology is, therefore, effective in treating wastewaters with low TDS content. When the water is diluted to an extent that the applied current is used to oxidize water into hydrogen and hydroxyl radicals. The generated hydrogen and hydroxyl radicals become useful to regenerate the ion exchange resins in-situ without additive chemicals (Fig. 5)[8, 50]. Electro deionization has been applied for produced water treatment in laboratory scale experiments and proved to be cost effective for TDS concentration less than 8000 mg/L. The energy consumption is found to be in the range of 0.14 -0.2 kWh/Ib NaCl removed and the water recovery can exceed 90% [50]

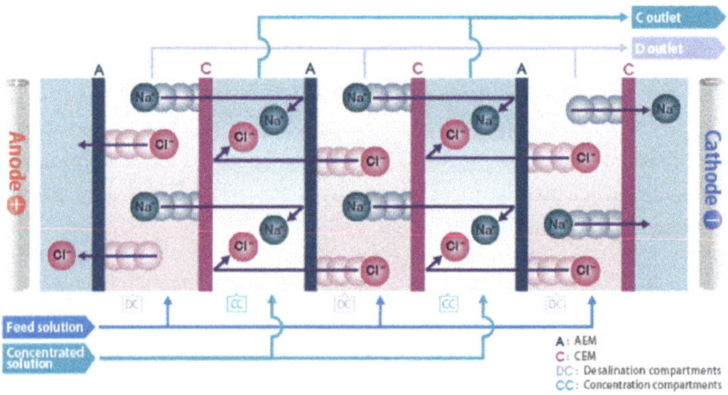


*Fig. 5. Electro-deionization cell [8].*

## 2.6 Capacitive deionization

Capacitive deionization (CDI) is an emerging produced water desalination technology that uses porous electrode material connected to positive and negative terminals instead of the regular metal electrodes. Porous electrodes allow the adsorption of salts onto the surface where the ions are mobilized as a result of the applied electric potential. Porous electrodes are easily regenerated by simply reversing the polarity or switching off the power supply, hence, no chemical regeneration is required (Fig. 6).

Carbon aerogel proved to be an optimum electrode material due to the high electrical conductivity, controllable pore size and high specific surface area [53-55]. Since CDI is an emerging PW technology, it was seldom reported in the literature for the treatment of PW. CDI was tested on both continuous lab scale and pilot scale experiments for PW treatment using 24 carbon aerogel electrodes (12 anodes, 12 cathodes) [56]. The results revealed that electrode deterioration was not observed and the process is cost effective at low TDS concertation (<3000 mg/L) [8, 50].

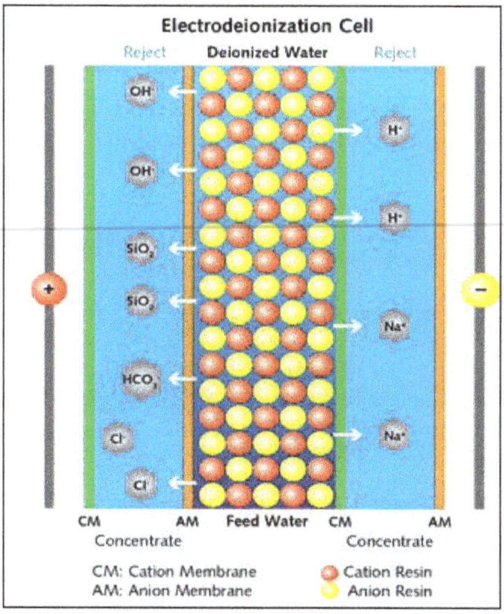


*Fig. 6. Capacitive deionization cell [8].*

### 3. Opportunities for produced water reuse

More than 1.2 billion people all over the world lack access to clean and safe water for domestic use according to world health organization report (WHO) [5]. According to the world business council for sustainable development report, fresh water accounts for less than 3% of the world's water; the remaining is seawater. About 2.5% of the 3% fresh water is frozen in the north and south poles and thus it is unavailable for a human being; therefore, the whole world's population is depending on 0.5% of the world's water [57]. Tremendous amounts of water are consumed every day all over the globe for daily life activities and in almost all industrial processes. Worldwide water consumption is split into 70% agricultural, 22% industrial and 8% domestic. Fig. 7 illustrates the water stress level in different parts of the world. It could be seen that the Gulf/ Middle East are the regions with critical water scarcity problems. These regions are also known to be major oil & gas producers, and hence are producers of large volumes of produced water. It is also worth mentioning that these areas have no natural fresh water resources other than ground water, which is becoming increasingly saline due to sea water intrusions. Therefore, Gulf area relies heavily on water desalination for their domestic/drinking water supply. This necessitates the concept of recycling and reusing produced water to cut down the fresh water consumption. Treated produced water could be used in various applications such as irrigation for landscaping or food crops, fire control, cooling water for chemical processes, wildlife-livestock, dust control and vehicle/ equipment washing [58-64].

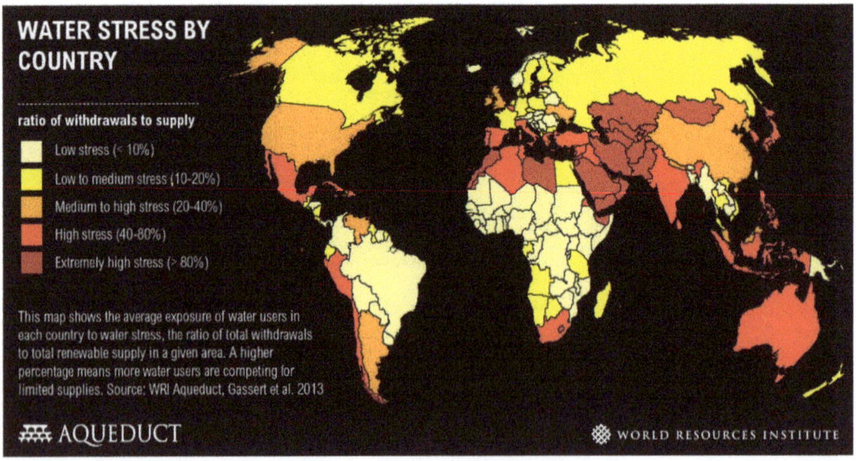


*Fig. 7. Water stress map in 2013 [57].*

## 4. Summary and potential for future research

The extraction of produced water as a byproduct of the oil and gas production operation is a challenge that cannot be eliminated but could be minimized to a certain extent.

Produced water is characterized by a complex composition that renders it toxic to surface water, soil, and groundwater. Treatment of produced water is becoming increasingly vital step not just due to environmental regulations, but also due to the water scarcity problem in many parts of the world. Proper treatment of produced water will not only help sustain resources and environment but will also limit the dependence on the rather limited fresh water resources. Currently, the only treatment of produced water before disposal is simple oil/ water separation, since oil content is the only regulated contaminant. Although a wide range of robust, well-established technologies are available for the treatment of produced water, these either fail to meet the PW reuse limits, or require hazardous chemical additives, generate toxic sludge, can't tolerate the salinity of produced water, or may require a large footprint, capital, or operating cost. These limitations lead to an increased interest in relatively new technologies known as electrochemical technologies, which are often referred to as green technologies due to the fact that they have a minimum or no impact on the environment. The status of electrochemical technologies for produced water treatment is not mature enough to be used as a centralized water treatment technologies. However, they proved to be effective for laboratory scale applications. Therefore, future research should focus on optimizing these technologies through studying the scale up, cost and kinetics. Moreover, several authors reported excellent results of integrating electrochemical technologies as pretreatment or post treatment of existing technologies to enhance the treatment efficiency, protect the existing equipment and achieving the PW reuse specifications. The feasibility of integrating electrochemical technologies with existing technologies is an important aspect that must be taken into consideration in future work [9, 56, 65-69].

## References


[1] Y. Deng, J.D. Englehardt, Electrochemical oxidation for landfill leachate treatment, Waste Management 27 (2007) 380-388. https://doi.org/10.1016/j.wasman.2006.02.004

[2] C.C.d. Almeida, P.R.F.d. Costa, M.J.d.M. Melo, E.V.d. Santos, C.A. Martínez-Huitle, Application of Electrochemical Technology for Water Treatment of Brazilian Industry Effluents, Journal of the Mexican Chemical Society 58 (2014) 276-286.

[3] R.T. Duraisamy, A.H. Beni, A. Henni, State of the Art Treatment of Produced Water, INTECH Open Access Publisher 9 (2013) 199-222.

[4] E.T. Igunnu, G.Z. Chen, Produced water treatment technologies, International Journal of Low-Carbon Technologies (2012) cts049.

[5] M. Nasiri, I. Jafari, Produced Water from Oil-Gas Plants: A Short Review on Challenges and Opportunities, Periodica Polytechnica Chemical Engineering (2016). https://doi.org/10.3311/PPch.8786

[6] J.A. Ahan, characterization of produced water from two offshore oil fields in Qatar, Qatar University, 2014.

[7] J. Veil, M. Puder, D. Elcock, R. Redweik Jr, A white paper describing produced water from production of crude oil, natural gas and coal bed methane. Argonne National Laboratory, Contract W-31-109-Eng-38 (2004).

[8] J.D. Arthur, B.G. Langhus, C. Patel, Technical summary of oil & gas produced water treatment technologies, All Consulting, LLC, Tulsa, OK (2005).

[9] P. Jain, M. Sharma, P. Dureja, P.M. Sarma, B. Lal, Bioelectrochemical approaches for removal of sulfate, hydrocarbon and salinity from produced water, Chemosphere 166 (2017) 96-108. https://doi.org/10.1016/j.chemosphere.2016.09.081

[10] J. Neff, K. Lee, E.M. DeBlois, Produced water: overview of composition, fates, and effects, Produced water, Springer2011, pp. 3-54. https://doi.org/10.1007/978-1-4614-0046-2_1

[11] A. Fakhru'l-Razi, A. Pendashteh, L.C. Abdullah, D.R.A. Biak, S.S. Madaeni, Z.Z. Abidin, Review of technologies for oil and gas produced water treatment, Journal of hazardous materials 170 (2009) 530-551. https://doi.org/10.1016/j.jhazmat.2009.05.044

[12] E.T. Igunnu, G.Z. Chen, Produced water treatment technologies, International Journal of Low-Carbon Technologies 9 (2012) 157-177. https://doi.org/10.1093/ijlct/cts049

[13] J.A. Veil, M.G. Puder, D. Elcock, R.J. Redweik Jr, A white paper describing produced water from production of crude oil, natural gas, and coal bed methane, Argonne National Laboratory, Technical Report (2004).

[14] J. Veil, Why are produced water discharge standards different throughout the world, Environmental Science Division Argonne National Laboratory. 13th IPEC, San Antonio, Texas, October 19th. ipec. utulsa. edu/Conf2006/Papers/Veil prodwater. pdf (2006).

[15] N. Gaurina-Međimurec, K. Simon, B. Pašić, Offshore Drilling and Environmental Protection, Energy and Environment (Energija i okoliš) 2006, 2006.

[16] R. Bernier, E. Garland, A. Glickman, F. Jones, H. Mairs, R. Melton, J. Ray, J. Smith, D. Thomas, J. Campbell, Environmental aspects of the use and disposal of non aqueous drilling fluids associated with offshore oil & gas operations, International Association of Oil & Gas Producers Report 342 (2003).

[17] H. Backer, J.-M. Leppänen, A.C. Brusendorff, K. Forsius, M. Stankiewicz, J. Mehtonen, M. Pyhälä, M. Laamanen, H. Paulomäki, N. Vlasov, HELCOM Baltic Sea Action Plan–A regional programme of measures for the marine environment based on the Ecosystem Approach, Marine pollution bulletin 60 (2010) 642-649. https://doi.org/10.1016/j.marpolbul.2009.11.016

[18] D.T. Moussa, M.H. El-Naas, M. Nasser, M.J. Al-Marri, A comprehensive review of electrocoagulation for water treatment: Potentials and challenges, Journal of Environmental Management 186 (2017) 24-41. https://doi.org/10.1016/j.jenvman.2016.10.032

[19] U.N.E.a.S.C.f.W. Asia, Waste-water Treatment Technologies: A General Review, United Nations, Economic and Social Commission for Western Asia2003. https://doi.org/10.1016/j.apcatb.2016.08.037

[20] F.C. Moreira, R.A.R. Boaventura, E. Brillas, V.J.P. Vilar, Electrochemical advanced oxidation processes: A review on their application to synthetic and real wastewaters, Applied Catalysis B: Environmental 202 (2017) 217-261. https://doi.org/10.1016/j.fuproc.2011.12.011

[21] J.H.B. Rocha, M.M.S. Gomes, N.S. Fernandes, D.R. da Silva, C.A. Martínez-Huitle, Application of electrochemical oxidation as alternative treatment of produced water generated by Brazilian petrochemical industry, Fuel Processing Technology 96 (2012) 80-87.

[22] E.V. dos Santos, S.F.M. Sena, D.R. da Silva, S. Ferro, A. De Battisti, C.A. Martínez-Huitle, Scale-up of electrochemical oxidation system for treatment of produced water generated by Brazilian petrochemical industry, Environmental Science and Pollution Research 21 (2014) 8466-8475. https://doi.org/10.1007/s11356-014-2779-x

[23] A.J.C. da Silva, E.V. dos Santos, C.C. de Oliveira Morais, C.A. Martínez-Huitle, S.S.L. Castro, Electrochemical treatment of fresh, brine and saline produced water generated by petrochemical industry using Ti/IrO2–Ta2O5 and BDD in flow reactor, Chemical Engineering Journal 233 (2013) 47-55. https://doi.org/10.1016/j.cej.2013.08.023

[24] E.V. dos Santos, J.H. Bezerra Rocha, D.M. de Araújo, D.C. de Moura, C.A. Martínez-Huitle, Decontamination of produced water containing petroleum hydrocarbons by electrochemical methods: a minireview, Environmental Science and Pollution Research 21 (2014) 8432-8441. https://doi.org/10.1007/s11356-014-2780-4

[25] B. Gargouri, O.D. Gargouri, B. Gargouri, S.K. Trabelsi, R. Abdelhedi, M. Bouaziz, Application of electrochemical technology for removing petroleum hydrocarbons from produced water using lead dioxide and boron-doped diamond electrodes, Chemosphere 117 (2014) 309-315. https://doi.org/10.1016/j.chemosphere.2014.07.067

[26] S.B. Dimitrijević, S.P. Dimitrijević, M.D. Vuković, Modern water treatment by electrochemical oxidation-a review.

[27] I.D. Santos, M. Dezotti, A.J.B. Dutra, Electrochemical treatment of effluents from petroleum industry using a Ti/RuO2 anode, Chemical Engineering Journal 226 (2013) 293-299. https://doi.org/10.1016/j.cej.2013.04.080

[28] J. Radjenovic, D.L. Sedlak, Challenges and Opportunities for Electrochemical Processes as Next-Generation Technologies for the Treatment of Contaminated Water, Environmental Science & Technology 49 (2015) 11292-11302. https://doi.org/10.1021/acs.est.5b02414

[29] O. Scialdone, Electrochemical oxidation of organic pollutants in water at metal oxide electrodes: A simple theoretical model including direct and indirect oxidation processes at the anodic surface, Electrochimica Acta 54 (2009) 6140-6147. https://doi.org/10.1016/j.electacta.2009.05.066

[30] M.R.G. Santos, M.O.F. Goulart, J. Tonholo, C.L.P.S. Zanta, The application of electrochemical technology to the remediation of oily wastewater, Chemosphere 64 (2006) 393-399. https://doi.org/10.1016/j.chemosphere.2005.12.036

[31] C.A. Martínez-Huitle, M.A. Rodrigo, I. Sirés, O. Scialdone, Single and coupled electrochemical processes and reactors for the abatement of organic water pollutants: a critical review, Chemical reviews 115 (2015) 13362-13407. https://doi.org/10.1021/acs.chemrev.5b00361

[32] S.A. Babu, S. Raja, S. Sibi, T. Sundaram, Direct And Indirect Electrochemical Oxidation Of Organic Pollutants From Industrially Polluted Water, I Control Pollution 2012 (2015).

[33] C.A. Martinez-Huitle, S. Ferro, Electrochemical oxidation of organic pollutants for the wastewater treatment: direct and indirect processes, Chemical Society Reviews 35 (2006) 1324-1340. https://doi.org/10.1039/B517632H

[34] C.L. De Silva, R.K. Garlapalli, J.P. Trembly, Removal of phenol from oil/gas produced water using supercritical water treatment with TiO2 supported MnO2 ctalyst, Journal of Environmental Chemical Engineering 5 (2017) 488-493. https://doi.org/10.1016/j.jece.2016.12.015

[35] J.H. Luong, K.B. Male, J.D. Glennon, Boron-doped diamond electrode: synthesis, characterization, functionalization and analytical applications, Analyst 134 (2009) 1965-1979.

[36] J. Iniesta, P.A. Michaud, M. Panizza, G. Cerisola, A. Aldaz, C. Comninellis, Electrochemical oxidation of phenol at boron-doped diamond electrode, Electrochimica Acta 46 (2001) 3573-3578. https://doi.org/10.1016/S0013-4686(01)00630-2

[37] M.Y.A. Mollah, P. Morkovsky, J.A.G. Gomes, M. Kesmez, J. Parga, D.L. Cocke, Fundamentals, present and future perspectives of electrocoagulation, Journal of Hazardous Materials 114 (2004) 199-210. https://doi.org/10.1016/j.jhazmat.2004.08.009

[38] M.Y.A. Mollah, R. Schennach, J.R. Parga, D.L. Cocke, Electrocoagulation (EC) — science and applications, Journal of Hazardous Materials 84 (2001) 29-41. https://doi.org/10.1016/S0304-3894(01)00176-5

[39] P.K. Holt, G.W. Barton, C.A. Mitchell, The future for electrocoagulation as a localised water treatment technology, Chemosphere 59 (2005) 355-367. https://doi.org/10.1016/j.chemosphere.2004.10.023

[40] P.K. Holt, G.W. Barton, M. Wark, C.A. Mitchell, A quantitative comparison between chemical dosing and electrocoagulation, Colloids and Surfaces A: Physicochemical and Engineering Aspects 211 (2002) 233-248. https://doi.org/10.1016/S0927-7757(02)00285-6

[41] E. Bazrafshan, L. Mohammadi, A. Ansari-Moghaddam, A.H. Mahvi, Heavy metals removal from aqueous environments by electrocoagulation process– a systematic review, Journal of Environmental Health Science and Engineering 13 (2015) 1-16. https://doi.org/10.1186/s40201-015-0233-8

[42] S. Zodi, J.-N. Louvet, C. Michon, O. Potier, M.-N. Pons, F. Lapicque, J.-P. Leclerc, Electrocoagulation as a tertiary treatment for paper mill wastewater:

Removal of non-biodegradable organic pollution and arsenic, Separation and Purification Technology 81 (2011) 62-68. https://doi.org/10.1016/j.seppur.2011.07.002

[43] J.N. Hakizimana, B. Gourich, M. Chafi, Y. Stiriba, C. Vial, P. Drogui, J. Naja, Electrocoagulation process in water treatment: A review of electrocoagulation modeling approaches, Desalination 404 (2017) 1-21. https://doi.org/10.1016/j.desal.2016.10.011

[44] E. Henry Ezechi, M. Hasnain Isa, S.R.b.M. Kutty, Z. Ahmed, Electrochemical removal of boron from produced water and recovery, Journal of Environmental Chemical Engineering 3 (2015) 1962-1973. https://doi.org/10.1016/j.jece.2015.05.015

[45] F.L. Lobo, H. Wang, T. Huggins, J. Rosenblum, K.G. Linden, Z.J. Ren, Low-energy hydraulic fracturing wastewater treatment via AC powered electrocoagulation with biochar, Journal of Hazardous Materials 309 (2016) 180-184. https://doi.org/10.1016/j.jhazmat.2016.02.020

[46] N. Esmaeilirad, K. Carlson, P. Omur Ozbek, Influence of softening sequencing on electrocoagulation treatment of produced water, Journal of Hazardous Materials 283 (2015) 721-729. https://doi.org/10.1016/j.jhazmat.2014.10.046

[47] E.H. Ezechi, M.H. Isa, S.R.M. Kutty, A. Yaqub, Boron removal from produced water using electrocoagulation, Process Safety and Environmental Protection 92 (2014) 509-514. https://doi.org/10.1016/j.psep.2014.08.003

[48] S. Zhao, G. Huang, G. Cheng, Y. Wang, H. Fu, Hardness, COD and turbidity removals from produced water by electrocoagulation pretreatment prior to Reverse Osmosis membranes, Desalination 344 (2014) 454-462. https://doi.org/10.1016/j.desal.2014.04.014

[49] E.T. Igunnu, Treatment of produced water by simultaneous removal of heavy metals and dissolved polycyclic aromatic hydrocarbons in a photoelectrochemical cell, University of Nottingham, 2014.

[50] J.E. Drewes, T.Y. Cath, P. Xu, J. Graydon, J. Veil, S. Snyder, An integrated framework for treatment and management of produced water, RPSEA Project (2009) 07122-07112.

[51] R.K. McGovern, A.M. Weiner, L. Sun, C.G. Chambers, S.M. Zubair, On the cost of electrodialysis for the desalination of high salinity feeds, Applied Energy 136 (2014) 649-661. https://doi.org/10.1016/j.apenergy.2014.09.050

[52] T. Sirivedhin, J. McCue, L. Dallbauman, Reclaiming produced water for beneficial use: salt removal by electrodialysis, Journal of Membrane Science 243 (2004) 335-343. https://doi.org/10.1016/j.memsci.2004.06.038

[53] C. Tsouris, R. Mayes, J. Kiggans, K. Sharma, S. Yiacoumi, D. DePaoli, S. Dai, Mesoporous carbon for capacitive deionization of saline water, Environmental science & technology 45 (2011) 10243-10249. https://doi.org/10.1021/es201551e

[54] S. Porada, R. Zhao, A. Van Der Wal, V. Presser, P. Biesheuvel, Review on the science and technology of water desalination by capacitive deionization, Progress in Materials Science 58 (2013) 1388-1442. https://doi.org/10.1016/j.pmatsci.2013.03.005

[55] S. Porada, L. Weinstein, R. Dash, A. Van Der Wal, M. Bryjak, Y. Gogotsi, P. Biesheuvel, Water desalination using capacitive deionization with microporous carbon electrodes, ACS applied materials & interfaces 4 (2012) 1194-1199. https://doi.org/10.1021/am201683j

[56] P. Xu, J.E. Drewes, D. Heil, G. Wang, Treatment of brackish produced water using carbon aerogel-based capacitive deionization technology, Water Research 42 (2008) 2605-2617. https://doi.org/10.1016/j.watres.2008.01.011

[57] U. Water, Wa ter, Facts and Trends, World Business Council for Sustainable Development (2006).

[58] M. Abdou, A. Carnegie, S.G. Mathews, K. McCarthy, M. O'Keefe, B. Raghuraman, W. Wei, C. Xian, Finding value in formation water, Oilfield Review 23 (2011) 24-35.

[59] G.W. Miller, Integrated concepts in water reuse: managing global water needs, Desalination 187 (2006) 65-75. https://doi.org/10.1016/j.desal.2005.04.068

[60] K.B. Gregory, R.D. Vidic, D.A. Dzombak, Water management challenges associated with the production of shale gas by hydraulic fracturing, Elements 7 (2011) 181-186. https://doi.org/10.2113/gselements.7.3.181

[61] D.L. Shaffer, L.H. Arias Chavez, M. Ben-Sasson, S. Romero-Vargas Castrillón, N.Y. Yip, M. Elimelech, Desalination and reuse of high-salinity shale gas produced water: drivers, technologies, and future directions, Environmental science & technology 47 (2013) 9569-9583.

[62] R. Kimball, Key considerations for frac flowback/produced water reuse and treatment, NJWEA Annual Conference, 2011, pp. 9-13.

[63] M.E. Mantell, Produced water reuse and recycling challenges and opportunities across major shale plays, Proceedings of the technical workshops for the hydraulic fracturing study: water resources management. EPA, 2011, pp. 49-57.

[64] Z. Khatib, P. Verbeek, Water to value-produced water management for sustainable field development of mature and green fields, Journal of Petroleum Technology 55 (2003) 26-28. https://doi.org/10.2118/0103-0026-JPT

[65] S. Zha, P. Gusnawan, J. Lin, G. Zhang, N. Liu, J. Yu, Integrating a novel TS-af-HFM NF process for portable treatment of oilfield produced water, Chemical Engineering Journal 311 (2017) 203-208. https://doi.org/10.1016/j.cej.2016.11.090

[66] S. Jiménez, M.M. Micó, M. Arnaldos, E. Ferrero, J.J. Malfeito, F. Medina, S. Contreras, Integrated processes for produced water polishing: Enhanced flotation/sedimentation combined with advanced oxidation processes, Chemosphere 168 (2017) 309-317. https://doi.org/10.1016/j.chemosphere.2016.10.055.

[67] O. Monzon, Y. Yang, J. Kim, A. Heldenbrand, Q. Li, P.J.J. Alvarez, Microbial fuel cell fed by Barnett Shale produced water: Power production by hypersaline autochthonous bacteria and coupling to a desalination unit, Biochemical Engineering Journal 117, Part A (2017) 87-91.

[68] A. Venkatesan, P.C. Wankat, Produced water desalination: An exploratory study, Desalination 404 (2017) 328-340. https://doi.org/10.1016/j.desal.2016.11.013

[69] Z.A. Stoll, C. Forrestal, Z.J. Ren, P. Xu, Shale gas produced water treatment using innovative microbial capacitive desalination cell, Journal of Hazardous Materials 283 (2015) 847-855. https://doi.org/10.1016/j.jhazmat.2014.10.015

# Keywords

# About the Editors

## Assistant Professor Inamuddin

**Dr. Inamuddin** is currently working as Assistant Professor in the Chemistry Department, Faculty of Science, King Abdulaziz University, Jeddah, Saudi Arabia. He obtained Master of Science degree in Organic Chemistry from Chaudhary Charan Singh (CCS) University, Meerut, India, in 2002. He received his Master of Philosophy and Doctor of Philosophy degrees in Applied Chemistry from AMU in 2004 and 2007, respectively. He has extensive research experience in multidisciplinary fields of Analytical Chemistry, Materials Chemistry, and Electrochemistry and, more specifically, Renewable Energy and Environment. He has worked on different research projects as project fellow and senior research fellow funded by University Grants Commission (UGC), Government of India, and Council of Scientific and Industrial Research (CSIR), Government of India. He has received Fast Track Young Scientist Award from the Department of Science and Technology, India, to work in the area of bending actuators and artificial muscles. He has completed four major research projects sanctioned by University Grant Commission, Department of Science and Technology, Council of Scientific and Industrial Research, and Council of Science and Technology, India. He has published 76 research articles in international journals of repute and twelve book chapters in knowledge-based book editions published by renowned international publishers. He has published six edited books with Springer, United Kingdom, three by Nova Science Publishers, Inc. U.S.A., one by CRC Press Taylor & Francis Asia Pacific, two by Trans Tech Publications Ltd., Switzerland and two by Materials Science Forum, U.S.A. He is the member of various editorial boards of the journals and also serving as associate editor for a journal Environmental Chemistry Letter, Springer Nature. He has attended as well as chaired sessions in various international and national conferences. He has worked as a Postdoctoral Fellow, leading a research team at the Creative Research Initiative Center for Bio-Artificial Muscle, Hanyang University, South Korea, in the field of renewable energy, especially biofuel cells. He has also worked as a Postdoctoral Fellow at the Center of Research Excellence in Renewable Energy, King Fahd University of Petroleum and Minerals, Saudi Arabia, in the field of polymer electrolyte membrane fuel cells and computational fluid dynamics of polymer electrolyte membrane fuel cells. He is a life member of the Journal of the Indian Chemical Society. His research interest includes ion exchange materials, a sensor for heavy metal ions, biofuel cells, supercapacitors and bending actuators.

**Professor Ali Mohammad**

Prof. Ali Mohammad is presently working as UGC-Emeritus Fellow in the Department of Applied Chemistry, Faculty of Engineering and Technology, Aligarh Muslim University, Aligarh, India after serving as chairman of the department of Applied Chemistry for 6 years. His scientific interests include physico-analytical aspects of solid-state reactions, micellar thin layer chromatography, surfactants analysis, and green chromatography. He is the author or coauthor of 247 scientific publications including research articles, reviews, and book chapters. He has supervised 53 students for Ph.D./M.Phil. and M.Tech. degrees. He has also served as Editor of Scientific Journal, "Chemical and Environmental Research" published from India since 1992 to 2012 and as the Associate Editor for Analytical Chemistry section of the Journal of Indian Chemical Society. He has published four edited books with Springer, the United Kingdom and two by Materials Science Forum, U.S.A. He has been the member of editorial boards of Acta Chromatographica, Acta Universitatis Cibiniensis Seria F. Chemia, Air Pollution, and Annals of Agrarian Science. He has attended as well as chaired sessions in various international and national conferences. He is the life member of several Indian Scientific and Chemical Societies. He has also served as Visiting Professor at King Saud University, Riyadh, Kingdom of Saudi Arabia. Dr. Mohammad obtained his M.Sc. (1972), M.Phil. (1975), Ph.D. (1978), and D.Sc. (1996) degrees from Aligarh Muslim University, Aligarh, India.

**Professor Abdullah M. Asiri**

Prof. Abdullah M. Asiri is the Head of the Chemistry Department at King Abdulaziz University since October 2009 and he is the founder and the Director of the Center of Excellence for Advanced Materials Research (CEAMR) since 2010 till date. He is the Professor of Organic Photochemistry. He graduated from King Abdulaziz University (KAU) with B.Sc. in Chemistry in 1990 and a Ph.D from University of Wales, College of Cardiff, U.K. in 1995. His research interest covers color chemistry, synthesis of novel photochromic and thermochromic systems, synthesis of novel coloring matters and dyeing of textiles, materials chemistry, nanochemistry and nanotechnology, polymers and plastics. Prof. Asiri is the principal supervisors of more than 20 M.Sc. and six Ph.D theses; He is the main author of ten books of different chemistry disciplines. Prof. Asiri is the Editor-in-Chief of King Abdulaziz University Journal of Science. A major achievement of Prof. Asiri is the discovery of tribochromic compounds, a class of compounds which change from slightly or colorless to deep colored when subjected to small pressure or when grind. This discovery was introduced to the scientific community as a new terminology published by IUPAC in 2000. This discovery was awarded a patent from European Patent office and from UK patent. Prof. Asiri involved in many committees at the KAU level and also on the national level, he took a major role in the advanced materials committee working for KACST to identify the National plan for science and technology in 2007. Prof. Asiri played a major role in advancing the chemistry education and research in KAU, he has been awarded the best Researchers from KAU for the past five years. He also awarded the Young Scientist award from the Saudi Chemical Society in 2009, and also the first prize for the distinction in science from the Saudi Chemical Society in 2012. He also received a recognition certificate from the American Chemical society (Gulf region Chapter) for the advancement of chemical science in the Kingdome. Also he received a Scopus certificate for the most Publishing Scientist in Saudi Arabia in chemistry in 2008. He is also a member of the Editorial Board of various journals of international repute. He is the Vice- President of Saudi Chemical Society (Western Province Branch). He holds four USA patents, more than 800 Publications in international journals, seven book chapters, and ten books.

www.ingramcontent.com/pod-product-compliance
Ingram Content Group UK Ltd.
Pitfield, Milton Keynes, MK11 3LW, UK
UKHW020659041026
15590UKWH00001B/360

9 781945 291340